Progress in Mathematics
Volume 143

Algorithms in Algebraic Geometry and Applications

Laureano González-Vega
Tomás Recio
Editors

Birkhäuser Verlag
Basel · Boston · Berlin

Editors:

Laureano González-Vega and
Tomás Recio
Departamento de Matemáticas
Facultad de Ciencias
Avenida Los Castros s/n
Santander 39071
Spain

A CIP catalogue record for this book is available from the Library of Congress, Washington D.C., USA

Deutsche Bibliothek Cataloging-in-Publication Data

Algorithms in algebraic geometry and applications / Laureano González-Vega ; Tomás Recio Ed. – Basel ; Boston ; Berlin : Birkhäuser, 1996
(Progress in mathematics ; Vol. 143)
ISBN-13:978-3-0348-9908-6 e-ISBN-13:978-3-0348-9104-2
DOI: 10.1007/978-3-0348-9104-2

NE: González-Vega, Laureano [Hrsg.]; GT

Softcover reprint of the hardcover 1st edition 1996

Printed on acid-free paper produced of chlorine-free pulp. TCF ∞

9 8 7 6 5 4 3 2 1

Table of Contents

Preface vii

Zeros, multiplicities, and idempotents for zero-dimensional systems
M. E. Alonso, E. Becker, M.-F. Roy, and T. Wörmann 1

On a conjecture of C. Berenstein and A. Yger
F. Amoroso 17

Computation of the splitting fields and the Galois groups of polynomials
H. Anai, M. Noro, and K. Yokoyama 29

How to compute the canonical module of a set of points
S. Beck and M. Kreuzer 51

Multivariate Bezoutians, Kronecker symbol and Eisenbud-Levine formula
E. Becker, J. P. Cardinal, M.-F. Roy, and Z. Szafraniec 79

Some effective methods in pseudo-linear algebra
M. Bronstein 105

Gröbner basis and characteristically nilpotent filiform Lie algebras of dimension 10
F. J. Castro and J. Núñez-Valdés 115

Computing multidimensional residues
E. Cattani, A. Dickenstein, and B. Sturmfels 135

The arithmetic of hyperelliptic curves
E. V. Flynn 165

Viro's method and T-curves
I. Itenberg 177

A computational method for diophantine approximation
T. Krick and L. M. Pardo 193

An effective method to classify nilpotent orbits
P. Littelmann 255

Some algebraic geometry problems arising in the field of mechanism theory
J.-P. Merlet 271

Enumeration problems in geometry, robotics and vision
B. Mourrain 285

Mixed monomial bases
P. Pedersen and B. Sturmfels 307

The complexity and enumerative geometry of aspect graphs of smooth surfaces
S. Petitjean 317

Aspect graphs of bodies of revolution with algorithms of real algebraic geometry
M.-F. Roy and T. Van Effelterre 353

Computational conformal geometry
M. Seppälä 365

An algorithm and bounds for the real effective Nullstellensatz in one variable
H. Warou 373

Solving zero-dimensional involutive systems
A. Zarkov 389

Preface

This volume arises from the contributions presented at the **MEGA**-94 Conference (**M**étodos **E**fectivos en **G**eometría **A**lgebraica = Effective Methods in Algebraic Geometry), held at the University of Cantabria (Santander, Spain) April 5–9, 1994. Previous sessions of this biannual conference had taken place in Castiglioncello (Livorno, Italy, 1990) and in Nice (France, 1992) and the corresponding proceedings have been published in the Birkhäuser series *Progress in Mathematics*, volumes no. 94 and 109, respectively.

The present collection consists of twenty articles involving miscellaneous topics concerning algorithms in algebra, algebraic geometry and related applications. Fourteen of these papers correspond to the contents of the Conference's regular scientific program and have been selected, by the MEGA Committee, from the submitted contributions after a very rigorous refereeing procedure entailing an average of three independent reports per paper and two Program Committee panel discussions before and after the Conference. The remaining six papers (by S. Beck & M. Kreuzer, M. Bronstein, E. V. Flynn, I. Itenberg, J.-P. Merlet and M. Seppälä) correspond to invited talks and have also been subject to a post-conference refereeing procedure.

In appreciation of this generous effort towards the achievement of excellence in the discipline, the editors of this volume are pleased to mention here the members of the MEGA-94 Committee:

B. Buchberger (Linz), A. M. Cohen (Eindhoven), A. Conte (Torino), J. H. Davenport (Bath), A. Galligo (Nice), D. Yu. Grigoriev (Leningrad), J. Heintz (Santander and Buenos Aires), W. Lassner (Leipzig), D. Lazard (Paris), H. M. Möller (Hagen), T. Mora (Genova), M. Pohst (Düsseldorf), T. Recio (Santander), J.-J. Risler (Paris), M.-F. Roy (Rennes), R. Schoof (Trento) and C. Traverso (Pisa).

We also wish to express our gratitude to all the 120 participants at the Conference; specially to those, listed below, who contributed to the regular program or to the informal sessions (on applications, algorithmics and software) but whose papers, for whatever reason, do not appear in this volume:

- J. Abbott: PoSSo: Past, present and future.
- C. Alonso: FRAC: A Maple package for computing in the rational function field $\mathbf{K}(X)$.
- F. Barkats: Computation of local cohomology modules of monomial ideals.
- C. Berenstein: Abel and Jacobi residue identities and their role in the effective Nullstellensatz.

- I. Bermejo and M. Lejeune: Quelques remarques sur la complexité de calcul des projections d'une courbe projective.
- F. Broglia, F. Acquistapace and P. Vélez-Melon: An algorithmic criterion for basicness in dimension 2.
- A. Capani and L. Robbiano: Some new features of CoCoA.
- L. Ceballos and L. González-Vega: Extending the capabilities of the Maple realroot function by using Thom's codes.
- M. Chardin: Multivariate subresultants.
- J.-P. Dedieu and J.-C. Yakoubshon: Exclusion methods in numerical analysis and symbolic computation.
- D. Dubhasi: (Real) Algebraic geometry and computational geometry.
- I. Duursma: Divisor class groups and weight distributions.
- P. Gimenez: Effective determination of the special fibre of some varieties.
- T. Gómez-Díaz: Jordan forms.
- M.-J. González-Lopez and T. Recio: Kovacs' Conjecture and the inverse kinematics of the general 6R manipulator.
- L. González-Vega and H. Lombardi: A continuous version of Thom's lemma and its applications.
- H. Gräbe: Algorithms in local algebra.
- D. Grigoriev: NC solving of a system of linear ordinary differential equations in several unknowns.
- M. Kreuzer: Computing 0-dimensional schemes with COP.
- B. Lopez: The Construction of a basis for $L(G)$ on the Drinfeld modular curve $X_1(I)$.
- M. McGettrick: Integrated computation: linking Axiom and NAG.
- J. L. Montaña, E. Morais, and L. M. Pardo: Sum of Betti numbers and parallel time.
- A. Montes and J. Castro: Solving the load flow problem using Gröbner basis.
- M. Niermann: Linear methods in Gröbner basis theory.
- D. O'Shea: Computing limiting normals to real surfaces.
- A. H. Park: An algorithmic proof of the Suslin's stability theorem.
- F. Piras: Some remarks on Macaulay's inverse systems.
- P. Pisón, E. Briales, J. Borrego, and M.-J. Pérez: Computing with ideals of semigroups.
- R. Rioboo: Infinitesimals and real closure.
- T. Sander: Effective algebraic geometry over fields not algebraically closed.
- J. Schmid: Complexity of the real Nullstellensatz in the 0-dimensional case.
- H. Schönemann: The computer algebra system SINGULAR: applications in algebraic geometry.
- T. Siebert: Effective implementation of standard bases and syzygies.
- A. Yger: Properness and residues.

They all contributed to a pleasant working atmosphere during the conference days in Santander, while the outside atmosphere was simply horrible: most participants will forever remember the hurling winds and pouring rain, unfortunately quite typical of the North of Spain during early spring. But the conference was also a landmark for other, more pleasant, reasons. Following an initiative of Prof. T. Mora, this event received, for the first time, support through the HCM (Human Capital and Mobility) Euroconference Program, contract number ERB-CHEC-CT93-0109. Under the acronym of "GAME", this contract allowed 22 young European researchers from seven different countries to attended the meeting. A second important fact was the organization of two scientific events of international character around the conference, namely the Workshop on Complexity organized the weekend after MEGA by the experts of the ESPRIT/BRA project PoSSo; and the business meeting and dinner that gathered principal investigators or their deputies for most European research projects related to the topics of the Conference, working towards the establishment of several permanent information exchange procedures.

The editors of this volume are, as is already traditional for MEGA conferences, also the local organizers. Therefore, it is a pleasure for us to thank the many national and local institutions and persons who supported financially and logistically the Conference: first of all our thanks to the Departamento de Matemáticas, Estadística y Computación and the Facultad de Ciencias and its Dean, Prof. M. Arrate, who helped so much with the logistics; next, to the financial help provided by the Vicerrectorado de Investigación de la Universidad de Cantabria through the STRIDE Program; and to the Ministerio de Educación y Ciencia and its grant DGICyT CO9-0095. And, last but not least, our gratitude goes to the enthusiastic team of young volunteer students who managed, as professionals, the logistics of the Conference during a vacation period.

The reader will have gained the impression, after reading the above, that many people and a lot of effort have been required for the final outcome of this volume. If the reader is not satisfied with what he/she finds, it is, perhaps, the fault of

The Editors,

Laureano González-Vega and Tomás Recio

Progress in Mathematics, Vol. 143, © 1996 Birkhäuser Verlag Basel/Switzerland

Zeros, multiplicities, and idempotents for zero-dimensional systems

M.-E. Alonso, E. Becker, M.-F. Roy, T. Wörmann*

1 Introduction

We want to propose alternative computational methods for dealing with the following three classical problems in the study of zero-dimensional systems, rephrased in the context of finite-dimensional algebras over a field k of characteristic zero. It is the main feature of our approach to adapt to the affine case the concept of the u-Chow form (or u-resultant) which was developed in the projective case (and has been used by several authors, e.g., [Ca] and [Re]).

Let $\mathcal{I}$ be a zero-dimensional ideal of $k[X_1, \ldots, X_n]$, and set

$$A = k[X_1, \ldots, X_n]/\mathcal{I}$$

and $V(\mathcal{I}) = \{\alpha \in \bar{k}^n | f(\alpha) = 0 \text{ for every } f \in \mathcal{I}\}$, where $\bar{k}$ denotes the algebraic closure of k.

1) Compute $V(\mathcal{I})$

In a widely known method this is done by computing the radical of $\mathcal{I}$ in the shape lemma form, i.e., expressing one coordinate as the zeros of a univariate polynomial and describing the other coordinates as polynomial functions of the first one. This can be done if a coordinate (or a linear combination of them) is found such that the projection of the variety onto that coordinate is injective. $V(\mathcal{I})$ is then said to be in general position with respect to that coordinate. All this is not only a time-consuming process, but also the output is very inconvenient (see the examples in section 6).

An inspection of several examples for the shape lemma shows that the coefficients of the univariate polynomial are usually of moderate size, whereas the coefficients of the other polynomials tend to be very large. How does one cope with this situation and arrive at a numerically nicer expression of the zeros is displayed in section 2. There are two main ingredients. First we introduce the u-Chow polynomial in section 2. Second, and important at least from a technical point, we slightly generalize the notion of being in general position: an element $u \in A$ is called separating if the function $V(\mathcal{I}) \to \bar{k}, \alpha \mapsto u(\alpha)$ is injective. A separating element is used to "encode" the points in $V(\mathcal{I})$ as the zeros of a univariate polynomial, e.g., the u-Chow polynomial $\chi(u: T)$ in section 2.

(*) Partially supported by POSSO, Esprit BRA 6846, 2nd and 4th author also acknowledge support from DFG

According to the multiplicities of the points, the variety $V(\mathcal{I})$ is split into subsets $V_\mu(\mathcal{I})$ of points of equal multiplicity μ. The u-Chow polynomial will then be used to express the coordinates of the points $\alpha \in V_\mu(\mathcal{I})$ as rational functions of the zeros of $\chi(u;T)$.

In section 3, the ideas above are used in a modified way. Disregarding the splitting of $V(\mathcal{I})$ according to multiplicities we obtain a global rational expression of zeros. Moreover, what we call the generalized shape lemma supplies a set of generators of the radical $\sqrt{\mathcal{I}}$ by somewhat explicit formulae.

2) Computing multiplicities
In section 2, the u-Chow polynomial is used to derive which multiplicities may occur. Section 3.1 contains an explicit description of the function $\mu : V(\mathcal{I}) \to \mathbf{N}$, $\alpha \mapsto \mu_\alpha =$ the multiplicity of α. In fact, we provide a polynomial (or rational) function on $V(\mathcal{I})$ representing μ. Consequently the multiplicity μ_α can be computed once a sufficiently good approximation for α is known.

3) Computing idempotents
Section 3.2 is concerned with the decomposition of A into its local components. Equivalently this amounts to computing the indecomposable idempotents in A. This can be achieved either by using the u-Chow polynomial of section 2.1 or by appealing to the minimal polynomial of section 2.3.

We apply our methods in two different contexts.
1) When the structure of the ring is explicitly given by a Gröbner basis, we shall follow the philosophy of several papers (e.g., [FGLM], [PRS], [BW2]) and explain how to make the computations starting from the multiplication tables in the ring A, with methods working in polynomial time in the output of the Gröbner basis computation.
2) When the ring is complete intersection, we shall make a deformation of the equations and perform the computations in a modified ring where the Gröbner basis is immediate. The complexity will be polynomial in d^n where d is the degree and n is the number of variables. We shall then explain how to obtain information in the original ring.

2 The Chow polynomial and expressing zeros

We keep the notations introduced above; let x_i denote the canonical images under the mapping $k[X_1, \ldots, X_n] \to A$. Let us denote by μ_α the multiplicity of a zero α of $\bar{A} = A \otimes \bar{k}$, that is, the dimension as a vector space of the local factor $\bar{A}_\alpha$ over $\bar{k}$.

2.1 The u-Chow polynomial

The following can be directly proved by passing to the scalar extension $\bar{A}$, decomposing $\bar{A}$ into its local factors $\bar{A}_\alpha$ of dimension μ_α and taking into account $\bar{A}_{red} = A_{red} \otimes \bar{k} \simeq \prod_{\alpha\in V(\mathcal{I})} \bar{k}_\alpha$, where $\bar{k}_\alpha = \bar{k}$ for every $\alpha \in V(\mathcal{I})$.
The induced mapping $A \to \prod_{\alpha\in V(\mathcal{I})} \bar{k}_\alpha$ is the evaluation map

$$u \mapsto (u(\alpha))_{\alpha\in V(\mathcal{I})}.$$

2.1 Definition: For a given $u \in A$, we define an equivalence relation on $V(\mathcal{I})$:

$$\alpha \sim \beta \;:\Longleftrightarrow u(\alpha) = u(\beta).$$

$[\alpha]$ will mean the equivalence class with respect to $\sim$. By $\chi(u;T)$ we mean the characteristic polynomial of the k-linear multiplication map

$$M(u) : A \to A, v \mapsto u \cdot v.$$

2.2 Lemma: Defining $\mu_{[\alpha]} := \sum_{\beta\in[\alpha]} \mu_\beta$ we have

$$\chi(u;T) = \prod_{[\alpha]} (T - u(\alpha))^{\mu_{[\alpha]}}.$$

Proof: The proof is an easy consequence of [PRS].
2.3 Corollary:
i) The trace $\mathrm{tr}(u)$ of $M(u)$ is

$$\sum_{\alpha\in V(\mathcal{I})} \mu_{[\alpha]} u(\alpha).$$

ii) The determinant of $M(u)$ is

$$\prod_{\alpha\in V(\mathcal{I})} u(\alpha)^{\mu_{[\alpha]}}.$$

Looking at the above formula for $\chi(u;T)$, we see that the multiplicities μ_α appear as the multiplicities of the eigenvalues only if all values $u(\alpha)$ are pairwise distinct. Therefore we introduce the following

2.4 Definition:
i) An element $u \in A$ is called separating if the mapping $V(\mathcal{I}) \to \bar{k}, \alpha \mapsto u(\alpha)$ is injective. Equivalently this means that under the canonical surjection $A \to A_{red}$, the image of u is a generating element for A_{red} if and only if u is a separating element (e.g., [BW2] for this and related matters below).
ii) If $u \in A$ is a separating element, the characteristic polynomial $\chi(u;T)$ is called the u-**Chow polynomial**.

2.5 Corollary: If u separates the zeros of $\mathcal{I}$, then the multiplicities of the roots of $\chi(u;T)$ are the multiplicities of the zeros of $\mathcal{I}$.

2.2 Expressing zeros according to multiplicity

Having defined the u-Chow polynomial and supposing we found a u separating the zeros, we can use the method described in [Re] to express the zeros as functions of u. We define $g(u;T)$ as the derivative of $\chi(u;T)$ with respect to T.

We consider a polynomial $\chi(u,v;T,S)$ which is the characteristic polynomial of the multiplication by $u+Sv$ in $A\otimes_k k(S)$, where S is a new indeterminate, and we define $g(u,v;T)$ as the derivative of $\chi(u,v;T,S)$ with respect to S taken in $S=0$.

2.6 Proposition: Let $V_\mu := \{\alpha \in V(\mathcal{I}) | \mu_\alpha = \mu\}, u \in A$ a separating element and $v \in A$ any element. Then we have the following properties:
i)

$$g^{(\mu-1)}(u;u(\alpha)) \neq 0 \ \forall \alpha \in V_\mu,$$

ii)

$$v = -\frac{g^{(\mu-1)}(u,v;u)}{g^{(\mu-1)}(u;u)} \text{ on } V_\mu,$$

on V_μ.
iii)

$$x_i = -\frac{g^{(\mu-1)}(u,x_i;u)}{g^{(\mu-1)}(u;u)} \text{ on } V_\mu.$$

Here $g^{(i)}(u;T)$ (resp. $g^{(i)}(u,v;T)$) is the i-th derivative of $g(u;T)$ with respect to T (resp. $g(u,v;T)$).

Proof of the proposition:
The proposition results clearly from the formulas

$$\chi(u;T) = \prod_{\alpha\in V(\mathcal{I})} (T-u(\alpha))^{\mu_\alpha}$$

and

$$\chi(u,v;T,S) = \prod_{\alpha\in V(\mathcal{I})} (T-(u(\alpha)+Sv(\alpha)))^{\mu_\alpha}$$

which implies

$$g^{(\mu_\alpha-1)}(u;T) =$$
$$= (\mu_\alpha!)\left[\left(\prod_{\beta\in V(\mathcal{I})\beta\neq\alpha} (u(\alpha)-u(\beta))^{\mu_\beta}\right) + (T-u(\alpha))\cdot h_1(T)\right]$$
$$g^{(\mu_\alpha-1)}(u,v;T) =$$
$$= -(\mu_\alpha!)v(\alpha)\left[\left(\prod_{\beta\in V(\mathcal{I})\beta\neq\alpha} (u(\alpha)-u(\beta))^{\mu_\beta}\right) + (T-u(\alpha))\cdot h_2(T)\right],$$

for some polynomials h_1, h_2. Substituting $u(\alpha)$ gives the announced result. □

2.3 The generalized shape lemma

If we are not interested in multiplicities we can use the concept of the u-Chow polynomial in a modified form to obtain another well-behaved description of the variety $V(\mathcal{I})$ through polynomials with small coefficients generating the ideal $\sqrt{\mathcal{I}}$. We not only have a look at the characteristic polynomial $\chi(u;T)$ of the multiplication by $u \in A$ but also at its minimal polynomial $m(u;T)$, which is also the minimal polynomial of u viewed as an element of the algebra. Even more important for our approach is what we call the reduced minimal polynomial $\bar{m}(u;T)$, by definition the minimal polynomial of $\bar{u} := u + Nil(A) \in A_{red} := A/Nil(A)$, where $Nil(A)$ denotes the (nilpotent) radical of A.

According to definition 2.4, an element $u \in A$ is separating if and only if the mapping $V(\mathcal{I}) \to \bar{k}, \alpha \mapsto u(\alpha)$ is injective. Note that

$$A_{red} = k[X_1, \ldots, X_n]/\sqrt{\mathcal{I}}$$

and, given $u, v \in A$, $\bar{u} = \bar{v}$ if and only if $u(\alpha) = v(\alpha)$ for every $\alpha \in V(\mathcal{I})$. Using the terminology developed above we get the following:

2.7. Lemma: For any $u \in A$ we have:
i)

$$m(u;T) = \prod_{[\alpha]} (T - u(\alpha))^{r_{[\alpha]}}, \quad 1 \leq r_{[\alpha]} \leq \mu_{[\alpha]}$$

ii)

$$\bar{m}(u;T) = \prod_{[\alpha]} (T - u(\alpha)).$$

We see that $\bar{m}(u;T)$ is the square-free kernel of $\chi(u;T)$. We will make use of the derivative

$$h(u;T) = \sum_{[\alpha]} \prod_{[\beta] \neq [\alpha]} (T - u(\beta))$$

of $\bar{m}(u;T)$. Substituting u, we get

$$h(u;u)(\gamma) = \left(\sum_{[\alpha]} \prod_{[\beta] \neq [\alpha]} (u - u(\beta)) \right) (\gamma) = \prod_{[\beta] : [\beta] \neq [\gamma]} (u(\gamma) - u(\beta)) \neq 0.$$

i.e., $h(u;\bar{u})$ is a unit in A_{red}. Therefore $h(u;u)$ is a unit in A since $Nil(A)$ is nilpotent.

We are next going to derive a formula which is suggested by the idea of the u-Chow polynomial and which is the key for our present approach. To this end we fix a separating element $u \in A$ and let $v \in A$ be arbitrary. Now, as in the case of the u-Chow polynomial, we pass to the base field extension $A \otimes_k k(S)$, where S is a new variable. Consider in the latter algebra the element $u + Sv$, which turns out to be separating as well. Therefore

$$\bar{m}(u + Sv; T) = \prod_{\alpha \in V(\mathcal{I})} (T - (u(\alpha) + Sv(\alpha))) .$$

We see that $\bar{m}(u + Sv; T)$ is a polynomial in $k[T, S]$ and we set

$$h(u, v; T) := \left(\frac{\partial}{\partial S} \bar{m}(u + Sv; T) \right)_{S=0} .$$

Hence,

$$h(u, v; T) = -\sum_{\alpha} v(\alpha) \prod_{\beta : \beta \neq \alpha} (T - u(\beta)) .$$

Substituting we get

$$(h(u, v; u))\,(\alpha) = h(u, v; u(\alpha)) = -v(\alpha)(h(u; u))(\alpha),$$

showing that $h(u, v; u)$ equals $-vh(u; u)$ as a function on $V(\mathcal{I})$ or, equivalently, that their sum lies in $Nil(A)$. We summarize:

2.8. Lemma: Given a separating element u and any $v \in A$ we can construct a polynomial $h(u, v; T)$ such that

$$v + \frac{h(u, v; u)}{h(u; u)} \in Nil(A)$$

or, equivalently, that

$$v = -\frac{h(u, v; u)}{h(u; u)}$$

as a function on $V(\mathcal{I})$.

The formula of the lemma provides a set of generators of the radical of a zero-dimensional ideal by an explicit formula. According to their shape we call the following result the

2.9. Generalized shape lemma: Let $\mathcal{I} \subset k[X_1, \ldots, X_n]$ be a zero-dimensional ideal and the characteristic of k equal to 0. Choose $U \in k[X_1, \ldots, X_n]$ such that $u := U + \mathcal{I}$ is a separating element. Then

$$\sqrt{(\mathcal{I})} = (f(U), f'(U) \cdot X_1 + h_1(U), \ldots, f'(U) \cdot X_n + h_n(U)) ,$$

where $deg(h_i) < deg(f), i = 1, \ldots, n,\ f(U) = \bar{m}'(u; U), f'(U) = h(u; U) = \bar{m}'(u; U)$ and $h_i(U) = h(u, x_i; U)$.

The proof uses lemma 3.2 and the fact that $\bar{m}(u; T)$ and $\bar{m}'(u: T)$ are coprime.

Remark: The theorem above also holds in characteristic $p > 0$. if A_{red} is separable.

If U is equal to one of the variables, then the above expression reduces to the usual shape lemma. In fact, assume $U = X_1, u = x_1$ and set $H(T) = f'(T) \cdot T + h_1(T)$. Since $H(x_1) \in Nil(A)$ we get $H(\bar{x_1}) = 0$ in A_{red}. Thus $H(X_1)$ can be dropped and we get

$$\sqrt{(\mathcal{I})} = (f(X_1), f'(X_1) \cdot X_2 + h_2(X_1), \ldots, f'(X_1) \cdot X_n + h_n(X_1)) .$$

Using ithe fact that f and f' are coprime, one finally derives the usual generators of type $X_i - k_i(X_1), i = 2, \ldots, n$. The generalized shape lemma is closely related to Rückert's parametrization in the local analytic case (see e.g., [Na], [Ru]) as was kindly pointed out to us by the referee.

2.4 Size of coefficients

We have found in 2.6. (resp. 2.8) descriptions for the coordinate functions as rational functions on the subvariety of points of equal multiplicity (resp. as rational functions on the whole set of zeros). The coefficients of the numerator and denominator obtained by these methods are clearly of the same order of size. One may calculate for each multiplicity the inverse of the denominator (since this denominator is coprime with the polynomial in the square free decomposition of $\chi(u; T)$ corresponding to roots of a given multiplicity) and derive a polynomial expression of v in function of u instead. Equally, in (2.8), $h(u; u)$ is a unit in A. Hence one may calculate its inverse and derives a polynomial expression of v instead.
The inverting of these denominators accounts for the appearance of high coefficients in the polynomial expression. Therefore it seems advisable to stay with the rational expressions as long as possible in a given context.

Let us explain heuristically, in the shape lemma case, why in a lexicographical Gröbner basis the coefficients of the univariate polynomial are small while the coefficients of the other polynomials are big, and why in our methods all coefficients are small. We start from a zero-dimensional ideal I. Suppose that X_1 is separating, that a Gröbner basis G of I is already computed (for any order we want), and denote by N the dimension of quotient A and by t the size of the coefficients in the multiplication tables by the variables in the basis of monomials associated to G. Then the size of the coefficients in the generalized shape lemma presentation is bounded by $O(tN)$. If we invert the denominators in order to get the shape lemma representation and thus get a lexicographical Gröbner basis, the size of coefficients gets bounded by $O(tN^2)$.

These estimates correspond to what we observed in the examples we computed: If the length of the long integers in the lexicographical Gröbner basis are of order T, the length of integers in the generalized shape lemma presentation is of order $T^{1/2}$.

An additional remark seems in order. In getting the zeros of $\chi(u;T)$ or $\bar{m}(u;T)$, one may appeal to numerical methods. The formula of the lemma yields simultaneous approximations of the coordinates with a control of the error of approximation. This plays a key role in section 3, where we show how to compute the multiplicity out of a good approximation of the zero.

Let us remark lastly that even if an ideal is already given in the usual shape lemma form, it can be transformed directly by a few computations to the generalized shape lemma form to get a smaller representation of the problem. This is how the examples shown in this paper were computed.

We shall discuss computational strategies in section 4. Efficient computational methods for making the computations proposed in 2.6 or 2.8 are important for practical purposes.

3 Computing idempotents and multiplicities

3.1 Multiplicities

In this part we will again use the notion of a separating element $u \in A$ to find the multiplicity of a root α in case that a numerical approximation of α or $u(\alpha)$ (i.e., a root of $\chi(u;T)$ or $\bar{m}(u;T)$) is known. Let d be the degree of the squarefree kernel $\bar{m}$ of the characteristic (or minimal) polynomial of $M(u)$: $A \to A, x \mapsto ux$. By $M_{red}(a) : A_{red} \to A_{red}, b \mapsto ab$ we mean the multiplication map on A_{red}. Accordingly, we consider the trace function $tr_{red} : A \to k$.

3.1 Lemma: We can construct an element $N \in A$ such that

$$N(\alpha) = \mu_\alpha.$$

Or, to be more precise, there exists a polynomial $F \in k[T]$ of degree at most $d-1$ such that

$$\mu_\alpha = F(u(\alpha)),$$

α being any point in $V(\mathcal{I})$.

Proof: We consider the system of equations

$$tr(u^j) = tr_{red}(N \cdot \bar{u}^j) \;\; \forall j = 0, \ldots, d-1.$$

If we put $N = \sum_{i=0}^{d-1} a_i \bar{u}^i$, we get equivalently the following system of linear equations:

$$\begin{pmatrix} tr_{red}(1) & \dots & tr_{red}(\bar{u}^{d-1}) \\ \vdots & \vdots & \vdots \\ tr_{red}(\bar{u}^{d-1}) & \dots & tr_{red}(\bar{u}^{2d-2}) \end{pmatrix} \cdot \begin{pmatrix} a_0 \\ \vdots \\ a_{d-1} \end{pmatrix} = \begin{pmatrix} tr(1) \\ \vdots \\ tr(u^{d-1}) \end{pmatrix}.$$

This system has a unique solution since the matrix is regular (see [BW2]), so we get an element $N \in A_{red}$. The traces inside the matrix can be computed easily since A_{red} is isomorphic to $k[T]/(\bar{m}(u;T))$ and hence the traces $tr_{red}(\bar{u}^j)$ are equal to the trace of the multiplications by $t^j := T^j + \bar{m}(u;T)$ in $k[T]/(\bar{m}(u;T))$. Using $tr(u^j) = tr_{red}(N \cdot \bar{u}^j)$ and

$$tr(u^j) = \sum_{\alpha \in V(\mathcal{I})} \mu_\alpha \cdot u(\alpha)^j,$$

$$tr_{red}(N \cdot \bar{u}^j) = \sum_{\alpha \in V(\mathcal{I})} N(\alpha) \cdot u(\alpha)^j$$

we get $N(\alpha) = \mu_\alpha$ since u is injective as a function on the variety.

Looking now at the element $Z := \sum_{i=0}^{d-1} a_i u^i \in A$ we see that $\bar{Z} = N$, and hence Z and N have the same values on the variety. Clearly $F(T) := \sum_{i=0}^{d-1} a_i T^i$ is the polynomial searched for.

Remark: If we want to describe μ as a rational function, we start with the system of equations

$$tr(h(u;u) \cdot u^j) = tr_{red}(N \cdot \bar{u}^j) \quad \forall j = 0, \dots, d-1$$

and proceed like in the proof above we get that on $V(\mathcal{I})$

$$\mu_\alpha = \frac{F(u(\alpha))}{h(u;u(\alpha))}.$$

3.2 Idempotents

We start by listing the following facts about idempotents in finite-dimensional k-algebras. Since $Nil(A)$ has finite nilpotence-index, i.e., $Nil(A)^r = 0$ for some r, we derive that the canonical epimorphism

$$A \to A_{red} = A/Nil(A)$$

induces a bijection between the idempotents of A and A_{red} where indecomposability of idempotents is preserved.

Next suppose that u is a separating element in A. Then

$$A_{red} = k[\bar{u}] \simeq k[T]/(\bar{m}(u;T))$$

and the indecomposable idempotents of A_{red} are in one-to-one correspondence with the irreducible factors of $\bar{m}(u;T)$.

Tracing this back to A we obtain that again the the indecomposable idempotents e of A correspond to the irreducible factors p of $\bar{m}(u;T)$ where e is assigned to p if and only if $\bar{e} \equiv 1 \mod p(\bar{u})$ and $\bar{e} \equiv 0 \mod q(\bar{u})$ for every irreducible factor q of $\bar{m}(u;T)$ different from p.
Next let e and p be related as just described. Then $B := Ae$ is a local component of A with residue field $k[T]/(p(T))$. After base field extension to the algebraic closure we get

$$B \otimes \bar{k} = \oplus \bar{A}_\alpha,$$

where α ranges over those zeros in $V(\mathcal{I})$ such that $p(u(\alpha)) = 0$. The local components are conjugate under the natural action of the Galois group $G(\bar{k}|k)$. Hence we deduce

3.2 Proposition: In the situation above

$$\mu_\alpha = \frac{dim_k(Ae)}{deg(p(T))},$$

where $p(u(\alpha)) = 0$ and the idempotent e belongs to p.

The considerations above can be turned into explicit constructions of idempotents. We start with a separating element $u \in A$ and a polynomial $f(T) \in k[T]$ annihilating u in A, i.e., $f(u) = 0$ in A (e.g., $f(T) = \chi(u;T)$ or $f(T) = m(u;T)$ may serve). Then, clearly, $\bar{m}(u;T)|f(T)$. Next let $p(T)$ be an irreducible factor of $\bar{m}(u;T)$. Hence $f(T) = p^r(T) \cdot f_1(T)$ where $p(T)$ does not divide $f_1(T)$. Writing down the Bezout identity between the coprime polynomials $p^r(T)$ and $f_1(T)$, we end up with an identity

$$1 = g(T) + h(T),$$

where $p^r(T)|g(T), f_1(T)|h(T)$. Substituting u we get $1 = g(u) + h(u)$. Furthermore, $f(T)$ divides $g(T) \cdot h(T)$ showing $g(u) \cdot h(u) = 0$ in A. Hence, $h(u) = h(u)^2$.

In view of the criterion above we see that $h(u)$ is the indecomposable idempotent belonging to $p(T)$. Then

$$A \cdot h(u) = ker M(g(u)),$$

and, as an algebra

$$A \cdot h(u) \simeq k[X_1, \ldots, X_n]/(\mathcal{I}, g(U)), \quad u = U + \mathcal{I}$$

is the corresponding local component. The multiplicities of zeros $\alpha \in V(\mathcal{I})$ with $p(u(\alpha)) = 0$ can easily be computed once $h(u)$ has been obtained.

4 Computational strategies

4.1 Finding separating elements

Since it is known that already an element of the form $u_1x_1 + \cdots + u_nx_n$, $u_1, \ldots, u_n \in k$ is separating (see [BW2]), we consider the linear combination

$$U := U_1x_1 + \cdots + U_nx_n \in A \otimes k(U_1, \ldots, U_n)$$

which is a separating element for $A \otimes k(U_1, \ldots, U_n), with U_1, \ldots, U_n$ being a new set of indeterminates. There are several methods for finding a vector u, a specialization of U, that separates the points of $\mathcal{I}$.

One method consists in specializing $(U_1, \ldots, U_n)$ to $(1, V, \ldots, V^{n-1})$ where V is a new variable, specializing U to u accordingly. Compute $\chi(u;T)$, then the sequence of subresultants of $\chi(u;T)$ (considered as a polynomial in T). Since V is a variable, u separates the points and the degree of the first non-zero subresultant is exactly $\sum(\mu_\alpha - 1)$. Choose a value v of V such that the first non zero of these subresultants is not identically 0 when v is substituted to V. Specialize now $(U_1, \ldots, U_n)$ to $(1, v, \ldots, v^{n-1})$. The accordingly obtained specialization u of U is a separating element.

Another method for finding a separating u is described in [Re].

4.2 The affine computations

We suppose now that we have a basis of the vector space A and the multiplication table of A in this basis. This will be the case as soon as we know a Gröbner basis of A (for any ordering). We are now able to compute a separating u, the u-Chow polynomial $\chi(u;T)$, and the polynomials $\chi(u, x_i; T, S)$ using, for example, the definitions through caracteristic polynomials. We can also compute the reduced minimal polynomial of u by, e.g., checking the successive powers $1, u, u^2, \ldots$ for linear dependency. Moreover we can compute the generalized shape lemma form by terms of the following

Algorithm GSL

Input: A set of generators of the ideal $\mathcal{I}$.
Output: $f, h_1, \ldots, h_n$ as in the generalized shape lemma.

- Compute a Gröbner basis of the ideal with respect to any admissible term ordering. Check whether the ideal is zero-dimensional; if not, leave the algorithm. Determine the multiplication table of A and $N = dim_k A$.
- Find a separating element $u \in A$ and determine the square-free kernel f of the characteristic or minimal polynomial of $M(u)$. Set $d := deg(f)$.

- Determine $h_1, \ldots, h_n$ subject to

$$tr((f'(u) \cdot x_i + h_i(u)) \cdot u^j) = 0 \ \ \forall j = 0, \ldots, d-1,$$

or if we write $h_i(X) = a_0^i + \cdots + a_{d-1}^i X^{d-1}$, we get the systems of linear equations

$$\begin{pmatrix} tr(1) & \ldots & tr(u^{d-1}) \\ \vdots & \vdots & \vdots \\ tr(u^{d-1}) & \ldots & tr(u^{2d-2}) \end{pmatrix} \begin{pmatrix} a_0^i \\ \vdots \\ a_{d-1}^i \end{pmatrix} = \begin{pmatrix} tr(-f'(u)x_i) \\ \vdots \\ tr(-f'(u) \cdot x_i \cdot u^{d-1}) \end{pmatrix}.$$

This means inverting one regular Hankel-matrix of size $d \times d$.

See [BW2] for a proof of correctness and complexity of the algorithm.

Also, having computed the idempotent associated to α, we can compute the structure of the local ring A_α by projecting over A_α, i.e., by multiplying the basis of the vector space A by $h(u)$. In particular, we can find a basis of the vector space A_α by doing linear algebra in A.

Note that all the computations in 4.2 can be done in polynomial time in the dimension D of the vector space A (and linear in the size t of the integers appearing in the multiplication tables of A).

4.3 The complete intersection case

We have a complete intersection zero-dimensional algebra A given by equations $(p_1, \ldots, p_n)$ of degree bounded by d and we suppose that the origin is one of its zeros. This situation is important, e.g., for computing topological degrees (see [LS]).

We shall not work with the original equations since the corresponding quotient ring A may have a complicated structure and computing its Gröbner basis might be expensive, but instead we shall consider the modified equations

$$\phi_1 = \epsilon X_1^N + p_1(X_1, \ldots, X_n), \ldots, \phi_n = \epsilon X_n^N + p_k(X_1, \ldots, X_n)$$

with N big enough (bigger for example than $(D+1)$ where D is the dimension of the vector space A) and ϵ small enough (that is, a new infinitesimal variable), and the ring

$$A' = K[X_1, \ldots, X_n]/(\phi_1, \ldots, \phi_n), \text{where } \epsilon \in K.$$

This idea has been used by several authors (see [CGr], [Ca], [Re]) in the global context. We use it here also for local informations as in [BCRS].

The properties of the ring A' are the following (see [BCRS]). The quotient structure of A' is easy to describe: The polynomials $\phi_1, \ldots, \phi_n$ form a Gröbner basis for the lexicographical ordering, and thus a linear basis of A' is given by the residue classes of the monomials $X^m = X_1{}^{m_1} \ldots X_n{}^{m_n}$ where $m_i \leq N-1$, the multiplication table of elements in A' being given by the reduction process modulo the Gröbner basis $(\phi_1, \ldots, \phi_n)$, the local rings A'_0 and A_0 are isomorphic.

Solving equations

How to solve the equations of A by this method has already being explained in [Ca] and [Re]: One performs the computations indicated in section 2 in the ring A', which are the quotients of polynomials in u, T, ϵ, and in order to get the coordinates of the original zeros in A, one "makes the variable ϵ tend to 0" as explained in [Re] or in [BPR]. Some zeros of A' tend to infinity, and the others tend to the zeros of A.

We obtain at the end the zeros of A in the announced form.

Computing idempotents and local structures

We use the fact that A_0 and A'_0 are isomorphic. We perform the computations indicated in section 3.2 in A' and we get the idempotent associated to 0 in A'. Using this idempotent we can deduce the local structure of A_0 which is isomorphic to A'_0, in particular a basis of linearly independent monomials for any chosen ordering. Note that this computation does not give a priori the idempotent in A, and that at the end the multiplication table of A stays unknown.

Note that all the computations in 4.3 are clearly polynomial in d^n (and linear in the size t of the integers appearing in the equations $p_1, \ldots, p_n$).

5 Examples

In the following example, chosen at random,

$$x^2+xy^2z-2xy+y^4+y^2+z^2, -x^3y^2+xy^2z+xyz^3-2xy+y^4, -2x^2y+xy^4+yz^4-3$$

the output of the lexicographical Gröbner basis is <u>450 lines</u> long (coefficient size is approximately 200 digits). The output given by our method is reduced to 10% compared to the output of the lexicographical Gröbner basis!

The following bivariate example comes from equations considered by H. Caprasse, who computed the lexicographical Gröbner basis. We exhibit the generalized shape lemma's form in the following way. The first entry in each example is the description of the ideal by a lexicographical Gröbner basis with respect to $X < Y$, the first polynomial being a polynomial f in X (with small coefficients), and the second a polynomial of degree 1 in Y of the form $cY - \bar{h}$, where c is an integer and $\bar{h}$ a polynomial in X (with much bigger coefficients). The following entry is the polynomials h obtained by our method, which give the description of the ideal in generalized shape lemma form, i.e.,

$(f(X), f'(X) \cdot Y + h(X))$. One can observe that the coefficients of h are much smaller than c and the coefficients of k.

$$
\begin{gathered}
(81621X^{13} + 675405X^{12} + 1078749X^{11} - 1209917X^{10} + 11263786X^9 \\
-6215992X^8 - 96012986X^7 + 234089060X^6 - 247966892X^5 \\
+161951496X^4 - 109108912X^3 + 85958560X^2 - 43381760X + 8744960
\end{gathered}
$$

$$
\begin{gathered}
-6479010026722119881345625461369103153481978568 1X^{12} \\
-568043526935731232207054677589796122492357375343X^{11} \\
-1126050073604155297184983842272188670719220083635X^{10} \\
+509289294372693186136985887117887720729227929479X^9 \\
-836113795531338528906570018623724326793142433845 6X^8 \\
+1197280905856553024292863271500299304506129153480X^7 \\
+7855580167622734231833177351693987569544607068521 8X^6 \\
-1447253546326254672317963387414506365468995701818 96X^5 \\
+1162903282124117639811580039742213640793141442284 92X^4 \\
-638936724477040253148998673390856236718517578431 04X^3 \\
+534405698549339901101478096718645054567235254367 20X^2 \\
-398018634130980065549501760671744833948890577617 92X \\
+92362848812063315428669629142102482394449631846 4Y \\
+1139710836712982987765092620983360587467425190912 0,
\end{gathered}
$$

$$
\begin{gathered}
1031067X^{12} + 2628006X^{11} - 12139567X^{10} - 9881580X^9 + 108725358X^8 \\
+164726272X^7 - 647777846X^6 + 340231008X^5 + 418343972X^4 \\
-436204688X^3 - 37169360X^2 + 167262720X - 50001920)
\end{gathered}
$$

References

[BPR] S. Basu, R. Pollack, M.-F. Roy: A new algorithm to find a point in every cell defined by a family of polynomials, quantifier elimination and cylindrical algebraic decomposition, B. Caviness, J. Johnson (Eds.), Springer-Verlag, to appear.

[BCRS] E. Becker, J.-P. Cardinal, M.-F. Roy, Z. Szafraniec: Multivariate bezoutians, Kronecker symbol and Eisenbud-Levine formula. In this volume.

[BW1] E. Becker, T. Wörmann: On the trace formula for quadratic forms, recent advances in real algebraic geometry and quadratic forms, Proceedings of the RAGSQUAD year, Berkeley 1990–1991, W. B. Jacob, T.-Y. Lam, R. O. Robson (editors), Contemporary Mathematics 155, pp. 271–291 (1994).

[BW2] E. Becker, T. Wörmann: Radical computations of zero-dimensional ideals and real root counting, to appear in Mathematics and Computers in Simulation.

[Ca88] J.F. Canny: Some algebraic and geometric computations in PSPACE, In Proc. Twentieth ACM Symp. on Theory of Computing. pp. 460-467 (1988).

[Cap] H. Caprasse, Departement d'Astronomie et d'Astrophysique. Institut de Physique B5, Sart Tilman, B-4000 LIEGE.

[CGr] A. Chistov, D. Grigor'ev: Subexponential time solving of systems of algebraic equations, Preprint LOMIE-9-83, Leningrad.

[FGLM] J.C. Faugére, P. Gianni, D. Lazard, T. Mora: Efficient computation of zero-dimensional Gröbner-Bases by change of orderings. unpublished manuscript (1989).

[LS] A. Lecki, Z. Szafraniec: Application of the Eisenbud-Levine's theorem to real algebraic geometry, Proceedings of MEGA 1992.

[Na] R. Narasimhan: Introduction to the theory of analytic spaces, Lecture Notes in Mathematics 25, Springer (1966).

[PRS] P. Pedersen, M.-F. Roy, A. Szpirglas: Counting real zeros in the multivariate case, Computational Algebraic Geometry, Frédéric Eyssette. André Galligo (editors), pp. 203–223 (1993), Birkhäuser.

[Re] J. Renegar: On the computational complexity and geometry of the first-order theory of the reals, parts I, II and III. Journal of Symbolic Computation, 13(3), pp. 255–352, 1992.

[Ru] J. M. Ruiz: The basic theory of power series, Advanced Lectures in Mathematics, Vieweg, Braunschweig/Wiesbaden (1993).

[SchSt] G. Scheja, U. Storch: Lehrbuch der Algebra, Band 2 (1988). Teubner.

M.-E. Alonso (`mariemi@sunal1.mat.ucm.es`)

Departamento de Álgebra, Facultad de Matemáticas. Universidad Complutense, Madrid, Spain.

E. Becker (`becker@emmy.mathematik.uni-dortmund.de`)

FB Mathematik, University of Dortmund. 44221 Dortmund (Germany).

M.-F. Roy (`costeroy@univ-rennes1.fr`)

IRMAR, University of Rennes I, 35042 Rennes Cedex (France).

T. Wörmann (`thorsten@emmy.mathematik.uni-dortmund.de`)

FB Mathematik, University of Dortmund. 44221 Dortmund (Germany).

Progress in Mathematics, Vol. 143, © 1996 Birkhäuser Verlag Basel/Switzerland

On a conjecture of C. Berenstein and A. Yger

F. Amoroso

1 Introduction

Let $d \geq 5$ and $k \geq 1$ be two integers, and let $n = 10k + 1$. A well-known example of E. Mayr and A. Meyer (see [MM]) shows that there are n polynomials $f_1, \dots, f_n \in \mathbf{C}[x_1, \dots, x_n]$ of degree $\leq d$ such that x_1 belongs to the ideal generated by $f_1, \dots, f_n$ that and each solution $a_1, \dots, a_n$ of the equation

$$x_1 = a_1 f_1 + \cdots + a_n f_n, \qquad a_1, \dots, a_n \in \mathbf{C}[x_1, \dots, x_n]$$

satisfies $\max \deg a_i > (d-2)^{2^{k-1}}$. In other words, the growth of the degrees of the polynomial coefficients in the representation problem for an ideal $\mathbf{a} \subseteq \mathbf{C}[x_1, \dots, x_n]$ is, in general, double-exponential.

Given an ideal $\mathbf{a} \subset \mathbf{K}[x_1, \dots, x_n] = \mathcal{R}$ and a positive integer d, we define $\phi_{\mathbf{a}}(d)$ as the minimum integer D such that for all systems of generators $\{f_1, \dots, f_m\}$ of $\mathbf{a}$ with $\deg f_i \leq d$ and for all $f \in \mathbf{a}$, we can find a representation

$$f = a_1 f_1 + \cdots + a_m f_m$$

with

$$\max_i \deg a_i \leq \deg f + D.$$

A classical result of Hermann (see [H]) shows that $\phi_{\mathbf{a}}(d) \leq 2(2d)^{2^{n-1}}$ for all ideals $\mathbf{a}$. In 1991, Krick and Logar (see [KL]), using effective linear algebra techniques, improved the previous bound to $\phi_{\mathbf{a}}(d) \leq d^{O(n^2 3^r)}$, where r is the dimension of $\mathbf{a}$. On the other hand, the quoted result of Mayr-Meyer gives an ideal $\mathbf{a}$ for which

$$\phi_{\mathbf{a}}(d) > (d-2)^{2^{(n-11)/10}}.$$

Therefore, it seems to be interesting to give algebraic conditions on $\mathbf{a}$ to avoid the double exponential growth of $\phi_{\mathbf{a}}$. Several authors have written papers on this subject, working with different techniques: elimination theory, complex analysis, algebraic geometry, Gröbner basis theory, algebraic complexity theory. The principal results we know are the following ones.

First, for $\mathbf{a} = \mathcal{R}$ the problem is reduced to the effective Nullstellensatz. In 1987, Brownawell (see [B]), using the theory of Chow forms and a theorem of Skoda, found $\phi_R(d) \leq 3n^2 d^n$ provided that car $\mathbf{K} = 0$. In 1988, Caniglia-Galligo-Heintz (see [CGH]), with simple algebraic arguments, proved the weaker result $\phi_R(d) \leq d^{n(n+3)/2}$ without any assumption on $\mathbf{K}$. The same year Kollár (see [K]), using algebraic geometry, proved $\phi_R(d) \leq d^n$ without any assumption on $\mathbf{K}$ but under the technical condition $d \geq 3$. A similar result was found by Philippon (see [P1]), who used the homology of Koszul complex instead of local cohomology, and by Fitchas-Galligo (see [FG]).

Also for zero-dimensional ideals and for complete intersection ideals, $\phi_{\mathbf{a}}$ grows exponentially. In 1990, Berenstein-Yger (see [BY1]) proved with analytic methods $\phi_{\mathbf{a}}(d) \leq 3(n+1)d^n$ for zero-dimensional ideals and $\phi_{\mathbf{a}}(d) \leq 6(n+1)^2(2k^{k+1}+n)d^k$ for complete intersection ideals of codimension k, again under the assumption car $\mathbf{K} = 0$. In 1991, Dickenstein-Fitchas-Giusti-Sessa (see [DFGS]) obtained from Gröbner basis theory $\phi_{\mathbf{a}}(d) \leq nd^{2n} + d^n + d$ (if dim $\mathbf{a} = 0$) and $\phi_{\mathbf{a}}(d) \leq d^k$ (if $\mathbf{a}$ is a complete intersection of codimension k). In 1989, the author (see [A1] and also [A2] for an errata-corrigé), using Kollár-Philippon's method, showed that $\phi_{\mathbf{a}}(d) \leq d^n + d - 1$ provided that $d \geq 3$ for all zero-dimensional ideals, and that $\phi_{\mathbf{a}}(d) \leq d^{k(n-k+1)} + d^k + 1$ $(d \geq 3)$ for unmixed, locally complete intersection ideals of codimension k.

A problem which is related but not equivalent to the previous one is the membership problem: For a given ideal $\mathbf{a} \subset \mathcal{R}$, decide whether f belongs to $\mathbf{a}$. Of course, a solution for the representation problem gives a solution for the membership problem, but the converse is not true. A strong result of Dickenstein-Fitchas-Giusti-Sessa (see [DFGS]) shows that the membership problem for unmixed ideals is solvable in single exponential time.

Last but not least, the analogous diophantine problem: We assume $f_1, \ldots, f_m \in \mathcal{R}' = \mathbf{Z}[x_1, \ldots, x_n]$ and we try to find a representation

$$\lambda f = a_1 f_1 + \cdots + a_m f_m, \qquad \lambda \in \mathbf{Z}\backslash\{0\},\ a_1, \ldots, a_m \in \mathcal{R}'$$

for an arbitrary $f \in (f_1, \ldots, f_m)$, with good bounds not only for deg a_i but also for λ and for $H(a_i)$ (=max |coefficients (a_i)|). In 1991, Berenstein-Yger (see [BY2] and [BY3]), combining analytic methods with a theorem of Philippon (see [P2]), solved this problem when $f_1, \ldots, f_m$ don't have common zeros in $\mathbf{C}^n$. Recently, Elkadi (see [E]) and Krick-Pardo (see [KP1] and [KP2]) independently have also found good bounds when $f_1, \ldots, f_m$ is a regular sequence.

Instead of looking at a representation of $f \in \mathbf{a}$, it seems interesting also to try a representation of "small" powers of f. Given two positive integers e and d, let us define $\phi_{\mathbf{a}}(e, d)$ as the minimum integer D such that for all systems of generators $\{f_1, \ldots, f_m\}$ of $\mathbf{a}$ with deg $f_i \leq d$ and for all $f \in \mathbf{a}$, we can find a representation

$$f^e = a_1 f_1 + \cdots + a_m f_m$$

with

$$\max_i \deg a_i \leq e \deg f + D.$$

Obviously, $\phi_{\mathbf{a}}(e, d) \leq \phi_{\mathbf{a}}(d)$. Moreover, the effective Nullstellensatz implies $\phi_{\mathbf{a}}(e, d) \leq d^n$, provided that $d \geq 3$ and $e \geq d^n$. The study of the applications of integral representation formulas (such as the Weil formula or the weighted Bochner-Martinelli formula) suggests that $\phi_{\mathbf{a}}(n, d)$ is bounded essentially by d^n for all ideals $\mathbf{a}$ (see [BY4]). Recently (see [A2]) the author has proved this conjecture for one-dimensional ideals (not necessarly unmixed): $\phi_{\mathbf{a}}(n, d) \leq d^n + d - 1$ for $d \geq 3$. The proof combined Kollár-Philippon's method with the theory of reduction of ideals (developed by Northcott and Rees), a theorem of Lipman-Tessier (see [LT]) on the integral closure of ideals, Bertini's theorem, and a theorem on Gruson-Lazarsfeld-Peskine (see [GPL]) on the regularity of the Hilbert's function of a reduced ideal of dimension 1. In this paper we generalize this result, obtaining $\phi_{\mathbf{a}}(3^n, d) \leq d^n + d$ ($d \geq 3$) for all ideals $\mathbf{a}$.

More precisely, let us define for an ideal $\mathbf{a}$ in a ring $\mathcal{R}$ the integral closure of $\mathbf{a}$ as the set $\bar{\mathbf{a}}$ of elements $g \in \mathcal{R}$ for which $g^n \in \mathbf{a}(\mathbf{a}, g)^{n-1}$ (notice that this set is an ideal of $\mathcal{R}$). We have the following result.

Theorem.
Let $\mathbf{K}$ *be a field of arbitrary characteristic,* $\mathbf{a} \subseteq \mathbf{K}[x_1, \ldots, x_n]$ *an ideal generated by polynomials* $f_1, \ldots, f_m$ *of degrees* $d_1 \geq d_2 \geq \cdots \geq d_m \geq 3$ *and set*

$$\eta = \frac{3}{8}(3^m - 1) + \frac{m^2}{4}, \qquad \gamma = d_1 \cdots d_m, \quad \text{if } m \leq n - 1.$$
$$\eta = \frac{3}{8}(3^{n-1} - 1) + \frac{(n-1)^2}{4} + 1, \qquad \gamma = d_1 \cdots d_n + d_1 \text{ if } m \geq n.$$

Then, for any $f \in (\bar{\mathbf{a}})^\eta$, *we can find polynomials* $a_1, \ldots, a_m \in \mathbf{K}[x_1, \ldots, x_n]$ *with*

$$\max \deg\, (a_i f_i) \leq \deg\, f + \gamma$$

such that

$$f = a_1 f_1 + \cdots + a_m f_m.$$

2 Superficial ideals

As in the proofs of the explicit Nullstellensatz (see [B] and [K]), the first step in order to prove our main theorem consists in replacing the sequence $f_1, \ldots, f_m$ of generators of $\mathbf{a}$ by $m' \leq \min\{m, n+1\}$ suitable linear combinations $g_1, \ldots, g_{m'} \in \mathbf{K}f_1 + \cdots + \mathbf{K}f_m$, in such a way that some "regularity" assumptions are satisfied. However, in our case, the polynomials g_i must be chosen more carefully. For this, we need some definitions coming from local algebra, namely the notions of superficial element and reduction of an ideal. A complete discussion about this theory for a local ring can be found in [ZS] and [NR], but for our purposes we must work on a finitely generated K-algebra and we also need some additional results. For the above reasons and also for completeness, we prefer to develop the results we need independently, even though there will be some overlap with the quoted papers.

We start from the following definitions. Let $\mathcal{R}$ be a noetherian ring and let $\mathbf{a} \subseteq \mathcal{R}$ be an ideal. An element $x \in \mathbf{a}$ is called a superficial element (for $\mathbf{a}$) if there exists a natural number c such that $\big(\mathbf{a}^n : (x)\big) \cap \mathbf{a}^c = \mathbf{a}^{n-1}$ for any sufficiently large n. An ideal $\mathbf{b} \subseteq \mathbf{a}$ is called superficial (for $\mathbf{a}$) if $\mathbf{a}^n \cap \mathbf{b} = \mathbf{b}\mathbf{a}^{n-1}$ for any sufficiently large n. Finally, a reduction of $\mathbf{a}$ is an ideal $\mathbf{b} \subseteq \mathbf{a}$ such that $\mathbf{a}^n = \mathbf{b}\mathbf{a}^{n-1}$ for some (and then for any sufficiently large) n.

Remarks.

(1) An ideal $\mathbf{b} \subseteq \mathbf{a}$ is superficial if and only if there exists a natural number c such that $\mathbf{a}^n \cap \mathbf{b}\mathbf{a}^c = \mathbf{b}\mathbf{a}^{n-1}$ for any sufficiently large n. Indeed, by a lemma of E. Artin and D.G. Rees ([ZS] Theorem 4′, p.254), there exists $k \in \mathbf{N}$ such that $\mathbf{a}^n \cap \mathbf{b} = \mathbf{a}^{n-k}(\mathbf{a}^k \cap \mathbf{b})$ for any $n \geq k$. Therefore, if $n \geq k + c$, we get $\mathbf{a}^n \cap \mathbf{b} \subseteq \mathbf{a}^n \cap \mathbf{b}\mathbf{a}^c$.

(2) Let $\mathbf{b} = (x) \subseteq \mathbf{a}$ be a principal ideal. If x is superficial, the ideal $\mathbf{b}$ is also superficial (apply the last remark) and the converse is true if x is a non-zero-divisor.

(3) Let $(\mathcal{R}, \mathbf{m})$ be a local ring and let $\mathbf{b} \subseteq \mathbf{a}$ be two open ideals (i.e., $\sqrt{\mathbf{a}} = \sqrt{\mathbf{b}} = \mathbf{m}$). Then $\mathbf{b}$ is a reduction of $\mathbf{a}$ if and only if $\mathbf{b}$ is superficial (indeed we have $\mathbf{a}^n \subseteq \mathbf{b}$ for any sufficiently large n).

Superficial ideals will be constructed inductively using the following lemma:

Lemma 1.
Let $\mathbf{b} \subseteq \mathbf{a} \subseteq \mathcal{R}$ be two ideals with $\mathbf{b}$ superficial, and let x be an element of $\mathbf{a}$ such that its image $\overline{x}$ in $\mathbf{a}/\mathbf{b}$ is a superficial element. Then $(\mathbf{b}, x)$ is superficial for $\mathbf{a}$.

Proof.
If $\overline{x} \in \mathbf{a}/\mathbf{b}$ is a superficial element, by definition there exists $c \in \mathbf{N}$ such that $\big((\mathbf{a}^n + \mathbf{b}) : (x)\big) \cap (\mathbf{a}^c + \mathbf{b}) = \mathbf{a}^{n-1} + \mathbf{b}$ for any $n \geq n_1$. Therefore, if $y \in \mathbf{a}^n \cap (\mathbf{b}, x)\mathbf{a}^c$ and n is sufficiently large, we have $y = \alpha x + \beta$ where $\alpha \in \mathbf{a}^{n-1}$ and $\beta \in \mathbf{b}\mathbf{a}^c$. Hence $\beta \in \mathbf{a}^n \cap \mathbf{b}\mathbf{a}^c = \mathbf{b}\mathbf{a}^{n-1}$, since $\mathbf{b}$ is superficial for $\mathbf{a}$, and so $y \in (\mathbf{b}, x)\mathbf{a}^{n-1}$. From the remark (1) above, we deduce that $(\mathbf{b}, x)$ is superficial for $\mathbf{a}$. □

If the ring $\mathcal{R}$ is a finitely generated $\mathbf{K}$-algebra ($\mathbf{K}$ being an infinite field), we can always find superficial elements $x \in \mathbf{a}$. More precisely, we have the following result (see also [ZS] pp. 286-287).

Lemma 2.
Let $\mathcal{R}$ be as above and let $\mathbf{a} = (x_1, \ldots, x_m) \subseteq \mathcal{R}$ be an ideal. Then there exists a finite number of proper linear subspaces $\mathbf{V}_1, \ldots, \mathbf{V}_u \subset \mathbf{K}^m$ such that for any vector $\lambda \in \mathbf{K}^m \setminus \bigcup_{i=1}^{u} \mathbf{V}_i$ [(1)] the element $x = \lambda_1 x_1 + \cdots + \lambda_m x_m \in \mathbf{a}$ is superficial for $\mathbf{a}$.

---

(1) This last set is non-empty because $\mathbf{K}$ is infinite.

Proof.

Let $G(\mathcal{R}) = \sum_{h=0}^{\infty} \mathbf{a}^h/\mathbf{a}^{h+1}$ be the graded associated ring and let

$$\mathbf{X} = \sum_{h=1}^{\infty} \mathbf{a}^h/\mathbf{a}^{h+1}$$

be the ideal of the elements of positive degree. We must find $x \in \mathcal{R}$ such that $\overline{x} \in \mathbf{a}/\mathbf{a}^2 \subseteq G(\mathcal{R})$ satisfies

$$\overline{x}y = 0, \quad y \in \mathbf{X}^c \Rightarrow y = 0$$

for some natural number c. Let $\wp_1, \ldots, \wp_s$ be the minimal primes of $G(\mathcal{R})$, and assume $\wp_1, \ldots, \wp_u \not\supseteq \mathbf{a}/\mathbf{a}^2$ and $\wp_{u+1}, \ldots, \wp_s \supseteq \mathbf{a}/\mathbf{a}^2$. Then, for $i = 1, \ldots, u$, the set

$$\mathbf{V}_i = \{\lambda \in \mathbf{K}^m \mid \lambda_1\overline{x}_1 + \cdots \lambda_m\overline{x}_m \in \wp_i\}$$

is a proper subspace of $\mathbf{K}^m$; choose $\lambda \in \mathbf{K}^m \setminus \bigcup_{i=1}^{u} \mathbf{V}_i$ and set $x = \lambda_1 x_1 + \cdots \lambda_m x_m \in \mathbf{a}$. We have

$$(0) = Q_1 \cap \cdots \cap Q_u \cap Q_{u+1} \cap \cdots \cap Q_s$$

where the Q_i's are $\wp_i$-primary. For some c, $\mathbf{X}^c \subseteq Q_{u+1} \cap \cdots \cap Q_s$: therefore, if $y \in \mathbf{X}^c$ and $\overline{x}y = 0$, we have $y \in Q_{u+1} \cap \cdots \cap Q_s$. Moreover, $y \in Q_1 \cap \cdots \cap Q_u$, since $\overline{x} \notin \wp_i$ for $i = 1, \ldots, u$, and so $y = 0$. □

3 Reduction to an $\mathcal{U}$-regular sequence

Let $\mathbf{a} \subseteq \mathcal{R}$ be an ideal and let us consider the Zariski open set $\mathcal{U} = \mathcal{U}_{\mathbf{a}} = \{\wp \in \mathrm{Spec}\mathcal{R}, \wp \not\supseteq \mathbf{a}\}$. A sequence $x_1, \ldots, x_k$ is called $\mathcal{U}$-regular if it is a regular sequence in the ring $\mathcal{R}_{\mathcal{U}} = \bigcap_{\wp \in \mathcal{U}} \mathcal{R}_{\wp}$. In other words, $x_1, \ldots, x_k$ is $\mathcal{U}$-regular if $(x_1, \ldots, x_k) \colon \langle \mathbf{a} \rangle = \bigcup_{r \in \mathbf{N}} (x_1, \ldots, x_k) \colon \mathbf{a}^r$ is a proper ideal and if x_i is a non-zero-divisor in $\mathcal{R}/(x_1, \ldots, x_{i-1}) \colon \langle \mathbf{a} \rangle$ for $i = 1, \ldots, k$. If $\mathcal{R}$ is a Cohen-Maculay ring, $\mathcal{R}_{\mathcal{U}}$ is also Cohen-Maculay and the purity theorem holds; therefore, if $\mathbf{b} \subseteq \mathcal{R}$ is an ideal generated by an $\mathcal{U}$-regular sequence of length k, then $\mathbf{b} \colon \langle \mathbf{a} \rangle$ is unmixed of rank k.

We also need the following corollary of Bertini's theorem:

Lemma 3.

Let $\mathcal{R}$ be a finitely generated algebra over an algebraically closed field $\mathbf{K}$ and let $\mathbf{a} = (x_1, \ldots, x_m) \subseteq \mathcal{R}$ be an ideal. Then there exists a non-empty Zariski open set $\mathbf{V} \subset \mathbf{K}^m$ such that for any $\lambda \in \mathbf{K}^m \backslash \mathbf{V}$ the ideal $(\lambda_1 x_1 + \cdots + \lambda_m x_m) \colon \langle \mathbf{a} \rangle$ is radical (i.e., it coincides with its radical).

Proof.

Apply [J], corollary 6.7, to $X = \mathrm{Spec}\mathcal{R}_{\mathcal{U}}$ and to the morphism $f\colon X \to \mathrm{Aff}^m_{\mathbf{K}}$ defined by $f = (x_1, \ldots, x_m)$. □

Proposition 1.

Let $f_1, \ldots, f_m$ be a system of generators of the ideal $\mathbf{a} \subseteq \mathcal{R} = K'$, where $\mathbf{K}$ is an algebraically closed field. Then we can find an integer $m' \leq \min\{m, n+1\}$ and m' linear combinations (with coefficients in $\mathbf{K}$) of $f_1, \ldots, f_m$, say $g_1, \ldots, g_{m'}$, such that $\sqrt{\mathbf{a}} = \sqrt{(g_1, \ldots, g_{m'})}$ and the following assertions hold for $i = 1, \ldots, m'$:

i) The ideal $\mathbf{b}_i = (g_1, \ldots, g_i)$ is superficial for $\mathbf{a}$;

ii) The ideal $\mathbf{b}'_i = \mathbf{b}_i \colon \langle \mathbf{a} \rangle$ is unmixed of rank i and g_i is a non-zero-divisor in $R/\mathbf{b}'_{i-1}$;

iii) $\mathbf{b}'_i$ is radical.

Proof.

We choose $g_1, \ldots, g_{m'}$ by induction. Let us assume that assertions i), ii), and iii) are satisfied for $i = 1, \ldots, h$. If $\sqrt{(g_1, \ldots, g_h)} = \sqrt{\mathbf{a}}$, we put $m' = h$; otherwise $\mathbf{b}_h \colon \langle \mathbf{a} \rangle$ is a proper ideal and so (by ii)) $g_1, \ldots, g_h$ is a $\mathcal{U}$-regular sequence and $\mathbf{b}_h \colon \langle \mathbf{a} \rangle$ is unmixed of rank h. Let $\wp_1, \ldots, \wp_r$ be its associated primes and let us consider the proper subspaces $\mathbf{W}_i = \{\lambda \in \mathbf{K}^m$ such that $\lambda_1 f_1 + \cdots + \lambda_m f_m \notin \wp_i\}$, $i = 1, \ldots, r$. We also denote by $\mathbf{V}_i$ $(i = 1, \ldots, s)$ and by $\mathbf{V}$ the subspaces and the Zariski open set of $\mathbf{K}^m$ obtained by respectively applying lemma 2 and lemma 3 to the $\mathbf{K}$-algebra $\mathcal{R}/\mathbf{b}_h$ and to the ideal $\mathbf{a}/\mathbf{b}_h$ generated by the images of $f_1, \ldots, f_m$. Then, if $\lambda \in \mathbf{K}^m$ lies outside $\mathbf{W}_i$, $\mathbf{V}_i$, and $\mathbf{V}$, the polynomial $g_{h+1} = \lambda_1 f_1 + \cdots + \lambda_m f_m$ has the required properties. □

Remark.

It is easy to see that we can choose g_i as a linear combination of $f_i, \ldots, f_m$.

Let us define for an ideal $\mathbf{a}$ in a ring $\mathcal{R}$ its integral closure $\bar{\mathbf{a}}$ as the set of elements $x \in \mathcal{R}$ for which $x^n \in \mathbf{a}(\mathbf{a}, x)^{n-1}$; notice that this set is an ideal of $\mathcal{R}$. We have

Corollary 1.

Let g_i, $\mathbf{b}_i$, and $\mathbf{b}'_i$ be as in the previous proposition. Then the ideals $\mathbf{b}'_{i-1} + (g_i)$ and $\mathbf{b}'_{i-1} + \mathbf{a}$ have the same integral closure in $\mathcal{R}_\wp$, for any prime ideal $\wp \supseteq \mathbf{b}'_{i-1} + \mathbf{a}$ of rank i.

Proof.
Let $\wp \supseteq \mathbf{b}'_{i-1} + \mathbf{a}$ be a prime ideal of rank i. The ideal $(\mathbf{b}'_{i-1}, g_i)\mathcal{R}_\wp$ is open (use ii) of the proposition above); hence there exists $c_1 \in \mathbf{N}$ such that $\mathbf{a}^{c_1}\mathcal{R}_\wp \subseteq (\mathbf{b}'_{i-1}, g_i)\mathcal{R}_\wp$. Moreover, there exists another constant $c_2 \in \mathbf{N}$ such that

$$\mathbf{a}^{c_2}\mathbf{b}'_{i-1} \subseteq \mathbf{b}_{i-1} \subseteq \mathbf{b}'_{i-1}.$$

Therefore, for any $n \geq c_1 + c_2$,

$$\mathbf{a}^n\mathcal{R}_\wp \subseteq \mathbf{a}^{c_1+c_2}\mathcal{R}_\wp \subseteq \mathbf{a}^{c_2}(\mathbf{b}'_{i-1}, g_i)\mathcal{R}_\wp \subseteq (\mathbf{b}_{i-1}, g_i)\mathcal{R}_\wp = \mathbf{b}_i\mathcal{R}_\wp.$$

Hence, for any sufficiently large n,

$$\begin{aligned}(\mathbf{b}'_{i-1} + \mathbf{a})^n\mathcal{R}_\wp &= \mathbf{b}'_{i-1}(\mathbf{b}'_{i-1} + \mathbf{a})^{n-1}\mathcal{R}_\wp + \mathbf{a}^n\mathcal{R}_\wp \\ &= \mathbf{b}'_{i-1}(\mathbf{b}'_{i-1} + \mathbf{a})^{n-1}\mathcal{R}_\wp + (\mathbf{a}^n \cap \mathbf{b}_i)\mathcal{R}_\wp \\ &= \mathbf{b}'_{i-1}(\mathbf{b}'_{i-1} + \mathbf{a})^{n-1}\mathcal{R}_\wp + \mathbf{b}_i\mathbf{a}^{n-1}\mathcal{R}_\wp \\ &\subseteq (\mathbf{b}'_{i-1}, g_i)(\mathbf{b}'_{i-1} + \mathbf{a})^{n-1}\mathcal{R}_p\end{aligned}$$

since $\mathbf{b}_i$ is superficial for $\mathbf{a}$. Hence $(\mathbf{b}'_{i-1}, g_i)\mathcal{R}_\wp$ is a reduction of $(\mathbf{b}'_{i-1}+\mathbf{a})\mathcal{R}_\wp$. To conclude the proof, we use the following lemma (see [NR], corollary of theorem 1, p. 155).

Lemma 4.
Let $\mathbf{b} \subseteq \mathbf{a}$ two ideals in a noetherian ring $\mathcal{R}$ and let us assume that $\mathbf{a}$ is not entirely composed by zero-divisors. Then, if $\mathbf{b}$ is a reduction of $\mathbf{a}$, these two ideals have the same integral closure.
Proof.
Let $x \in \mathbf{a}$. From $\mathbf{a}\mathbf{a}^{n-1} = \mathbf{b}\mathbf{a}^{n-1}$ we see from a determinant argument that there exists $\phi \in x^n + \mathbf{b}(\mathbf{b}, x)^{n-1}$ such that $\phi\mathbf{a}^{n-1} = 0$. Hence $\phi = 0$ and $x^n \in \mathbf{b}(\mathbf{b}, x)^{n-1}$. □

Now we quote the following theorem of J. Lipman and B. Tessier:
Theorem. (see [LT])
Let $\mathcal{R}$ be a regular local ring of dimension d and let $\mathbf{a} \subseteq \mathcal{R}$ be an open ideal. Then $\overline{\mathbf{a}}^d \subseteq \mathbf{a}$.

From this theorem and from the corollary above, we deduce that

$$\overline{\mathbf{a}}^i \subseteq (\mathbf{b}'_{i-1} + (g_i))\mathcal{R}_\wp \cap \mathcal{R} \tag{1}$$

for any prime $\wp \supseteq b'_{i-1} + \mathbf{a}$ of rank i. The previous statement is one of the crucial points in the proof of our main result.

4 Proof of the main theorem

We shall work over the homogeneous ring $\mathcal{A} = \mathbf{K}[x_0, \ldots, x_n]$; for $i = 1, \ldots, m'$, let $G_i = {}^h g_i$ be the homogenization of the polynomials g_i, given by proposition 1, and let $J_i = {}^h b'_i$ be the homogenization of the ideal b'_i; we also put $J_0 = (0)$. We denote by I_{i+1} the intersection of the isolated components of (J_i, G_{i+1}) whose radicals contain $I = {}^h\mathbf{a}$ but not x_0. Similarly, let K_{i+1} be the intersection of the isolated components whose radicals contain x_0 and let L_{i+1} be the intersection of the embedded ones. Then

$$(J_i, G_i) = J_{i+1} \cap I_{i+1} \cap K_{i+1} \cap L_{i+1}$$

and $x_0 \cdot I \subseteq \sqrt{L_{i+1}}$, since $G_1, \ldots, G_{m'}$ is an $\mathcal{U}$-regular sequence in the Zariski open set $\mathcal{U} = \{\wp \in \mathrm{Spec}\mathcal{A} \text{ such that } x_0 \cdot I \not\subseteq \wp\}$.

We want to give explicitly two integers, γ_{i+1} and η_{i+1}, such that

$$x_0^{\gamma_{i+1}} \overline{I}^{\eta_{i+1}} J_{i+1} \subseteq (J_i, G_{i+1}).$$

To this end, let $\delta_{i+1} = \deg \mathcal{A}/K_{i+1}$, which is easily estimated by Bezout's theorem. From (1), we find that

$$x_0^{\delta_{i+1}} \overline{I}^{i+1} J_{i+1} \subseteq J_{i+1} \cap I_{i+1} \cap K_{i+1}. \tag{2}$$

Thus, we must only control the embedded components with a suitable modification of Kollar's method in Patrice Philippon's algebraic version (see [P1] and [P2]).

Definition.
Let $J \subseteq \mathcal{A}$ be a homogeneous ideal. We say that J has type (ε, ρ) if for any

$$\alpha \in \mathcal{F} = \{(\alpha_1, \ldots, \alpha_s) \subseteq \mathcal{A}^s \mid x_0 \cdot I \subseteq \sqrt{(\alpha_1, \ldots, \alpha_s)}\}$$

and for any

$$\sigma < \dim \mathcal{A}/J - \dim \mathcal{A}/(\alpha)$$

we have $x_0^{\varepsilon} \cdot I^{\rho} \cdot H_{s-\sigma}(\alpha|\mathcal{A}/J) = 0$ (2).

Lemma 5.
Let us assume that J_i has type (ε_i, ρ_i). Then

$$x_0^{\varepsilon_i + \delta i+1} \overline{I}^{\rho_i + i+1} J_{i+1} \subseteq (J_i, G_{i+1}).$$

(2) The $H_{s-\sigma}(\alpha|M)$ are the homological modules associated with the Koszul complex $K(\alpha|\mathcal{A}|M)$ (see [N] §8.2).

Proof.
Let $\alpha = (\alpha_1, \dots, \alpha_s) \in \mathcal{F}$ be a family of generators of L_{i+1} and put

$$M = (J_{i+1} \cap I_{i+1} \cap K_{i+1})/(J_i, G_{i+1}).$$

From the inclusion $M \hookrightarrow \mathcal{A}/(J_i, G_{i+1})$ we get

$$M = H_s(\alpha|M) \hookrightarrow H_s(\alpha|\mathcal{A}/(J_i, G_{i+1})).$$

On the other hand, the exact sequence

$$0 \longrightarrow \mathcal{A}/J_i \overset{\times G_{i+1}}{\longrightarrow} \mathcal{A}/J_i \longrightarrow \mathcal{A}/(J_i, G_{i+1}) \longrightarrow 0 \tag{3}$$

gives

$$H_s(\alpha|\mathcal{A}/(J_i, G_{i+1})) \hookrightarrow H_{s-1}(\alpha|\mathcal{A}/J_i)$$

(since any associated prime of J_i contains neither x_0 nor I, we have $H_s(\alpha|\mathcal{A}/J_i) = 0:_{\mathcal{A}/J_i} L_{i+1} = 0$). Therefore, composing the two last inclusions we get

$$M \hookrightarrow H_{s-1}(\alpha|\mathcal{A}/J_i).$$

Now, taking into account that $\dim \mathcal{A}/J_i - \dim \mathcal{A}/L_{i+1} > 1$ and our definition of type, the last line implies $x_0^{\varepsilon_i} \overline{I}^{\rho_i} M = 0$. From this and (2) we obtain our claim. □

Lemma 6.
Let us assume that J_i has type (ε_i, ρ_i). Then J_{i+1} has type $(3\varepsilon_i + \delta_{i+1}, 3\rho_i + i + 1)$.
Proof.
Let $\alpha \in \mathcal{F}$ and let $\sigma < \dim \mathcal{A}/J_{i+1} - \dim \mathcal{A}/(\alpha) < \dim \mathcal{A}/J_i - \dim \mathcal{A}/(\alpha)$. From (3) we get

$$x_0^{2\varepsilon_i} \overline{I}^{2\rho_i} \cdot H_{s-\sigma}(\alpha|\mathcal{A}/(J_i, G_{i+1})) = 0. \tag{4}$$

On the other hand, the exact sequence

$$0 \longrightarrow J_{i+1}/(J_i, G_{i+1}) \longrightarrow \mathcal{A}/(J_i, G_{i+1}) \longrightarrow \mathcal{A}/J_{i+1} \longrightarrow 0$$

gives rise to

$$H_{s-\sigma}(\alpha|\mathcal{A}/(J_i, G_{i+1})) \rightarrow H_{s-\sigma}(\alpha|\mathcal{A}/J_{i+1}) \rightarrow H_{s-\sigma-1}(\alpha|J_{i+1}/(J_i, G_{i+1})).$$

From lemma 5 we know that $x_0^{\varepsilon_i + \delta i+1} \overline{I}^{\rho_i + i + 1}$ kills $J_{i+1}/(J_i, G_{i+1})$, and hence also

$$H_{s-\sigma-1}(\alpha|J_{i+1}/(J_i, G_{i+1})).$$

Taking into account (4), we find

$$x_0^{3\varepsilon_i + \delta i+1} \overline{I}^{3\rho_i + i + 1} \cdot H_{s-\sigma}(\alpha|\mathcal{A}/J_{i+1})) = 0.$$ □

The ideal J_0 has type $(0,0)$; hence we easily obtain from the lemma above that the ideal J_i has type (ε_i, ρ_i), where

$$\varepsilon_i = \sum_{h=1}^{i} 3^{i-h}\delta_h, \qquad \rho_i = \sum_{h=1}^{i} 3^{i-h}h.$$

Now repeated applications of lemma 6 give

$$x_0^{\gamma_i}\overline{I}^{\eta_i} J_i \subseteq (G_1, \dots, G_i) \tag{5}$$

where

$$\begin{aligned} \eta_i &= \sum_{h=1}^{i} h + \sum_{h=1}^{i-1} \rho_h = \frac{3}{8}(3^i - 1) + \frac{i^2}{4} \\ \gamma_i &= \sum_{h=1}^{i} \delta_h + \sum_{h=1}^{i-1} \varepsilon_{h-1} \leq \sum_{h=1}^{i} 3^{i-h}\delta_h \\ &\leq \sum_{h=1}^{i} d_{k+1} \cdots d_i \delta_k \leq d_1 \cdots d_i - \deg \mathcal{A}/J_i, \end{aligned} \tag{6}$$

the last inequality coming from Bezout's theorem. We distinguish two cases.

- **First case** $m' \leq n-1$.

  We take $i = m' \leq \min\{m, n-1\}$ in (5) and (6).
- **Second case** $m' \geq n$ (and so $m \geq n$)

  The ideal J_{n-1} is a homogeneous radical ideal of rank $n-1$ (use iii) of proposition 1) and we apply the following two lemmas (for the proofs, see [A2] lemma 5 and lemma 4 respectively):

Lemma 7.

Let $c_1 = d_n \deg \mathcal{A}/J_{n-1} - \deg \mathcal{A}/J_n \cap I_n$ and $c_2 = \deg \mathcal{A}/J_n \cap I_n + 1$. Then

$$x_0^{c_1}(J_n \cap I_n)_\nu \subseteq (J_{n-1}, G_n)$$

for any integer $\nu \geq c_2$.

Lemma 8.

Let $F_i = {}^h f_i$, $i = 1, \dots, m$. Then $I_\nu \subseteq (J_n \cap I_n, F_1, \dots, F_m)$, for $\nu \geq c_3 =$ $\deg \mathcal{A}/J_n \cap I_n + d_1$.

Combining these two results, we get

$$x_0^{c_1 + \max\{c_2, c_3\}} I \subseteq (J_{n-1}, G_n, F_1, \dots, F_m).$$

Taking into account (5) with $i = n-1$, this gives

$$x_0^{\gamma_n'}\overline{I}^{\eta_n'} J_{n-1} \subseteq (G_1, \dots, G_{n-1}, G_n, F_1, \dots, F_m).$$

where

$$\gamma'_n = \gamma_{n-1} + d_n \deg \mathcal{A}/J_{n-1} + d_1 \leq d_1 \cdots d_n + d_1$$
$$\eta'_n = \eta_{n-1} + 1 = \frac{3}{8}(3^{n-1} - 1) + \frac{(n-1)^2}{4} + 1.$$

Our theorem follows.

References

[A1] F. AMOROSO, "Tests d'appartenance d'après un théorème de J. Kollár". C.R. Acad. Sc. Paris, 309, Série I, pp. 691–694, (1989).

[A2] F. AMOROSO, "Membership problem", in "Approximation Diophantiennes et Nombres Transcendants. Luminy 1990", ed. P. Philippon, Walter de Gruyter, Berlin, pp. 1–13 (1992).

[B] W.D. BROWNAWELL, "Bounds for the degrees in the Nullstellensatz", Annals of Mathematics 126, pp. 577–591 (1987).

[BY1] C.A. BERENSTEIN and A. YGER, "Bounds for the degrees in the division problem", Mich. Math. J. 37, No.1, pp. 25–43 (1990).

[BY2] C.A. BERENSTEIN and A. YGER, "Effective Bezout identities in $\mathbf{Q}[x_1, \ldots, x_n]$", Acta Math. 166, 1991, pp. 69–120 (1991).

[BY3] C.A. BERENSTEIN and A. YGER, "Une formule de Jacobi et ses conséquences", Ann. Sci. E.N.S., 4^{ieme} série, 24, pp. 363–377 (1991).

[BY4] C.A. BERENSTEIN and A. YGER, "Représentations intégrales et division", in "Approximation Diophantiennes et Nombres Transcendants. Luminy 1990", ed. P. Philippon, Walter de Gruyter, Berlin pp. 15–37 (1992).

[CGH] L. CANIGLIA, A. GALLIGO and J. HEINTZ, "Borne simple exponentielle pour les degrés dans le théorème des zéros sur un corps de caractéristique quelconque", C.R. Acad. Sc. Paris, t.307, Série I, pp. 255–258, (1988).

[DFGS] A. DICKENSTEIN, N. FITCHAS, M. GIUSTI and C. SESSA, "The membership problem for unmixed polynomial ideals is solvable in single exponential time", Discrete Appl. Math. 33, pp. 73–94 (1991).

[E] M. ELKADI, "Bornes pour le degré et les hauteurs dans le problème de division", to appear in Michigan Math. Journal.

[FG] N. FITCHAS and A. GALLIGO, "Nullstellensatz effectif et conjecture de Serre (Théorème de Quillen-Suslin) pour le Calcul Formel". Math. Nachr. 149, pp. 231–253 (1990).

[GPL] L. GRUSON, R. LAZARSFELD and C. PESKINE, "On a theorem of Castelnuovo and the equations defining space curves". Inv. Math. 72, pp. 491–506 (1983).

[H] G. HERMANN, "Die Frage der endlich vielen Schritte in der Theorie der Polynomideale", Math. Ann. 95, pp. 736–788 (1926).

[KL] T. KRICK and A. LOGAR, "Membership problem, Representation problem and the computation of the radical for one-dimensional ideals", in Effective Methods in Algebraic Geometry, Proc. Intern. Conf. MEGA 90, Castiglioncello 1990, T. Mora and C. Traverso, eds., Progress in Math. 94, pp. 169-193 (1991).

[KP] T. KRICK and L.M. PARDO, "Une approche informatique pour l'approximation diophantienne", C.R. Acad. Sc. Paris, t.318, Série I, pp. 407–412, (1994).

[KP] T. KRICK and L.M. PARDO, "A Computational Method for Diophantine Approximation", in this volume.

[J] J. -P. JOUANOLOU, "Théorèmes de Bertini et applications", Progress in Math., Birkhäuser, Boston, (1983).

[K] J. KOLLÁR, "Sharp effective Nullstellensatz", Journal of the American Math. Soc. 1, No.4, pp. 963–975 (1988).

[MM] E. MAYR and A. MEYER, "The complexity of the word problem for commutative semigroup and polynomial ideals", Advances in Math. 46, pp. 305–329 (1982).

[N] D.G. NORTHCOTT , "Lessons on rings, modules and multiplicities", Cambridge University Press 1968.

[NR] D.G. NORTHCOTT and D. REES, "Reduction of ideals in local rings", Proc. Camb. Phil. Soc., pp. 145–158 (1954).

[LT] J. LIPMAN and B. TEISSIER, "Pseudo-rational local rings and a theorem of Briançon-Skoda about integral closures of ideals", Michigan Math. J., 28, pp. 97–116 (1981).

[P1] P. PHILIPPON, "Théorème des zéros effectif, d'après J.Kollár", dans "Problèmes Diophantiens 1987/88", Publications de l'Université de Paris VI No.88 (1988).

[P2] P. PHILIPPON, "Dénominateurs dans le théorème des zéros de Hilbert", Acta Arith. 58, pp. 1–25 (1991).

[ZS] O. ZARISKI and P. SAMUEL, "Commutative Algebra", Vol.2, Van Nostrand, Princeton 1960; Springer-Verlag, Berlin-Heidelberg-New York 1986.

F. Amoroso (`amoroso@dm.unipi.it`)
Dip. di Matematica, University of Pisa, Via F. Buonarroti, 2. 56100 Pisa (Italy).

Progress in Mathematics, Vol. 143, © 1996 Birkhäuser Verlag Basel/Switzerland

Computation of the splitting fields and the Galois groups of polynomials

H. Anai, M. Noro, K. Yokoyama

1 Introduction

This study is a continuation of Yokoyama et al. [22], which improved the method by Landau and Miller [11] for the determination of solvability of a polynomial over the integers. In both methods, the solvability of a polynomial is reduced, in polynomial time, to that of polynomials, each of which is constructed so that its Galois group acts primitively on its roots. Then, by virtue of Pálfy's bound [14], solvability of polynomials with primitive Galois groups can be determined in polynomial time. An effective method, thus, exists in theory. For practical computation, however, the most serious problem remains: How to determine solvability of each polynomial with primitive Galois group.

We present a practical method to compute the Galois group, not necessarily primitive, by which solvability is decided. The method applies to polynomials over an arbitrary finite extension field of the rationals. In this paper, however, we shall focus on integral polynomials. There are two kinds of approaches, we think, to determine the Galois group of a given integral polynomial:

(1) In one approach, every root of the given polynomial is first presented explicitly as an element in a certain algebraic extension field. Then, the Galois group is presented as a permutation group on the roots.
(2) In the other approach, roots are not presented explicitly. The Galois group is not presented as a permutation group on the roots but determined to belong to a class of permutation groups to within relabeling of the roots. (This class is the conjugacy class of the Galois group in the symmetric group).

Significant work was done for the second approach by using classification tables of all transitive subgroups in the symmetric groups (see Soicher [16], Kolesova and Mckay [9], and Ford and Mckay [6]). Here, we call such methods *table-based* methods. On the other hand, we call the first one the *direct approach* and methods in the first one *direct methods*. *Table-based* methods can determine Galois groups of polynomials very efficiently both in theory and in practice as far as tables for the degrees of input polynomials are known. We mention that, as a variant of table-based methods, there is another way using group-theoretical knowledge, where the roots of an input polynomial are assigned to their approximate numerical values and functions on the roots associated with transitive subgroups in the symmetric group are computed, from which the Galois group is determined. Stauduhar [17] used resolvent polynomials to compute such functions.

Here, we present a direct method which seems very efficient compared to direct methods. The following explains the reasons why we employ the direct approach.

(i) Our final goal is to express every root of a given polynomial by radicals when it is solvable. For this goal, we have to express all roots as elements of the splitting field and present the Galois group as a concrete permutation group on the roots.
(ii) In the computational view point, it is useful to seek a method applicable for polynomials with arbitrary degree. In both theory and practice, it is very important to give a complete procedure for the problem. At present, we cannot employ the table-based method since such tables are not known for arbitrary degrees.
(iii) The *modified* Trager algorithm for algebraic factorization proposed by Noro and Takeshima [13] has been extended and applied effectively to computation of splitting fields of polynomials. By this extension, we can compute the splitting field of a given polynomial very efficiently.
(iv) From a practical point of view, we think, there are cases where the direct approach works well such as a case where no table is known for the degree of an input polynomial but its splitting field is considerably small. Moreover, for the purpose described in (i), we have only to deal with polynomials whose Galois groups are primitive. Then, Pálfy's bound will make the direct method very practical.

Thus, we employ the direct approach. Now we explain several important points in the direct approach briefly. From the description of the approach, it can be seen that efficiency of the method depends on splitting field computation and expression of the Galois group as a permutation group.

For example, methods in which the splitting field is represented by a primitive element are already described in the literature (cf. [11]). In those methods, computation of the Galois group requires not only factorization of the given polynomial but also factorization of the minimal polynomial of the primitive element employed. This fact may reduce total efficiences (see remark 11).

Meanwhile, as an alternate way, we employ the description of the splitting field by *successive extension*, where the splitting field is obtained by adjoining several roots of the given polynomial. By employing this description, we can extend the modified Trager algorithm to an algorithm for computation of splitting fields effectively. By making good use of this efficient algorithm for splitting fields, we obtain an algorithm to represent the Galois group as a permutation group on the roots, where a set of generators of the Galois group is computed. Moreover, the computed set of generators forms a *strong generating set* which is suited for further computation of group properties, such as commutators, normal closures, *etc.*

As for efficiency of the proposed method, we did not state its theoretical complexity partly because we employed non-polynomial-time algorithms for factorization of univariate polynomials. But we examined its efficiency by experiment. We implemented the method on the Risa/Asir computer algebra system [13], and computed a number of examples. From its performance listed in Appendix, we are convinced that the method is very practical.

This paper is organized as follows. In Section 2 we provide notations and the mathematical basis for the expression of splitting fields of polynomials as algebraic extension fields. In Section 3 we present a new method for direct computation of Galois groups, and in Section 4 we present a new method for computation of splitting fields. We discuss practical computations in Section 5. Finally, in Section 6, we give concluding remarks. In the Appendix, we list several bench-mark results of the method implemented on the Risa/Asir system.

2 Representation of the splitting field

Here, we provide some notions on the splitting fields which can be found in textbooks on Galois theory.

Let $f(x)$ be an irreducible and monic polynomial in x over the rational number field $\mathbb{Q}$. We denote the set of all roots of $f(x)$ in the algebraic closure $\bar{\mathbb{Q}}$ of $\mathbb{Q}$ by Ω_f and thus, $\Omega_f = \{\alpha_1, \ldots, \alpha_n\}$, where $n = \deg(f(x))$. Then, the splitting field K_f of $f(x)$ is the finite extension field $\mathbb{Q}(\Omega_f)$ by adjoining all roots. We denote by N the extension degree of K_f over $\mathbb{Q}$. The Galois group G_f of $f(x)$ is a transitive and faithful permutation group acting on the set Ω_f and its order coincides with N. As mentioned in Section 1, computation of the Galois group G_f depends on the description of all roots, or that of the splitting field.

In general, it is very convenient to represent a finite extension field K over $\mathbb{Q}$ by a residue class ring $\mathbb{Q}[x_1, \ldots, x_r]/I$ of a polynomial ring over $\mathbb{Q}$ by its maximal ideal I. Unique representation of elements in the residue class ring is obtained easily by the notion of Gröbner basis and normal forms (see Buchberger [3] and Yokoyama *et al.* [21]). Among these representations, the simplest one is a residue class ring of the ring of univariate polynomials. In this case, the indeterminate x corresponds to a primitive element α of K, and every element of the extension field is represented by a univariate polynomial in x modulo the minimal polynomial $m(x)$ of α.

A description of the splitting field K_f is obtained by factoring the original polynomial f into its linear factors. This factorization, in general, requires a number of factorizations of factors of f over algebraic extension fields obtained by adding roots of f successively. In more detail, the procedure is described as follows.

SPLITTING FIELD GENERAL PROCEDURE:
A subfield K_i of K_f is defined by $K_i = K_{i-1}(\alpha_i) = \mathbb{Q}(\alpha_1, \ldots, \alpha_i)$, where $K_0 = \mathbb{Q}$ and $\alpha_1, \ldots, \alpha_i$ are distinct roots of f and α_k does not belong to K_{k-1} for $k = 1, \ldots, i$. K_i is represented by a residue class ring.

Over the field K_i, $f(x)$ is factorized into its irreducible factors. And if there is a non-linear irreducible factor h, then fix one root of h as α_{i+1} and proceed to factorization over K_{i+1}. Otherwise, terminate the procedure.

The result of the *SPLITTING FIELD GENERAL PROCEDURE* depends on the method of algebraic factorization employed there. Among several ways to represent the splitting field, we choose the following one which is derived quite naturally from *SPLITTING FIELD GENERAL PROCEDURE.*

2.1 Representation by successive extension

Suppose that K_f is obtained by adding ℓ distinct roots in the *SPLITTING FIELD GENERAL PROCEDURE.* By letting each α_k correspond to x_k for $k = 1, \ldots, \ell$, K_ℓ is represented by

$$K_\ell \equiv \mathbb{Q}[x_1, \ldots, x_\ell] / < f_1, \ldots, f_\ell >,$$

where $< f_1, \ldots, f_\ell >$ is the ideal generated by polynomials $f_1, \ldots, f_\ell$ and each polynomial f_k, $k = 1, \ldots, \ell$, is a non-linear irreducible factor of f over $K_{k-1} \equiv \mathbb{Q}[x_1, \ldots, x_{k-1}] / < f_1, \ldots, f_{k-1} >$. Thus, $f_1(x_1) = f(x_1)$ and the set $\{f_1, .., f_\ell\}$ forms a "triangular" set. The following details the expression:

- α_1 is a fixed root of $f(x)$, i.e., $f_1(\alpha_1) = 0$,
- α_2 is a fixed root of $f_2(\alpha_1, x)$, i.e., $f_2(\alpha_1, \alpha_2) = 0$, where $f_2(x_1, x)$ is a non-linear irreducible factor of $f(x)$ over $\mathbb{Q}[x_1]/f_1(x_1)$,

$\vdots$

- α_ℓ is a fixed root of $f_\ell(\alpha_1, \ldots, \alpha_{\ell-1}, x)$, i.e., $f_\ell(\alpha_1, \ldots, \alpha_\ell) = 0$, where $f_\ell(x_1, \ldots, x_{\ell-1}, x)$ is a non-linear irreducible factor of $f(x)$ over $\mathbb{Q}[x_1, \ldots, x_{\ell-1}] / < f_1(x_1), \ldots, f_{\ell-1}(x_1, \ldots, x_{\ell-1}) >$,
- $\alpha_{\ell+1}$ is expressed as a polynomial $A_{\ell+1}(\alpha_1, \ldots, \alpha_\ell)$,

$\vdots$

- α_n is expressed as a polynomial $A_n(\alpha_1, \ldots, \alpha_\ell)$.

Moreover, by letting $f_i(x_1, \ldots, x_\ell, x_i) = x_i - A_i(x_1, \ldots, x_\ell)$ for defining α_i, $i = \ell + 1, \ldots, n$, we have more convenient representation of K_f:

$$K_f \equiv \mathbb{Q}[x_1, \ldots, x_n]/J,$$

where J is the ideal generated by polynomials $f_1, \ldots, f_n$. J is the ideal of relations between the roots $\alpha_1, \ldots, \alpha_n$. i.e.,

$$J = \{r \in \mathbb{Q}[x_1, \ldots, x_n] \mid r(\alpha_1, \ldots, \alpha_n) = 0\}.$$

Definition 1 *We call ℓ the* length of the representation *of K_f with respect to $\alpha_1, \ldots, \alpha_n$. For each i, $i = 1, \ldots, n$, we call $f_i(x_1, \ldots, x_i)$ the* defining polynomial *of α_i, and we call J the* defining ideal *of K_f. Moreover, for each i, $i = \ell + 1, \ldots, n$, we call $A_i(x_1, \ldots, x_\ell)$ the* polynomial expression *of α_i with respect to $\alpha_1, \ldots, \alpha_\ell$.*

Remark 2 *Let $n_i = \deg_{x_i}(f_i)$ for each i. Then, $N = n_1 \cdots n_\ell = n_1 \cdots n_n$. A set $\{x_1^{e_1} \cdots x_\ell^{e_\ell} | 0 \leq e_1 < n_1, \ldots, 0 \leq e_\ell < n_\ell\}$ forms a basis of $\mathbb{Q}[x_1, \ldots, x_n]/J$ as a vector space over $\mathbb{Q}$.*

3 Computation of Galois groups

Now, we show how we can compute the Galois group G_f of a given polynomial f on the basis of the representation of the splitting field K_f by successive extension. First, we give a procedure to determine whether a permutation on the roots belongs to the Galois group or not; second, we give an efficient procedure finding a set of generators of the Galois group among all permutations on the roots. We begin with defining several necessary notations.

Definition 3 *For an element g in G_f, the image α_i^g of α_i by the action of g coincides with some α_j. Simply we denote this by $i^g = j$. (So we identify $\{1, \ldots, n\}$ with $\{\alpha_1, \ldots, \alpha_n\}$.) Moreover, we denote the permutation representation of g by $(1^g, \ldots, n^g)$, where $\alpha_i^g = \alpha_{i^g}$ for $i = 1, \ldots, n$.*

We also use the same notation for an arbitrary permutation g on $\{1, \ldots, n\}$. That is, g is expressed by $(1^g, \ldots, n^g)$. We call $(1^g, \ldots, t^g)$ the first t-part *of g for t, $1 \leq t \leq n$.*

3.1 Membership of permutations to a Galois group

Now, we present a new method to determine the membership of permutations to the Galois group. The mathematical basis of the method described below may form a refinement of very old works by Mertens and others (cf. Tschebotaröw and Schwerdtfeger [19]). To devise an efficient algorithm, we use the length ℓ and the first ℓ-part of permutations.

Suppose the splitting field K_f and all roots are described as in the previous section. So, $K_f \equiv \mathbb{Q}[x_1, \ldots, x_n]/J$, where J is the defining ideal generated by the defining polynomials $f_1, \ldots, f_n$, and where $A_{\ell+1}, \ldots, A_n$, where ℓ is the length, are the polynomial expressions for $\alpha_{\ell+1}, \ldots, \alpha_n$. Then, the following holds.

Lemma 4 *An element g in G_f is determined uniquely by its first ℓ-part $(1^g, \ldots, \ell^g)$.*

Lemma 4 is shown by the fact that the stabilizer of $\alpha_1, \ldots, \alpha_\ell$ coincides with the unit group.

Theorem 5 *Let $C = \{c_1, \ldots, c_t\}$ be a subset, with t elements, of $\{1, \ldots, n\}$. Then:*

(1) *There is an element g in G_f having $(c_1, \ldots, c_t)$ as its first t-part, if and only if $\alpha_{c_1}, \ldots, \alpha_{c_t}$ satisfy the first t defining polynomials, that is, $f_k(\alpha_{c_1}, \ldots, \alpha_{c_k}) = 0$ for $k = 1, \ldots, t$. If $t \geq \ell$, the existence is unique.*

(2) *Assume that there exists an element g in G_f having $(c_1, \ldots, c_\ell)$ as its first ℓ-part. Then, for $i = \ell+1, \ldots, n$, i^g is determined as an integer k such that $f_i(\alpha_{c_1}, \ldots, \alpha_{c_\ell}, \alpha_k) = \alpha_k - A_i(\alpha_{c_1}, \ldots, \alpha_{c_\ell}) = 0$.*

Proof. (1) We show *if* part. It is well-known that for a field K, any embedding of K into its algebraic closure $\bar{K}$ can be extended to an automorphism of $\bar{K}$. The fact that $\alpha_{c_1}, \ldots, \alpha_{c_t}$ satisfy the first t defining polynomials implies that there is a ring isomorphism τ from $\mathbb{Q}(\alpha_1, \ldots, \alpha_t)$ to $\mathbb{Q}(\alpha_{c_1}, \ldots, \alpha_{c_t})$. So, there is an automorphism $\bar{\tau}$ of $\bar{\mathbb{Q}}$ which is an extension of τ, and the restriction of $\bar{\tau}$ to the splitting field K_f becomes an automorphism of K_f, i.e., $\bar{\tau}$ is an element of G_f. *Only if* part is shown by the fact that every element g of the Galois group G_f fixes $\mathbb{Q}$, that is, $0 = f_k(\alpha_1, \ldots, \alpha_k) = f_k(\alpha_1, \ldots, \alpha_k)^g = f_k(\alpha_1^g, \ldots, \alpha_k^g)$.
(2) As $f_i(x_1, \ldots, x_i) = x_i - A_i(x_1, \ldots, x_\ell)$, the second statement holds. Q.E.D.

Since the splitting field K_f is expressed by the residue class ring, we can check zero-relations among elements in K_f as the membership of the polynomials derived from zero-relations to the ideal.

Corollary 6 *We use the same notation as in theorem 5. Then:*

(1) *There is an element g in G_f having $(c_1, \ldots, c_t)$ as its first t-part, if and only if $f_k(x_{c_1}, \ldots, x_{c_k})$ belongs to the ideal J for $k = 1, \ldots, t$.*

(2) *In the case (2) in theorem 5, for $i = \ell + 1, \ldots, n$, i^g is determined as an integer k such that $x_k - A_i(x_{c_1}, \ldots, x_{c_\ell})$ belongs to J.*

Since $f_1, \ldots, f_n$ forms a Gröbner basis for the ideal J, the membership problem of polynomials to J can be solved easily by using the normal forms. Practical computation will be described later.

Thus, for a permutation g such that only first t-part is known, we can check whether there exists such an element g in G_f or not, and if so, we can construct an element having the given first t-part.

Procedure $CHECK(DP, C)$

Input: a set of defining polynomials $DP = \{f_1, \ldots, f_n\}$ and the first t-part $C = [c_1, \ldots, c_t]$ of g, where $t \leq \ell$.
Output: the first ℓ-part of g if g exists; $\emptyset$, otherwise.

```
    for i from 1 by 1 to t do
        h ← NF(f_i(x_{c_1}, ..., x_{c_i}), DP)
        if h ≠ 0 then return ∅
    if t < ℓ then
        for k from t + 1 by 1 to ℓ do
            for each c in {1, ..., n} \ C do
                h ← NF(f_k(x_{c_1}, ..., x_{c_{k-1}}, x_c), DP)
                if h = 0 then
                    C ← APPEND(C, c)
                    break
    return C
end
```

<u>*Procedure COMPLETE(DP, C)*</u>

Input: a set of defining polynomials DP and the first ℓ-part $C = [c_1, \ldots, c_\ell]$ of g.
Output: the complete expression of g.

```
    for i from ℓ + 1 by 1 to n do
        h ← NF(A_i(x_{c_1}, ..., x_{c_ℓ}), DP)
            for k from 1 by 1 to n do
                if x_k = h then
                    C ← APPEND(C, k)
    return C
end
```

$APPEND([a_1, \ldots, a_r], a_{r+1})$ returns the list $[a_1, \ldots, a_{r+1}]$. $NF(p, B)$ calculates the normal form of a polynomial p with respect to a set of polynomials B under some fixed term-ordering.

Thus, by checking the membership for all permutations, we can construct the Galois group G_f. But it is not an efficient way, because the number of all permutations is $n!$. So we employ some strategy for generation of permutations.

3.2 Finding a set of generators of a Galois group

First, we introduce an important notion *strong generators*. (See Butler [4].)

Definition 7 *Let G be a permutation group on the set $\{1, \ldots, n\}$. By $G_{1,\ldots,i}$ or simply $G_{(i)}$, we denote the pointwise stabilizer of $1, \ldots, i$ in G, that is, $G_{(i)} = \{g \in G | j^g = j \text{ for } j = 1, \ldots, i\}$. Moreover set $G_{(0)} = G$. Then, we have a chain of stabilizers:*

$$G = G_{(0)} \supset G_{(1)} \supset G_{(2)} \supset \cdots \supset G_{(n)} = 1.$$

Let k be the smallest integer such that $G_{(k)} = 1$. We call a sequence $[1, 2, \ldots, k]$

a basis *for G. For each i in the basis $[1,\dots,k]$, we denote by S_i a set of all (right) coset representatives of $G_{(i)}$ in $G_{(i-1)}$. Therefore, by setting $S_i = \{s_1^{(i)},\dots,s_{t_i}^{(i)}\}$,*

$$G_{(i-1)} = G_{(i)}s_1^{(i)} \cup G_{(i)}s_2^{(i)} \cup \cdots \cup G_{(i)}s_{t_i}^{(i)}.$$

Then, the union $S = \cup_{i=1}^k S_i$ generates G. We call S a strong generating set *and elements of S* strong generators.

Remark 8 *A strong generating set has a useful property: Every element g in G can be written uniquely as $g = g_1g_2\cdots g_k$, where g_i belongs to S_i for each i, $i = 1,\dots,k$. Moreover, the number of elements in S is bounded by $k(2n-k+1)/2$.*

Now, let us consider how to construct a strong generating set S for G_f. Let $[1,\dots,k]$ be the basis for G_f and let S_i be a set of all coset representatives of $(G_f)_{(i)}$ in $(G_f)_{(i-1)}$ for i, $1 \le i \le k$. P_i be the set of all integers which correspond to the roots of the defining polynomial f_i of α_i, that is $P_i = \{j | f_i(\alpha_1,\dots,\alpha_{i-1},\alpha_j) = 0\}$. Then, P_i is contained in $\{i, i+1,\dots,n\}$ and it is the unique suborbit of $(G_f)_{(i-1)}$ containing i. From this, we have the following.

Lemma 9 *The length ℓ coincides with k. Moreover, for the coset representatives S_i for each i, we can choose t permutations $s_1,\dots,s_t$ such that $t = |P_i|$ and each s_k transports i to $p_k^{(i)}$, where $P_i = \{p_1^{(i)}(=i), p_2^{(i)},\dots,p_t^{(i)}\}$. Especially, we can choose the unit $1 = (1,\dots,n)$ for s_1.*

Thus, applying *CHECK* and *COMPLETE* for a coset representative g having $(1,\dots,i,p)$ as its first $i+1$ part, where $p \in P_i$, we obtain a strong generating sets. By this, the number of trials for checking the existence is bounded in a small amount.

Procedure STRONG_GENERATORS(DP)

Input: a set of defining polynomials $DP = \{f_1,\dots,f_n\}$.
Output: a strong generating set.

```
L ← {}
S ← {}
for i from 1 by 1 to n do
    if i = 1 then
        for j from 1 by 1 to ℓ − 1 do
            C ← {1,...,j}
            for each k from {1,...,n} \ C do
                g ← CHECK([1,...,j,k],DP)
```

```
                if g ≠ ∅ then L ← L ∪ {g}
        else
            g ← CHECK([i], DP)
            L ← L ∪ {g}
    for each g ∈ L do
        h ← COMPLETE(g, DP)
        S ← S ∪ {h}
    return S
end
```

We will give a bound on the number of normal form computations with respect to the ideal J. Since the total complexity of the procedure is the product of the number of polynomials for which we compute the normal forms and the cost of one normal form computation, the following bound shows the efficiency of the procedure.

Lemma 10

(1) *The length ℓ is bounded by $min\{log_2(N), n\}$. If G_f is primitive and solvable, then ℓ is bounded by $O(log(n))$.*

(2) *The number of normal form computations is bounded by $O(n^2\ell^2)$ and so $O(n^4)$.*

Proof. (1) Since $N = n_1 \cdots n_\ell$, where $n_i = \deg_{x_i}(f_i) > 1$ for $i = 1, \ldots, \ell$, we have $2^\ell < N$ and so $\ell < log_2(N)$. If G_f is primitive and solvable, we have $log_2(N) = O(log(n))$ by [14].

(2) Consider the case where we find a permutation g with its first 1-part (c_1). Since α_{c_1} is a root of f_1, we need not compute the normal form of $f_1(\alpha_{c_1})$. Then, we seek for c_2 from $\{1, \ldots, n\} \setminus \{c_1\}$ such that $\alpha_{c_1}, \alpha_{c_2}$ satisfy f_2. So, in this step, at most $n-1$ normal form computations are necessary. After finding such an integer c_2, we next seek for c_3 and at most $n-2$ normal form computations are required. Thus, for finding $c_2, \ldots, c_\ell$, at most $(n-1)+(n-2)+\cdots+(n-\ell+1)$ normal form computations are required. After finding the first ℓ-part, the remaining part is determined by $n-\ell$ normal form computations of A_i's. Thus, to find such an element g, at most $(2n-\ell-1)\ell/2$ normal form computations are necessary. Thus, to complete a set S_1 of representatives of $(G_f)_{(1)}$ in $(G_f)_{(0)}$, at most $(n-1)n\ell$ normal form computations are necessary. By an induction argument, to complete a set S_i of representatives of $(G_f)_{(i)}$ in $(G_f)_{(i-1)}$, at most $(n_i-1)n(\ell-i+1)$ normal form computations are necessary, where $n_i = \deg_{x_i}(f_i) < n$. Thus, in the whole, at most $n^2\ell(\ell+1)/2$ normal form computations are necessary. Q.E.D.

Remark 11 *We describe a method using a primitive element briefly. K_f is represented by $\mathbb{Q}[z]/ < m(z) >$, where new indeterminate z corresponds to a primitive element α and $m(z)$ is the minimal polynomial of α over $\mathbb{Q}$. So each root α_i is expressed as a polynomial $A_i(z)$ in z over $\mathbb{Q}$ with degree less than $N = \deg(m)$.*

The Galois group G_f is computed by using the regularity of the action of G_f on the roots $\beta_1, \ldots, \beta_N$ of $m(x)$. Here, we need the factorization of $m(x)$. Since K_f is a Galois extension, $m(x)$ is factorized into linear factors over K_f and so each root β_i is expressed by a polynomial $B_i(z)$ in z over $\mathbb{Q}$. Then, each element of G_f is determined uniquely by the action on $z = B_1$, that is, $G_f = \{g_1, \ldots, g_N\}$, where $z^{g_i} = B_1^{g_i} = B_i(z)$. The permutation representation of g_k on Ω_f is determined by the following:

$$j = i^{g_k} \text{ if and only if } A_i(B_k(z)) \equiv A_j(z) \text{ in } \mathbb{Q}[z]/ < m(z) >.$$

To complete the permutation representation of g_k, where $z^{g_k} = B_k(z)$, we have to compute the normal forms $A_i(B_k(z))$ for all i, $i = 1, \ldots, n$. Therefore, the number of normal form computations is Nn. By employing tricks for avoiding unnecessary normal form computations, the number of normal form computations can be reduced to some extent.

4 Computation of splitting fields

Employing successive extension to compute splitting fields is not a new idea and it is already implemented on Maple. The dominant part is univariate factorization over algebraic number fields. For practical computation, improvements for this part are necessary. For factorization over algebraic number fields, several algorithms are proposed. We refer to Abott [1] for the history. Here, we propose a new method for the problem based on the modified Trager algorithm in [13].

We first review Trager's algorithm presented in [18]. Let $\alpha_1, \alpha_2, \ldots, \alpha_\ell$ be algebraic numbers such that the minimal polynomial of each α_i is defined over $\mathbb{Q}(\alpha_1, \ldots, \alpha_{i-1})$. By applying his method directly to factorization over extension fields represented by successive extension, we obtain the following procedure.

<u>*Procedure $AFACTOR\_TRAGER(AS, f)$*</u>

Input: a set of algebraic numbers $AS = \{\alpha_i, \ldots, \alpha_1\}$ and a square-free univariate polynomial f in x over $\mathbb{Q}(AS)$.
Output: the set of irreducible factors of f over $\mathbb{Q}(AS)$.

```
    if AS = ∅ then return FACTOR(f)
    α ← FIRST(AS)
    (shift, r) ← SQFR_NORM(f, α)
    GS ← AFACTOR_TRAGER(AS \ {α}, r)
    Factors ← {}
    for each member g in GS do
        h ← GCD(f, g(x + shift · α))
        f ← f/h
        Factors ← Factors ∪ {h}
    return Factors
end
```

Here, $FACTOR$ is a factorizer of univariate polynomials over $\mathbb{Q}$, $FIRST(AS)$ gives α_i having the largest index in AS, and $SQFR\_NORM$ finds an integer $shift$ such that $Norm_{\mathbb{Q}(AS)/\mathbb{Q}(AS\setminus\{\alpha\})}(f(x-shift\cdot\alpha))$ is square-free and returns the pair of the integer and the norm. The correctness of this algorithm is deduced by the following.

Lemma 12 (Theorem 2.1 in [18]) *Let $f(x)$ be an irreducible polynomial over $K(\alpha)$. Then, $Norm_{K(\alpha)/K}(f)$ is a power of an irreducible polynomial over K.*

In [13], we proposed a modification of Trager's algorithm for univariate factorization over algebraic number fields. (Wang [20] also proposed a modified Trager type algorithm, which uses the characteristic set method extensively.) In [13] only the outline of the algorithm was stated for a simple extension over the rationals, but we can apply it to more general cases.

The key of our algorithm is the *use of non square-free norms*. In the original Trager's algorithm [18] and $AFACTOR\_TRAGER$, non square-free norms are discarded by reason that there is no guarantee that they produce irreducible factors. But if a non square-free norm has multiple irreducible factors, we can split the input polynomial into proper factors.

Lemma 13 *Let $f(x)$ be a squarefree polynomial over $K(\alpha)$ and let*

$$Norm_{K(\alpha)/K}(f) = \prod_{i=1}^{r} f_i^{m_i}$$

be the decomposition of $Norm_{K(\alpha)/K}(f)$ into its irreducible factors over K, where each f_i is irreducible. Then,

$$f = \prod_{i=1}^{r} GCD(f, f_i)$$

and $GCD(f, f_i)$ is a non-constant factor of f for each f_i. Especially $GCD(f, f_i)$ is irreducible if $m_i = 1$ for a factor f_i.

Proof. By the definition of the norm, f is a factor of $Norm_{K(\alpha)/K}(f)$ over $K(\alpha)$. Since f is square-free, we have $f = \prod_{i=1}^{r} GCD(f, f_i)$. By lemma 12, each f_i contains some irreducible factors of f. Therefore, $GCD(f, f_i)$ is not a constant. If $m_i = 1$, f_i must be the norm of an irreducible factor h of f. This implies that $GCD(f, f_i)$ is irreducible. Q.E.D.

$AFACTOR\_TRAGER$ is a recursive procedure. At the bottom of the recursion, factorizer for integral polynomials is called. When it is called, the degree of a polynomial to be factorized tends to be larger. The following lemma reduces the cost of trial division in factorization of integral polynomials appeared in the procedure.

Lemma 14 *Let L/K be an algebraic field extension of degree n and $f(x)$ be an element of $L[x]$, and suppose that the degree of each irreducible factor of f is a multiple of an integer d. Then, the degree of an irreducible factor of multiplicity m of $Norm_{L/K}(f)$ is a multiple of $dn/GCD(m, dn)$.*

Proof. Let f_i, $1 = 1, \ldots, k$, be irreducible factors of f such that $Norm_{L/K}(f_i) = g^{m_i}$ for same irreducible polynomial g. By the assumption for the degrees of f_i, there is an integer t_i such that $\deg(f_i) = dt_i$ and $m_i \deg(g) = ndt_i$. Then, $\deg(g) = (dn \sum_{i=1}^{k} t_i)/\sum_{i=1}^{k} m_i$. Let $m = \sum_{i=1}^{k} m_i$ and $t = \sum_{i=1}^{k} t_i$. Then, m is the multiplicity of g in $Norm_{L/K}(f)$ and $\deg(g)m = dnt$. From this, $\deg(g) \times (m/GCD(m, dn)) = (dn/GCD(m, dn)) \times t$ and so $(dn/GCD(m, dn))$ must divide $\deg(g)$ because $GCD(dn/GCD(m, dn), m/GCD(m, dn)) = 1$. Q.E.D.

Thus, we have the following procedure.

Procedure $AFACTOR(AS, f, shift, hint)$

Input: a set of algebraic numbers $AS = \{\alpha_i, \ldots, \alpha_1\}$, a square-free univariate polynomial f in x over $\mathbb{Q}(AS)$, an integer $shift$ and an integer $hint$.
Restriction: the degree of each irreducible factor of f is a multiple of $hint$.
Output: the set of irreducible factors of f over $\mathbb{Q}(AS)$.

```
    if DEG(f) = 1 then return {f}
    if AS = ∅ then return UFACTOR_HINT(f, hint)
    α ← FIRST(AS)
    r ← Norm_{Q(AS)/Q(AS\{α})}(f(x − shift · α))
    Sq ← SQFR(r)
    Factors ← {}
    for each member {g, m} in Sq do
        d ← DEG(DEFPOLY(α))
        T ← AFACTOR(AS \ {α}, g, 1, hint · d/GCD(m, hint · d))
        for each member h in T do
            p ← GCD(f, h(x + shift · α))
            f ← f/p
            if m = 1 then
                Factors ← Factors ∪ {p}
            else
                shift1 ← NEXT_SHIFT(shift)
                Factors1 ← AFACTOR(AS, p, shift1, hint)
                Factors ← Factors ∪ Factors1
    return Factors
end
```

In $AFACTOR$, $DEFPOLY$ returns the minimal polynomial of an algebraic number. The square-free factorizer $SQFR$ returns the set of pairs of a factor and its multiplicity. $NEXT\_SHIFT(s)$ returns $-s$ for $s > 0$, $-s+1$ for

$s \leq 0$. $UFACTOR\_HINT$ is a variation of Berlekamp-Hensel type factorizer for univariate polynomials. This can be used when it is known a priori that the degree of each irreducible factors is a multiple of *hint*. In the trial division stage, only candidates whose degrees are multiples of *hint* are actually tested. Though the complexity of this modified factorizer is still exponential with respect to the number of modular factors, it greatly reduces the practical computing time when *hint* is large. By lemma 14, the propagation of *hint* stated in the procedure is correct.

Remark 15 *Except for the use of non square-free norm, $AFACTOR$ is essentially the same as $AFCTOR\_TRAGER$. Especially, if the norm is square-free, only the hint will take effects. But in the computation of splitting fields there are several cases that non square-free norms appear. We will discuss this behavior later.*

In the following, the procedure $SPLITTING\_FIELD$ is described which can also deal with reducible polynomials over $\mathbb{Q}$. In this procedure, $AFACTOR$ is called with $shift = hint = 1$. The initial value of $shift$ is chosen to be 1 because the value 0 makes the norm in $AFACTOR$ useless.

Procedure $SPLITTING\_FIELD(f)$

Input: a univariate polynomial f in x over $\mathbb{Q}$.
Output: algebraic numbers required to construct the splitting field and the roots of f represented by the above algebraic numbers.

$AS \leftarrow \{\}$
$Factors \leftarrow$ irreducible factors of f over $\mathbb{Q}$
do
  $T \leftarrow$ non-linear factors in $Factors$
  if $T = \emptyset$ then return $\{AS, Factors\}$
  $defpoly \leftarrow$ an element of T
  $\alpha \leftarrow$ a root of $defpoly$
  $AS \leftarrow \{\alpha\} \cup AS$
  $Factors1 \leftarrow \{x - \alpha, defpoly/(x - \alpha)\}$
  $Factors \leftarrow (Factors \backslash \{defpoly\}) \cup Factors1$
  $Factors \leftarrow \bigcup_{g \in Factors} AFACTOR(AS, g, 1, 1)$
end

Remark 16 *In $AFACTOR$, each irreducible factor of*

$$Norm_{\mathbb{Q}(AS)/\mathbb{Q}(AS\backslash\{\alpha\})}(f(x - shift \cdot \alpha))$$

corresponds with the minimal polynomial of an algebraic number $\beta = \gamma + shift \cdot \alpha$, where γ is a root of the corresponding irreducible factor of $f(x)$ over $\mathbb{Q}(AS)$. If β is not a primitive element of the extension field $\mathbb{Q}(AS \cup \{\gamma\})$ over $\mathbb{Q}(AS \backslash$

$\{\alpha\}$) or if β is algebraically conjugate to $\beta' = \gamma' + shift \cdot \alpha$ over $\mathbb{Q}(AS \setminus \{\alpha\})$, where γ' is a root of another irreducible factor of f over $\mathbb{Q}(AS)$, the corresponding irreducible factor is a non square-free factor (see [21]).

In SPLITTING_FIELD, such an algebraic number β can be expressed as $\beta = a_{s+1}\alpha_{s+1} + \cdots + a_{s+t}\alpha_{s+t}$, where $0 \leq s < s+t \leq n$ and $a_{s+1}, \ldots, a_{s+t}$ are integers used for $shifts$, and its minimal polynomial over $\mathbb{Q}(\alpha_1, \ldots, \alpha_s)$ is considered. Then, a non square-free norm appears if the pointwise stabilizer $G_{\alpha_1,\ldots,\alpha_s,\beta}$ of the point $\alpha_1, \ldots, \alpha_s, \beta$ does not coincide with that of $\alpha_1, \ldots, \alpha_{s+t}$ or if $G_{\alpha_1,\ldots,\alpha_s,\beta}$ is conjugate to $G_{\alpha_1,\ldots,\alpha_s,\beta'}$ in $G_{\alpha_1,\ldots,\alpha_s}$, where $\beta' = a_{s+1}\alpha'_{s+1} + \cdots + a_{s+t}\alpha'_{s+t}$ for another t roots $\alpha'_{s+1}, \ldots, \alpha'_{s+t}$. For example, if the pointwise stabilizer $G_{\alpha_1,\ldots,\alpha_t}$ does not coincide with the setwise stabilizer $G_{\{\alpha_1,\ldots,\alpha_t\}}$, $\beta = \alpha_1 + \cdots + \alpha_t$ with $a_1 = \cdots = a_t = 1$ gives a non square-free norm. Moreover, in the case where f is irreducible and G_f is small, there are many linear relations between the roots which tend to give non square-free norms. By this observation, we are convinced that the modification is very suited for computation of splitting fields.

5 Discussion on practical computation

Here, we report our efforts to realize our new method as an efficient one on a computer algebra system. The method requires heavy algebraic operations such as algebraic factorization, algebraic GCD, *etc.* Therefore, how we choose algorithms for algebraic operations properly and how we combine those effectively are very serious issues. The following explains not only our current efforts to these issues but also difficulties which must be overcome in the future.

5.1 Practical computation of splitting fields

There are three major time consuming parts in *AFACTOR*: *1. univariate factorization over $\mathbb{Q}$; 2. norm computation; 3. polynomial GCD computation over algebraic number fields.*

5.1.1 Univariate factorization over $\mathbb{Q}$

By using *UFACTOR_HINT*, computing times are greatly reduced for several cases, but this is still a major obstacle, we think, especially when the Galois group is large. Currently, the order approximately 200 is an upper limit for one day computation.

In more detail, consider the case where the input polynomial f is integral and irreducible. In the main procedure *SPLITTING_FIELD*, each univariate polynomial g to be factored by *UFACTOR_HINT* is a factor of the norm of $f(x - c_1\alpha_1 - \cdots - c_s\alpha_s)$, where $s \leq \ell$ and $c_1, \ldots, c_s$ are integers used for $shifts$. Then, g is a factor of $RES_{v_1}(\cdots RES_{v_s}(f(x - c_1v_1 - \cdots - c_sv_s), f(v_s)), \cdots, f(v_1))$, where $RES_v(F, G)$ is the resultant of F and G with respect to v. So, it can be easily shown that for each prime p, the image $\bar{g}$ of g over the finite field $GF(p)$ also splits over the splitting field L of the image $\bar{f}$ of f over $GF(p)$. Therefore, the degree of each irreducible factor of $\bar{g}$ is a divisor of the extension degree of L. Since the extension degree of L is the LCM of degrees of irreducible factors of $\bar{f}$, $\bar{g}$ tends to have many irreducible factors which do not correspond with true factors of g. For example, when $\deg(f)$ is a prime q, and $\bar{f}$ is also irreducible, $\bar{g}$ is factorized into linear factors and factors with degree q. This behavior is well-known. (See Abott et al. [2] and Kaltofen et al. [8].) Thus, a factorizer based on the Berlekamp-Hensel algorithm often suffers *extraneous factor problem* in this case. In our method, only *hint* has an effect on this disaster. The lattice algorithm, we think, is a promising one to solve this problem.

5.1.2 Norm computation

Norm is computed by resultant. Although there are several methods to compute resultant, it is difficult to choose an algorithm which may work well for various inputs. The subresultant algorithm and the Chinese remainder algorithm are two major algorithms for resultant computation. However, we don't know the practical condition which one to choose. So, we are now experimenting with both methods.

5.1.3 GCD computation over algebraic number fields

GCD computation by an Euclidean algorithm often causes serious intermediate expression swells. For simple extension fields over $\mathbb{Q}$, the modular methods seem to be practically best and are proposed by several authors; Geddes et al. [7], Smedley [15], and Langemyr and McCallum [11]. In our current implementation, no modular methods are employed but Gröbner basis computation is used to compute GCD over algebraic number fields. Modular methods will be tested in future works. From our experiments, the method based on Gröbner basis computation is more efficient than ordinary Euclidean algorithms for many cases. The reason is that coefficient growths are smaller in M-reductions than in pseudo-remainder computations. In our implementation, the content of a normal form is always removed to make coefficient growth as small as possible.

5.2 Polynomial GCD computation over algebraic number fields

How efficiently we execute the practical computation of the Galois groups depends on how efficiently we compute the normal forms because, as mentioned before, a simplification of an expression containing algebraic numbers is considered to be a normal form computation in a polynomial ring.

There are several methods to compute normal form. By treating the set of defining polynomials as the reduced Gröbner basis, normal forms are computed by M-reductions. And also by treating it as the characteristic set, M-reductions gives the same result as pseudo-divisions, since each defining polynomial has a rational leading coefficient with respect to its main variable. Practically, the total cost of normal form computation depends heavily on the choice of reducer. For example, if the selection strategy of reducer is that the first one with respect to the *natural* order is selected, the first reducer reduces the polynomial repeatedly and the result may be very large before other reducers are used. So reductions by lower-order reducers must be executed to make the intermediate expressions smaller. Appearance of fractional coefficients also makes the cost very high because many integer GCD computations are required. To avoid that, when successive reductions by non-monic reducers are executed, computations of pseudo-remainder are first repeated, and finally the result is divided by the product of the numbers which were used to make the divisions integral.

In addition to the above general observations, another technique based on the property of the algorithm is used. It is the use of *history*. The main part of the algorithm is polynomial substitution and reduction, and the reducers and the values for substitution are fixed throughout the execution of the algorithm. So, once we make tables of powers of the values modulo the reducers, we can use repeatedly the tables to compute reduced values of substituted polynomials. By these techniques, the practical efficiency has been much improved.

As a future work, modular computation will be examined. Because, in the computation of Galois groups, normal form computation is used to determine the coincidences with 0. So it seems very effective for avoiding coefficient swell.

6 Concluding remarks

We presented a method for computing Galois groups using successive extension representation. We aim at *direct* computation of Galois groups which is more efficient both in theory and practice. Although we pointed out that the new method has a small number of normal form computations, we did not give the total complexity of the method. It is because the difficulty to estimate the complexity of the splitting field computation. (The complexity shall vary with algorithms for algebraic factorization. And if we had employed lattice algorithms for factorization, the arguments in Chistov and Grigoryev [5] or Landau [10] could be used to estimate the total complexity.) Instead, we examined the efficiency of the method by experiment with a number of examples. In Appendix, Table 1 shows the comparison of two methods for direct computation and Table 2 shows that Galois groups with small orders (< 200) are computable in time. So, we are convinced that the method is applicable to polynomials with small splitting field.

Finally, we comment on the comparison of two methods, a table-based one and our direct one. As Table 3 in the appendix shows, a table-based method on Maple is more efficient than our direct method for polynomials with small degree, to which classification tables for transitive subgroups are known. But, as described before, for arbitrary degrees, there are no tables for them. From this, we think that both approaches are important for further study of the subject. Especially, usage of a certain knowledge of the Galois group will make direct methods more efficient. Anyway, since both approaches use factorization as their parts, progress of factorization will make them more efficient.

Appendix

Risa/Asir [13] is a computer algebra system, which provides a programming system Asir with several subroutine libraries which can also be used as the parts of other programs. It runs on major UNIX workstations, Macintoshes and MS-DOS machines under DOS-Extenders. Though its capability of doing various computations is currently limited, its efficiency for fundamental algebraic computations is sufficiently high. Especially the univariate factorizer is much tuned, which was very suited to implement our algorithm for the computation of splitting fields and Galois groups.

All the procedures stated in this paper were written in the Asir user language. Timings (given in seconds) of Asir and Maple were measured for 37 polynomials listed below on SPARCstation 2 (40MHz SPARC, 64Mbytes of main memory). In the tables, GC times are excluded for Asir. For Maple, the timings were measured by `showtime()`, but we don't know whether the timing reported includes GC time or not.

For all tables, ℓ denotes the length of the splitting field in successive representation, and "–" means that we could not get the result in 3600 seconds in Table 1 and in 30000 seconds in Table 2 and Table 3.

Table 1 shows the comparison of two methods for computation of splitting fields, a method using a primitive element and one using successive extension.

Table 2 shows the times for splitting field computations and total times to construct strong generating sets of Galois groups. For the sample (20), its splitting field was obtained in 4.5 hours by employing Maple GCD computation. But, the original procedure in Asir might take more than 6 hours and so $x > 21600$.

Table 3 shows the comparison of two methods, a table-based one on Maple and our direct one on Asir. Names of Galois groups are obtained by `galois(f)` in Maple. Since `galois(f)` is not applicable to polynomials with degree more than 7, we compare them only for samples (1) $\sim$ (29).

Remark 17 *By the proposed direct method we can obtain the Galois group as a strong generating set, whose elements are permutations on all roots of the given polynomial. Of course, the name of Galois group cannot be known directly, but to know it is easy for small degrees.*

(1) $x^5+x^4-4x^3-3x^2+3x+1$ (2) x^5-3x^2+2x+1
(3) x^5-2 (4) $x^5-2x^4+10x^3-10x^2-10x-10$
(5) x^5-x^2-2x-3 (6) x^5-x+1
(7) x^6+x^3+1 (8) $x^6+2x^3+9x^2-6x+2$
(9) x^6-3 (10) $x^6+9x^4-4x^2-4$
(11) x^6+x^3+7 (12) x^6-3x^4+1
(13) x^6+x^4-9 (14) x^6+6x^2+4
(15) x^6-2x^3-2 (16) $x^6+6x^4+2x^3+9x^2+6x-4$
(17) x^6+x^4-8 (18) $x^6-9x^3+6x^2+9x+2$
(19) $x^6+x^4-x^2+5x-5$
(20) $x^6+10x^5+55x^4+140x^3+175x^2-3019x+25$
(21) $x^6-9x^3+3x^2-6x+1$ (22) x^6+x+1
(23) $x^7+x^6-12x^5-7x^4+28x^3+14x^2-9x+1$
(24) $x^7+7x^3+7x^2+7x-1$ (25) $x^7-14x^5+56x^3-56x+22$
(26) x^7-2 (27) x^7-7x+3
(28) $x^7+7x^4-7x^3-9$ (29) x^7+x+1
(30) x^8-2 (31) x^9-2
(32) $x^9-15x^6-87x^3-125$ (33) $x^{10}-2$
(34) $x^{11}-2$ (35) $x^{12}-2$
(36) $x^{15}-2$ (37) $x^{16}-2$

Table 1 : Comparison of two methods for splitting fields (seconds)

	ℓ	successive	primitive
(2)	2	4.20	220.23
(3)	2	3.50	427.93
(9)	2	0.96	10.48
(10)	2	2.32	58.92
(11)	2	1.98	317.73
(12)	2	2.17	—
(26)	2	21.40	—

Table 2 : Timing statistics for constructing strong generating sets (seconds)

	order	ℓ	split	total
(1)	5	1	1.18	1.56
(2)	10	2	4.20	5.06
(3)	20	2	3.50	4.26
(4)	20	2	24.02	50.73
(5)	60	3	7415.04	7430.39
(6)	120	4	2059.06	2060.25
(7)	6	1	1.61	1.91
(8)	6	1	1.86	2.56
(9)	12	2	0.96	2.03
(10)	12	2	2.32	4.01
(11)	18	2	1.98	3.56
(12)	24	3	2.17	3.17
(13)	24	2	2.48	4.09
(14)	24	2	26.49	28.45
(15)	36	3	13.51	14.82
(16)	36	3	364.55	369.91
(17)	48	3	15.20	16.62
(18)	60	3	953.42	1029.02
(19)	72	4	92.03	95.24
(20)	120	3	x	x+162.38
(21)	360	4	—	—
(22)	720	5	—	—
(23)	7	1	6.43	7.68
(24)	14	2	17.46	27.00
(25)	21	2	50.18	65.41
(26)	42	2	21.4	25.06
(27)	168	3	25157.98	25497.72
(28)	2520	5	—	—
(29)	5040	6	—	—
(30)	16	2	2.33	4.93
(31)	54	2	18.04	23.38
(32)	18	2	9.79	27.55
(33)	40	2	18.39	26.54
(34)	110	2	2738.16	2744.18
(35)	48	2	7.91	20.51
(36)	120	2	225.78	240.06
(37)	64	2	89.92	119.74

Table 3 : Comparison of table-based method and direct one (seconds)

	Group	ℓ	table	direct
(1)	+Z5	1	0.68	1.56
(2)	+D5	2	0.82	5.06
(3)	F20	2	0.37	4.26
(4)	F20	2	0.78	50.73
(5)	+A5	3	0.12	7430.39
(6)	S5	4	0.13	2060.25
(7)	Z6	1	2.02	1.91
(8)	S3	1	1.75	2.56
(9)	D6	2	1.55	2.03
(10)	+A4	2	3.07	4.01
(11)	3.S3	2	6.73	3.56
(12)	2.A4	2	2.02	3.17
(13)	+S4/V4	2	1.97	4.09
(14)	S4/Z4	2	17.65	28.45
(15)	3^2.2^2	3	2.37	14.82
(16)	+3^2.4	3	1.88	369.91
(17)	2.S4	3	1.68	16.62
(18)	+PSL2(5)	3	1.78	1029.02
(19)	3^2.D4	4	1.43	95.24
(20)	PGL2(5)	3	24.25	x+162.38
(21)	+A6	4	0.26	—
(22)	S6	5	0.27	—
(23)	+Z7	1	1.28	7.68
(24)	D7	2	1.57	27.00
(25)	+F21	2	5.33	65.41
(26)	F42	2	6.41	25.06
(27)	+PSL3(2)	3	5.73	25497.72
(28)	+A7	5	0.42	—
(29)	S7	6	0.25	—

References

[1] Abott, J. A., *On the factorisation of polynomials over algebraic fields*, Ph.D thesis, School of Math. Sci., University of Bath (1989).

[2] Abott, J. A., Bradford, R. J., Davenport, J. H., *Factorisation of polynomials: Old ideas and recent results*, in "Trends in Computer Algebra," L. N. Comp. Sci. **296**, pp. 81–91 (1987).

[3] Buchberger, B., *Gröbner bases: an algorithmic method in polynomial ideal theory*, in "Multidimensional System Theory," Reidel Publ. Comp., pp. 184–232 (1985).

[4] Butler, G., "Fundamental algorithms for permutation groups," L. N. Comp. Sci. **559** (1991).

[5] Chistov, A. L., Grigoryev, D. Yu, *Polynomial-time factoring of the multivariable polynomials over a global field*, preprint, LOMI E-5-82, Leningrad (1982).

[6] Ford, D. J., Mckay, J., *Computation of Galois groups from polynomials over the rationals*, in "Computer Algebra," L. N. Pure Appl. Math. **113**, pp. 145–150 (1989).

[7] Geddes, K. O., Gonnet, G. H., Smedley, T. J., *Group theoretical methods for operations with algebraic numbers*, in "Proc. ISSAC '88," ACM Press, pp. 475–480 (1988).

[8] Kaltofen, E., Musser, D. R., Saunders, B. D., *A generalized class of polynomials that are hard to factor*, SIAM J. Comput. **12**, pp. 473–483 (1983).

[9] Kolesova, G., Mckay, J., *Practical strategies for computing Galois groups*, in "Computational Group Theory", Academic Press, pp. 297–299 (1984).

[10] Landau, S., *Factoring polynomials over algebraic number fields*, SIAM J. Comput. **14**, pp. 184–195 (1985).

[11] Landau, S., Miller, G. L., *Solvability by radicals is in polynomial time*, J. Comput. System Sci. **30**, pp. 179–208 (1985).

[12] Langemyr, E. G., Langemyr, L., McCallum, S., *The computation of polynomial greatest common divisors over an algebraic number field*, J. Symbolic. Comp. **8**, pp. 429–448 (1989).

[13] Noro, M., Takeshima, T, *Risa/Asir – a computer algebra system*, in "Proc. ISSAC '92," ACM Press, pp. 387–396 (1992).

[14] Pálfy, P., *A polynomial bound for the orders of primitive solvable groups*, J. Algebra, pp. 127–137 (1982).

[15] Smedley, T. J., *A new modular algorithm for computation of algebraic number polynomial gcds*, in Proc. ISSAC '89," ACM Press, pp. 91–94 (1989).

[16] Soicher, L. H., *An algorithm for computing Galois groups*, in "Computational Group Theory," Academic Press, pp. 291–296 (1984).

[17] Stauduhar, R. P., *The determination of Galois groups*, Math. Comp. **27**, pp. 981–996 (1973).

[18] Trager, B. M., *Algebraic factoring and rational function integration*, in Proc. SYMSAC '76," ACM Press, pp. 219–226 (1976).

[19] Tschebotaröw, N., Schwerdtfeger, H., "Grundzüge der Galois'schen Theorie," P. Noordhoff, 1950.

[20] Wang, D. M., *A method for factorizing multivariate polynomials over successive algebraic extension fields*, preprint, RISC-LINZ, 1992.

[21] Yokoyama, K., Noro, M., Takeshima, T., *Computing primitive elements of extension fields*, J. Symb. Comp. **8**, pp. 553–580 (1989).

[22] Yokoyama, K., Noro, M., Takeshima, T., *On determining the solvability of polynomials*, in "Proc. ISSAC '90," ACM Press, pp. 127–134 (1990).

H. Anai (`anai@iias.flab.fujitsu.co.jp`)

M. Noro (`noro@iias.flab.fujitsu.co.jp`)

K. Yokoyama (`momoko@iias.flab.fujitsu.co.jp`)

ISIS, Fujitsu Laboratories, 140 Miyamoto, Numazu-shi, Shizuoka, 410-03 (Japan).

Progress in Mathematics, Vol. 143,

How to compute the canonical module of a set of points

S. Beck, M. Kreuzer

1 Introduction

Suppose K is a computable field and $\mathbb{X} \subseteq \mathbb{P}^n_K$ is a set of points (i. e., a zero-dimensional, reduced subscheme with K-rational support) given by their coordinates. The first object of interest in this situation is, of course, the homogeneous vanishing ideal $I_{\mathbb{X}} \subseteq K[X_0, \ldots, X_n]$ of $\mathbb{X}$. A naïve and inefficient approach to find this ideal would be to form the homogeneous prime ideals of height n corresponding to those points and to compute their intersection. It has been shown to be much faster, and in fact of polynomial complexity in n and the number of points, to use the *Buchberger-Möller algorithm* for this task (cf. [MMM]). Based on the results of this algorithm, many elementary properties of $\mathbb{X}$ and its homogeneous coordinate ring R like its Hilbert function, Cohen-Macaulay type, a minimal system of generators of $I_{\mathbb{X}}$, the separators (cf. [GKR], sect. 2), etc. can be easily computed.

What is the canonical module?

In this paper we want to deal computationally with deeper geometric properties of $\mathbb{X}$ resp. algebraic properties of R. Our main object of interest is the *canonical module* of R. This finitely generated, graded R-module historically was first introduced in *duality theory* via

$$\omega_R \cong \underline{\mathrm{Hom}}_K(H^1_{\mathfrak{m}}(R), K).$$

Here $\mathfrak{m} = \oplus_{i>0} R_i$ is the homogeneous maximal ideal of R and $\underline{\mathrm{Hom}}$ means "graded homomorphisms", i.e., finite sums of homogeneous linear maps. The module ω_R plays a key rôle in this theory because of the

Graded duality theorem. *(cf. [GW], 2.1.6) For every finitely generated, graded R-module M there is an isomorphism*

$$\underline{\mathrm{Hom}}_R(M, \omega_R) \cong \underline{\mathrm{Hom}}_K(H^1_{\mathfrak{m}}(M), K)$$

of graded R-modules which is functorial in M.

Why are we interested in the canonical module?

Later it was realized that the multiplicative structure of this module carries a large amount of information about the geometry of the points of $\mathbb{X}$ (cf. [GKR], [K1], [K2]). In particular, it is possible to characterize the Cayley-Bacharach property, various uniformity conditions, linearly and higher-order general position properties solely in terms of the structure of the multiplication maps from one homogeneous component of ω_R to another.

Third, considering the *adjunction formula*

$$\omega_R \cong \underline{\mathrm{Ext}}^n_{K[X_0,\ldots,X_n]}(R, K[X_0,\ldots,X_n])(-n-1),$$

it is clear that knowing ω_R may be useful for computing the minimal graded free $P := K[X_0,\ldots,X_n]$-resolution of R. Some hints in this direction were already given in [K1], sect. 5, but let us be more precise. If

$$0 \longrightarrow \bigoplus_{i=1}^{\beta_n} P(-\alpha_{ni}) \xrightarrow{\Phi_n} \cdots \longrightarrow \bigoplus_{i=1}^{\beta_1} P(-\alpha_{1i}) \xrightarrow{\Phi_1} P \longrightarrow R \longrightarrow 0$$

is the minimal graded free P-resolution of R, then

$$0 \to P(-n-1) \xrightarrow{\check{\Phi}_1} \bigoplus_{i=1}^{\beta_1} P(\alpha_{1i}-n-1) \longrightarrow \cdots \xrightarrow{\check{\Phi}_n}$$
$$\xrightarrow{\check{\Phi}_n} \bigoplus_{i=1}^{\beta_n} P(\alpha_{ni}-n-1) \to \omega_R \to 0$$

is the minimal graded free P-resolution of ω_R, since R is a Cohen-Macaulay ring and ω_R is given by the adjunction formula (cf. sect. 7). Thus the resolution of ω_R is dual to the resolution of R. Now, often it happens that the Betti numbers β_i or the shifts α_{ij} we are interested in are the ones near the end of the resolution of R. Clearly, if we know a minimal graded presentation of ω_R and start resolving it, we can find those numbers much more easily.

A case in point is the situation when $\mathbb{X}$ has generic Hilbert function, e.g., if the points of $\mathbb{X}$ are chosen randomly. In section 7 we shall demonstrate that it suffices in this case to compute either the linear part of the resolution of R or of ω_R, in order to get all numbers α_{ij}, β_i. Consequently, we are able to compute those numbers very efficiently, and we apply our results to solve the *Minimal Resolution Conjecture* computationally in many cases.

How can we calculate a minimal graded presentation of ω_R?

This is the main topic of the present paper. We want to give a solution in the spirit of the Buchberger-Möller algorithm, i.e., avoiding the potentially expensive calculation of Gröbner bases via Buchberger's algorithm. In view of this strategy, the naïve method of using the description of ω_R given by the Adjunction Formula is totally out of the question: Computing the whole resolution of R requires computing n Gröbner bases of its syzygy modules, the sizes of the middle ones of which are known to be large because of the exponential growth of the Betti numbers.

A much better idea is to change coordinates such that $\mathbb{X} \subseteq D_+(X_0) = \mathbb{P}^n_K \setminus \mathcal{V}(X_0)$ and to use the description of the canonical module

$$\omega_R \cong \underline{\mathrm{Hom}}_{K[x_0]}(R, K[x_0])(-1)$$

derived from the *Noetherian Normalization* $K[x_0] \subseteq R$. Here x_0, the image of X_0 in R, is a linear non-zero-divisor. The elements of ω_R are then viewed as $K[x_0]$-linear maps $\varphi : R \longrightarrow K[x_0]$, and its R-module structure is given by $(r\varphi)(r') = \varphi(rr')$ for $r, r' \in R$.

Now we give a more detailed outline of the contents of the individual sections. In section 2 we introduce a generalization of the usual theory of graded and filtered rings and modules. We allow rings with filtrations indexed by abelian groups and modules with filtrations indexed by sets on which those abelian groups operate suitably. This allows us to treat the cases of filtrations of free modules (cf. Example 2.6) and dual modules (cf. Example 2.7) simultaneously. Then, in section 3, we review and adapt some standard computer algebra techniques which are useful for our purposes. In particular, we recall how one can deal computationally with zero-dimensional subschemes of $\mathbb{P}^n_K$, and we provide some reduction techniques for syzygies (cf. propositions 3.7 and 3.8).

Sections 4 to 6 form the central part of this paper and contain the main algorithm. In section 4 we start by describing a Gröbner basis of ω_R with respect to a suitable filtration. The idea for constructing it comes from looking at the order ideal $\mathcal{O}_\sigma(J_\mathbb{X})$ of the affine ideal $J_\mathbb{X}$ of $\mathbb{X} \subseteq D_+(X_0)$: Each monomial there has a multiple which is a *socle monomial*, i.e., all of whose multiples are outside $\mathcal{O}_\sigma(J_\mathbb{X})$. In the language of projections $R \longrightarrow K[x_0]$ to those monomials this says that the projections to socle monomials generate ω_R as an R-module. In fact, if we filter ω_R suitably, those projections actually form a minimal Gröbner basis.

Thus we proceed by studying the module of syzygies of those projections. Again, by looking at the above order ideal, we see that one has to consider two kinds of syzygies (cf. propositions 5.1 and 5.2), and we construct from them a Gröbner basis of the syzygy module of all projections (cf. proposition 5.3). Then we reduce those syzygies to a Gröbner basis of the syzygy module of the minimal Gröbner basis found before (cf. proposition 5.4). Finally, we apply in section 6 several minimalization steps to get the desired minimal graded presentation of ω_R.

The paper concludes with two sections about applications of the canonical module. First, in section 7, we explain in detail how one can use ω_R to speed up the computation of minimal graded free resolutions of points with generic Hilbert function. In particular, this method is applied to produce proofs or counterexamples to the Minimal Resolution Conjecture for numerous cases with small n and moderate numbers of points.

In the last section, we give some applications to checking higher uniformities. For small uniformities, we can give a fast direct method (cf. Remark

8.2), whereas for high uniformities we can apply our knowledge of ω_R to reduce the problem to checking biinjectivity of certain explicitly computable bilinear maps of finite-dimensional K-vector spaces. For that problem we offer two approaches (cf. Remarks 8.3 and 8.4), both of which, however, require standard Gröbner basis computations.

Acknowledgments. The authors are indebted to E. Kunz (Regensburg) for his continuing interest in this project, and for the opportunity to lecture about it in his seminar. They would also like to thank T. Mora (Genova), L. Robbiano (Genova), and F. Schreyer (Bayreuth) for stimulating discussions about the subject. Last but not least, thanks to the organizers of the conference "MEGA 94" (Santander / Spain) for enabling the second author to present this work to a wider audience.

Implementation. The algorithms described in this paper have been implemented by the first author using $C++$ in a program called "COP" (Computation Of Points). This program runs on PC's and Unix platforms equipped with the GNU compiler, and it is available from the authors upon request.

2 Filtered modules

For our purposes, it will prove useful to introduce the following generalization of the usual theory of graded and filtered rings and modules (cf. [R1], [PS]). We start with an abelian group $(G, +)$ which operates additively on a set H.

Definition 2.1. A *G-grading* on a ring R is a set of additive subgroups $R_g \subseteq R$ for $g \in G$ such that $R = \oplus_{g\in G} R_g$ and $R_g \cdot R_{g'} \subseteq R_{g+g'}$ for $g, g' \in G$. An *H-grading* on an R-module M is a set of R-submodules $M_h \subseteq M$ for $h \in H$ such that $M = \oplus_{h\in H} M_h$ and $R_g \cdot M_h \subseteq M_{g+h}$ for $g \in G$, $h \in H$.

Example 2.2. If R is a G-graded ring, $g \in G$, and M an H-graded R-module, then $M(g)$ is the H-graded R-module such that $M(g)_h = M_{g+h}$. We call $M(g)$ the *shift* of M by g.

Now let $(G, <_\sigma)$ and $(H, <_\tau)$ be totally ordered, and assume that the operation $+ : G \times H \longrightarrow H$ is compatible with those orderings, i.e., $h <_\tau h'$ implies $g + h <_\tau g + h'$, and $g <_\sigma g'$ implies $g + h <_\tau g' + h$ for $g, g' \in G$ and $h, h' \in H$.

Definition 2.3. An (ascending, exhaustive) *G-filtration* $\mathcal{G}$ of an R-algebra S is a set of R-submodules $\mathcal{G}_g \subseteq S$ for $g \in G$ such that

a) $\mathcal{G}_g \subseteq \mathcal{G}_{g'}$ for $g, g' \in G$ with $g <_\sigma g'$,
b) $\mathcal{G}_g \cdot \mathcal{G}_{g'} \subseteq \mathcal{G}_{g+g'}$ for $g, g' \in G$,
c) $1 \in \mathcal{G}_0$,
d) $\cup_{g\in G} \mathcal{G}_g = S$.

If $(S/R, \mathcal{G})$ is a G-filtered algebra and M an S-module, then an (ascending, exhaustive) *H-filtration* $\mathcal{H}$ of M is a set of R-submodules $\mathcal{H}_h \subseteq M$ for $h \in H$ such that

a) $\mathcal{H}_h \subseteq \mathcal{H}_{h'}$ for $h, h' \in H$ such that $h <_\tau h'$,
b) $\mathcal{G}_g \cdot \mathcal{H}_h \subseteq \mathcal{H}_{g+h}$ for $g \in G$, $h \in H$,
c) $\cup_{h \in H} \mathcal{H}_h = M$.

In the sequel, we let $(S/R, \mathcal{G})$ be a G-filtered algebra and $(M, \mathcal{H})$ an H-filtered S-module. As usual, letting $\mathcal{H}_h^< = \cup_{h'<h} \mathcal{H}_{h'}$ and $(\mathrm{gr}_\mathcal{H} M)_h = \mathcal{H}_h / \mathcal{H}_h^<$ for $h \in H$, we obtain an R-module $\mathrm{gr}_\mathcal{H} M = \oplus_{h \in H} (\mathrm{gr}_\mathcal{H} M)_h$ which we call the *associated graded module* of $(M, \mathcal{H})$. By $(s + \mathcal{G}_g^<) \cdot (m + \mathcal{H}_h^<) := sm + \mathcal{H}_{g+h}^<$ for $g \in G$, $h \in H$, $s \in \mathcal{G}_g$, $m \in \mathcal{H}_h$, it has the structure of an H-graded module over the associated G-graded ring $\mathrm{gr}_\mathcal{G} S = \oplus_{g \in G} \mathcal{G}_g / \mathcal{G}_g^<$ of $(S/R, \mathcal{G})$.

Definition 2.4. (cf. [PS], 2.19 and 4.1) The H-filtration $\mathcal{H}$ of M is called *valued*, if for any $m \in M \setminus \{0\}$ the *order* $\mathrm{ord}_\mathcal{H} m = \min\{h \in H \mid m \in \mathcal{H}_h\}$ exists. Furthermore, $\mathcal{H}$ is called a *Gröbner filtration*, if it is valued and the set $\{\mathrm{ord}_\mathcal{H} m \mid m \in M \setminus \{0\}\}$ is well ordered with respect to $<_\tau$.

From now on, all filtrations will be assumed to be Gröbner filtrations. For $m \in M \setminus \{0\}$, the element $L_\mathcal{H} m = m + \mathcal{H}^<_{\mathrm{ord}_\mathcal{H} m} \in \mathrm{gr}_\mathcal{H} M$ is called the *leading form* of m. If $m = 0$, we let $L_\mathcal{H} m := 0$. Note that with M, also every S-submodule $N \subseteq M$, resp. every quotient module M/N, is canonically H-filtered via $\{N \cap \mathcal{H}_h\}_{h \in H}$, resp. $\{\mathcal{H}_h / (N \cap \mathcal{H}_h)\}_{h \in H}$. The following proposition collects some basic properties of Gröbner filtrations. Its proof is completely analogous to the classical case (cf. [R1], sect. 3, and [PS], sect. 4) and will be omitted.

Proposition 2.5. *Let $(M, \mathcal{H})$ be an H-filtered S-module and $N \subseteq M$ an S-submodule.*

a) *The induced filtrations on N and M/N are Gröbner filtrations again.*
b) *If $\mathrm{gr}_\mathcal{H} N = \sum_{i=1}^r \mathrm{gr}_\mathcal{G} S \cdot L_\mathcal{H} m_i$ for some elements $m_1, \ldots, m_r \in N$, then $N = \sum_{i=1}^r S m_i$. In this case, $\{m_1, \ldots, m_r\}$ is called a* Gröbner basis *of N.*
c) *A set $\{m_1, \ldots, m_r\}$ is a Gröbner basis of N if and only if every element $n \in N \setminus \{0\}$ has a representation $n = \sum_{i=1}^r s_i m_i$ with $s_i \in S$ and $\mathrm{ord}_\mathcal{G} s_i + \mathrm{ord}_\mathcal{H} m_i \le \mathrm{ord}_\mathcal{H} n$ whenever $s_i \neq 0$.*

Example 2.6. *(Canonical filtration of a free module)*

For a G-filtered algebra $(S/R, \mathcal{G})$, we want to define an H-filtration $\mathcal{H}$ of $M = S^r$ with $r \geq 1$ such that $H = G \times \{1, \ldots, r\}$, and the standard basis vectors $e_1, \ldots, e_r$ satisfy $\mathrm{ord}_\mathcal{H} e_i = (a_i, i)$ with preassigned weights $a_i \in G$. We start by defining an operation of G on H by $g + (g', i) := (g + g', i)$ and a total ordering $<_\tau$ on H by

$$(g, i) <_\tau (g', j) :\Leftrightarrow \begin{cases} g - a_i <_\sigma g' - a_j & \text{or} \\ g - a_i = g' - a_j & \text{and } i < j \end{cases}$$

for $g, g' \in G$, $i, j \in \{1, \ldots, r\}$. Then we let $\mathcal{H} = \{\mathcal{H}_h\}_{h \in H}$ with $\mathcal{H}_{(g,i)}$ being the R-submodule of S^r generated by all elements of the form se_j with $s \in S$

and $(\mathrm{ord}_{\mathcal{G}}\, s + a_j, j) <_\tau (g, i)$. It is easy to check that $(S^r, \mathcal{H})$ is an H-filtered S-module, $\mathcal{H}$ is a Gröbner filtration, and we have a canonical isomorphism of H-graded $\mathrm{gr}_{\mathcal{G}}\, S$-modules

$$\mathrm{gr}_{\mathcal{H}}\, S^r \longrightarrow \bigoplus_{i=1}^{r} (\mathrm{gr}_{\mathcal{G}}\, S)(-a_i)$$

which maps $L_{\mathcal{H}} e_i$ to the i-th standard basis vector of the right-hand side.

Example 2.7. *(Canonical filtration of the dual module)*

For a G-filtered algebra $(S/R, \mathcal{G})$, the dual module $\mathrm{Hom}_R(S, R)$ carries a natural S-module structure via $s \cdot \varphi : S \longrightarrow R$ $(s' \mapsto \varphi(ss'))$. Suppose $\mathcal{G}$ is *bounded from below*, i.e., there is a $g_0 \in G$ such that $\mathcal{G}_{g_0} = 0$. We define a G-filtration $\mathcal{H} = \{\mathcal{H}_g\}_{g \in G}$ on $\mathrm{Hom}_R(S, R)$ by setting $\mathcal{H}_g := \{\varphi \in \mathrm{Hom}_R(S, R) \mid \varphi(\mathcal{G}^{<}_{-g}) = 0\}$ for $g \in G$. It is easy to check that this makes $(\mathrm{Hom}_R(S, R), \mathcal{H})$ a G-filtered S-module, $\mathcal{H}$ is a Gröbner filtration, and we have a canonical isomorphism of G-graded $\mathrm{gr}_{\mathcal{G}}\, S$-modules

$$\begin{array}{ccc} \mathrm{gr}_{\mathcal{H}}\, \mathrm{Hom}_R(S, R) & \longrightarrow & \mathrm{Hom}_R(\mathrm{gr}_{\mathcal{G}}\, S, R) \\ L_{\mathcal{H}} \varphi & \longmapsto & \left(L_{\mathcal{G}} s \mapsto \begin{cases} \varphi(s) & \text{if } \mathrm{ord}_{\mathcal{G}}\, s = -\,\mathrm{ord}_{\mathcal{H}}\, \varphi, \\ 0 & \text{otherwise.} \end{cases} \right) \end{array}$$

3 Some useful computer algebra techniques

Later in this paper we want to calculate a minimal homogeneous presentation of a certain graded module over an explicitly given standard graded algebra. For this we review concrete methods to deal computationally with the material explained in the previous section. The techniques presented subsequently are elaborated from the introductory papers [R2] and [MR].

Let us start with the simplest algebraic objects available for computer manipulation, namely *terms* (sometimes also called *monomials*), i.e., objects of the form $X^\nu := X_1^{\nu_1} \cdots X_n^{\nu_n}$ with variables $X_1, \ldots, X_n$, an exponent $\nu = (\nu_1, \ldots, \nu_n) \in \mathbb{N}^n$, and $n \geq 1$. The set of all terms is denoted by $\mathcal{T}^n$. A subset $\mathcal{M} \subseteq \mathcal{T}^n$ is called a *monoideal*, if $t \in \mathcal{M}$ implies $tt' \in \mathcal{M}$ for all $t' \in \mathcal{T}^n$. A subset $\mathcal{O} \subseteq \mathcal{T}^n$ is called an *order ideal*, if $t \in \mathcal{O}$ implies $t' \in \mathcal{O}$ for all $t' \in \mathcal{T}^n$ dividing t. Clearly, if $\mathcal{M} \subseteq \mathcal{T}^n$ is a monoideal, then $\mathcal{O} = \mathcal{T}^n \setminus \mathcal{M}$ is an order ideal, and vice versa. The set $\mathcal{B} := \{t \in M \mid t = X_i t' \text{ for some } t' \in \mathcal{O}, i \in \{1, \ldots, n\}\}$ is called the *border* of $\mathcal{M}$. Obviously $\mathcal{O} \cup \mathcal{B}$ is an order ideal, too. A subset $\mathcal{M}' \subseteq \mathcal{M}$ is called a *system of generators* of $\mathcal{M}$, if any term in $\mathcal{M}$ is a multiple of some term in $\mathcal{M}'$. The following proposition helps us to compute a minimal system of generators of $\mathcal{M}$ (cf. [R2], sect. 3).

Proposition 3.1. *Every monoideal $\mathcal{M} \subseteq \mathcal{T}^n$ has a unique, finite, minimal system of generators, namely $\mathcal{M}_0 = \{t \in \mathcal{M} \mid t/X_i \notin \mathcal{M}$ for all i such that $X_i \mid t\}$.*

A *term ordering* $<_\sigma$ on $\mathcal{T}^n$ is a total ordering such that $t <_\sigma t'$ implies $st <_\sigma st'$ for all $s, t, t' \in \mathcal{T}^n$, and such that $1 <_\sigma t$ for all $t \in \mathcal{T}^n \setminus \{1\}$. By looking at the exponents only, we can regard $<_\sigma$ as an ordering on $\mathbb{N}^n$. Then we can extend $<_\sigma$ uniquely to $\mathbb{Z}^n$ by defining $\nu <_\sigma \mu$ for $\nu, \mu \in \mathbb{Z}^n$ if and only if $\nu = \nu' - \nu''$, $\mu = \mu' - \mu''$ with $\nu', \nu'', \mu', \mu'' \in \mathbb{N}^n$, and $\nu' + \mu'' <_\sigma \mu' + \nu''$.

Now we extend those methods to deal with arbitrary polynomials. From now on, we will always work over a computable base field K. We let $A = K[X_1, \dots, X_n]$ be the polynomial ring, and $<_\sigma$ a term ordering on $\mathcal{T}^n$.

Definition 3.2. The $\mathbb{Z}^n$-filtration $\mathcal{F}^\sigma = \{\mathcal{F}^\sigma_\nu\}_{\nu \in \mathbb{Z}^n}$ given by

$$\mathcal{F}^\sigma_\nu = \begin{cases} \langle \{t \in \mathcal{T}^n \mid t \leq_\sigma X^\nu\} \rangle_K & \text{if } \nu \in \mathbb{N}^n, \\ 0 & \text{otherwise,} \end{cases}$$

is called the *term filtration* of A associated with $<_\sigma$.

It is easy to verify that $\mathcal{F}^\sigma$ is in fact a Gröbner filtration of A (cf. [R2], 3.5). The associated graded ring $\mathrm{gr}_{\mathcal{F}^\sigma} A$ is nothing but A, together with its canonical $\mathbb{Z}^n$-grading. Every polynomial $f \in A \setminus \{0\}$ has a unique representation $f = \sum_{i=1}^r c_i t_i$ with $c_i \in K \setminus \{0\}$, $t_i \in \mathcal{T}^n$, and $t_1 >_\sigma \cdots >_\sigma t_r$. We call $\mathrm{lt}_\sigma(f) := t_1$ the *leading term* of f, and $\mathrm{lc}_\sigma(f) := c_1$ the *leading coefficient* of f.

Next we extend our methods again and deal with ideals and residue class rings of A. For an ideal $I \subseteq A$, we call $\mathrm{lt}_\sigma(I) := (\mathrm{lt}_\sigma(f))_{f \in I}$ the *ideal of leading terms* of I. Clearly, this is a monoideal in $\mathcal{T}^n$. Its associated order ideal $\mathcal{T}^n \setminus \mathrm{lt}_\sigma(I)$ will be denoted by $\mathcal{O}_\sigma(I)$. This order ideal contains information about the arithmetic of A/I (cf. [R2], sect. 8).

Proposition 3.3. *Let $I \subseteq A$ be an ideal. Then every polynomial $f \in A$ has a unique decomposition $f = g + h$ such that $g \in I$ and $h \in \langle \mathcal{O}_\sigma(I) \rangle_K$. In particular, we have isomorphisms of K-vector spaces $A \cong I \oplus \langle \mathcal{O}_\sigma(I) \rangle_K$ and $A/I \cong \langle \mathcal{O}_\sigma(I) \rangle_K$.*

For a decomposition $f = g + h$ as above, we call $\mathrm{can}(f, I) := h$ the *normal form* of f with respect to I and $<_\sigma$. Our goal now is to show how to compute $\mathrm{can}(f, I)$ and thus compute in A/I. Recall that with A, also I is $\mathbb{Z}^n$-filtered by $\{I \cap \mathcal{F}^\sigma_\nu\}_{\nu \in \mathbb{Z}^n}$. From proposition 2.5 we obtain the following characterization of Gröbner bases.

Proposition 3.4. *For an ideal $I \subseteq A$ and $\{f_1, \dots, f_r\} \subseteq I \setminus \{0\}$, the following conditions are equivalent.*

a) *The set $\{f_1, \dots, f_r\}$ is a Gröbner basis of I, i.e.,*

$$\mathrm{gr}_{\mathcal{F}^\sigma} I = (L_{\mathcal{F}^\sigma} f_1, \dots, L_{\mathcal{F}^\sigma} f_r).$$

b) *The monoideal $\mathrm{lt}_\sigma(I)$ is generated by $\{\mathrm{lt}_\sigma(f_1), \dots, \mathrm{lt}_\sigma(f_r)\}$.*

c) *Every $f \in I$ has a representation $f = \sum_{i=1}^r g_i f_i$ such that $g_i \in A$, and such that $\mathrm{lt}_\sigma(g_i)\,\mathrm{lt}_\sigma(f_i) \leq_\sigma \mathrm{lt}_\sigma(f)$ whenever $g_i \neq 0$.*

Suppose we are given a Gröbner basis $\{f_1, \ldots, f_r\}$ of I with respect to $<_\sigma$. Then $\mathrm{can}(f, I)$ can be computed for any $f \in A$ by the *division algorithm* (cf. [R2], sect. 4, and [MR], sect. 1.1). From that Gröbner basis we also get a minimal system of generators $\{t_1, \ldots, t_s\}$ of $\mathrm{lt}_\sigma(I)$ by applying proposition 3.1. Now proposition 3.4 yields the following situation.

Corollary 3.5. *a) The set* $\{t_1 - \mathrm{can}(t_1, I), \ldots, t_s - \mathrm{can}(t_s, I)\}$ *is a Gröbner basis of* I. *It is uniquely determined by* I *and* $<_\sigma$ *and is called the* reduced Gröber basis *of* I *with respect to* $<_\sigma$.

b) For $i = 1, \ldots, s$, *the leading term of* $t_i - \mathrm{can}(t_i, I)$ *with respect to* $<_\sigma$ *is* $t_i \in \mathrm{lt}_\sigma(I)$, *and all other terms appearing in this polynomial are from* $\mathcal{O}_\sigma(I)$.

Thus we are able to compute in the coordinate ring of a closed subscheme of $\mathbb{A}^n_K$, if we are given a Gröbner basis of its vanishing ideal. Most of the time, however, we are interested in closed subschemes of the projective space $\mathbb{P}^n_K$ and their homogeneous coordinate rings. Hence we have to extend our methods once again.

The homogeneous coordinate ring of $\mathbb{P}^n_K$ itself is $P := K[X_0, \ldots, X_n]$. For $t = X^\nu = X_1^{\nu_1} \cdots X_n^{\nu_n} \in \mathcal{T}^n$, we call $\deg t := \nu_1 + \cdots + \nu_n$ the *degree* of t. If $f \in A \setminus \{0\}$ is in its standard form $f = \sum_{i=1}^r c_i t_i$ with $c_i \in K \setminus \{0\}$, $t_i \in \mathcal{T}^n$, and $t_1 >_\sigma \cdots >_\sigma t_r$, then $\deg f := \max_{i=1..r}\{\deg t_i\}$ is called the *degree* of f, and ${}^h f := \sum_{i=1}^r c_i t_i X_0^{\deg f - \deg t_i} \in P$ is called the *homogenization* of f. For an ideal $I \subseteq A$, its *homogenization* ${}^h I$ is defined by ${}^h I := ({}^h f)_{f \in I} \subseteq P$. Here we let ${}^h 0 := 0$.

Conversely, if we start with a homogeneous polynomial $g \in P$ and an ideal $J \subseteq P$, we have their *dehomogenizations* ${}^a g := g(1, X_1, \ldots, X_n) \in A$ and ${}^a J := ({}^a g)_{g \in J} \subseteq A$. The processes of homogenization and dehomogenization are almost inverses of each other, in the sense that ${}^a({}^h f) = f$ for $f \in A$ and $X_0^\epsilon \cdot {}^h({}^a g) = g$ for $g \in P$ homogeneous and $\epsilon = \max\{i \in \mathbb{N} \mid X_0^i \text{ divides } g\}$.

A similar procedure can be applied to term orderings $<_\sigma$ of $\mathcal{T}^n$. We extend $<_\sigma$ to an ordering on $\mathcal{T}^{n+1} = \{X_0^{\nu_0} X^\nu \mid (\nu_0, \nu) \in \mathbb{N}^{n+1}\}$ by defining

$$X_0^i X^\nu <_\tau X_0^j X^\mu \Leftrightarrow \begin{cases} \nu <_\sigma \mu & \text{or} \\ \nu = \mu & \text{and } i < j. \end{cases}$$

We call $<_\tau$ the *projective extension* of $<_\sigma$. Clearly, $<_\tau$ is again a term ordering, and its associated term filtration $\mathcal{F}^\tau$ is a $\mathbb{Z}^{n+1}$-filtration of the K-algebra P. The basic properties of the described processes are summarized by the following proposition (cf. [MR], sect. 1.1).

Proposition 3.6. *Suppose* $<_\sigma$ *is a* degree-compatible *term ordering on* $\mathcal{T}^n$, *i.e., all* $t, t' \in \mathcal{T}^n$ *with* $\deg t < \deg t'$ *satisfy* $t <_\sigma t'$. *Let* $<_\tau$ *be the projective extension of* $<_\sigma$, *let* $I \subseteq A$ *be an ideal, and let* $\{f_1, \ldots, f_r\}$ *be a Gröbner basis of* I *with respect to* $<_\sigma$.

a) *For all* $f \in A$ *we have* $\mathrm{lt}_\sigma(f) = \mathrm{lt}_\tau({}^h f)$.

b) *The set* $\{{}^h f_1, \ldots, {}^h f_r\}$ *is a Gröbner basis of* ${}^h I$ *with respect to* $<_\tau$.

c) *We have* $\mathcal{O}_\tau({}^h I) = \cup_{i\geq 0} X_0^i \cdot \mathcal{O}_\sigma(I)$.
d) *For* $g \in P$, *we have* $\operatorname{can}(g, {}^h I) = X_0^{\deg g - \deg {}^a g} \cdot {}^h \operatorname{can}({}^a g, I)$. *In particular, if* $f \in A$ *is homogeneous, then* $\operatorname{can}(f, {}^h I) = {}^h \operatorname{can}(f, I)$.
e) *If* $\{f_1, \dots, f_r\}$ *is the reduced Gröbner basis of* I *with respect to* $<_\sigma$, *then* $\{{}^h f_1, \dots, {}^h f_r\}$ *is the reduced Gröbner basis of* ${}^h I$ *with respect to* $<_\tau$.

Consequently, given a closed subscheme $\mathbb{X} \subseteq \mathbb{P}^n_K$ and a Gröbner basis of the vanishing ideal I of $\mathbb{X} \cap D_+(X_0)$ in $D_+(X_0) \cong \mathbb{A}^n_K$ with respect to a degree-compatible term ordering, we can effectively compute in the homogeneous coordinate ring $R = P/{}^h I$ of $\mathbb{X}$ in $\mathbb{P}^n_K$. In the present paper, this will be applied to compute in the homogeneous coordinate ring of a set of points, i.e., a reduced zero-dimensional subscheme of $\mathbb{P}^n_K$ consisting of K-rational points, which we can (and shall) assume to be entirely contained in $D_+(X_0)$. In this case, the *Buchberger-Möller algorithm* (cf. [BM] and [MR], sect. 1.2) allows us to compute the desired Gröbner basis in a very efficient and fast way. As a byproduct, we are given also the order ideal $\mathcal{O}_\sigma(I)$ of $\mathbb{X}$ in $\mathbb{A}^n_K$.

Finally, for one last time, we have to extend the scope of our methods in order to be able to deal with a finitely generated, graded module M over $R = P/J$, where J is a homogeneous ideal of P. Consider M as a graded P-module, and let M be given via a homogeneous presentation

$$P^q \xrightarrow{\ \delta\ } P^r \xrightarrow{\ \epsilon\ } M \longrightarrow 0,$$

i.e., if we let $m_i := \epsilon(\tilde{e}_i)$ and $s_j = (s_{j1}, \dots, s_{jr}) := \delta(e_j)$ for $i = 1, \dots, r$ and $j = 1, \dots, q$, then $\{m_1, \dots, m_r\}$ is a homogeneous system of generators of M, and the matrix of δ is the transpose of the matrix $\mathfrak{M}$ whose rows are the syzygies $s_1, \dots, s_q$.

Recalling that for any term filtration $\mathcal{F}^\tau$ of P, we have canonical Gröbner filtrations of P^r (cf. Example 2.6) and M (cf. proposition 2.5.a), it is clear what the notions leading term, Gröbner basis, etc. mean in this context. It is also clear, how one can adapt the previously explained material to effectively compute in M, if one is given a Gröbner basis of the syzygy module $\ker \epsilon$. We conclude by showing how one can extract a minimal homogeneous system of generators of M from the given one, and how one can compute its module of syzygies. The first step is an extension of the case of ideals (cf. [MMM] and [MR], sect. 2.1), and follows easily from the graded version of Nakayama's Lemma.

Proposition 3.7. *Let* $\overline{\mathfrak{M}} = (\bar{s}_{ij})$ *be the matrix of constants* $\bar{s}_{ij} = s_{ij}(0, \dots, 0)$ *of* $\mathfrak{M}$. *Let* u *be the rank of* $\overline{\mathfrak{M}}$, *and let* $\{j_1, \dots, j_u\}$ *be the numbers of the columns involved in some nonzero maximal minor of* $\overline{\mathfrak{M}}$. *Then* $\Gamma := \{m_j \mid j \in \{1, \dots, r\} \setminus \{j_1, \dots, j_u\}\}$ *is a minimal homogeneous system of generators of* M.

Now we permute the columns of $\mathfrak{M}$ (i.e., the copies of P in P^r) such that $j_1 = 1, \dots, j_u = u$. We also permute the rows of $\mathfrak{M}$ (i.e., the copies of P in

P^q) such that the rows involved in our maximal minor of $\overline{\mathfrak{M}}$ have numbers $\{1,\dots,u\}$. By applying suitable K-linear operations on those rows, we can assume that $(\bar{s}_{ij})_{i,j=1,..,u}$ is the identity matrix. Since $\mathfrak{M}$ is homogeneous, then even $s_{ii} = 1$ for $i = 1,\dots,u$. Next we use Gaussian elimination to annihilate every entry in $\mathfrak{M}$ below this partial diagonal by homogeneous P-linear row operations. Notice that the entries $s_{11} = \cdots = s_{uu} = 1$ do not change, since all operations are homogeneous and all elements above this partial diagonal have constant term zero. We end up with a matrix $\mathfrak{N}$ of the form

$$\mathfrak{N} = \left[\begin{array}{ccc|c} 1 & & * & \\ & \ddots & & * \\ 0 & & 1 & \\ \hline & & & s'_1 \\ & 0 & & \vdots \\ & & & s'_p \end{array}\right]$$

where $p = q - u$ and $s'_1, \dots, s'_p \in P^{r-u}$. Obviously $\mathfrak{N}$ has same row space as $\mathfrak{M}$. The elementary proof of our last result in this section is left to the reader.

Proposition 3.8. *Using the above notations, the elements $s'_1, \dots, s'_p$ form a homogeneous system of generators of the module of syzygies of Γ. If $\{s_1, \dots, s_q\}$ is a minimal system of generators of $\ker \epsilon$, then $\{s'_1, \dots, s'_p\}$ is also minimal.*

4 A Gröbner basis of the canonical module

For the remainder of this paper we will use the following notations. Let $\mathbb{X}$ be a zero-dimensional, reduced, K-rational subscheme of (i.e., a set of points in) some projective space $\mathbb{P}^n_K$ over a computable field K. Let $s := \deg \mathbb{X}$ be the number of those points. The homogeneous vanishing ideal of $\mathbb{X}$ in $P := K[X_0, \dots, X_n]$ is denoted by $I_{\mathbb{X}}$. Then $R = P/I_{\mathbb{X}}$ is the homogeneous coordinate ring of $\mathbb{X}$. We always assume that $\mathbb{X}$ is entirely contained in the affine space $D_+(X_0) \cong \mathbb{A}^n_K$. Then the image x_0 of X_0 in R is a linear non-zero-divisor, and we can identify $K[X_0]$ with its image under the canonical epimorphism $P \twoheadrightarrow R$. The affine ideal of $\mathbb{X}$ is $J_{\mathbb{X}} = (I_{\mathbb{X}} + (X_0 - 1))/(X_0 - 1) \subseteq A := K[X_1, \dots, X_n]$.

Let $<_\sigma$ be a degree-compatible term ordering on $\mathfrak{T}^n$. We suppose that a Gröbner basis of $J_{\mathbb{X}}$ with respect to $<_\sigma$ and $\mathcal{O}_\sigma(J_{\mathbb{X}})$ have been computed from the coordinates of the points via the Buchberger-Möller algorithm. As laid out in section 3, we can then effectively compute in $A/J_{\mathbb{X}}$ and $R = P/I_{\mathbb{X}}$. Let $\mathcal{O}_\sigma(J_{\mathbb{X}}) = \{T_1, \dots, T_s\}$ with $T_i = X^{\nu_i} = X_1^{\nu_{i1}} \cdots X_n^{\nu_{in}}$ and $\nu_i = (\nu_{i1}, \dots, \nu_{in}) \in \mathbb{N}^n$ for $i = 1, \dots, s$, and let $t_i = T_i + I_{\mathbb{X}} \in R$. Without loss of generality let $\nu_1 <_\sigma \cdots <_\sigma \nu_s$. Then the degrees $n_i := \deg T_i$ satisfy $n_1 \le \cdots \le n_s$.

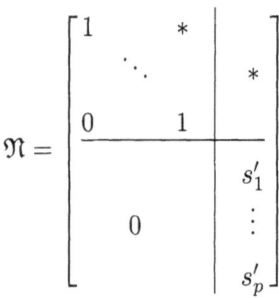

Algebraically, $R = \oplus_{i \ge 0} R_i$ is a standard graded K-algebra (i.e., $R_0 = K$, $\dim_K R_1$ is finite, and $R = K[R_1]$) which is a one-dimensional Cohen-Macaulay ring with Noetherian normalization $K[x_0] \subseteq R$. In particular, R is a free $K[x_0]$-module of rank s with $K[x_0]$-basis $\{t_1, \ldots, t_s\}$ (cf. propositions 3.3 and 3.6).

Definition 4.1. The graded R-module

$$\omega_R := \underline{\mathrm{Hom}}_{K[x_0]}(R, K[x_0])(-1)$$

is called the *canonical module* of R (cf. [K1]). Its R-module structure is given by $r \cdot \varphi : R \longrightarrow K[x_0]$ $(r' \mapsto \varphi(rr'))$, and $\underline{\mathrm{Hom}}$ means graded homomorphisms, i.e., in each degree $i \in \mathbb{Z}$ we have $(\omega_R)_i = \mathrm{Hom}_{K[x_0]}(R, K[x_0])_{i-1}$.

Our goal in this section is to find a Gröbner basis of ω_R with respect to a suitable filtration. If we consider $P = K[X_0][X_1, \ldots, X_n]$ as a polynomial ring over $K[X_0]$, then $<_\sigma$ induces a $\mathbb{Z}^n$-filtration $\mathcal{E} = \{\mathcal{E}_\nu\}_{\nu \in \mathbb{Z}^n}$ of P with $\mathcal{E}_\nu = \langle \{X^\mu \mid \mu \leq_\sigma \nu\} \rangle_{K[X_0]}$ for $\nu \in \mathbb{Z}^n$. Let $\mathcal{F} = \{\mathcal{F}_\nu\}_{\nu \in \mathbb{Z}^n}$ be the induced $\mathbb{Z}^n$-filtration of $R/K[x_0]$, i.e., for $\nu \in \mathbb{Z}^n$ we have $\mathcal{F}_\nu = \langle \{x^\mu \mid \mu \leq_\sigma \nu\} \rangle_{K[x_0]}$, where x^μ denotes the image of X^μ in R. Using Example 2.7, we obtain a $\mathbb{Z}^n$-filtration $\mathcal{G} = \{\mathcal{G}_\nu\}_{\nu \in \mathbb{Z}^n}$ of $\underline{\mathrm{Hom}}_{K[x_0]}(R, K[x_0])$ which is given by $\mathcal{G}_\nu = \langle \{\varphi : R \rightarrow K[x_0]$ homogeneous, $K[x_0]$-linear $\mid \varphi(\mathcal{F}^<_{-\nu}) = 0\} \rangle_{K[x_0]}$ for $\nu \in \mathbb{Z}^n$. Then ω_R is $\mathbb{Z}^n$-filtered by $\mathcal{G}$, too, and we are looking for a Gröbner basis of ω_R with respect to this filtration. We start by studying the filtration $\mathcal{F}$ more closely.

Proposition 4.2.

a) *Let $\nu \in \mathbb{Z}^n$. If $\nu <_\sigma \nu_1$, then $\mathcal{F}_\nu = \{0\}$, and if $\nu \geq_\sigma \nu_1$, then $\mathcal{F}_\nu = \mathcal{F}_{\nu_i}$ for some $i \in \{1, \ldots, s\}$. Thus the filtration $\mathcal{F}$ consists only of the $s+1$ different $K[x_0]$-modules*

$$\{0\} \subsetneq \mathcal{F}_{\nu_1} \subsetneq \cdots \subsetneq \mathcal{F}_{\nu_s} = R.$$

b) *The set $\{L_\mathcal{F} t_1, \ldots, L_\mathcal{F} t_s\}$ is a $K[x_0]$-basis of $\mathrm{gr}_\mathcal{F} R$. In particular, $\mathrm{gr}_\mathcal{F} R$ is a one-dimensional Cohen-Macaulay ring with Noetherian normalization*

$$K[x_0] \subseteq \mathrm{gr}_\mathcal{F} R$$

and canonical module $\omega_{\mathrm{gr}_\mathcal{F} R} \cong \underline{\mathrm{Hom}}_{K[x_0]}(\mathrm{gr}_\mathcal{F} R, K[x_0])(-1)$.

Proof: Part a) is clear, since $\{t_1, \ldots, t_s\}$ is a $K[x_0]$-basis of R and $\mathrm{ord}_\mathcal{F} t_i = \nu_i$ for $i = 1, \ldots, s$. Choose $\nu_0 \in \mathbb{Z}^n$ such that $\nu_0 <_\sigma \nu_1$. Then $\mathrm{gr}_\mathcal{F} R = \oplus_{i=1}^s \mathcal{F}_{\nu_i}/\mathcal{F}_{\nu_{i-1}}$ shows that $\{L_\mathcal{F} t_1, \ldots, L_\mathcal{F} t_s\}$ is a $K[x_0]$-basis of $\mathrm{gr}_\mathcal{F} R$. The additional claims follow from this observation using [M], 19.1 and [GW], 2.2.9. □

Next we consider the $K[x_0]$-linear maps $p_i : R \longrightarrow K[x_0]$ which are given by $p_i(t_j) = \delta_{ij}$ for $i, j = 1, \ldots, s$, i.e., $\{p_1, \ldots, p_s\}$ is the dual basis of $\{t_1, \ldots, t_s\}$. Obviously, p_i is homogeneous of degree $\deg p_i = -n_i$ in $\underline{\mathrm{Hom}}_{K[x_0]}(R, K[x_0])$, of degree $-n_i + 1$ in ω_R, and we have $\mathrm{ord}_\mathcal{G} p_i = -\nu_i$ for $i = 1, \ldots, s$. The following proposition describes the module structure of ω_R in terms of those projections.

Proposition 4.3.

a) Suppose $t_i t_k = \sum_{j=1}^{s} c_{ijk} t_j$ *with* $c_{ijk} \in K[x_0]$ *for* $i, j, k = 1, \ldots, s$. *Then*

$$t_i p_j = \sum_{k=1}^{s} c_{ijk} p_k = \begin{cases} \sum\limits_{\substack{k=1,\ldots,s \\ T_i T_k >_\sigma T_j}} c_{ijk} p_k & \text{if } T_i \nmid T_j, \\ p_m + \sum\limits_{k=m+1}^{s} c_{ijk} p_k & \text{if } T_j = T_i T_m. \end{cases}$$

b) The set $\{L_{\mathcal{G}} p_1, \ldots, L_{\mathcal{G}} p_s\}$ *is a* $K[x_0]$*-basis of* $\operatorname{gr}_{\mathcal{G}} \underline{\operatorname{Hom}}_{K[x_0]}(R, K[x_0])$. *In particular,* $\{p_1, \ldots, p_s\}$ *is a Gröbner basis of* $\underline{\operatorname{Hom}}_{K[x_0]}(R, K[x_0])$ *and of* ω_R.

Proof: From $t_i p_j(t_k) = p_j(t_i t_k) = \sum_{l=1}^{s} p_j(c_{ilk} t_l) = c_{ijk} = (\sum_{l=1}^{s} c_{ijl} p_l)(t_k)$ we conclude that $t_i p_j = \sum_{k=1}^{s} c_{ijk} p_k$. Furthermore, from proposition 4.2.b and $t_i t_k = \sum_{j=1}^{s} c_{ijk} t_j$, it follows that $c_{ijk} = 0$ whenever $T_i T_k <_\sigma T_j$, and $c_{ijm} = 1$ whenever $T_i T_m = T_j$. Thus claim a) is proved, and we turn to showing b). Proposition 4.2.a implies that the filtration $\mathcal{G}$ is completely described by

$$\{0\} = \mathcal{G}_{-\nu_{s+1}} \subsetneq \mathcal{G}_{-\nu_s} \subsetneq \cdots \subsetneq \mathcal{G}_{-\nu_1} = \underline{\operatorname{Hom}}_{K[x_0]}(R, K[x_0]),$$

where $\nu_{s+1} \in \mathbb{Z}^n$ is any tuple such that $\nu_{s+1} >_\sigma \nu_s$. Thus we have an isomorphism of $K[x_0]$-algebras

$$\operatorname{gr}_{\mathcal{G}} \underline{\operatorname{Hom}}_{K[x_0]}(R, K[x_0]) \cong \oplus_{i=1}^{s} \mathcal{G}_{-\nu_i} / \mathcal{G}_{-\nu_{i+1}}.$$

If $L_{\mathcal{G}} \varphi \in \mathcal{G}_{-\nu_i} / \mathcal{G}_{-\nu_{i+1}}$ for some $\varphi \in \underline{\operatorname{Hom}}_{K[x_0]}(R, K[x_0])$, then $\varphi(t_j) = 0$ for $j < i$ and $\varphi(t_i) \neq 0$. Hence $\varphi = \sum_{j \geq i} \varphi(t_j) p_j$ and $L_{\mathcal{G}} \varphi = \varphi(t_i) L_{\mathcal{G}} p_i$. Since $\{L_{\mathcal{G}} p_1, \ldots, L_{\mathcal{G}} p_s\}$ is obviously $K[x_0]$-linearly independent, the claim follows.□

For computational purposes, the Gröbner basis $\{p_1, \ldots, p_s\}$ of ω_R we just found is much too large and has too many syzygies. Therefore we want to find a subset which is a *minimal Gröbner basis*, i.e., whose leading forms are a minimal system of generators of the $\operatorname{gr}_{\mathcal{F}} R$-module $\operatorname{gr}_{\mathcal{G}} \omega_R$.

The elements of the set $\operatorname{soc} \mathcal{O}_\sigma(J_{\mathbb{X}}) = \{T_i \in \mathcal{O}_\sigma(J_{\mathbb{X}}) \mid X_j T_i \notin \mathcal{O}_\sigma(J_{\mathbb{X}}) \text{ for all } j = 1, \ldots, n\}$ are called the *socle monomials* of $\mathbb{X}$. Let $t \geq 1$ and $\mu_1, \ldots, \mu_t \in \{1, \ldots, s\}$ such that $\operatorname{soc} \mathcal{O}_\sigma(J_{\mathbb{X}}) = \{T_{\mu_1}, \ldots, T_{\mu_t}\}$. Actually, from the proof of our next proposition it is clear that t is the Cohen-Macaulay type of the ring $\operatorname{gr}_{\mathcal{F}} R$.

Proposition 4.4. *The set* $\{p_{\mu_1}, \ldots, p_{\mu_t}\}$ *is a minimal Gröbner basis of the* R*-module* $\underline{\operatorname{Hom}}_{K[x_0]}(R, K[x_0])$*, and therefore also of* ω_R.

Proof: We apply the isomorphism

$$\operatorname{gr}_{\mathcal{G}} \underline{\operatorname{Hom}}_{K[x_0]}(R, K[x_0]) \cong \underline{\operatorname{Hom}}_{K[x_0]}(\operatorname{gr}_{\mathcal{F}} R, K[x_0])$$

described in Example 2.7, and denote for $i = 1, \ldots, s$ the image of $L_{\mathcal{G}} p_i$ by π_i. From the definition of p_i and this isomorphism we get $\pi_i(L_{\mathcal{F}} t_j) = \delta_{ij}$ for $i, j = 1, \ldots, s$. Now we use the multiplication table

$$L_{\mathcal{F}} t_i \cdot L_{\mathcal{F}} t_k = \begin{cases} L_{\mathcal{F}}(t_i t_k) & \text{if } \ T_i T_k \in \mathcal{O}_\sigma(J_{\mathbb{X}}), \\ 0 & \text{otherwise,} \end{cases}$$

of $\operatorname{gr}_{\mathcal{F}} R$ and conclude that the multiplication table of $\underline{\operatorname{Hom}}_{K[x_0]}(\operatorname{gr}_{\mathcal{F}} R, K[x_0])$ is given by

$$L_{\mathcal{F}} t_i \cdot \pi_j = \begin{cases} 0 & \text{if } T_i \nmid T_j, \\ \pi_m & \text{if } T_j = T_i T_m. \end{cases}$$

In view of this table it is immediately clear that $\{\pi_{\mu_1}, \ldots, \pi_{\mu_t}\}$ is a minimal system of generators of the $\operatorname{gr}_{\mathcal{F}} R$-module $\underline{\operatorname{Hom}}_{K[x_0]}(\operatorname{gr}_{\mathcal{F}} R, K[x_0])$. □

5 A Gröbner basis of the syzygy module

As explained in section 3, in order to compute effectively in ω_R, we need a Gröbner basis of the syzygy module $\ker \epsilon_t$ of the graded epimorphism $\epsilon_t : \oplus_{i=1}^t P(n_{\mu_i} - 1) \longrightarrow \omega_R$ given by $e_i \mapsto p_{\mu_i}$ for $i = 1, \ldots, t$. This is achieved by finding generators for the module of syzygies $\ker \epsilon_s$ of the epimorphism $\epsilon_s : \oplus_{i=1}^s P(n_i - 1) \longrightarrow \omega_R$ ($e_i \mapsto p_i$) first, and then reducing them to syzygies of $\{p_{\mu_1}, \ldots, p_{\mu_t}\}$ using the process described at the end of section 3.

Recall that $<_\sigma$ was a degree-compatible term ordering on $\mathcal{T}^n$. Let $<_\tau$ be the projective extension of $<_\sigma$. Then $(P, \mathcal{F}^\tau)$ is a $\mathbb{Z}^{n+1}$-filtered K-algebra. By setting $H := \mathbb{Z}^{n+1} \times \{1, \ldots, s\}$ and applying Example 2.6, we can obtain an H-filtration $\mathcal{H}$ of P^s such that $\operatorname{ord}_{\mathcal{H}} = ((0, \operatorname{ord}_{\mathcal{G}} p_i), i) = ((0, -\nu_i), i)$ and $\operatorname{gr}_{\mathcal{H}} P^s \cong \oplus_{i=1}^s P(0, \nu_i)$. The ordering on H will be denoted by $<_\eta$. Recall also that the multiplication table of R is given by $t_i t_k = \sum_{j=1}^s c_{ijk} t_j$ with $c_{ijk} \in K[x_0]$ for $i, j, k = 1, \ldots, s$. Clearly, we can choose $c_{ijk} \in K[x_0]$ homogeneous of degree $n_i - n_j + n_k$, and we can identify c_{ijk} with its preimage in $K[X_0]$. Thirdly, recall that $\mathcal{O}_\sigma(J_{\mathbb{X}}) = \{T_1, \ldots, T_s\}$ is an order ideal. For $i, j \in \{1, \ldots, s\}$ we thus find $g(i,j), v(i,j) \in \{1, \ldots, s\}$ such that $T_{g(i,j)} = \gcd(T_i, T_j)$ and $T_i = T_{g(i,j)} T_{v(i,j)}$.

Proposition 5.1. *For $i, j \in \{1, \ldots, s\}$ with $i < j$, the element*

$$S_{ij} = T_{v(i,j)} e_i - T_{v(j,i)} e_j - \sum_{k=g(i,j)+1}^{s} (c_{v(i,j)ik} - c_{v(j,i)jk}) e_k$$

of P^s is a syzygy of $p_1, \ldots, p_s$ with leading form $L_{\mathcal{H}} S_{i,j} = T_{v(i,j)} e_i$. Considered as an element of $\ker \epsilon_s \subseteq \oplus_{k=1}^s P(n_k - 1)$, it is homogeneous of degree $-n_{g(i,j)} + 1$.

Proof: By proposition 4.3.a, we have

$$t_{v(i,j)} p_i = p_{g(i,j)} + \sum_{k=g(i,j)+1}^{s} c_{v(i,j)ik} p_k$$

and

$$t_{v(j,i)}p_j = p_{g(i,j)} + \sum_{k=g(i,j)+1}^{s} c_{v(j,i)jk}p_k.$$

Consequently, S_{ij} is a syzygy of $p_1, \dots, p_s$. Since $\nu_{v(i,j)} - \nu_i = \nu_{v(j,i)} - \nu_j = -\nu_{g(i,j)}$ for $i < j$, we get the inequality $\operatorname{ord}_{\mathcal{H}} T_{v(i,j)}e_i = ((0, \nu_{v(i,j)} - \nu_i), i) >_\eta \operatorname{ord}_{\mathcal{H}} T_{v(j,i)}e_j = ((0, \nu_{v(j,i)} - \nu_j), j)$ and the equation

$$\deg T_{v(i,j)}e_i = \deg T_{v(j,i)}e_j = -n_{g(i,j)} + 1.$$

Because of $\nu_{v(i,j)} - \nu_i = -\nu_{g(i,j)} >_\sigma -\nu_k$ for $k > g(i,j)$, we have $\operatorname{ord}_{\mathcal{H}} T_{v(i,j)}e_i = ((0, \nu_{v(i,j)} - \nu_i), i) >_\eta \operatorname{ord}_{\mathcal{H}}(c_{v(i,j)ik} - c_{v(j,i)jk})e_k = ((-n_{g(i,j)} + n_k, -\nu_k), k)$ and $\deg(c_{v(i,j)ik} - c_{v(j,i)jk})e_k = -n_{g(i,j)} + 1$. Therefore the leading form of S_{ij} is $T_{v(i,j)}e_i$, and S_{ij} is homogeneous of degree $-n_{g(i,j)} + 1$. □

Recall that, for $i = 1, \dots, s$, we have $T_i = X^{\nu_i} = X_1^{\nu_{i1}} \cdots X_n^{\nu_{in}}$ with $\nu_{ij} \in \mathbb{N}$. Clearly, $\{X_1^{\nu_{i1}+1}, \dots, X_n^{\nu_{in}+1}\}$ is a set of generators of the monoideal $\{T \in \mathcal{T}^n \mid T \nmid T_i\}$. Those elements can be used to construct more syzygies of $p_1, \dots, p_s$.

Proposition 5.2. *For $i \in \{1, \dots, s\}$ and $j \in \{1, \dots, n\}$ we write*

$$\operatorname{can}(X_j^{\nu_{ij}+1}T_l, I_{\mathbb{X}}) = \sum_{k=1}^{s} d_{ijkl}T_k$$

with homogeneous elements $d_{ijkl} \in K[X_0]$ of degree $\deg d_{ijkl} = \nu_{ij} + 1 - n_k + n_l$. The element

$$S'_{ij} = X_j^{\nu_{ij}+1}e_i - \sum_{k=1}^{s} d_{ijik}e_k$$

of P^s is a syzygy of $p_1, \dots, p_s$ with leading form $L_{\mathcal{H}}S'_{ij} = X_j^{\nu_{ij}+1}e_i$. Considered as an element of $\ker \epsilon_s \subseteq \oplus_{k=1}^{s} P(n_k - 1)$, it is homogeneous of degree $\nu_{ij} - n_i + 2$.

Proof: For $l = 1, \dots, s$ we have

$$(X_j^{\nu_{ij}+1}p_i)(t_l) = p_i(X_j^{\nu_{ij}+1}t_l) = p_i(\sum_{k=1}^{s} d_{ijkl}t_l) = d_{ijil} = (\sum_{k=1}^{s} d_{ijik}p_k)(t_k).$$

Consequently, S'_{ij} is a syzygy of $p_1, \dots, p_s$. If $d_{ijik} \neq 0$ for some $k \in \{1, \dots, s\}$, then $X_j^{\nu_{ij}+1}T_k \geq_\sigma T_i$, since the division algorithm only reduces leading forms. In this case we even have $X_j^{\nu_{ij}+1}T_k >_\sigma T_i$, since $X_j^{\nu_{ij}+1}$ is not a divisor of T_i. Therefore $\nu_{ij} + 1 - \nu_i >_\sigma -\nu_k$ and $\operatorname{ord}_{\mathcal{H}} X_j^{\nu_{ij}+1}e_i = ((0, \nu_{ij} + 1 - \nu_i), i) >_\eta \operatorname{ord}_{\mathcal{H}} d_{ijik}e_k = ((\nu_{ij} + 1 - n_i + n_k, -\nu_k), k)$. Hence we have $L_{\mathcal{H}}S'_{ij} = X_j^{\nu_{ij}+1}e_i$, and from $\deg X_j^{\nu_{ij}+1}e_i = \nu_{ij} - n_i + 2 = \deg d_{ijik}e_k$ we get the last claim. □

After constructing sufficiently many syzygies of $p_1, \ldots, p_s$, we show that they actually form a Gröbner basis of $\ker \epsilon_s$.

Proposition 5.3. *The set* $\mathfrak{S}_s = \{S_{ij} \mid i,j \in \{1,\ldots,s\}, i < j\} \cup \{S'_{ij} \mid i \in \{1,\ldots,s\}, j \in \{1,\ldots,n\}\}$ *is a Gröbner basis of* $\ker \epsilon_s$ *with respect to the filtration* $\mathcal{H}$. *In particular,* $\mathfrak{S}_s$ *is a system of generators of* $\ker \epsilon_s$.

Proof: Let $S = (F_1, \ldots, F_s) \in (\ker \epsilon_s)_\alpha$ with $F_i \in P_{n_i - 1 + \alpha}$ be a homogeneous syzygy of $p_1, \ldots, p_s$ of degree α, and let $L_{\mathcal{H}} S = X_0^m X^\nu e_i$ for $m \geq 0$, $\nu \in \mathbb{Z}^n$, and $i \in \{1, \ldots, s\}$. We have to show $L_{\mathcal{H}} S \in \langle L_{\mathcal{H}} S' \rangle_{S' \in \mathfrak{S}_s}$. For each term $cX_0^{m'} X^{\nu'} e_{i'}$ with $c \in K \setminus \{0\}$ appearing in S we have $((m', \nu'), i') \leq_\eta ((m, \nu), i)$, implying $(m', \nu' - \nu_{i'}) \leq_\tau (m, \nu - \nu_i)$, and therefore $\nu' - \nu_{i'} \leq_\sigma \nu - \nu_i$. Since $\deg F_i = n_i - 1 + \alpha = m + \deg X^\nu$ and $\deg F_{i'} = n_{i'} - 1 + \alpha = m + \deg X^{\nu'}$, and since the degree-compatibility of $<_\sigma$ yields $\deg X^{\nu'} - n_{i'} \leq \deg X^\nu - n_i$, we obtain $m' \geq m$. Thus S is divisible by X_0^m, and we can assume $m = 0$. Now we distinguish two cases.

Case 1: $X^\nu \nmid T_i$. Then we have $X_j^{\nu_{ij}+1} \mid X^\nu$ for some $j \in \{1, \ldots, n\}$. This yields that $L_{\mathcal{H}} S$ is a multiple of $L_{\mathcal{H}} S'_{ij} = X_j^{\nu_{ij}+1} e_i$.

Case 2: $X^\nu \mid T_i$. We set $x^\nu := X^\nu + I_{\mathbb{X}} \in R$ and $f_j := F_j + I_{\mathbb{X}} \in R$ for $j = 1, \ldots, s$. Since $X^\nu \mid T_i$, we have $x^\nu p_i \neq 0$, and thus $\operatorname{ord}_{\mathcal{G}}(f_i p_i) = \nu - \nu_i$. Now $\sum_{j=1}^s f_j p_j = 0$, together with $\operatorname{ord}_{\mathcal{F}} f_j \leq_\sigma \nu - \nu_i + \nu_j$ and $\operatorname{ord}_{\mathcal{G}} p_j = -\nu_j$, shows $\sum_{j \in L} L_{\mathcal{G}}(f_j p_j) = \sum_{j \in L} L_{\mathcal{F}} f_j L_{\mathcal{G}} p_j = 0$ for $L = \{j \in \{1, \ldots, s\} \mid \operatorname{ord}_{\mathcal{G}}(f_j p_j) = \nu - \nu_i\}$. Hence there is an index $j \in L \setminus \{i\}$ such that $L_{\mathcal{F}} f_j L_{\mathcal{G}} p_j \neq 0$.

Write $L_{\mathcal{F}} f_j = X_0^{m'} X^{\nu'}$ with $m' \geq 0$ and $\nu' \in \mathbb{N}^n$ such that $\nu' \leq_\sigma \nu$ and $m' + \deg X^{\nu'} = \deg X^\nu$. From $X^{\nu'} L_{\mathcal{G}} p_j \neq 0$ we conclude that $X^{\nu'} \mid T_j$ and $\nu' - \nu_j = \nu - \nu_i$. Hence $T_i = X^\nu T$ and $T_j = X^{\nu'} T$ for some $T \in \mathcal{O}_\sigma(J_{\mathbb{X}})$, and henceforth $T_{v(i,j)} \mid X^\nu$. Since $\nu' \leq_\sigma \nu$, we also get $j > i$. Altogether, $L_{\mathcal{H}} S = X^\nu e_i$ is a multiple of $L_{\mathcal{H}} S_{ij} = T_{v(i,j)} e_j$.

Having established that $\mathfrak{S}_s$ is a Gröbner basis of $\ker \epsilon_s$, we conclude from proposition 2.5.b that $\mathfrak{S}_s$ is also a system of generators of $\ker \epsilon_s$, since $\mathcal{H}$ is a Gröbner filtration by Example 2.6. □

Next we proceed with our task to find a Gröbner basis of $\ker \epsilon_t$. For $i \in \{1, \ldots, s\} \setminus \{\mu_1, \ldots, \mu_t\}$ choose $\lambda(i) \in \{1, \ldots, s\}$ such that $T_{\lambda(i)}$ is the largest multiple of T_i with respect to $<_\sigma$ which is in $\mathcal{O}_\sigma(J_{\mathbb{X}})$. Clearly, $T_{\lambda(i)}$ is a socle monomial, i.e., $\lambda(i) \in \{\mu_1, \ldots, \mu_t\}$. Now we form the matrix $\mathfrak{M}$ whose rows are the syzygies $S_{i\lambda(i)}$ for $i \in \{1, \ldots, s\} \setminus \{\mu_1 \ldots, \mu_t\}$, S_{ij} for $i, j \in \{\mu_1, \ldots, \mu_t\}$ with $i < j$, and S'_{ij} for $i \in \{\mu_1, \ldots, \mu_t\}$, $j \in \{1, \ldots, n\}$.

We apply the process described at the end of section 3. For $i \in \{1, \ldots, s\} \setminus \{\mu_1, \ldots, \mu_t\}$ we have $T_i \mid T_{\lambda(i)}$ by construction, so $T_{v(i,\lambda(i))} = 1$. Thus the syzygy $S_{i\lambda(i)}$ has leading form $L_{\mathcal{H}} S_{i\lambda(i)} = 1 \cdot e_i$, and after a suitable permutation of columns, the matrix $\overline{\mathfrak{M}}$ is of the form

$$\overline{\mathfrak{M}} = \left[\begin{array}{ccc|c} 1 & & * & \\ & \ddots & & * \\ * & & 1 & \\ \hline & * & & * \end{array}\right]$$

with the partial diagonal consisting of ones being of size $s-t$. As explained in section 3, we can then use that partial diagonal and homogeneous P-linear row operations to reduce $\overline{\mathfrak{M}}$ to a matrix $\mathfrak{N}$ of the form

$$\mathfrak{N} = \left[\begin{array}{ccc|c} 1 & & * & \\ & \ddots & & * \\ 0 & & 1 & \\ \hline & 0 & & s_{ij} \\ & 0 & & s'_{kl} \end{array}\right]$$

where $i,j \in \{\mu_1,\ldots,\mu_t\}$ with $i<j$, $k \in \{\mu_1,\ldots,\mu_t\}$, $l \in \{1,\ldots,n\}$, and $s_{ij}, s'_{kl} \in \oplus_{i=1}^t P(n_{\mu_i}-1)$ are homogeneous syzygies of $p_{\mu_1},\ldots,p_{\mu_t}$. We end this section by showing that those syzygies solve our problem.

Proposition 5.4.

a) *Consider the injection* $\iota : \oplus_{i=1}^t P(n_{\mu_i}-1) \longrightarrow \oplus_{j=1}^s P(n_j-1)$ *given by* $e_i \mapsto e_{\mu_i}$ *for* $i=1,\ldots,t$, *and equip* $\oplus_{i=1}^t P(n_{\mu_i}-1)$ *with the induced filtration* $\mathcal{H}' = \{\iota^{-1}(\mathcal{H}_\nu \cap \operatorname{im}\iota)\}_{\nu\in H}$. *Then* $L_{\mathcal{H}'} s_{ij} = T_{v(i,j)} e_i$ *for* $i,j \in \{\mu_1,\ldots,\mu_t\}$ *with* $i<j$, *and* $L_{\mathcal{H}'} s'_{kl} = X_l^{\nu_{kl}+1} e_k$ *for* $k \in \{\mu_1,\ldots,\mu_t\}$, $l \in \{1,\ldots,n\}$.

b) *The set*

$$\mathfrak{S}_t = \{s_{ij} \mid i,j \in \{\mu_1,\ldots,\mu_t\}, i<j\} \cup \{s'_{kl} \mid k \in \{\mu_1,\ldots,\mu_t\}, l \in \{1,\ldots,n\}\}$$

is a Gröbner basis of $\ker \epsilon_t$ *with respect to* $\mathcal{H}'$. *In particular,* $\mathfrak{S}_t$ *is a set of generators of* $\ker \epsilon_t$.

Proof: All syzygies from

$$\{S_{ij} \mid i,j \in \{\mu_1,\ldots,\mu_t\}, i<j\} \cup \{S'_{kl} \mid k \in \{\mu_1,\ldots,\mu_t\}, l \in \{1,\ldots,n\}\}$$

have their leading form with respect to $\mathcal{H}$ in one of their last t columns (cf. Propositions 5.1 and 5.2). Thus this leading form is not changed during the elimination process and part a) follows. The proof of part b) is completely analogous to the proof of proposition 5.3, and therefore left to the reader. □

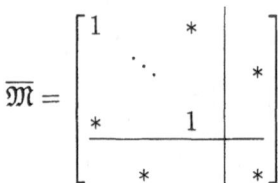

6 A presentation of the canonical module

In the previous sections we found a presentation of ω_R of the form

$$P^{\binom{t}{2}+nt} \longrightarrow P^t \longrightarrow \omega_R \longrightarrow 0.$$

Of course we could use the elements of $\mathfrak{S}_t$ now in order to find a minimal system of generators of ω_R via proposition 3.7, and a set of generators of their syzygy module via proposition 3.8. But as we are interested in a minimal homogeneous presentation of ω_R, we would then have to compute a Gröbner basis of that syzygy module to be able to minimize its set of generators. Since this could become quite costly, we find it easier to proceed as follows.

First we renumber the elements of $\mathfrak{S}_t$ such that $\mathfrak{S}_t = \{s_1, \ldots, s_q\}$ with $q = \binom{t}{2} + nt$. The module $\oplus_{i=1}^t P(n_{\mu_i} - 1)$ is H-filtered by $\mathcal{H}'$. As, in fact, all elements have orders in the subset of H, which is the image of $H' = \mathbb{Z}^{n+1} \times \{1, \ldots, t\}$ under the injection $H' \hookrightarrow H$, $(\nu, i) \mapsto (\nu, \mu_i)$, we can consider $\mathcal{H}'$ as an H'-filtration. The induced ordering on H' is denoted by $<_{\eta'}$. For $i = 1, \ldots, q$, we let $\operatorname{ord}_{\mathcal{H}'} s_q = (\psi_i, \varrho_i)$ with $\psi_i \in \mathbb{Z}^{n+1}$ and $\varrho_i \in \{1, \ldots, t\}$ given by proposition 5.4.a. The free P-module P^q has an $I := \mathbb{Z}^{n+1} \times \{1, \ldots, q\}$-filtration $\mathfrak{J} = \{\mathfrak{J}_{(\nu,i)}\}_{(\nu,i)\in I}$ such that $\operatorname{ord}_{\mathfrak{J}} e_i = (\psi_i, i)$ for $i = 1, \ldots, q$. Our next goal is to find a Gröbner basis of the syzygy module of $\mathfrak{S}_t$ with respect to this filtration $\mathfrak{J}$.

For $i, j = 1, \ldots, q$, we let $k(i,j) \in \mathbb{Z}^{n+1}$ be the componentwise maximum of ψ_i and ψ_j, and $w(i,j) := k(i,j) - \psi_i \in \mathbb{N}^{n+1}$. Setting $\underline{X}^v = X_0^{v_0} \cdots X_n^{v_n}$ for $v = (v_0, \ldots, v_n) \in \mathbb{Z}^{n+1}$, we can interpret $\underline{X}^{k(i,j)}$ as the least common multiple of $\underline{X}^{\psi_i}$ and $\underline{X}^{\psi_j}$, although some components of ψ_i and ψ_j may be negative. Obviously, for $i, j \in \{1, \ldots, q\}$ we have $\operatorname{ord}_{\mathcal{H}'}(\underline{X}^{w(i,j)} s_i) = (k(i,j), \varrho_i)$. Now we are ready to describe the desired Gröbner basis.

Proposition 6.1. *Let $i, j \in \{1, \ldots, q\}$ such that $i < j$ and $\varrho_i = \varrho_j$.*

a) *There are homogeneous polynomials $E_{ijk} \in P$ of degree $\deg E_{ijk} = \deg \underline{X}^{w(i,j)} + \deg s_i - \deg s_k$ such that*

$$\underline{X}^{w(i,j)} s_i - \underline{X}^{w(j,i)} s_j = \sum_{k=1}^{q} E_{ijk} s_k$$

and such that $\operatorname{ord}_{\mathcal{H}'} E_{ijk} s_k <_{\eta'} (k(i,j), \varrho_i)$ for all $k \in \{1, \ldots, q\}$ with $E_{ijk} \neq 0$.

b) *The element $\Sigma_{ij} = \underline{X}^{w(i,j)} e_i - \underline{X}^{w(j,i)} e_j - \sum_{k=1}^q E_{ijk} e_k$ of P^q is a homogeneous syzygy of $s_1, \ldots, s_q$ with leading form $L_{\mathfrak{J}} \Sigma_{ij} = \underline{X}^{w(i,j)} e_i$.*

c) *The set $\mathfrak{S}_q = \{\Sigma_{ij} \mid i, j \in \{1, \ldots, q\}, \varrho_i = \varrho_j\}$ is a Gröbner basis of the module of syzygies $\ker \epsilon_q$ of $s_1, \ldots, s_q$ with respect to the filtration $\mathfrak{J}$. Here ϵ_q is the homomorphism $\epsilon_q : P^q \longrightarrow P^t$ given by $e_i \mapsto s_i$ for $i = 1, \ldots, q$.*

Proof: Part a) follows immediately from the observation that $\operatorname{ord}_{\mathcal{H}'}(\underline{X}^{w(i,j)} s_i - \underline{X}^{w(j,i)} s_j) <_{\eta'} (k(i,j), \varrho_i)$ and the fact that $\{s_1, \ldots, s_q\}$ is a homogeneous Gröbner basis of $\ker \epsilon_t$ by proposition 5.4.b. Part b) is then clear, and the proof of part c) is completely analogous to the proof of proposition 5.3 and therefore omitted. □

Using this proposition, we are able to compute the desired minimal presentation. In the first step, we apply Propsition 3.7 and use our knowledge of $\mathfrak{S}_q$ to find a minimal system of generators $\mathfrak{S}' \subseteq \mathfrak{S}_t$ of $\ker \epsilon_t$. Then we apply proposition 3.7 for a second time and find a minimal system of generators $\mathfrak{G} \subseteq \{p_{\mu_1}, \ldots, p_{\mu_t}\}$ of ω_R. Let β be the minimal number of generators of ω_R, i.e., the Cohen-Macaulay type of R, and let $m_1, \ldots, m_\beta \in \{\mu_1, \ldots, \mu_t\}$ such that $\mathfrak{G} = \{p_{m_1}, \ldots, p_{m_\beta}\}$. Also, let $\alpha_i = \deg p_{m_i}$ for $i = 1, \ldots, \beta$. In the third step, we apply the second part of proposition 3.8 and compute from $\mathfrak{S}'$ a minimal system of generators $\mathfrak{S}$ of the module of syzygies of $\mathfrak{G}$. Let $\mathfrak{S} = \{s'_1, \ldots, s'_{\beta'}\}$ with $\beta' \geq 0$, and let $\alpha'_i = \deg s'_i$ for $i = 1, \ldots, \beta'$. All together, we arrive at the following result.

Proposition 6.2. *The exact sequence*

$$\begin{array}{ccccccc} \bigoplus_{i=1}^{\beta'} P(-\alpha'_i) & \longrightarrow & \bigoplus_{i=1}^{\beta} P(-\alpha_i) & \longrightarrow & \omega_R & \longrightarrow & 0 \\ e_i & \longmapsto & s'_i & e_i & \longmapsto & p_{m_i} & \end{array}$$

is a minimal homogeneous presentation of the graded P-module ω_R.

7 Application to minimal resolutions

Starting again with a zero-dimensional, reduced subscheme $\mathbb{X} \subseteq \mathbb{P}^n_K$, consisting of K-rational points, with homogeneous coordinate ring $R = P/I_{\mathbb{X}}$, and with canonical module ω_R, we are going to examine the minimal graded free resolutions of $I_{\mathbb{X}}$ and ω_R, considered as graded modules over $P = K[X_0, \ldots, X_n]$. The graded free P-resolution of $I_{\mathbb{X}}$ is of the form

$$0 \longrightarrow \bigoplus_{i=1}^{\beta_n} P(-\alpha_{ni}) \xrightarrow{\Phi_n} \cdots \longrightarrow \bigoplus_{i=1}^{\beta_1} P(-\alpha_{1i}) \xrightarrow{\Phi_1} I_{\mathbb{X}} \longrightarrow 0$$

with $\alpha_{ij}, \beta_i \in \mathbb{N}$, and homogeneous P-linear maps Φ_i of degree zero whose matrices $\mathfrak{A}_i$ consist of homogeneous polynomials of positive degree.

Since R is a one-dimensional Cohen-Macaulay ring, we have $\underline{\operatorname{Ext}}^i_P(R, P) = 0$ for $i = 0, \ldots, n-1$ (see [M], 17.1, or [G], IV, § 4, for an elementary approach) and $\omega_R \cong \underline{\operatorname{Ext}}^n_P(R, P)(-n-1)$ (cf. [GW], 2.1.6). Therefore, if we dualize and shift the above resolution, we obtain a minimal graded free P-resolution of ω_R of the form

$$0 \to P(-n-1) \xrightarrow{\check{\Phi_1}} \bigoplus_{i=1}^{\beta_1} P(\alpha_{1i} - n - 1) \longrightarrow \cdots \xrightarrow{\check{\Phi_n}}$$

$$\xrightarrow{\check{\Phi_n}} \bigoplus_{i=1}^{\beta_n} P(\alpha_{ni} - n - 1) \to \omega_R \to 0.$$

The numbers α_{ij}, β_i are invariants of $\mathbb{X} \subseteq \mathbb{P}^n_K$ and thus interesting to compute. For two of them we reserve special names: $\alpha_{\mathbb{X}} = \min\{\alpha_{11}, \dots, \alpha_{1\beta_1}\} = \min\{i \in \mathbb{N} \mid (I_{\mathbb{X}})_i \neq 0\}$ and $\sigma_{\mathbb{X}} = \max\{\alpha_{n1} - n - 1, \dots, \alpha_{n\beta_n} - n - 1\} = -\min\{i \in \mathbb{Z} \mid (\omega_R)_i \neq 0\}$.

It is known (cf. [S], Kor. 1) that $\sigma_{\mathbb{X}} \geq \alpha_{\mathbb{X}} - 2$ holds all the time, and that the numbers α_{ij}, β_i are uniquely determined, if $\sigma_{\mathbb{X}} = \alpha_{\mathbb{X}} - 2$ (cf. [S], Thm. A). The next case is of particular interest.

Definition 7.1. The *Hilbert function* of $\mathbb{X}$ is the function $H_{\mathbb{X}} : \mathbb{Z} \longrightarrow \mathbb{N}$ given by $i \mapsto \dim_K R_i$. We say that $\mathbb{X}$ has *generic Hilbert function*, if $H_{\mathbb{X}} = \min\{H_{\mathbb{P}^n}, \deg \mathbb{X}\}$.

It follows from [K1], Lemma 1.3.c, that $\sigma_{\mathbb{X}} = \max\{i \in \mathbb{Z} \mid H_{\mathbb{X}}(i) < \deg \mathbb{X}\}$, and therefore $\mathbb{X}$ has generic Hilbert function, if and only if $\sigma_{\mathbb{X}} \leq \alpha_{\mathbb{X}} - 1$. The reason for the choice of this name is that, in case of an algebraically closed field K, a generic set of points $\mathbb{X} \subseteq \mathbb{P}^n_K$ has generic Hilbert function. Here "generic set of points" means that there is a nonempty open subset of the symmetric power $(\mathbb{P}^n_K)^{(s)}$ such that all sets of s points in $\mathbb{P}^n_K$ corresponding to closed points of this open set have the stated property.

For the remainder of this section, we shall assume that K is algebraically closed and $\mathbb{X} \subseteq \mathbb{P}^n_K$ is a set of points with $\sigma_{\mathbb{X}} = \alpha_{\mathbb{X}} - 1$. Our goal is to compute the invariants α_{ij}, β_i. Without loss of generality we can assume that $\alpha_{i1} \leq \dots \leq \alpha_{i\beta_i}$ for $i = 1, \dots, n$. Since all entries in the matrices $\mathfrak{A}_1, \dots, \mathfrak{A}_n$ have positive degrees, we have $\alpha_{11} < \dots < \alpha_{n1}$. Since their transposed matrices have the same property, we can look at the above resolution of ω_R and get $\alpha_{1\beta_1} < \dots < \alpha_{n\beta_n}$. Thus $\sigma_{\mathbb{X}} = \alpha_{\mathbb{X}} - 1$ implies that in the chain of inequalities $\sigma_{\mathbb{X}} = \alpha_{n\beta_n} - n - 1 \geq \alpha_{n1} - n - 1 \geq \alpha_{n-1\,1} - n \geq \dots \geq \alpha_{11} - 2 = \alpha_{\mathbb{X}} - 2$ we have exactly one step where there is a strict decrease by one, and otherwise we have equality. In other words, the resolution of $I_{\mathbb{X}}$ is of the form $(*)$

$$0 \to P(-\alpha_{\mathbb{X}} - n + 1)^{a_n} \oplus P(-\alpha_{\mathbb{X}} - n)^{b_n} \to \cdots \to$$
$$\to P(-\alpha_{\mathbb{X}})^{a_1} \oplus P(-\alpha_{\mathbb{X}} - 1)^{b_1} \to I_{\mathbb{X}} \to 0$$

and we want to compute the numbers a_i, b_i. Clearly, if $a_i = 0$ for some $i \in \{1, \dots, n\}$, then $a_i = \dots = a_n = 0$, and if $b_i = 0$, then $b_1 = \dots = b_i = 0$. For the numbers a_i, b_i there is the following intriguing conjecture (cf. [L], sect. 2).

Minimal resolution conjecture (MRC). If $\mathbb{X} \subseteq \mathbb{P}^n_K$ is a generic set of points and $m := \lceil \alpha_{\mathbb{X}}(\binom{\alpha_{\mathbb{X}}+n}{n} / \deg \mathbb{X} - 1) \rceil$, then $a_{m+1} = b_{m-1} = 0$.

In fact, we shall see in a moment that if $\mathbb{X}$ satisfies MRC, then all numbers a_i, b_i are uniquely determined. Since $\sigma_{\mathbb{X}} = \alpha_{\mathbb{X}} - 1$, we have $\binom{n+\alpha_{\mathbb{X}}-1}{n} = H_{\mathbb{P}^n}(\alpha_{\mathbb{X}} - 1) < \deg \mathbb{X} < H_{\mathbb{P}^n}(\alpha_{\mathbb{X}})$. The number $\Delta_{\mathbb{X}} = \deg \mathbb{X} - \binom{n+\alpha_{\mathbb{X}}-1}{n}$ is called the *last difference* of $\mathbb{X}$, as it is the last nonzero value of the first difference function of $H_{\mathbb{X}}$. Recall that the minimal graded free resolution of $\binom{n+\alpha_{\mathbb{X}}-1}{n}$ points $\mathbb{Y} \subseteq \mathbb{P}^n_K$ is given by

$$0 \longrightarrow P(-\alpha_{\mathbb{X}} - n + 1)^{c_n} \longrightarrow \cdots \longrightarrow P(-\alpha_{\mathbb{X}})^{c_1} \longrightarrow I_{\mathbb{Y}} \longrightarrow 0$$

with $c_i = \binom{n+\alpha_{\mathbb{X}}-1}{i+\alpha_{\mathbb{X}}-1}\binom{i+\alpha_{\mathbb{X}}-2}{\alpha_{\mathbb{X}}-1}$ for $i = 1, \dots, n$ (cf. [S], Thm. A.b). Our next proposition gives a strong restriction on the numbers a_i, b_i of $\mathbb{X}$ and implies the description of the conjectured generic minimal resolution in [L], 2.3. (A different proof is given in [Eh], sect. 4.)

Lemma 7.2. *For natural numbers α, i, n with $i \leq n$ we have*

$$\sum_{j=0}^{i}(-1)^j\binom{n+i-j}{n}\binom{n+\alpha-1}{\alpha+j}\binom{\alpha-1+j}{\alpha-1} = \binom{n+\alpha+i}{n} - \binom{n+\alpha-1}{n}.$$

Proof: In view of the Gould inverse relations given by $a_i = \sum_{j=0}^{i}\binom{n+1}{i-j}b_j$ and $b_i = \sum_{j=0}^{i}(-1)^{i+j}\binom{n+i-j}{i-j}a_j$ (cf. [Ri], p. 52), we see that it suffices to show the equation

$$\sum_{j=0}^{i}(-1)^j\binom{n+1}{i-j}\binom{n+\alpha+j}{n} - \sum_{j=0}^{i}(-1)^j\binom{n+1}{i-j}\binom{n+\alpha-1}{n} = \binom{n+\alpha-1}{\alpha+i}\binom{\alpha-1+i}{\alpha-1}.$$

Using [Ri], p. 33, ex. 9, this boils down to showing that

$$\sum_{j=0}^{i}(-1)^j\binom{n+1}{i-j}\binom{n+\alpha+j}{n} = \binom{n+\alpha}{\alpha+i}\binom{\alpha-1+i}{\alpha-1}.$$

By applying [Ri], ch. 1.3, eqs. (3), (5a), and (11), we find

$$\begin{aligned}
\sum_{j=0}^{i}(-1)^j\binom{n+1}{i-j}\binom{n+\alpha+j}{n} &= \sum_{j=0}^{i}(-1)^j\binom{n+1}{i-j}\sum_{k\geq 0}\binom{n+\alpha}{n-k}\binom{j}{k}\\
&= \sum_{k\geq 0}(-1)^i\binom{n+\alpha}{n-k}\sum_{j=0}^{i}(-1)^j\binom{n+1}{j}\binom{i-j}{k}\\
&= \sum_{k\geq 0}(-1)^i\binom{n+\alpha}{n-k}\binom{i-n-1}{i-k} = \sum_{k\geq 0}(-1)^k\binom{n+\alpha}{n-k}\binom{n-k}{n-i}\\
&= \sum_{k\geq 0}(-1)^{n-k}\binom{k}{n-i}\binom{n+\alpha}{k} = \binom{\alpha+i-1}{i}\binom{n+\alpha}{n-1},
\end{aligned}$$

as desired. □

Proposition 7.3. *Let $\mathbb{X} \subseteq \mathbb{P}^n_K$ be a set of points with $\sigma_{\mathbb{X}} = \alpha_{\mathbb{X}} - 1$.*

a) We have $a_1 = \binom{n+\alpha_{\mathbb{X}}}{n} - \deg\mathbb{X}$, $b_n = \Delta_{\mathbb{X}}$, and $a_i - b_{i-1} = c_i - \Delta_{\mathbb{X}}\binom{n}{i-1}$ for $i = 2, \dots, n$.

b) If $\mathbb{X}$ satisfies MRC, then

$$a_i = \begin{cases} c_i - \Delta_{\mathbb{X}}\binom{n}{i-1} & \text{if } c_i - \Delta_{\mathbb{X}}\binom{n}{i-1} \geq 0,\\ 0 & \text{otherwise, and}\end{cases}$$

$$b_{i-1} = \begin{cases} -c_i + \Delta_{\mathbb{X}}\binom{n}{i-1} & \text{if } c_i - \Delta_{\mathbb{X}}\binom{n}{i-1} \leq 0,\\ 0 & \text{otherwise.}\end{cases}$$

Proof: In order to prove a), we compute the dimensions of the homogeneous components of $(*)$ and get

$$\dim_K(I_{\mathbb{X}})_{\alpha_{\mathbb{X}}+i} = a_1\binom{n+i}{n} + \sum_{j=1}^{i}(-1)^j(a_{j+1}-b_j)\binom{n+i-j}{n}$$

for $i=0,\dots,n-1$. Thus $a_1 = \dim_K(I_{\mathbb{X}})_{\alpha_{\mathbb{X}}} = \binom{n+\alpha_{\mathbb{X}}}{n} - \deg\mathbb{X}$ and $a_2 - b_1 = a_1\binom{n+1}{n} - \dim_K(I_{\mathbb{X}})_{\alpha_{\mathbb{X}}+1} = (n+1)\binom{n+\alpha_{\mathbb{X}}}{n} - n\deg\mathbb{X} - \binom{n+\alpha_{\mathbb{X}}+1}{n} = \alpha_{\mathbb{X}}\binom{n+\alpha_{\mathbb{X}}-1}{\alpha_{\mathbb{X}}+1} - n\deg\mathbb{X} + n\binom{n+\alpha_{\mathbb{X}}-1}{n} = c_2 - \Delta_{\mathbb{X}}\binom{n}{1}$. Letting $\alpha := \alpha_{\mathbb{X}}$, remembering

$$\dim_K(I_{\mathbb{X}})_{\alpha+i} = \binom{n+\alpha+i}{n} - \deg\mathbb{X},$$

and using Lemma 7.2, we get for $i=1,\dots,n-1$ inductively

$$\begin{aligned}
a_{i+1} - b_i =& (-1)^i\dim_K(I_{\mathbb{X}})_{\alpha+i} + (-1)^{i+1}\binom{n+i}{n}a_1 \\
&+ (-1)^{i+1}\sum_{j=1}^{i-1}(-1)^j(a_{j+1}-b_j)\binom{n+i-j}{n} \\
=& (-1)^i\binom{n+\alpha+i}{n} + (-1)^{i+1}\deg\mathbb{X} + (-1)^{i+1}\binom{n+i}{n}\binom{n+\alpha}{n} \\
&+ (-1)^i\binom{n+i}{n}\deg\mathbb{X} + (-1)^i\sum_{j=1}^{i-1}(-1)^j\binom{n+i-j}{n}\binom{n}{j}\deg\mathbb{X} \\
&+ (-1)^{i+1}\sum_{j=1}^{i-1}(-1)^j\binom{n+i-j}{n}\binom{n}{j}\binom{n+\alpha-1}{\alpha-1} \\
&+ (-1)^{i+1}\sum_{j=1}^{i-1}(-1)^j\binom{n+i-j}{n}\binom{n+\alpha-1}{\alpha+j}\binom{\alpha-1+j}{\alpha-1} \\
=& \binom{n+\alpha-1}{\alpha+i}\binom{\alpha-1+i}{\alpha-1} - \binom{n}{i}\deg\mathbb{X} + \binom{n}{i}\binom{n+\alpha-1}{\alpha-1} = c_{i+1} - \Delta_{\mathbb{X}}\binom{n}{i}.
\end{aligned}$$

Part b) follows immediately from a) and $a_{m+1} = b_{m-1} = 0$. □

Let us illustrate this proposition with an example.

Example 7.4. Let K be an algebraically closed field of characteristic zero, and let $\mathbb{X} \subseteq \mathbb{P}^6_K$ be a set of 11 generic points. The Hilbert function of $\mathbb{X}$ is $H_{\mathbb{X}}$: 1 7 11 11 $\cdots$, so $\sigma_{\mathbb{X}} = 1$, $\alpha_{\mathbb{X}} = 2$, and $\Delta_{\mathbb{X}} = 4$. The minimal resolution of 7 generic points in $\mathbb{P}^6_K$ is given by the table

i	1	2	3	4	5	6
c_i	21	70	105	84	35	6

Subtracting $\Delta_{\mathbb{X}} = 4$ times the 6th row of Pascal's triangle yields

i	1	2	3	4	5	6	7
$c_i - 4\binom{6}{i-1}$	17	46	45	4	-25	-18	-4

Hence the minimal resolution of $I_{\mathbb{X}}$ expected by MRC is given by the table

i	1	2	3	4	5	6
a_i	17	46	45	4	0	0
b_i	0	0	0	25	18	4

In fact, in concrete examples the minimal resolution of $\mathbb{X}$ is really given by the table

i	1	2	3	4	5	6
a_i	17	46	45	5	0	0
b_i	0	0	1	25	18	4

still in accordance with part a) of proposition 7.3, but violating MRC.

How can we use the material presented in sections 3 to 6 and proposition 7.3 to calculate minimal resolutions, in particular of schemes $\mathbb{X} \subseteq \mathbb{P}^n_K$ with $\sigma_{\mathbb{X}} = \alpha_{\mathbb{X}} - 1$? The following remark collects what we know so far.

Remark 7.5.

a) Given $\mathbb{X} \subseteq \mathbb{P}^n_K$, the Buchberger-Möller algorithm allows us to compute $I_{\mathbb{X}} \subseteq K[X_0, \ldots, X_n]$ in a fast (polynomial-time) way.

b) The algorithm described in sections 4 to 6 allows us to compute a minimal presentation of ω_R in a fast way. The resolution of ω_R is dual to the resolution of $I_{\mathbb{X}}$.

c) If we want to know the numbers a_i, b_i for $\mathbb{X} \subseteq \mathbb{P}^n_K$ with $\sigma_{\mathbb{X}} = \alpha_{\mathbb{X}} - 1$, it suffices by proposition 7.3.a to compute either $a_1, \ldots, a_n$ or $b_1, \ldots, b_n$. As an additional help, we can use $a_i = 0 \Rightarrow a_i = \cdots = a_n = 0$ and $b_i = 0 \Rightarrow b_1 = \cdots = b_i = 0$.

Now we consider the following notion.

Definition 7.6. Let $P = K[X_0, \ldots, X_n]$, let M be a finitely generated, graded P-module, and let $\alpha = \min\{i \in \mathbb{Z} \mid M_i \neq 0\}$. Furthermore, let

$$0 \longrightarrow \bigoplus_{i=1}^{\beta_n} P(-\alpha_{ni}) \longrightarrow \cdots \longrightarrow \bigoplus_{i=1}^{\beta_0} P(-\alpha_{0i}) \longrightarrow M \longrightarrow 0$$

be the minimal graded free P-resolution of M, where $\alpha + i \leq \alpha_{i1} \leq \cdots \leq \alpha_{i\beta_i}$ for $i = 0, \ldots, n$. Then the subcomplex

$$0 \longrightarrow \bigoplus_{\{i \mid \alpha_{ni} = \alpha + n\}} P(-\alpha - n) \longrightarrow \cdots \longrightarrow \bigoplus_{\{i \mid \alpha_{0i} = \alpha\}} P(-\alpha) \longrightarrow M \longrightarrow 0$$

is called the *linear part* of the resolution of M.

In our situation, the linear part of the resolution of $I_{\mathbb{X}}$ is given by

$$0 \longrightarrow P(-\alpha_{\mathbb{X}} - n + 1)^{a_n} \longrightarrow \cdots \longrightarrow P(-\alpha_{\mathbb{X}})^{a_1} \longrightarrow I_{\mathbb{X}} \longrightarrow 0,$$

and the linear part of the resolution of ω_R is given by

$$0 \longrightarrow P(\sigma_{\mathbb{X}} - n + 1)^{b_1} \longrightarrow \cdots \longrightarrow P(\sigma_{\mathbb{X}})^{b_n} \longrightarrow \omega_R \longrightarrow 0.$$

Therefore we can compute the numbers a_i, b_i by finding the linear parts of the resolutions of $I_{\mathbb{X}}$ and ω_R. This is done as follows.

Remark 7.7. Let $M \subseteq \oplus_{i=1}^r P(-\alpha_i)$ be a graded submodule, let $\alpha = \min\{i \in \mathbb{Z} \mid M_i \neq 0\}$, let $<_\sigma$ be a degree-compatible term ordering of $\mathcal{T}^{n+1}$, and let $\mathcal{G}$ be the $\mathbb{Z}^{n+1} \times \{1, \dots, r\}$-filtration induced by $\mathcal{F}^\sigma$ on $\oplus_{i=1}^r P(-\alpha_i)$ such that $\operatorname{ord}_{\mathcal{G}} e_i = (m_i, i)$ for some $m_1, \dots, m_r \in \mathbb{Z}^{n+1}$. Furthermore, let $\{g_1, \dots, g_s\}$ be a K-basis of M_α such that $L_{\mathcal{G}} g_i = T_i e_{\mu(i)}$ with $T_1, \dots, T_s \in \mathcal{T}^{n+1}$ pairwise distinct, and with $\mu(i) \in \{1, \dots, r\}$. Thus we have $\deg T_1 + \alpha_{\mu(1)} = \cdots = \deg T_s + \alpha_{\mu(s)}$. Then we form the set

$$S = \{X_\beta g_i - X_\gamma g_j \mid i, j \in \{1, \dots, s\},\ \beta, \gamma \in \{0, \dots, n\},\ X_\beta T_i = X_\gamma T_j\},$$

and we reduce each element of this set against $\{g_1, \dots, g_s\} \setminus \{g_i, g_j\}$ (resp. the part of the Gröbner basis of M with respect to $\mathcal{G}$ found so far) by successively eliminating leading forms until we either find a linear syzygy of $g_1, \dots, g_s$ or a new element of the Gröbner basis of M with respect to $\mathcal{G}$. It is clear that in this way we obtain a system of generators of the linear part of the syzygy module of M which we can then interreduce again to continue the process.

By repeatedly applying this remark, we can quickly calculate the linear part of the resolution of a given module, in particular of $I_{\mathbb{X}}$ and ω_R. In view of the next proposition, this allows us to prove MRC in many cases computationally.

Proposition 7.8. *Let K be an algebraically closed field. If there exists a set of s points in $\mathbb{P}_K^n$ satisfying MRC, then the generic set of s points in $\mathbb{P}_K^n$ satisfies MRC.*

Proof: Without loss of generality we can assume that $\sigma_{\mathbb{X}} = \alpha_{\mathbb{X}} - 1$, as in case $\sigma_{\mathbb{X}} = \alpha_{\mathbb{X}} - 2$ there is only one possible set of invariants $c_1, \dots, c_n$. Let $\mathbb{H}$ be the Hilbert scheme parametrizing zero-dimensional subschemes $\mathbb{X} \subseteq \mathbb{P}_K^n$ with $\deg \mathbb{X} = s$. From [BG], 2.15, it follows that there is an open subset $\mathbb{U} \subseteq \mathbb{H}$ whose closed points correspond to schemes $\mathbb{X}$ with $H_{\mathbb{X}} = \min\{H_{\mathbb{P}^n}, \deg \mathbb{X}\}$ and minimal Betti numbers $a_1 + b_1, \dots, a_n + b_n$. Since $a_1, a_2 - b_1, \dots, a_n - b_{n-1}, b_n$ are fixed by proposition 7.3.a, this means that such schemes have minimal numbers a_i, b_i. Clearly, the values of a_i, b_i given in proposition 7.3.b are the least possible, so if there exists a set of points $\mathbb{X}$ having those invariants (i.e., satisfying MRC), then the same is true for every zero-dimensional scheme corresponding to a closed point of $\mathbb{U}$.

Let $\mathbb{V}$ be the open subscheme of the Hilbert scheme whose closed points correspond to reduced zero-dimensional schemes. By assumption, $\mathbb{U} \cap \mathbb{V}$ is not empty. Now we apply the *Hilbert-Chow morphism* $\mathbb{H} \longrightarrow (\mathbb{P}_K^n)^{(s)}$ (cf. [F], [MF]). This morphism induces an isomorphism of $\mathbb{V}$ with the open subset $\mathbb{W}$ of $(\mathbb{P}_K^n)^{(s)}$ parametrizing cycles of s distinct points of $\mathbb{P}_K^n$. Thus the image of $\mathbb{U} \cap \mathbb{V}$ in $\mathbb{W}$ is a nonempty open subset whose closed points correspond to sets of points in $\mathbb{P}_K^n$ satisfying MRC. □

Using this proposition and the computational techniques described before, the authors were able to prove the following result by computer calculation.

Proposition 7.9. *Let K be an algebraically closed field of characteristic zero. Then the Minimal Resolution Conjecture is true for generic sets of s points in $\mathbb{P}^n_K$, as long as $n \le 9$, $s \le 50$, and $(n,s) \notin \{(6,11),(7,12),(8,13)\}$.*

What about the remaining three cases? In 1992, F. Schreyer pointed out that 12 generic points in $\mathbb{P}^7_{\mathbb{C}}$ and 13 generic points in $\mathbb{P}^8_{\mathbb{C}}$ appear to yield counterexamples to MRC. Using "random" points in $\mathbb{P}^n_{\mathbb{Q}}$, the authors found for 12 generic points in $\mathbb{P}^7_{\mathbb{C}}$ the resolution table

i	1	2	3	4	5	6	7
a_i	24	84	126	84	4	0	0
b_i	0	0	0	4	36	21	4

and for 13 generic points in $\mathbb{P}^8_{\mathbb{C}}$ it is given by

i	1	2	3	4	5	6	7	8
a_i	32	136	266	280	140	2	0	0
b_i	0	0	0	0	10	49	24	4

The apparent failure of MRC for 11 generic points in $\mathbb{P}^6_{\mathbb{C}}$ was reported by M. Boij and R. Fröberg in 1993 (see Example 7.4).

How can one disprove MRC in those particular cases using a computer? In view of [BG], 2.15, and proposition 7.3.a, it suffices to compute the numbers a_i, b_i for the set of points $\mathbb{X} = \{P_1, \dots, P_s\}$ with $P_i = (1 : Y_{i1} : \cdots : Y_{in})$ over the field $K = \mathbb{Q}(Y_{ij})_{i=1,\dots,s, j=1,\dots,n}$. Programming fast and efficient arithmetics for large fields of rational functions, however, seems to be a difficult problem that still awaits an answer.

8 Application to checking uniformity conditions

Let K be an algebraically closed field. For a zero-dimensional, reduced subscheme $\mathbb{X} \subseteq \mathbb{P}^n_K$ consisting of $s = \deg \mathbb{X}$ K-rational points, we have the following notion of uniformity (cf. [K1]).

Definition 8.1. For $i \ge 0$, we say that $\mathbb{X}$ is *i-uniform*, if every subset $\mathbb{Y} \subseteq \mathbb{X}$ of degree $\deg \mathbb{Y} \ge s - i$ has Hilbert function $H_{\mathbb{Y}} = \min\{H_{\mathbb{X}}, \deg \mathbb{Y}\}$. Equivalently, $\mathbb{X}$ is i-uniform, if and only if any two subsets $\mathbb{Y}, \mathbb{Y}'$ of degree $s - i \le \deg \mathbb{Y} = \deg \mathbb{Y}' \le s$ have the same Hilbert function $H_{\mathbb{Y}} = H_{\mathbb{Y}'}$.

The two extremal cases have special names. A 1-uniform set of points is also called a *Cayley-Bacharach scheme* (for short, CB-scheme, cf. [GKR]), and an $(s-1)$-uniform scheme is said to be in *uniform position* (cf. [H]).

In this section, we want to give a solution to the problem of checking those uniformity conditions computationally, if we are given the set of points $\mathbb{X}$ via their coordinates. The naïve approach is, of course, to calculate all Hilbert functions of the subsets $\mathbb{Y} \subseteq \mathbb{X}$ under consideration. For instance, in the case of checking uniform position, this entails the calculation of 2^s Hilbert functions. Already for moderate numbers of points s, this is clearly impracticable. Therefore we shall exhibit a different approach.

For small uniformities, and in particular for checking the Cayley-Bacharach property, the solution is provided by the following remark based on [K1], 3.4. Recall that, given $\mathbb{X} = \{P_1, \dots, P_s\} \subseteq \mathbb{P}^n_K$, we assumed $\mathbb{X} \subseteq D_+(X_0)$, we let $P = K[X_0, \dots, X_n]$ and $R = P/I_{\mathbb{X}}$, and we defined the invariants $\sigma_{\mathbb{X}} = -\min\{i \in \mathbb{Z} \mid (\omega_R)_i \neq 0\}$ and $\Delta_{\mathbb{X}} = H_{\mathbb{X}}(\sigma_{\mathbb{X}} + 1) - H_{\mathbb{X}}(\sigma_{\mathbb{X}})$.

Remark 8.2. Using the Buchberger-Möller algorithm, we can calculate a Gröbner basis of $I_{\mathbb{X}}$ with respect to a degree-compatible term ordering $<_\tau$ from the coordinates of $P_1, \dots, P_s$. As a byproduct, this algorithm yields polynomials $f_1, \dots, f_s \in P_{\sigma_{\mathbb{X}}+1}$ such that $f_i(P_j) = \delta_{ij}$ for $i, j = 1, \dots, s$ (cf. [MR], sect. 1.2, and [GKR], sect. 2). Here we write $P_i = (1 : p_{i1} : \cdots : p_{in})$, and we set $f(P_i) = f(1, p_{i1}, \dots, p_{in})$ for a homogeneous polynomial $f \in P$. The polynomials $f_1, \dots, f_s$ are called the *separators* of $\mathbb{X}$ and are represented in normal form with respect to $I_{\mathbb{X}}$. Their images in $R_{\sigma_{\mathbb{X}}+1}$ form a K-basis of that vector space (cf. [GKR], 1.13.a).

Next we compute $\bar{f}_i(X_1, \dots, X_n) = f_i(0, X_1, \dots, X_n)$ for $i = 1, \dots, s$. As the polynomials $f_1, \dots, f_s$ were in their normal form with respect to $I_{\mathbb{X}}$, the elements $\bar{f}_1, \dots, \bar{f}_s$ are K-linear combinations of the $\Delta_{\mathbb{X}}$ monomials of $\mathcal{O}_\tau(J_{\mathbb{X}})$ of degree $\sigma_{\mathbb{X}} + 1$ (cf. sect. 3). Let $T_1, \dots, T_\Delta$ be those monomials, and let $\bar{f}_i = \sum_{j=1}^{\Delta} c_{ij} T_j$ with $c_{ij} \in K$ for $i = 1, \dots, s$. Then [K1], 3.4, says that for any $i \in \{1, \dots, \Delta_{\mathbb{X}}\}$ the following conditions are equivalent:

a) The scheme $\mathbb{X}$ is i-uniform.
b) Any i columns of the matrix (c_{ij}) are K-linearly independent.

Of course, condition b) is easy to check for small i, and we are done.

A much more difficult situation arises for higher uniformities. It appears to be impossible to check those directly from the Gröbner basis of $I_{\mathbb{X}}$ or the separators. Again the results from [K1] come to the rescue: The relevant information is encoded in the multiplication maps of the canonical module ω_R! More precisely, [K1], 3.2, says that for $i = s - H_{\mathbb{X}}(j)$ and $j \in \{0, \dots, \sigma_{\mathbb{X}}\}$ the following conditions are equivalent:

a) The scheme $\mathbb{X}$ is i-uniform.
b) For each $k \in \{j, \dots, \sigma_{\mathbb{X}}\}$, the multiplication map $\mu : R_k \otimes (\omega_R)_{-k} \longrightarrow (\omega_R)_0$ is biinjective.

Here "biinjective" means that $\mu(r \otimes \varphi) = 0$ implies $r = 0$ or $\varphi = 0$. Let us see how one can check condition b) computationally. From sections 4 to 6 we know how to get a minimal graded presentation

$$\bigoplus_{i=1}^{\beta'} P(-\alpha'_i) \longrightarrow \bigoplus_{i=1}^{\beta} P(-\alpha_i) \longrightarrow \omega_R \longrightarrow 0.$$

Thus we can find elements $\varphi_1, \ldots, \varphi_m \in (\bigoplus_{i=1}^{\beta} P(-\alpha_i))_{-k}$ whose images form a K-basis of $(\omega_R)_{-k}$. Here $k \in \{j, \ldots, \sigma_{\mathbb{X}}\}$ and $m = \dim_K(\omega_R)_{-k} = s - H_{\mathbb{X}}(k)$. In the same way, we can find elements $\psi_1, \ldots, \psi_{s-1} \in (\bigoplus_{i=1}^{\beta} P(-\alpha_i))_0$ whose images form a K-basis of $(\omega_R)_0$. Moreover, let $r_1, \ldots, r_l \in P_k$ with $l = H_{\mathbb{X}}(k)$ be those monomials from $\mathcal{O}_\tau(J_{\mathbb{X}})$ whose residue classes form a K-basis of R_k. Then we write $r_\lambda \varphi_\mu$ in normal form with respect to the syzygy module of ω_R, and we get representations $r_\lambda \varphi_\mu = \sum_{\nu=1}^{s-1} c_{\lambda\mu\nu} \psi_\nu$ with $c_{\lambda\mu\nu} \in K$ for $\lambda = 1, \ldots, l$ and $\mu = 1, \ldots, m$. With this setup, we have at our disposal the following two methods to check whether the bilinear map given by $(c_{\lambda\mu\nu})$ is biinjective.

Remark 8.3. After introducing new variables $Y_1, \ldots, Y_l$ and $Z_1, \ldots, Z_m$, we form the quadratic polynomials $f_\nu = \sum_{\lambda,\mu} c_{\lambda\mu\nu} Y_\lambda Z_\mu$ for $\nu = 1, \ldots, s-1$. Since we want to check whether $(a_1 r_1 + \cdots + a_l r_l)(b_1 \varphi_1 + \cdots + b_m \varphi_m) = 0$ implies that $a_1 = \cdots = a_l = 0$ or $b_1 = \cdots = b_m = 0$, we need to find out whether $f_1(a_1, \ldots, a_l, b_1, \ldots, b_m) = \ldots = f_{s-1}(a_1, \ldots, a_l, b_1, \ldots, b_m) = 0$ yields the same implication. Thus we have to check whether the ideal $(Y_1, \ldots, Y_l) \cap (Z_1, \ldots, Z_m) = (Y_\lambda Z_\mu)_{\lambda=1,\ldots,l, \mu=1,\ldots,m}$ is contained in the radical of $(f_1, \ldots, f_{s-1})$, a condition which can be decided by standard computer algebra techniques (cf., e.g., [GTZ], sect. 9).

Our second method is based on a result of D. Eisenbud (cf. [Ei], 1.1).

Remark 8.4. For $\nu = 1, \ldots, s-1$ we let $\mathfrak{C}_\nu = (c_{\lambda\mu\nu})_{\lambda=1,\ldots,l, \mu=1,\ldots,m}$, and we denote by M the subspace of $\mathrm{Hom}_K(K^m, K^l)$ generated by $\mathfrak{C}_1, \ldots, \mathfrak{C}_{s-1}$. Then we let $M^\perp$ be the subspace of $\mathrm{Hom}_K(K^l, K^m)$ which is dual to M with respect to the trace map, i.e., the space

$$M^\perp = \{\mathfrak{D} \in \mathrm{Hom}_K(K^l, K^m) \mid \mathrm{tr}(\mathfrak{D}\mathfrak{C}_1) = \ldots = \mathrm{tr}(\mathfrak{D}\mathfrak{C}_{s-1}) = 0\}.$$

Obviously $M^\perp$ can be computed by solving a system of linear equations. Let $\mathfrak{D}_1, \ldots, \mathfrak{D}_{lm-s-1}$ be a K-basis of $M^\perp$. We introduce new variables $Y_1, \ldots, Y_{lm-s-1}$ and let $\mathfrak{M} = \mathfrak{D}_1 Y_1 + \cdots + \mathfrak{D}_{lm-s-1} Y_{lm-s-1}$ be the matrix of linear forms associated to $M^\perp$ (cf. [Ei], sect. 1). By [Ei], 1.1, we have to check whether $(M^\perp)_1 = 0$, i.e., whether the ideal of 2×2-minors of $\mathfrak{M}$ has radical $(Y_1, \ldots, Y_{lm-s-1})$. Again this is a straightforward task using standard techniques.

Thus we are able to check i-uniformity in case i is of the form $i = s - H_{\mathbb{X}}(j)$ with $j \in \{0, \ldots, \sigma_{\mathbb{X}}\}$. Notice that this means, in particular, that we can check the uniform position property. Finally, we are left with the general case. Let $i \in \{\Delta_{\mathbb{X}}, \ldots, s-1\}$, and write $i = s - H_{\mathbb{X}}(j) + k$ with $j \in \{0, \ldots, \sigma_{\mathbb{X}}\}$ and $k \in \{0, \ldots, H_{\mathbb{X}}(j) - H_{\mathbb{X}}(j-1) - 1\}$. Using Remark 8.3 or Remark 8.4, we first check whether $\mathbb{X}$ is $(s - H_{\mathbb{X}}(j))$-uniform. Suppose this is the case. Then, for each subscheme $\mathbb{Y} \subseteq \mathbb{X}$ consisting of $s - H_{\mathbb{X}}(j)$ points, we have $H_{\mathbb{Y}} = \min\{H_{\mathbb{X}}, \deg \mathbb{Y}\}$, and we have to check whether $\mathbb{Y}$ is k-uniform. Since the Hilbert function of $\mathbb{Y}$ is $H_{\mathbb{Y}} = \min\{H_{\mathbb{X}}, H_{\mathbb{X}}(j)\}$, we have $\Delta_{\mathbb{Y}} = H_{\mathbb{X}}(j) - H_{\mathbb{X}}(j-1)$ and $k < \Delta_{\mathbb{Y}}$, so that we can check $\mathbb{Y}$ for k-uniformity by applying Remark 8.2.

Let us end this section by showing how one can apply the described methods in a concrete example.

Example 8.5. Let $\mathbb{X} \subseteq \mathbb{P}^3_K$ be a set of 24 randomly chosen points, where $K = \overline{\mathbb{Q}}$ and "randomly" means that the coordinates of the points of $\mathbb{X}$ are random integers between, say, 0 and 100. Suppose a calculation of the Hilbert function $H_{\mathbb{X}}$ yields the expected result $H_{\mathbb{X}}$: 1 4 10 20 24 24 $\cdots$, and we want to check the uniformity of $\mathbb{X}$. Of course, we expect $\mathbb{X}$ to be in uniform position.

For checking $\Delta_{\mathbb{X}} = 4$-uniformity, it suffices to find the separators of $\mathbb{X}$ via the Buchberger-Möller algorithm, and to compute the $\binom{24}{4} = 10.626$ maximal minors of the resulting 4×24-matrix (cf. Remark 8.2). It follows that the multiplication map $R_3 \otimes (\omega_R)_{-3} \longrightarrow (\omega_R)_0$ is biinjective (cf. [K1], 3.1). Furthermore, we can easily check whether $\mathbb{X}$ is in linearly general position (cf. [K1], sect. 4) by computing another $\binom{24}{4} = 10,626$ determinants of 4×4-matrices made up from the coordinates of the points of $\mathbb{X}$. By [K1], 4.2, we then know that the multiplication map $R_1 \otimes (\omega_R)_{-1} \longrightarrow (\omega_R)_0$ is biinjective.

Thus we are left with checking the biinjectivity of the multiplication map $R_2 \otimes (\omega_R)_{-2} \longrightarrow (\omega_R)_0$. After finding the minimal homogeneous presentation of ω_R as outlined in sections 4 to 6, we calculate the $(\dim_K R_2)(\dim_K (\omega_R)_{-2})(s-1) = 3220$ coefficients $c_{\lambda\mu\nu}$ as explained above. From those we get 23 quadratic polynomials $f_\nu(Y_1, \ldots, Y_{10}, Z_1, \ldots, Z_{14})$, and we have to check whether the 140 polynomials $Y_\lambda Z_\mu$ are in the radical of $(f_1, \ldots, f_{23})$. This is achieved by computing a Gröbner basis of $(f_1, \ldots, f_{23})$ degree by degree and checking at each stage which of the polynomials $(Y_\lambda Z_\mu)^\nu$ are already contained in it.

The reader may compare this with the naïve method, where we would have to find $2^{24} = 16,777,216$ Hilbert functions of subsets of $\mathbb{X}$.

References

[BG] M. Boratynski and S. Greco, Hilbert functions and Betti numbers in a flat family, Ann. di Mat. Pura e Appl. (IV) **142** (1986), 277–292

[BM] B. Buchberger and H. M. Möller, The construction of multivariate polynomials with preassigned zeros, Computer Algebra, EUROCAM '82, Conf. Marseille / France 1982, Lect. Notes Comput. Sci. **144** (1982), 24–31

[Eh] S. Ehbauer, Graduierte Syzygienzahlen von Punkten im projektiven Raum und Anwendungen, Dissertation, Universität Bayreuth 1993

[Ei] D. Eisenbud, Linear sections of determinantal varieties, Amer. J. Math. **110** (1988), 541–575

[F] J. Fogarty, Algebraic families on an algebraic surface, Amer. J. Math. **90** (1968), 511–521

[G] W. Gröbner, Algebraische Geometrie II, B.I. Hochschultaschenbücher **737**, Bibliographisches Institut, Mannheim 1970

[GKR] A. V. Geramita, M. Kreuzer, and L. Robbiano, Cayley-Bacharach schemes and their canonical modules, Trans. Amer. Math. Soc. **339** (1993), 163–189

[GW] S. Goto and K. Watanabe, On graded rings, I, J. Math. Soc. Japan **30** (1978), 179–213

[GTZ] P. Gianni, B. Trager, and G. Zacharias, Gröbner bases and primary decomposition of polynomial ideals, J. Symbolic Comput. **6** (1988), 149–167

[H] J. Harris, The genus of space curves, Math. Ann. **249** (1980), 191–204

[K1] M. Kreuzer, On the canonical module of a zero-dimensional scheme, Can. J. Math. (to appear)

[K2] M. Kreuzer, Some applications of the canonical module of a zero-dimensional scheme, in: Proc. Conf. Zero-Dimensional Schemes, Ravello / Italy 1992 (to appear)

[L] A. Lorenzini, The minimal resolution conjecture, J. Algebra **156** (1993), 5–35

[M] H. Matsumura, Commutative ring theory, Cambridge University Press, Cambridge 1986

[MF] D. Mumford and J. Fogarty, Geometric invariant theory, Ergebnisse der Mathematik **34**, Springer, Berlin-Heidelberg-New York 1982

[MMM] M. G. Marinari, H. M. Möller, and F. Mora, Gröber bases of ideals defined by functionals with an application to ideals of projective points, Appl. Algebra Eng. Commun. Comput. **4** (1993)

[MR] F. Mora and L. Robbiano, Points in affine and projective spaces, in: Proc. Conf. Computational Algebraic Geometry and Commutative Algebra, Cortona / Italy 1991, Symp. Math. **34**, Cambridge University Press, Cambridge 1993

[PS] D. Portelli and W. Spangher, Rees algebras and Gröbner bases, Comm. Algebra **18** (1990), 2177–2197

[Ri] J. Riordan, Combinatorial identities, J. Wiley & Sons, New York 1968

[R1] L. Robbiano, On the theory of graded structures, J. Symbolic Comput. **2** (1986), 139–170

[R2] L. Robbiano, Introduction to the theory of Gröbner bases, in: The Curves Seminar at Queen's, Vol. V, Queen's Papers in Pure and Appl. Math. **80**, Queen's University, Kingston 1988

[S] P. Schenzel, Über die freien Auflösungen extremaler Cohen-Macaulay Ringe, J. Algebra **64** (1980), 93–101

S. Beck
Fakultät für Math., University of Regensburg, 93040 Regensburg (Germany).

M. Kreuzer (`kreuzer@vax1.rz.uni-regensburg.d400.de`)
Fakultät für Math., University of Regensburg, D-93040 Regensburg (Germany).

Progress in Mathematics, Vol. 143, © 1996 Birkhäuser Verlag Basel/Switzerland

Multivariate Bezoutians, Kronecker symbol and Eisenbud-Levine formula

E. Becker[1,2], J.P. Cardinal[1], M.-F. Roy[1], Z. Szafraniec[3]

1 Introduction

In the case of univariate polynomials, the *Bezoutian* $\frac{(P(x) - P(y))}{(x-y)}$ defines a quadratic form of maximal rank whose signature is 1 when the degree is odd and 0 when it is even (*). More generally the expression $\frac{(Q(y)P(x) - Q(x)P(y))}{(x-y)}$ defines a quadratic form whose signature is the Cauchy index of the rational function Q/P. The *Kronecker symbol* or global residue is the linear form ℓ associating to f (reduced modulo P) its coefficient of degree $d-1$ (where d is the degree of P). When the polynomial P has only simple roots the Kronecker symbol (or global residue) of f is the number$\sum f(p)/P'(p)$ and the signature of the quadratic form $\ell(Qf^2)$ is again the Cauchy index of Q/P.

In the multivariate case, the mathematical framework adapted for a purely algebraic generalization of this univariate situation is the theory of Frobenius algebras, which we explain in the second section of the paper. Following [15] and [6], we develop with elementary proofs (in the zero-dimensional case) the theory of duality. We study in section three the particular case of complete intersections by elementary methods and give direct methods for computing Bezoutians and Kronecker symbols from equations. We prove by an easy algebraic method two important classical results (see for example in [2]):

- a zero-dimensional complete intersection ring is a Frobenius algebra, and
- the image of the Jacobian of the equations is non-zero in every local factor.

Our method for proving these results is based on an interplay between the local and global properties. The technique is as follows: We modify the equations without changing a given localization in such a way that the properties we look for become easy in the new global ring we consider.

In the last section, we give a new and easy proof of the celebrated Eisenbud-Levine theorem (cf [7]), which is the generalization of (*). This last section relies heavily on the results and techniques in the first section.

Our results are mostly classical (most of them appear for example in [11], [15], [2], [7]), but our methods of proofs are new.

The theory of multivariate Bezoutians and Kronecker symbol has deep connections with the multivariate residues, but we do not discuss them here.

(1) Partially supported by POSSO, Esprit BRA 6846

(2) Deutsche Forschungsgemeinschaft

(3) Supported by grant KBN 2/1125/91/01

2 Frobenius algebras

In this section we review some facts about Frobenius algebras which are needed in this paper. Most of the results can be found scattered over the literature, cf., e.g., [15], [16], [13] Appendix, [8], [10]. It is our intention to give straightforward proofs and to add some results which are especially needed for this paper.

Let k be a field and A any finite-dimensional commutative k-algebra. We set $A^* = \operatorname{Hom}_k(A, k)$ for the dual space of the k-vectorspace A. A^* is an A-module via the composition

$$A \times A^* \to A^*, \ (a \cdot \lambda) : b \mapsto \lambda(ab).$$

With this A-module structure, A^* is called the *canonical module* of A. Now pick any linear form $\lambda \in A^*$. It gives rise to a symmetric k-bilinear form Φ_λ on A:

$$\Phi_\lambda(a, b) := \lambda(ab)$$

λ is called a *dualizing linear form* if Φ_λ is non degenerate, i.e., $\Phi_\lambda(a, A) = 0$ implies $a = 0$.

Proposition 2.1 *The following statements are equivalent:*

(i) λ *is dualizing,*

(ii) λ *generates the* A*-module* A^**,*

(iii) $A \to A^*$, $a \mapsto a \cdot \lambda$ *is an* A*-module isomorphism.*

Proof. The non degeneracy of Φ_λ is equivalent to the statement that the mapping $A \to A^*$, $a \mapsto (x \mapsto \Phi_\lambda(a, x))$ is an isomorphism. In this case $\Phi_\lambda(a, x) = (a \cdot \lambda)(x)$. From this and the fact $\dim_k A = \dim_k A^* < \infty$, the claim follows. □

Definition 2.2 *A is called a Frobenius algebra if it admits a dualizing linear form* λ.

To give reference to the linear form λ we also call (A, λ) a Frobenius algebra (or Frobenius, for short) if λ is a dualizing form. By 2.1, A is a Frobenius algebra if and only if A^* is a free A-module of rank 1, and the dualizing linear forms are just the (free) generators of A^*. Two bases $\{a_i\}$ and $\{b_i\}$ of the vector space A are *dual with respect to* λ if

$$\lambda(a_i b_j) = \delta_{ij}, \ i, j = 1, \dots, n.$$

From the non degeneracy of Φ_λ, one deduces

$$\forall a \in A, a = \sum \lambda(a_i) b_i.$$

Let (A, λ) be Frobenius and $\mu \in A^*$ be arbitrary.

Then there exists a unique element $J_\lambda(\mu) \in A$ with

$$\mu = J_\lambda(\mu) \cdot \lambda.$$

To describe $J_\lambda(\mu)$, let $\{a_i | i = 1, \ldots, n\}$ and $\{b_i | i = 1, \ldots, n\}$ be a pair of bases dual with respect to λ. With this choice of bases we get

2.3 $J_\lambda(\mu) = \sum_{i=1}^{n} \mu(a_i) b_i.$

In particular, this applies to the trace function $tr_{A|k} : A \to k$. In this case we denote $J_\lambda(tr_{A|k})$ by J_λ and get

2.4 $tr_{A|k} = J_\lambda \cdot \lambda$ *where* $J_\lambda = \sum tr(a_i) b_i = \sum_{i=1}^{n} a_i b_i.$

This follows from (2.3) and the identity

$$aa_i = \sum_{k=1}^{n} \lambda(aa_i b_k) a_k,$$

which in turn yields

$$tr_{A|k}(a) = \sum_{i=1}^{n} \lambda(aa_i b_i) = \lambda(a \times \sum_{i=1}^{n} a_i b_i).$$

$$J_\lambda = \sum_{i=1}^{n} a_i b_i.$$

From the point of view of ramification theory, the element J_λ should be called a Dedekind different, cf. [13], p. 229; [11], Appendix G; [15]. In our present context of algebras over fields, it is also appropriate to call J_λ the *generalized Jacobian relative to* λ. This is motivated by the special case of complete intersection algebras as presented in the next section.

Let (A, λ) be a Frobenius algebra, $\mathfrak{a}$ any ideal of A. $\pi : A \to A/\mathfrak{a}$ the canonical epimorphism and $\mu \in (A/\mathfrak{a})^*$. Then $\mu \circ \pi \in A^*$ and let $J_{\lambda,\mu}$ be defined by

$$\mu \circ \pi = J_{\lambda,\mu} \cdot \lambda.$$

Proposition 2.5 *If* (A, λ) *and* $(A/\mathfrak{a}, \mu)$ *are Frobenius algebras then*

$$Ann(\mathfrak{a}) := \{x \in A | x\mathfrak{a} = 0\}$$

is a free $A/\mathfrak{a}$*-module with basis* $J_{\lambda,\mu}$.

Proof. $Ann(\mathfrak{a})$ is clearly an $A/_{\mathfrak{a}}$-module.

Choose $a \in \mathfrak{a}$, then $a \cdot (\mu \circ \pi) = 0$. Hence, $(a \cdot J_{\lambda\mu}) \cdot \lambda = 0$ resulting in $aJ_{\lambda\mu} = 0$, and $J_{\lambda\mu} \in Ann(\mathfrak{a})$. Conversely, assume $x \cdot \mathfrak{a} = 0$. Then, by defining $\Phi_\mu(\overline{y}) := \lambda(xy)$ for any $\overline{y} = \pi(y)$, $y \in A$, we have got $\Phi_\mu \in A'$. By assumption, $\Phi_\mu = \pi(z) \cdot \mu$ for some $z \in A$. Then $x \cdot \lambda = \Phi_\mu \circ \pi = z \cdot (\mu \circ \pi) = zJ_{\lambda\mu} \cdot \lambda$, i.e., $x = zJ_{\lambda\mu}$. So far, $Ann(\mathfrak{a}) = A \cdot J_{\lambda\mu}$ has been proved. It remains to show that $xJ_{\lambda\mu} = 0$ implies $x \in \mathfrak{a}$. In fact, $xJ_{\lambda\mu} = 0$ leads to $\pi(x) \cdot \mu = 0$, and hence $\pi(x) = 0$, i.e., $x \in \mathfrak{a}$. □

The last proposition is essentially a result of Nakayama, cf. [8], p. 96, Th. (6.2).

We will apply the result above in two situations which, in each case, then yields a characterization of Frobenius algebras. We start with following fact:

2.6 *If* (A, λ) *and* (B, μ) *are Frobenius algebras then* $(A \otimes_k B, \lambda \otimes_k \mu : A \otimes B \to k \otimes_k k = k)$ *is a Frobenius algebra as well.*

Indeed, the linear form $\lambda \otimes \mu$: $a \otimes b \mapsto \lambda(a)\mu(b)$ induces the bilinear form $\Phi_\lambda \otimes \Phi_\mu = \Phi_{\lambda \otimes \mu}$ which is non degenerate under this assumption.

Given any algebra A we will consider the multiplication map

$$\delta : A \otimes_k A \mapsto A, \ a \otimes b \mapsto ab.$$

Furthermore, $A \otimes A$ will be understood as an A-module via $a \cdot (x \otimes y) := (a \otimes 1)(x \otimes y) = ax \otimes y$. Set $I := ker(\delta)$. One readily checks that I, as an A-module, is generated by the elements $a \otimes 1 - 1 \otimes a$, $a \in A$, and that $A \otimes A$ decomposes directly as an A-module

$$A \otimes A = I \oplus A \cdot (1 \otimes 1).$$

We now apply 2.5 to $(A \otimes A, \lambda \otimes \lambda)$ and $(A \otimes A)/_I = A$ and $\lambda \in A^*$, where we *assume* (A, λ) *to be a Frobenius algebra.* We obtain $B_\lambda \in A \otimes A$ such that the following holds:

2.7 $\lambda \circ \delta = B_\lambda \cdot \lambda$, *$Ann(I)$ is a free A-module with basis B_λ.*

Note that the $(A \otimes A)/_I$-module structure is exactly the A-module structure we have just defined.

Definition 2.8 *Let (A, λ) be a Frobenius algebra; then $B_\lambda \in A \otimes A$ is called the* generalized Bezoutian *of λ.*

This terminology comes from the fact that these objects generalize the classical Bezoutian

$$\frac{(P(X) - P(Y))}{(X - Y)}$$

where

$$P = X^d + \sum_{i=1,\ldots,d} a_i X^{d-i}$$

is a univariate polynomial of degree d. The classical Bezoutian is obtained from the linear form ℓ sending the basis

$$\{1, X, \ldots, X^{d-2}, X^{d-1}\}$$

of $A = K[X]/(P)$ to $0, \ldots, 0, 1$, since it is easy to see that the dual basis of $\{1, X, \ldots, X^{d-2}, X^{d-1}\}$ for ℓ is the Horner basis

$$\{X^{d-1} + \sum_{i=1,\ldots,d-1} a_i X^{d-i-1}, X^{d-2} + \sum_{i=1,\ldots,d-2} a_i X^{d-i-2}, \ldots, 1\}.$$

To derive a presentation of B_λ in the general case, we start with a pair of dual bases $\{a_i\}$, $\{b_i\}$ for λ, as before 2.3. Then $\{b_i \otimes a_j | i, j\}$ and $\{a_i \otimes b_j | i, j\}$ are dual bases for $\lambda \otimes \lambda$. According to 2.3, 2.4, and noting that the roles of $\{a_i\}$ and $\{b_i\}$ may be interchanged, we get

2.9 $B_\lambda = \sum_{i=1}^{n} a_i \otimes b_i = \sum_{i=1}^{n} b_i \otimes a_i$, $\delta(B_\lambda) = J_\lambda$.

We now derive one of the main results in this section the statement and proof of which are inspired by the work of Scheja/Storch cf. [15], [17].

Theorem 2.10

(i) A finite-dimensional k-algebra is a Frobenius algebra if and only if $Ann(I)$ is a free A-module of rank 1.

(ii) The assignment $\lambda \mapsto B_\lambda$ is a bijection between the set of dualizing forms on A and the set of free generators of $Ann(I)$ as an A-module.

(iii) Let $\{a_1, \ldots, a_n\}$ be any k-basis of A and present $B \in A \otimes_k A$ in the form $B = \sum_{i=1}^{n} a_i \otimes b_i$ with (uniquely determined) $b_1, \ldots, b_n \in A$. Then, B is a free generator of $Ann(I)$ if and only if

1) $B \in Ann(I)$, 2) $\{b_1, \ldots, b_n\}$ is a basis of A.

Proof. To prove the claims above we make use of the k-linear isomorphism

$$\rho : \begin{cases} A \otimes_k A & \to & Hom_k(A^*, A) \\ x & \mapsto & (\lambda \mapsto (\lambda \otimes id)(x)) \end{cases}$$

where we identify $k \otimes A$ and A in the natural way. Recall that we regard $A \otimes A$ and A^* as A-modules. One readily verifies

$$\rho(a \cdot x)(\varphi) = \rho(x)(a\varphi), \ \rho((1 \otimes a)x)(\varphi) = a \cdot [\rho(x)(\varphi)].$$

Using that ρ is bijective we derive the equivalence

$$x \in Ann(I) \Leftrightarrow \rho(x) \quad A\text{-linear}.$$

Now consider any $x \in Ann(I)$. Then, as $\rho(x)$ is A-linear, $\rho(x)(A^*)$ is an ideal of A. A is a finite-dimensional k-algebra, hence a zero-dimensional noetherian ring. In any such ring R, the annihilitor $Ann(\mathfrak{a})$ of any ideal $\mathfrak{a} \neq R$ is non-zero (e.g. use the fact $(0) = \prod_1^l \mathcal{M}_i^{r_i}$, $\mathcal{M}_1, \ldots, \mathcal{M}_l$ the maximal ideals of R). In our situation we obtain:

2.11 *$\rho(x)$ not an A-isomorphism $\Leftrightarrow$ $\rho(x)$ not surjective $\Leftrightarrow$ $\exists a \in A\backslash\{0\}$ s.t. $a \cdot \rho(x)(A^*) = \rho(a \cdot x)(A^*) = 0$, i.e., $\rho(a \cdot x) = 0$, or equivalently, $a \cdot x = 0$.*

If A is Frobenius then 2.7 states that $Ann(I)$ is a free A-module of rank 1. To study the converse let $B = \sum_1^n a_i \otimes b_i$; $\{a_i | i = 1, \ldots, n\}$ a k-basis of A, be any element of $Ann(I)$. Pick $\mu \in A^*$. Then $\rho(B)(\mu) = \sum_1^n \mu(a_i)b_i$. Denoting by $\{a_1^*, \ldots, a_n^*\}$ the dual basis of A^* relative to $\{a_1, \ldots, a_n\}$ we derive $\rho(B)(a_i^*) = b_i$. Hence, the ideal $\rho(B)(A^*)$ of A is generated by $\{b_1, \ldots, b_n\}$ as a k-vectorspace.

If B is a free generator of $Ann(I)$, it is clearly a non-torsion element. Thus, by 2.11, $\rho(B)$ is surjective, i.e., $\{b_1, \ldots, b_n\}$ a k-basis of A. To prove the converse start with the assumptions 1), 2) of iii). By 2.11, $\rho(B) : A^* \to A$ is an A-module isomorphism, and A is Frobenius by 2.1 with dualizing form $\lambda_B := \rho(B)^{-1}(1)$. It remains to show that the assignments $\lambda \mapsto B_\lambda$, $B \mapsto \lambda_B$ are inverse to each other. From 2.9 one readily checks the injectivity of $\lambda \mapsto B_\lambda$. To prove surjectivity, set $\lambda = \lambda_B$ and note $\rho(B)(a \cdot \lambda) = a$, since $\rho(B)$ is A-linear. This means $a = \sum_1^n \lambda(aa_i)b_i$ for every $a \in A$. Pluging in $a = b_j$ we derive $\lambda(a_i b_j) = \delta_{ij}$, using that $\{b_1, \ldots, b_n\}$ form a k-basis. Hence $B = B_\lambda$. □

Remarks 2.12

1) *In the proof we have seen that if B and λ are associated to each other then*

   i) *If $\lambda(a_i b_j) = \delta_{ij}$, then $B = B_\lambda = \sum a_i \otimes b_i$. ,*

   ii) *If $B = \sum_i a_i \otimes b_i$, then $\lambda(a_i b_j) = \delta_{ij}$, and $\{b_i\}$ is the dual basis of $\{a_i\}$.*

   iii) *For any $c \in A$, $c = (\lambda \otimes id)((1 \otimes c)B) = \sum \lambda(ca_i)b_i$; in particular, $1 = (\lambda \otimes id)(B) = \sum \lambda(a_i)b_i$.*

Since $\rho(B)$ is A-linear we get in addition

iv) $\mu = J_\lambda(\mu) \cdot \lambda,\ J_\lambda(\mu) = (\mu \otimes id)(B) = \sum \mu(a_i) b_i$

in accordance with 2.3.

2) $A \otimes A$ admits the involution $\tau : A \otimes A \to A \otimes A$. $a \otimes b \mapsto b \otimes a$. In 2.9 we have seen $\tau(B_\lambda) = B_\lambda$. Hence, by 2.10, every generator of $Ann(I)$ is symmetric as well.

From all this we can deduce the following. Let $a = \{a_i\}$ and $b = \{b_i\}$ be dual basis and let $\mu \in A^*$. Then the symmmetric matrix of the quadratic form Φ_μ in the basis b_i coincide with the coefficients of $J_\lambda(\mu) \cdot B_\lambda = (J_\lambda(\mu) \otimes 1) B_\lambda$ in the basis $a_i \otimes a_j$.

Let us prove it. The (i,j)-th entry of the matrix of Φ_μ in the basis b is by definition $\mu(b_i b_j) = \lambda(J_\lambda(\mu) b_i b_j)$. On the other hand we have

$$J_\lambda(\mu) b_i = \sum_j \lambda(J_\lambda(\mu) b_i b_j) a_j$$

$$(J_\lambda(\mu) b_i) \otimes a_i = \sum_j (\lambda(J_\lambda(\mu) b_i b_j) a_j) \otimes a_i$$

$$(J_\lambda(\mu) \otimes 1) B_\lambda = \sum_{i,j} \lambda(J_\lambda(\mu) b_i b_j) a_i \otimes a_j$$

In the sequel we are going to apply 2.5 to A and appropriate factor algebras $A/\mathfrak{a}$. Every maximal ideal $\mathcal{M}$ of A gives rise to the field extension $K = A/_{\mathcal{M}}$ over k. Clearly, being a field, K is a Frobenius algebra over k (with every $\lambda \neq 0$ as a dualizing form). Moreover, since a finite direct product of Frobenius algebras is again Frobenius, as can be deduced directly from the definition. the algebra

$$A_{red} := A/_{Nil(A)} = \prod_{\mathcal{M}} A/_{\mathcal{M}},\ \mathcal{M} \text{ ranging over the maximal ideals}$$

with $Nil(A) = \{\text{nilpotent elements of } A\}$ is Frobenius as well.

Any finite-dimensional k-algebra A is a zero-dimensional Noetherian ring. Therefore, the canonical maps $\pi_{\mathcal{M}} : A \to A_{\mathcal{M}}$, $\mathcal{M}$ any maximal ideal, induces a k-isomorphism

$$\pi : A \to \prod_{\mathcal{M}} A_{\mathcal{M}},\ \mathcal{M} \text{ ranging over the finitely many maximal ideals.}$$

This gives rise to an internal decomposition

2.13 $A = \oplus A e_{\mathcal{M}}$, *$e_{\mathcal{M}}$ idempotent satisfying $\pi_{\mathcal{M}}(A e_{\mathcal{M}'}) = 0$ if $\mathcal{M} \neq \mathcal{M}'$*, $\pi_{\mathcal{M}} : A e_{\mathcal{M}} \xrightarrow{\sim} A_{\mathcal{M}}$, $\pi_{\mathcal{M}}(e_{\mathcal{M}}) = 1$.

Proposition 2.14 *Let A be a finite-dimensional k-algebra. Then the following statements are equivalent.*

(i) A is Frobenius.

(ii) $Ann(\mathcal{M})$ is a $A/_{\mathcal{M}}$-vectorspace of dimension 1 for every maximal ideal $\mathcal{M}$.

(iii) $A_{\mathcal{M}}$ is Frobenius for every maximal ideal $\mathcal{M}$,

(iv) $Ann(Nil(A))$ is a free A_{red}-module of rank 1.

Proof. i) $\Rightarrow$ ii) follows from 2.5 and the considerations above. To prove ii) $\Rightarrow$ iii), set $Ann(\mathcal{M}) = J_{\mathcal{M}} \cdot A$. Since $\mathcal{M} = \bigoplus_{\mathcal{M}' \neq \mathcal{M}} Ae_{\mathcal{M}'} \oplus \mathcal{M}e$, we derive $J_{\mathcal{M}} = \sum_{\mathcal{M}'} J_{\mathcal{M}} \cdot e_{\mathcal{M}'} = J_{\mathcal{M}} \cdot e_{\mathcal{M}}$ and, consequently, that $\pi_{\mathcal{M}}(J_{\mathcal{M}})$ is a free generator of $Ann(\mathcal{M}A_{\mathcal{M}})$ as an $A_{\mathcal{M}}/_{\mathcal{M}A_{\mathcal{M}}} = A/_{\mathcal{M}}$-module. Choose any linear form $\lambda_{\mathcal{M}} : A_{\mathcal{M}} \to k$ which satisfies $\lambda_{\mathcal{M}}(\pi_{\mathcal{M}}(J_{\mathcal{M}})) = 1$. We now work inside $A_{\mathcal{M}}$ and set $\lambda = \lambda_{\mathcal{M}}$, $J := \pi_{\mathcal{M}}(J_{\mathcal{M}})$. $A_{\mathcal{M}}$ is a finite-dimensional local k-algebra. Therefore there is $e \in \mathbf{N}$ such that $\mathcal{M}^{e-1} \neq 0$, $\mathcal{M}^e = 0$. If $x \in A_{\mathcal{M}}$ is given, $x \neq 0$, we find $i \in \mathbf{N}$ satisfying $x\mathcal{M}^{i-1} \neq 0$, $x\mathcal{M}^i = 0$. Hence $x\mathcal{M}^{i-1} \subseteq A_{\mathcal{M}} \cdot J$, in particular: $xy = a \cdot J \neq 0$ for some $y \in \mathcal{M}^{i-1}$, $a \in A$. Clearly, $a \notin \mathcal{M}$, i.e., a is a unit, and we get $x(ya^{-1}) = c$, $\lambda(x \cdot ya^{-1}) = 1$. This proves Φ_λ to be non degenerate. That iii) $\Rightarrow$ i) holds is clear in view of $A \xrightarrow{\sim} \prod A_{\mathcal{M}}$. To prove ii) $\Rightarrow$ iv) one directly checks that $C := \sum J_{\mathcal{M}}$ is a free generator of $Ann(Nil(A))$ over A_{red}, by noticing $Nil(A) = \cap \mathcal{M} = \bigoplus_{\mathcal{M}} \mathcal{M} \cdot e_{\mathcal{M}}$. The remaining implication iv) $\Rightarrow$ i) can be deduced from the following fact: $x = \sum x_{\mathcal{M}}$, $x_{\mathcal{M}} \in Ae_{\mathcal{M}}$, lies in $Ann(Nil(A))$ if and only if $x_{\mathcal{M}} \in Ann(\mathcal{M})$ for every maximal ideal $\mathcal{M}$. □

Remarks 2.15 *As a special case of 2.14, we have proved that a* local *k-algebra A with maximal ideal $\mathcal{M}$ is Frobenius if and only if $Ann(\mathcal{M})$, often referred to as the socle of A, is a one-dimensional $A/_{\mathcal{M}}$-vectorspace. This property characterizes also the local Gorenstein algebras of Krull dimension zero. Hence, Frobenius algebras are exactly the finite-dimensional global Gorenstein algebras, see also [11], Appendix E, E.16 and [16], §4.*

The following considerations are partially needed in §3. In addition, we want to point out the importance of Frobenius algebras for the computational study of zero-dimensional polynomial systems. To this end we write

$$A = k[X_1, \ldots, X_n]/\mathfrak{a}, \quad \mathfrak{a} \lhd k[X_1, \ldots, X_n]$$

and set $V := V(\mathfrak{a}) = \{p = (x_1, \ldots, x_n) \in \overline{k}^n | p \text{ zero of } \mathfrak{a}\}$, where $\overline{k}$ denotes the algebraic closure of k. As said at the beginning, every linear form $\mu \in A^*$ gives rise to a bilinear form Φ_μ on A. It is the topic of this paper, as well as many others, to get information on V by studying the bilinear form Φ_λ, e.g., its rank, signature etc., for appropriate choices of λ.

If $\lambda = tr_{A|k}$ then Φ_λ can be used to count the number of all zeros or of all real zeros in V, cf. [14], [4]. This method, often referred to as the Hermite method, dates back to the last century, at least as far as the principles are concerned.

In the case of a Frobenius algebra A one may choose other linear forms λ with a meaningful Φ_λ. In particular, the Eisenbud-Levine formula in §4 is concerned with Φ_λ for a suitable choice of a dualizing form λ.

In order to expose the structure of the algebra A and to study the bilinear form Φ_λ we have to pass to appropriate base field extensions $L \otimes_k A$. Note that these extended algebras $L \otimes_k A$ are still Frobenius algebras with the extension λ_L of λ as a dualizing form. By 2.9, the generalized Bezoutian of λ is also the generalized Bezoutian of λ_L. The algebraic closure $\bar{k}$ is a first choice for L. However, for the sake of rationality questions, we introduce the following more general hypothesis which is trivially satisfied if char $k = 0$ or if we have passed to a base field extension A_L provided L contains the coordinates of every zero of $\mathfrak{a}$ (e.g. $L = \bar{k}$).

2.16 *For every maximal $\mathcal{M}$, the field $A/_{\mathcal{M}}$ is separable extension of k; equivalently, A_{red} is a separable algebra over k.*

Let $\bar{\pi} : A \to A_{red}$ denote the canonical epimorphism and consider the trace $tr_{A_{red}} : A_{red} \to k$. By abuse of notation, we call

$$tr_{red} := tr_{A_{red}} \circ \bar{\pi} : A \to k$$

the *reduced trace* on A.

By scalar extension to the algebraic closure and setting $\overline{A} = \bar{k} \otimes_k A$ we arrive at the following situation:

2.17 *For $p \in V$ set $\bar{\pi}_p : \overline{A} \to \bar{k}$, $f \mapsto f(p)$ and $\mathcal{M}_p = ker\ \pi_p$. Then $\{\mathcal{M}_p | p \in V\}$ is the set of maximal ideals of $\overline{A}$. The (external) decomposition*

$$\overline{A} \xrightarrow{\sim} \prod_p A_p, \ \textit{where}\ \overline{A}_{\mathcal{M}_p} =: A_p,$$

is reflected by in internal decomposition $\overline{A} = \oplus \overline{A} e_p$, $e_p^2 = e_p$, etc., as in 2.13. With the obvious change in notation, $\pi_p : A \to A_p$ induces an isomorphism $Ae_p \xrightarrow{\sim} A_p$ whereas $\pi_p(Ae_q) = 0$ if $q \neq p$.

We define $\mu_p := \dim_k A_p$ to be the *multiplicity* of p. Then

$$\textbf{2.18} \quad \begin{aligned} tr(f) &= \sum \mu_p f(p) \\ tr_{red}(f) &= \sum f(p). \end{aligned}$$

As shown in [1], (3.1) there is an element $N \in A$, the *dimension element*, such that $N(p) = \mu_p$ for every $p \in V$. Clearly, N is a unit of A and satisfies

2.19 $tr_{A|k} = N \cdot tr_{red}.$

We next assume (A, λ) to be Frobenius. The algebra A_{red} is separable, which implies that its trace function is dualizing. Applying 2.5 to tr_{red} and λ yields in view of 2.12, 1) ii)

2.20

$$\begin{cases} tr_{red} = \overline{J}_\lambda \cdot \lambda,\ \overline{J}_\lambda = (tr_{red} \otimes id)(B_\lambda) \\ Ann(Nil(A)) = A \cdot \overline{J}_\lambda \end{cases}$$

As a consequence we obtain

2.21 $J_\lambda = N \cdot \overline{J}_\lambda.$

To facilitate the notation we set $B = B_\lambda$, $\overline{J} = \overline{J}_\lambda$ in the following. The decomposition $\overline{A} = \oplus \overline{A} e_p$, $x = \sum x_p$, $x_p \in \overline{A} e_p$ gives rise to

$$\overline{A} \otimes \overline{A} = \oplus_{p,q} \overline{A} e_{p,q} \text{ where } e_{p,q} = e_p \otimes e_q.$$

Then $Be_{p,q} = B(e_p e_q \otimes 1)$ since $B \cdot I = 0$. Hence $Be_{p,q} = 0$ if $p \neq q$. Thus, setting $B_p := B(e_p \otimes e_p) = e_p \cdot B$, we have $B = \sum_p B_p$, $B_p \in (\overline{A} \otimes \overline{A}) e_p \otimes e_p$. Then $J = \delta(B) = \sum_p \delta(B_p)$ from which $J_p = \delta(B_p)$ follows because of $\delta(B_p) \subseteq Ae_p$. Similarily, $\overline{J} = \sum (\pi_p \otimes id)(B)$ implies $\overline{J}_p = (\pi_p \otimes id)(B) = (\pi_p \otimes id)(B_p)$. By 2.5 and 2.12, 1) $\overline{J}_p$ generates $Ann(\mathcal{M}_p)$. This means $(f - f(p))\overline{J}_p = 0$, i.e., $f \cdot \overline{J}_p = f(p)\overline{J}_p$ for every $f \in A$. Using $N(p) = \mu_p$ we get $J_p = (N\overline{J})e_p = N\overline{J}_p = \mu_p \overline{J}_p$. Finally, $\lambda(\overline{J}_p) = \lambda(\overline{J} e_p) = tr_{red}(e_p) = 1$.

We summarize the above properties:

2.22

$$\begin{aligned} &\delta(B_p) = J_p = \mu_p \overline{J}_p, \\ &\overline{J}_p = (\pi_p \otimes id)(B) = (\pi_p \otimes id)(B) \text{ generates } Ann(\mathcal{M}_p),\ \overline{J}_\lambda = \sum_p \overline{J}_p, \\ &f\overline{J}_p = f(p)\overline{J}_p,\ \lambda(\overline{J}_p) = 1. \end{aligned}$$

These facts are interesting for numerical methods solving polyomial systems (see for example [3] or [6]), which are looking for common eigenvectors for all multiplications in $\overline{A}$, $A = k[X]/\mathfrak{a}$ any finite-dimensional k-algebra. It is known that A is Frobenius if and only if $\overline{A}$ is Frobenius, cf., e.g., [8], 1.11. These eigenspaces are in one-to-one correspondence with the zeros of $\mathfrak{a}$ in $\overline{k}$; they are exactly $Ann(\mathcal{M}_p)$, p any zero. By 2.14, $\overline{A}$ is Frobenius if and only if all these eigenspaces are one-dimensional.

Moreover the eigenspace associated to p is generated by $\overline{J}_p$. If additionally every p is regular, i.e., $A_p = \overline{k}$, then $\overline{A}$ admits the basis $\{\overline{J}_p\}$.

The fact that, in the Frobenius case, all these eigenspaces are one-dimensional is the explanation for the good behaviour of numerical methods since fixed-point techniques converging to the eigenspace can be applied. We shall see in next section that every complete zero-dimensional intersection is Frobenius, so that the numerical methods based on finding the eigenspaces can be applied to *any complete intersection.*

Let us remark that for each $p \in V$ the local generalized Jacobian $\pi_p(J) = \pi_p(J_p)$ is not zero in A_p provided $char k = 0$ or $char\ k = p,\ p \not| \mu_p$.

Interpreting the bilinear forms Φ_μ under the aspect of getting geometric information on V seems difficult in general.

In the univariate case, this task is easy since we recover the theory of Cauchy index [6]. The Eisenbud-Levine formula we give in section 4 is just one example of such a geometric interpretation. Much simpler is this task if μ is related to the trace or reduced trace.

First note that Ae_p and Ae_q are orthogonal relative to μ if $p \neq q$. On Ae_p we choose a basis $\{a_1e_p, a_2e_p, \ldots, a_re_p\}$, $r = \mu_p$, $a_1 = 1, a_2, \ldots, a_n \in \mathcal{M}_p$. Further assume $\mu = D\overline{J} \cdot \lambda$ for some $D \in A$. Then $\Phi_\mu(a_ie_p, a_je_p) = \lambda(D \cdot \overline{J}a_ie_pa_je_p) = 0$ if not $i = j = 1$. Therefore $\Phi_\mu \simeq \perp_p < \lambda(D\overline{J}e_p), 0, \ldots, 0 >$ in diagonal presentation. We have $\lambda(D\overline{J}e_p) = tr_{red}(De_p) = D(p)$. Hence,

2.23 *rank* $\Phi_\mu = \{p \in V | D(p) \neq 0\}$ *if* $\mu = D \cdot tr_{red}$.

In case k is an ordered field with real closure R then, as in [14], [4]:

2.24

$$\begin{aligned} signature\ \Phi_\mu &= \#\{p \in V \cap R^n | D(p) > 0\} \\ &\quad -\#\{p \in V \cap R^n | D(p) < 0\},\ where\ \mu = D \cdot tr_{red} \end{aligned}$$

If we set $D = J$ and $D = J^2$, one can count in this way the number of regular solutions in $\overline{k}$ and of regular solutions in R, respectively. Note that $J(p) \neq 0$ if and only if p is regular.

3 The complete intersection case and the multivariate Bezoutian

By definition, a finite-dimensional k-algebra A is called a *complete intersection* if it has a presentation

$$A = k[X_1, \ldots, X_n]/_{(P_1, \ldots, P_n)}.$$

In this section we are going to prove the following

Theorem 3.1 *A finite-dimensional complete intersection k-algebra*

$$A = k[X_1, \ldots, X_n]/_{(P_1, \ldots, P_n)}$$

is a Frobenius algebra.

To be more precise, we will derive directly from the generators $P_1, \ldots, P_n$ a dualizing form ℓ by writing down a generator B of $Ann(I)$ over A. How to get a dualizing form out of a presentation of A was first shown by Tate, cf. [13], Appendix and [15] in the general context of complete intersection algebras over rings. In this paper we restrict ourselves to algebras over fields and use as a main new idea, a deformation trick which is inspired by Arnold's treatment of the Eisenbud-Levine formula [2], I, §5. By combining some of the algebraic arguments of the last section with this deformation idea we arrive at a simpler and direct proof.

Let $\bar{\pi} : k[X] \to A$ denote the canonical epimorphism. Set $x_i = \bar{\pi}(X_i)$, $x = (x_1, \ldots, x_n)$, hence $A = \{f(x) | f \in k[X]\}$. The assignments $a \mapsto a \otimes 1$ and $a \mapsto 1 \otimes a$ are injective maps $A \to A \otimes A$. We therefore also write x_i for $x_i \otimes 1$; in addition we set $y_i = 1 \otimes x_i$. In this way, $x_1, \ldots, x_n$ together with $y_1, \ldots, y_n$ form a system of generators of $A \otimes A$ as an k-algebra: $A \otimes A = \{f(x, y) | f \in k[X, Y]\}$.

The natural k-isomorphism $k[X] \otimes k[X] \xrightarrow{\sim} k[X, Y]$, $X_i \otimes 1 \mapsto X_i$, $1 \otimes X_i \mapsto Y_i$, induces a k-epimorphism

$$\tilde{\pi} : k[X, Y] \to A \otimes A, X_i \mapsto x_i, Y_i \mapsto y_i$$

with kernel

$$ker\ \tilde{\pi} = (\alpha(X), \alpha(Y)) = (P_1(X), \ldots, P_n(X), P_1(Y), \ldots, P_n(Y)).$$

We now introduce, relative to $P_1, \ldots, P_n$,

$$B(X, Y) := det(P_{ij}) \text{ with}$$

$$P_{ij} = \frac{P_i(Y_1, \ldots, Y_{j-1}, X_j, X_{j+1}, \ldots, X_n) - P_i(Y_1, \ldots, Y_{j-1}, Y_j, X_{j+1}, \ldots, X_n)}{X_j - Y_j}$$

and then call

$$Bez(x, y) := \tilde{\pi}(B(X, Y)) \in A \otimes A$$

the *Bezoutian* of $P_1, \ldots, P_n$. The *Jacobian* of $P_1, \ldots, P_n$ coincides with $Bez(x, x) = \delta(Bez(x, y))$, where $\delta(a \otimes b) = ab$.

In the univariate case, i.e., $n = 1$, we obtain $B(X, Y) = \dfrac{P(X) - P(Y)}{X - Y}$.

Theorem 3.2 *The complete intersection algebra* $A = k[X_1, \ldots, X_n]/_{(P_1, \ldots, P_n)}$ *admits a dualizing form* ℓ *with the Bezoutian* $Bez(x, y)$ *as its generalized Bezoutian* B_ℓ.

Theorem 3.1 is an immediate corollary of theorem 3.2.

This statement can also be deduced from Tate's construction, loc. cit., or from the Scheja/Storch approach. Our dualizing form is that of Tate or Scheja/Storch. In the Scheja/Storch case, this follows from the approach in §2. Concerning Tate's construction via the Koszul complex, one compares the properties (A.4)–(A.6) in [13], Appendix with the statements in 2.12 and takes 3.4 below into account. It coincides also with the global Residue or Kronecker symbol (see [2]).

The proof of this theorem covers the main part of this section. It proceeds in several steps.

To prove that $Bez(x,y) \in Ann(I)$. Because of our identification, this amounts to showing

3.3 $f(x)Bez(x,y) = f(y)Bez(x,y)$ *for every* $f(X) \in k[X]$, $X = (X_1,\ldots,X_n)$.

It is enough to prove 3.3 for the cases $f = X_j$, $j = 1,\ldots,n$. Let B_j denote the j-th column of (P_{ij}). Then

$$\begin{aligned}(X_j - Y_j)B(X,Y) &= det(B_1,\ldots,B_{j-1},(X_j-Y_j)B_j,B_{j+1},\ldots,B_n)\\ &= det(B_1,\ldots,B_{j-1},\sum_k(X_k-Y_k)B_k,B_{j+1},\ldots,B_n).\end{aligned}$$

Now,

$$\textbf{3.4}\quad \sum_k (X_k - Y_k)B_k = \begin{pmatrix} P_1(X) & - & P_1(Y)\\ \vdots & & \vdots \\ P_n(X) & - & P_n(Y)\end{pmatrix}.$$

Developing the last determinant relative to the j-th column we see that $(X_j - Y_j)B(X,Y) \in (\alpha(X),\alpha(Y))$. Hence $x_jBez(x,y) = y_jBez(x,y)$.

After this first step it remains to check the second condition of theorem 2.10, iii). To achieve this we pass to the algebraic closure $\overline{k}$ and the scalar extension $\overline{A} = \overline{k}\otimes A$. If $Bez(x,y)$ satisfies the condition 2) in 2.10, iii), over $\overline{k}$, it does the same over k as well.

Therefore we assume in the next step: k *is algebraically closed.*

Under this assumption we have the decomposition $A = \bigoplus_{p\in V} Ae_p$, where this time

$$V = \{p \in k^n | P_1(p) = \cdots = P_n(p) = 0\}.$$

The canonical map $\pi_p : A \to A_p$ induces an isomorphism $\pi_p : Ae_p \to A_p$, whereas $\pi_p(Ae_q) = 0$, if $q \neq p$. As shown in section 2 before 2.22, any $B \in Ann(I)$ must lie in $\sum(A\otimes A)(e_p\otimes e_p)$. We set $B := Bez(x,y)$. Then $B = \sum B_p$, $B_p = B(e_p\otimes e_p) = e_p\cdot B$. Assume that we can prove that $(\pi_p\otimes\pi_p)(B_p) \in A_p\otimes A_p$ is a generalized Bezoutian of a dualizing form $\lambda_p : A_p \to k$, i.e., $(\lambda_p \otimes id)(\pi_p \otimes \pi_p)(B_p) = 1$ in A_p. Noting $B_p \in Ae_p \otimes Ae_p$ and understanding $\lambda_p\pi_p$ as a linear form on Ae_p, we derive $(\lambda_p j_p\otimes id)(B_p) = e_p$. From $B_q = B\cdot(e_q\otimes e_q) = (e_q\otimes 1)B$ one deduces $(\lambda_p j_p)(e_q \otimes 1)B = 0$ if $q \neq p$. Thus $(\lambda_p j_p \otimes id)(B) = e_p$ and

$$1 = [(\sum_p \lambda_p j_p) \otimes id](B).$$

Setting $\lambda = \sum_p \lambda_p j_p$, i.e., $\lambda(\sum x_p) = \sum(\lambda_p j_p)(x_p)$, we have obtain that λ is the dualizing form associated to B.

As a consequence, *we are reduced to the local study* of $(\pi_p \otimes \pi_p)(B) = (\pi_p \otimes \pi_p)(B_p)$ in $A_p \otimes A_p$, $p \in V$. The epimorphism $k[X] \to A$ attachs to $\mathcal{M}_p$ a maximal ideal $\overline{\mathcal{M}}_p$ of $k[X]$ subject to $\overline{\mathcal{M}}_p/\mathfrak{a} = \mathcal{M}_p$. We next obtain the commutative diagram

$$\begin{array}{ccc} k[X] & \xrightarrow{i_p} & k[X]_{\mathcal{M}_p} =: \mathcal{O}_p \\ \downarrow \pi & /// & \downarrow \overline{\pi} \\ A & \xrightarrow{j_p} & A_p \end{array}$$

where i_p is the canonical map. We have

$$ker\ \overline{\pi} = \mathfrak{a}\mathcal{O}_p.$$

To study the case of $p \in V$ we may normalize the situation, set $p = 0$, $\pi = \pi_0$, $i = i_0$ and have to deal with $(i \otimes i)\ B(X,Y)$ and $(\pi \otimes \pi)Bez(x,y) = \overline{\pi}(i \otimes i)B(X,Y)$ and $\mathcal{O} = \mathcal{O}_0$ and A_0 as the rings in questions.

In the next step, *the deformation of the system* $P_1 = \cdots = P_n = 0$, we will define a new system $P_1' = \cdots = P_n' = 0$ with zero $p = 0$ yielding a modified algebra $A' = k[X]/\mathfrak{a}'$, $\mathfrak{a}' = (P_1', \ldots, P_n')$, a corresponding element $B'(X,Y)$, and Bezoutian $B' = Bez(x,y)$ of $P_1', \ldots, P_n'$ in $A' \otimes A'$. But, and this is the main point:

Fact 1) A' is Frobenius with Bezoutian $B' = B_{\ell'}$,

Fact 2) $A_0' = A_0$,

Fact 3) $(\pi' \otimes \pi')(B') = (\pi \otimes \pi)(B)$ where $B = Bez(x,y)$.

These three facts ensure that A is Frobenius too: From the local study in §2, if we know that $(\pi' \otimes \pi')(B')$ is the generalized Bezoutian of a dualizing form ℓ_0', then this holds for $(\pi \otimes \pi)(B)$.

We now define A' and prove Facts 2 and 3 above. Choose $M \in \mathbf{N}$ and set:

$$P_i'(X) = X_i^{M+1} + P_i(X),\ i = 1, \ldots, n.$$

M will be specified later on. We see that $p = 0$ is a zero of both systems. We have:

$$A_0 = \mathcal{O}/\mathfrak{a}\mathcal{O},\ A_0' = \mathcal{O}/\mathfrak{a}'\mathcal{O}.$$

If M is suitably large we want to show $\mathfrak{a}\mathcal{O} = \mathfrak{a}'\mathcal{O}$. Since A_0 is a finite-dimensional local algebra with maximal ideal $(x_1, \ldots, x_n)$, there is $e \in \mathbf{N}$ such that $(X_1, \ldots, X_n)^e \subseteq \mathfrak{a}\mathcal{O}$. Choose any $M \geq e$. Then for each $i = 1, \ldots, n$:

$X_i^{M+1} \in (X_iP_1, \ldots, X_iP_n)$ in $\mathcal{O}$, i.e., there are $a_{ij} \in \mathcal{O}$ satisfying $X_i^{M+1} = \sum_{j=1}^{n} a_{ij}X_iP_j$ entailing

$$P_i'(X) = \sum_{j=1}^{n}(a_{ij}X_i + \delta_{ij})P_j, \; i = 1, \ldots, n.$$

Since the determinant of $(a_{ij}X_i + \delta_{ij})$ is a unit in $\mathcal{O}$ we conclude $\mathfrak{a}\mathcal{O} = \mathfrak{a}'\mathcal{O}$ provided $M \geq e$.

We next show that $\pi' \otimes \pi'(B') = \pi \otimes \pi(B)$ if $M \geq 2e$. In fact, from $P_i'(X) = X_i^{M+1} + P_i(X)$, $i = 1, \ldots, n$ we first derive

$$P_{ij}' = \begin{cases} P_{ij} & \text{if } i \neq j \\ \frac{X_i^{M+1} - Y_i^{M+1}}{X_i - Y_i} + P_{ii} & \text{if } i = j. \end{cases}$$

We have $\pi \otimes \pi(B) = \overline{\pi}(i \otimes i)B(X,Y)$, $\pi' \otimes \pi'(B') = \overline{\pi} \circ (i \otimes i)(B'(X,Y))$. Thus, $i \otimes i(B(X,Y))$ and $i \otimes i(B'(X,Y))$ have to be compared. Note that $(i \otimes i)(f(X)) = i(f(x)) \otimes 1$ and $(i \otimes i)(f(Y)) = 1 \otimes i(f(x))$. Clearly

$$i \otimes i(B(X,Y)) = det((i \otimes i)(P_{rs})), \text{ etc.}$$

Either $i\otimes i(P_{rs}) = i\otimes i(P_{rs}')$ or $r = s$, $i\otimes i(P_{rr}') = i\otimes i(P_{rr}) + i\otimes i(\sum_{k=0}^{n} X_r^kY_r^{M-k})$. If $M \geq 2e$ then either $k \geq e$ or $M - k \geq e$. It follows $i\otimes i(\sum_k X_r^kY_r^{M-k}) \in \mathfrak{a}\mathcal{O} \otimes \mathcal{O} + \mathcal{O} \otimes \mathfrak{a}\mathcal{O} \subseteq ker\ \overline{\pi}$ (note that $\mathfrak{a}\mathcal{O} = \mathfrak{a}'\mathcal{O}$). Thus $\pi \otimes \pi(B) = \pi' \otimes \pi'(B')$ is proved.

It remains to prove Fact 1: A' is Frobenius and $B' = B_{\ell'}$ for some dualizing form.

To apply the following arguments in more general situations we don't assume $P_1, \ldots, P_n$ to define a complete intersection. They are arbitrarily chosen polynomials in $k[X_1, \ldots, X_n]$. Nevertheless, the system $P_1' = 0, \ldots, P_n' = 0$ is zero-dimensional as can be seen by Gröbner Bases techniques e.g., cf. [5], p. 274, Th. 6.54.

We already know $B' \in Ann(I')$, $I' = ker(\delta : A' \otimes A' \to A')$. It remains to check the second condition in 2.10, iii). This will be done rather explicitly by studying appropriate k-bases of A' and $A' \otimes A'$. We choose $M \geq max\{deg\ P_1, \ldots, deg\ P_n\}$. The system $P_1'(X), \ldots, P_n'(X)$ is then a Gröbner basis relative to a total-degree lexicographical term ordering in $k[X]$, and the system $P_1'(X), \ldots, P_n'(X)$, $P_1'(Y), \ldots, P_n'(Y)$ is a Gröbner bases for the corresponding term ordering in $k[X,Y]$, apply [5], p. 223, Th. 5.68, e.g. The highest terms are $X_1^{M+1}, \ldots, X_n^{M+1}$ and $Y_1^{M+1}, \ldots, Y_n^{M+1}$, respectively. We denote by x_i (respectively y_i) the image of X_i (respectively Y_i) in $A' \otimes A'$. Since the mapping $A' \to A' \otimes A'$, $a' \mapsto a' \otimes 1$ is injective, x_i is also the image in A'. The theory of Gröbner bases yields that

$$\mathcal{B} = \left\{ \prod_1^n x_i^{\alpha_i} | 0 \le \alpha_i \le M \right\} \text{ is a } k\text{-basis of } A'$$

and

$$\mathcal{B} \times \mathcal{B} = \left\{ \prod_1^n x_i^{\alpha_i} \cdot \prod_1^n y_i^{\beta_i} | 0 \le \alpha_i, \beta_i \le M \right\} \text{ is a } k\text{-basis of } A' \otimes A'.$$

In particular, $\dim_k A' = (M+1)^n$.

We now turn to the structure of $B'(X,Y) = det(P'_{ij})$. For convenience, we set

$$S_i = \sum_{k=0}^{M} X_i^k Y_i^{M-k}, \; d = \max_i \{deg\ P_i\}.$$

Then $B'(X,Y) = S_1 S_2 \cdots S_n +$ sums of products of the type $S_{i_1} \cdots S_{i_r} Q_{r+1} \cdots Q_n$, $r < n$ where the Q_l's are among the polynomials $\{P_{ij}\}_{i,j}$. We have $deg(S_1 \cdots S_n) = nM$ whereas

$$deg(S_{i_1} \cdots S_{i_r} Q_{r+1} \cdots Q_n) \le r \cdot M + (n-r)(d-1) < nM$$

if $r < n$. Each term $F = S_{i_1} \cdots S_{i_r} Q_{r+1} \cdots Q_n$ will now be reduced to its normal form $N(F)$ relative to the Gröbner basis $P'_1(X), \ldots, P'_n(Y)$. One readily checks: $deg\ N(F) \le deg\ F$; $deg_{X_i} F$, $deg_{Y_i} F \le M$ for every i.

For $B' = \tilde{\pi}(B'(X,Y)) \in A' \otimes A'$ this means

$$B' = \prod_{i=1}^{n} \left(\sum_{k=0}^{n} x_i^k y_i^{M-k} \right) + \text{ linear combinations of terms}$$

$$\prod_1^n x_i^{\alpha_i} \prod_1^n y_i^{\beta_i} \text{ where all } \alpha_i, \beta_i \le M \text{ and } \sum (\alpha_i + \beta_i) < nM.$$

For convenience, we write $\prod x_i^{\alpha_i} = x^\alpha$, $\alpha = (\alpha_1, \ldots, \alpha_n)$; the same applies to the $y_1, \ldots, y_n$. Furthermore $\mathbf{M} := (M, \ldots, M)$, $|\alpha| = |(\alpha_1, \ldots, \alpha_n)| = \sum \alpha_i$, and we can write

$$B' = \sum_{x^\alpha \in \mathcal{B}} x^\alpha \left(y^{\mathbf{M}-\alpha} + \sum_{\substack{y^\beta \in \mathcal{B} \\ |\alpha|+|\beta|<nM}} a_{\alpha,\beta} y^\beta \right), a_{\alpha,\beta} \in k.$$

The total-degree lexicographical term ordering on $k[X,Y]$ induces a total order on $\mathcal{B} \times \mathcal{B}$. Since $y^{\mathbf{M}-\alpha}$ has degree $nM - |\alpha|$, we see that $y^{\mathbf{M}-\alpha}$ is the leading term of the factor of x^α. Using this total order and the fact that the elements

y^β, $0 \le \beta_i \le M$ form a basis of A' $(= k[Y]/_{\mathfrak{a}'})$, we see the second condition of 2.10, iii) satisfied.

Let us describe what is the associated dualizing form ℓ'. Define z_α by

$$B' = \sum_{x^\alpha \in \mathcal{B}} x^\alpha (y^{\mathbf{M}-\alpha} + \text{ remainder}) =: \sum_{x^\alpha \in \mathcal{B}} x^\alpha z_\alpha.$$

For $\alpha = \mathbf{M} = (M, \ldots, M)$ we must have $z_\alpha = 1$. Therefore

$$\ell'(x^\alpha) = \begin{cases} 1 & \alpha = \mathbf{M}, \\ 0 & \alpha \neq \mathbf{M}. \end{cases}$$ □

We now want to make explicit the theory of the preceeding section in the case of a complete intersection

$$P_1 = \cdots = P_n = 0.$$

Again, let V denote the zeros of this system over the algebraic closure $\overline{k}$ of k. Intuitively, we regard the elements of $A = k[X]/_{(P_1,\ldots,P_n)}$, respectively of $A \otimes A$, as functions on V, respectively on $V \times V$.

The Bezoutian $Bez(x,y) \in A \otimes A$ gives rise to a dualizing form which we call, following [17], [12], [10], the *Global Residue* or, following [2], *Kronecker symbol* denoted by ℓ, of this system of equations. From §2 we carry over the relation between $Bez(x,y)$ and ℓ: if $Bez(x,y) = \sum_{i=1}^{n} e_i(x) f_i(y)$ then $e = \{e_i(x)\}$ and $f = \{f_i(x)\}$ are a basis of A. The coefficients of 1 in the basis $\{f_i(x)\}$ are the images of the $e_i(x)$ under ℓ. This allows a method for the *calculation of the Kronecker symbol ℓ for a general complete intersection:*

Suppose we are given a basis e of the vector space A (through a Gröbner basis computation, for example). Compute the Bezoutian and get the basis dual f. The value $\ell(g)$ is obtained as follows: Write g in the basis e, 1 in the dual basis f, and make the inner product of the vector of coefficients respectively obtained.

$$\text{if } g = \sum_1^n c_i e_i(x) \text{ and } 1 = \sum_1^n d_i f_i(x),$$

$$\text{then } \ell(f) = \sum_1^n c_i d_i.$$

We have already seen a concrete and basic description of the Kronecker symbol in a particular case in the proof of theorem 3.2. Assume

$$P_i(X) = X_i^{M+1} + Q_i(X),\ i = 1, \ldots, n \qquad \text{with } M \ge \max_i\{deg\ Q_i\}.$$

As already mentioned, the system $P_1 = \cdots = P_n = 0$ is zero-dimensional, hence a complete intersection. As seen in the proof of the theorem,

$$Bez(x,y) = \sum_{x^\alpha \in \mathcal{B}} x^\alpha (y^{\mathbf{M}-\alpha} + \text{remainder}) =: \sum_{x^\alpha \in \mathcal{B}} x^\alpha z_\alpha$$

$$\ell(x^\alpha) = \begin{cases} 1 & \alpha = (M, M, \ldots, M), \\ 0 & \alpha \neq (M, M, \ldots, M). \end{cases}$$

The next proposition concerning this particular case will be useful in section 4.

Proposition 3.5 *If K is a real closed field and $M+1$ is even then*

$$signature\, \Phi_\ell = 0.$$

Proof. It is clear since $\Phi_\ell(X^\alpha, X^{\mathbf{M}-\alpha}) = 1, \Phi_\ell(X^\alpha, X^\beta) = 0$ for $\alpha + \beta < M$ □

As a very special case we assume $n = 1$, i.e., we are dealing with a univariate k-algebra:

$$A = k[X]/_{P(X)},\ deg\ P = M+1,\ Bez(x,y) = \text{ image of } \frac{P(X) - P(Y)}{X - Y}$$

$$\ell(1) = \cdots = \ell(x^{M-1}) = 0,\ \ell(x^M) = 1.$$

Again back in the general case of an arbitrary complete intersection we first turn to the study of the generalized Jacobian J_ℓ attached to ℓ and then to local considerations.

The multiplication $\delta : A \otimes A \to A$ fits into the following diagram

$$\begin{array}{ccc} k[X,Y] & \overset{X \mapsto X, Y \mapsto X}{\longrightarrow} & k[X] \\ \downarrow \tilde{\pi} & /// & \downarrow \pi \\ A \otimes A & \overset{\delta}{\longrightarrow} & A \end{array}$$

Hence,

3.6 $J_\ell =$ Image of the Jacobian of $P_1, \ldots, P_n$ *in* A.

We therefore write Jac instead of J_ℓ.

For each $p \in V$, the map $\overline{\pi}_p : \overline{A} \to \overline{k}$ is just the evaluation map $f \mapsto f(p)$. Since B is symmetric we have $(\overline{\pi}_p \otimes id)(B) = (id \otimes \overline{\pi}_p)(B)$. We define for $p \in V$:

$$Bez(x,p) := (id \otimes \overline{\pi}_p)(B) = \sum_{i=1}^{n} e_i(x) f_j(p) \in \overline{A}.$$

From 2.22 we infer

3.7 *$f(x)Bez(x,p) = f(p)Bez(x,p)$ for every $f \in A$,*

3.8 $\ell(Bez(x,p)) = 1$,

3.9 $\ell_p(Jac_p) = \mu_p$.

In 3.9, we understand Jac_p as the image of Jac in A_p under π_p, and ℓ_p is the component of ℓ defined by $\ell_p\pi_p = \ell|_{Ae_p}$. From 2.22 we also get:

$$Jac_p = \mu_p \cdot \text{ image of } Bez(x,p).$$

This gives in the case of $char\ k = 0$ or $char\ k \nmid \mu_p$:

Proposition 3.10 *For every $p \in V$, the image of Jac in the local ring A_p is not zero.*

There is an immediate interpretation of ℓ_p if p is a simple root, i.e., $\mu_p = 1$ or, equivalently, $A_p = \overline{k}$.

Assume $\mu_p = 1$, then $Jac_p = Jac(p) = \pi_p(Bez(x,p)) \neq 0$. If $f \in A_p$ is given then $f(p) = f$ and

$$\ell_p(f) = \frac{f(p)}{Jac(p)}\ell_p(Jac(p)) = \frac{f(p)}{Jac(p)}.$$

In the case that all roots are simple, we have for $f \in A$

$$\ell(f) = \sum_p \ell_p\pi_p(f) = \sum_p \frac{f(p)}{Jac(p)}.$$

We summarize this in

Proposition 3.11

(i) If p is simple then $\ell_p(f) = \dfrac{f(p)}{Jac(p)}$,

(ii) if every zero p is simple then

$$\ell(f) = \sum_p \frac{f(p)}{Jac(p)}.$$ □

4 Proof of the Eisenbud and Levine theorem

In this section we present a new proof of the Eisenbud and Levine formula, based on the techniques and results of the preceeding sections, i.e., we shall prove that the local topological degree of a mapping germ $F : (R^n, \mathbf{0}) \longrightarrow (R^n, \mathbf{0})$ is equal to the signature of bilinear symmetric form associated to a dualizing λ with $\lambda(Jac_0) > 0$.

The main idea of our proof is to replace F by a polynomial mapping $G : (R^n, \mathbf{0}) \longrightarrow (R^n, \mathbf{0})$ being close to F near the origin and having only non degenerate zeros (except at the origin) and then to give the proof for G based on facts proved in previous sections. Proofs of this theorem were given by Eisenbud and Levine [7], Khimshiashvili [9] and Arnold, Varchenko, and Gusein-Zade [2].

Let K denote either the field R of real numbers or the field C of complex numbers, and let Γ_K denote the space of all n-tuples $(h_1, \ldots, h_n)$, where every $h_i : K^n \longrightarrow K$ is a homogeneous polynomial of degree s. Then Γ_C is the complexification of Γ_R. For $h = (h_1, \ldots, h_n) \in \Gamma_K$ and $f_1, \ldots, f_n \in K[x] = K[x_1, \ldots, x_n]$, write

$$G_C^h = (f_1 + h_1, \ldots, f_n + h_n) : C^n \longrightarrow \mathbf{C}^n$$

and denote $h_i = \sum h_\alpha^i x^\alpha$, where $x^\alpha = x_1^{\alpha_1} \cdots x_n^{\alpha_n}$.

The next theorem has a technical character; in cases $s = 0, 1$ it can be derived from the Sard theorem.

Theorem 4.1 *Assume that $f_1, \ldots, f_n \in R[x]$. Then for each positive integer s there is a dense semialgebraic set $U \subset \Gamma_R$ such that*

$$Jac\, G_C^h(z) \neq 0 \text{ at each } z \in (G_C^h)^{-1}(\mathbf{0}) - \{\mathbf{0}\}$$

for every $h \in U$.

Lemma 4.2 *Let $f = f(x_1, \ldots, x_n) : C^n \longrightarrow C$ be holomorphic, let $\tilde{z} = (\tilde{z}_1, \ldots, \tilde{z}_n) \in C^n - \{\mathbf{0}\}$, and let $L : C^n \longrightarrow C^n$ be a linear mapping with $L(1, 0, \ldots, 0) = \tilde{z}$. Define $h = f \circ L$. Then*

$$\frac{\partial h}{\partial x_1}(1, 0, \ldots, 0) = \frac{\partial f}{\partial x_1}(\tilde{z})\tilde{z}_1 + \cdots + \frac{\partial f}{\partial x_n}(\tilde{z})\tilde{z}_n.$$

Proof. Since $L(x_1, 0, \ldots, 0) = x_1 L(1, 0, \ldots, 0) = x_1 \tilde{z} = (x_1\tilde{z}_1, \ldots, x_1\tilde{z}_n)$, we have $h(x_1, 0, \ldots, 0) = f(x_1\tilde{z}_1, \ldots, x_1\tilde{z}_n)$, and it is enough to apply the chain rule. □

Proof of theorem 4.1. If all $f_1, \ldots, f_n$ are homogeneous of degree s then the proof is obvious because there is a dense semialgebraic U such that $(G_C^h)^{-1}(\mathbf{0}) = \{\mathbf{0}\}$ for every $h \in U$.

From now on we shall assume that f_1 is not homogeneous of degree s. Let

$$X = \{z \in C^n : \frac{\partial f_1}{\partial x_1}(z)z_1 + \cdots + \frac{\partial f_1}{\partial x_n}(z)z_n = s\, f_1(z)\}.$$

It is well-known that $X \neq C^n$, so $\dim_C X$ (i.e., the complex dimension of X) is $n-1$, and so $\operatorname{codim}_C X = 1$.

Define

$$\Delta = \{(z,h) \in C^n \times \Gamma_C : G_C^h(z) = \mathbf{0} \text{ and } \operatorname{Jac} G_C^h(z) = 0\}.$$

Obviously Δ is an algebraic subset of $C^n \times \Gamma_C$ defined by real polynomials. Let $\pi : \Delta \longrightarrow C^n$ denote the projection on the first factor. For each $z \in C^n$, $\pi^{-1}(z)$ is an algebraic subset of $\{z\} \times \Gamma_C \cong \Gamma_C$. Let $\operatorname{codim}_C \pi^{-1}(z)$ denote the codimension of $\pi^{-1}(z)$ in Γ_C. We shall prove

$$\operatorname{codim}_C \pi^{-1}(z) \geq n \text{ at each } z \in C^n - \{\mathbf{0}\}. \tag{1}$$

The linear change of coordinates preserves zeros, critical points, and the degree of homogeneous polynomials, so it is enough to consider the case $z = (1,0,\ldots,0)$. Since $h_i(1,0,\ldots,0) = h^i_{(s,0,\ldots,0)}$, then $\pi^{-1}((1,0,\ldots,0)) \subset Y = \{h \in \Gamma_C : G_C^h(1,0,\ldots,0) = \mathbf{0}\} = \{h \in \Gamma_C : h^i_{(s,0,\ldots,0)} = -f_i(1,0,\ldots,0) \text{ for } i = 1,\ldots,n\}$. Since $Y \subset \Gamma_C$ is an algebraic subset and $\operatorname{codim}_C Y = n$ we have $\operatorname{codim}_C \pi^{-1}((1,0,\ldots,0)) \geq n$.

Now we shall prove

$$\operatorname{codim}_C \pi^{-1}(z) \geq n+1 \text{ at each } z \in C^n - X. \tag{2}$$

From lemma 4.2, we may assume that $z = (1,0,\ldots,0)$ in some coordinate system and $\frac{\partial f_1}{\partial x_1}(1,0,\ldots,0) \neq s\, f_1(1,0,\ldots,0)$. Clearly:

$$\frac{\partial h_1}{\partial x_1}(1,0,\ldots,0) = s\, h^1_{(s,0,\ldots,0)},$$

$$\frac{\partial h_2}{\partial x_2}(1,0,\ldots,0) = h^2_{(s-1,1,0,\ldots,0)}, \ldots, \frac{\partial h_n}{\partial x_n}(1,0,\ldots,0) = h^n_{(s-1,0,\ldots,0,1)}.$$

If $h \in \pi^{-1}((1,0,\ldots,0))$, then $h^i_{(s,0,\ldots,0)} = -f_i(1,0,\ldots,0)$ for $i = 1,\ldots,n$ and

$$\begin{aligned} 0 &= \operatorname{Jac} G_C^h(1,0,\ldots,0) = \\ &= \left(\tfrac{\partial f_1}{\partial x_1}(1,0,\ldots,0) + s\, h^1_{(s,0,\ldots,0)}\right) h^2_{(s-1,1,0,\ldots,0)} \cdots h^n_{(s-1,0,\ldots,0,1)} + \ldots = \\ &= \left(\tfrac{\partial f_1}{\partial x_1}(1,0,\ldots,0) - s\, f_1(1,0,\ldots,0)\right) h^2_{(s-1,1,0,\ldots,0)} \cdots h^n_{(s-1,0,\ldots,0,1)} + \ldots . \end{aligned}$$

Since

$$\frac{\partial f_1}{\partial x_1}(1,0,\ldots,0) - s\, f_1(1,0,\ldots,0) \neq 0$$

then $\pi^{-1}((1,0,\ldots,0))$ is a proper subset of an irreducible Y and then

$$\operatorname{codim}_C \pi^{-1}((1,0,\ldots,0)) \geq \operatorname{codim}_C Y + 1 = n+1.$$

Both $\pi^{-1}(X - \{\mathbf{0}\})$ and $\pi^{-1}(C^n - X)$ are constructible subsets of $C^n \times \Gamma_C$. Hence, from (1) and (2), $\operatorname{codim}_C \pi^{-1}(X - \{\mathbf{0}\}) \geq \operatorname{codim}_C(X - \{\mathbf{0}\}) + n = 1 + n$

and $\text{codim}_C \pi^{-1}(C^n - X) \geq \text{codim}_C(C^n - X) + 1 + n = 1 + n$. Since $\Delta - \pi^{-1}(\mathbf{0}) = \pi^{-1}(X - \{\mathbf{0}\}) \cup \pi^{-1}(C^n - X)$ so $\text{codim}_C(\Delta - \pi^{-1}(\mathbf{0})) \geq 1 + n$.

Let $\eta : C^n \times \Gamma_C \longrightarrow \Gamma_C$ denote the projection on the second factor. Then $Z = \eta(\Delta - \pi^{-1}(\mathbf{0}))$ is a constructible subset of Γ_C defined by polynomials having real coefficients. Since $\dim_C \eta^{-1}(h) = \dim_C C^n \times \{h\} = n$ for every $h \in \Gamma_C$, $\text{codim}_C Z \geq 1$. Then $U = \Gamma_R - Z$ is a dense semialgebraic subset of Γ_R and if $h \in U$ then $\eta^{-1}(h) \cap (\Delta - \pi^{-1}(\mathbf{0})) = \emptyset$ and that means that $\text{Jac}\, G_C^h(z) \neq 0$ at each $z \in (G_C^h)^{-1}(\mathbf{0}) - \{\mathbf{0}\}$. □

Assume that $p \in C^n - R^n$. Let $\mathcal{F}_{C,p}$ denote the space of all functions $\{p, \overline{p}\} \longrightarrow$ C, where $\overline{p}$ is the complex conjugate. We shall write $f = (f(p), f(\overline{p}))$ for every $f \in \mathcal{F}_{C,p}$. Let $\tilde{A}_p = \{f \in \mathcal{F}_{C,p} : f(\overline{p}) = \overline{f(p)}\}$. Clearly $\tilde{A}_p$ is an R-algebra, $\dim_R \tilde{A}_p = 2$, and if $f = (f(p), f(\overline{p})) \in \tilde{A}_p$ and $\alpha \in$ C, then $\alpha * f = (\alpha f(p), \overline{\alpha} f(\overline{p})) \in \tilde{A}_p$ too. Assume that there is an R-linear form $\ell_p : \tilde{A}_p \longrightarrow$ R. Let $\Phi_{\ell p} : \tilde{A}_p \times \tilde{A}_p \longrightarrow$ R be the bilinear symmetric form given by $\Phi_{\ell p}(f, g) = \ell_p(fg)$.

Lemma 4.3 *signature* $\Phi_{\ell_p} = 0$.

Proof. Assume that there is $f \in \tilde{A}_p$ such that $\Phi_{\ell p}(f, f) = \ell_p(f^2) > 0$. Then $\sqrt{-1} * f \in \tilde{A}_p$ and $\Phi_{\ell p}(\sqrt{-1} * f, \sqrt{-1} * f) = \ell_p(-f^2) < 0$. Thus the dimension of the subspace of $\tilde{A}_p$ on which $\Phi_{\ell p}$ is positive definite is equal to that on which $\Phi_{\ell p}$ is negative definite. Hence signature $\Phi_{\ell_p} = 0$. □

Assume that $F : (R^n, \mathbf{0}) \longrightarrow (R^n, \mathbf{0})$ is a continuous mapping defined in some neighbourhood of the origin such that $\mathbf{0}$ is isolated in $F^{-1}(\mathbf{0})$. There is $r > 0$ such that the intersection of the ball $B_r = \{x \in R^n : \| x \| \leq r\}$ with $F^{-1}(\mathbf{0})$ is $\{\mathbf{0}\}$. Let $S_r = \partial B_r$ and let $\deg_0(F)$ denote the local topological degree of F at $\mathbf{0}$, i.e., the topological degree of the mapping

$$S_r \ni x \longmapsto F(x) / \| F(x) \| \in S^{n-1}.$$

If $F = (f_1, \ldots, f_n)$ is analytic and the R-algebra

$$A = \mathrm{R}[[x_1, \ldots, x_n]]/(f_1, \ldots, f_n)$$

is finite-dimensional, then it is known that $\mathbf{0}$ is isolated in $F^{-1}(\mathbf{0})$ and there is a positive integer N such that the residue class of any formal power series consisting of monomials of degree $\geq N$ is zero in A.

Theorem 4.4 (Eisenbud and Levine formula) *Assume that*

$$F = (f_1, \ldots, f_n) : (R^n, \mathbf{0}) \longrightarrow (R^n, \mathbf{0})$$

is an analytic mapping defined in some neighbourhood of $\mathbf{0}$ *such that* $\dim_R A < \infty$. *Let* Jac_O *denote the residue class of* $Jac\, F$ *in* A. *Then (i)* $Jac_O \neq 0$.

Let $\lambda : A \longrightarrow R$ be a linear form such that $\lambda(Jac_O) > 0$ and let $\Phi_\lambda : A \times A \longrightarrow R$ be the symmetric bilinear form given by $\Phi_\lambda(f, g) = \lambda(fg)$. Then (ii) Φ_λ is non degenerate and

$$deg_0(F) = signature\, \Phi_\lambda \ .$$

Proof. (i) Has been proved in 3.10.
(ii) It is easy to prove the following conditions:

(i) For every analytic germs $r_1, \dots, r_n : (R^n, \mathbf{0}) \longrightarrow (R, 0)$ consisting of monomials of degree $\geq N+1$, the mapping $\tilde{F} = (f_1 + r_1, \dots, f_n + r_n)$ has an isolated zero at $\mathbf{0}$, $\deg_0(F) = \deg_0(\tilde{F})$, $A = \mathrm{R}[[x_1, \dots, x_n]]/(\tilde{f}_1, \dots, \tilde{f}_n)$ and $Jac_O = \tilde{Jac}_O$ in A.

Thus we may assume that $f_1, \dots, f_n$ are polynomials of degree $\leq N$. From theorem 4.1, there are homogeneous polynomials $h_1, \dots, h_n : R^n \longrightarrow R$ of degree $N+1$ such that the mapping $G_C : (C^n, \mathbf{0}) \longrightarrow (C^n, \mathbf{0})$, where $G = (g_1, \dots, g_n) = (x_1^{2N} + f_1 + h_1, \dots, x_n^{2N} + f_n + h_n) : (R^n, \mathbf{0}) \longrightarrow (R^n, \mathbf{0})$, has only non degenerate zeros in $G_C^{-1}(\mathbf{0}) - \{\mathbf{0}\}$.

Let $A' = R[x_1, \dots, x_n]/(g_1, \dots, g_n)$ and let $h_i = f_i + q_i$. Then $g_i = x_i^{2N} + h_i$ and

$$2N - 1 > \max_{i=1}^{n} \deg(h_i)\}.$$

Let $\ell : A' \longrightarrow R$ be the Kronecker symbol (see section 3) associated to the equations and let $\Phi_\ell : A' \times A' \longrightarrow \mathrm{R}$ be the symmetric bilinear form given by $\Phi_\ell(f, g) = \ell(fg)$.

The set $G_C^{-1}(\mathbf{0})$ is finite so we may assume that

$$G_C^{-1}(\mathbf{0}) = \{\mathbf{0}, p_1, \dots, p_r, p_{r+1}, \overline{p_{r+1}}, \dots, p_s, \overline{p_s}\}.$$

where $\{\mathbf{0}, p_1, \dots, p_r\} = G_C^{-1}(\mathbf{0}) \cap R^n$ and $\{p_{r+1}, \overline{p_{r+1}}, \dots, p_s, \overline{p_s}\} = G_C^{-1}(0) \cap (C^n - R^n)$. The algebra $A'_C = C[x_1, \dots, x_n]/(g_1, \dots, g_n)$ is naturally isomorphic to

$$A'_{C,0} \times \prod_{i=1}^{r} A'_{C,p_i} \times \prod_{i=r+1}^{s} (A'_{C,p_i} \times A'_{C,\overline{p_i}}) = A'_{C,O} \times \prod_{i=1}^{r} A'_{C,p_i} \times \prod_{i=r+1}^{s} B_{C,p_i} \ ,$$

where $A'_{C,p}$ denotes the corresponding local ring at p. Let π_i denote either the projection of A' on A'_{C,p_i} for $i = 1, \dots, r$ or on B_{C,p_i} for $i = r+1, \dots, s$. Take $f \in A'$. Since $\mathrm{Jac}\, G_C(p_i) \neq 0$ for $i = 1, \dots, s$ then $\pi_i(f) = f(p_i)$ in the first case and $\pi_i(f) = (f(p_i), f(\overline{p_i}))$ in the second one.

Clearly $A' = \{f \in A'_C : \overline{f(\overline{z})} = f(z)\}$. Denote $A'_p = \{f \in A'_{C,p} : \overline{f(\overline{z})} = f(z)\}$ for $p \in \{\mathbf{0}, p_1, \dots, p_r\}$ and $\tilde{A}_p = \{f \in B_{C,p} : \overline{f(\overline{z})} = f(z)\}$ for $p \in \{p_{r+1}, \dots, p_s\}$. Then $A'_0 = \mathrm{R}[[x_1, \dots, x_n]]/(g_1, \dots, g_n)$, $A'_{p_i} = \{f \in A'_{C,p_i} : f(p_i) \in \mathrm{R}\}$ and $\tilde{A}'_{p_i} = \{f \in B_{C,p_i} : f(\overline{p_i}) = \overline{f(p_i)}\}$. Thus there is a natural isomorphism of algebras

$$A' \cong A'_0 \times \prod_{i=1}^{r} A'_{p_i} \times \prod_{i=r+1}^{s} \tilde{A}'_{p_i} \ .$$

From the condition (i), $A'_0 = A$ and $Jac_O = \mathrm{Jac}\, G$ in A'_0. Let ℓ_0 denote the natural restriction of ℓ to A'_0. As we proved in 3.9, $\ell_0(Jac_O) = \ell_0(\mathrm{Jac}\, G) = \mu_0$. From proposition 3.3, Φ_ℓ is non degenerate. Hence the symmetric bilinear form $\Phi_{\ell 0}$ on $A'_0 \times A'_0$ given by $\Phi_{\ell 0} = \ell_0(fg)$ is non degenerate too.

Let ℓ_i for $i = 1, \ldots, r$ denote the natural restriction of ℓ to A'_{p_i}. Since $\mathrm{Jac}\, G(p_i) \neq 0$ then, according to 3.9, $\ell_i(f) = f(p_i)/\mathrm{Jac}\, G(p_i)$. Then the symmetric bilinear form $\Phi_{\ell i}$ on $A'_{p_i} \times A'_{p_i}$ given by $\Phi_{\ell i}(f,g) = \ell_i(fg)$ is non degenerate and signature $\Phi_{\ell i} =$ sign $\mathrm{Jac}\, G(p_i)$.

Let ℓ_i for $i = r+1, \ldots, s$ denote the natural restriction of ℓ to $\tilde{A}_{p_i}$ and let $\Phi_{\ell i}$ be the symmetric bilinear form on $\tilde{A}_{p_i} \times \tilde{A}_{p_i}$ given by $\Phi_{\ell i}(f,g) = \ell_i(fg)$. From lemma 4.3, signature $\Phi_{\ell i} = 0$.

Clearly

$$signature\, \Phi_\ell = signature\, \Phi_{\ell 0} + \sum_{i=1}^{s} signature\, \Phi_{\ell i} \ .$$

From proposition 3.5 and from last remarks we get

$$0 = signature\, \Phi_{\ell 0} + \sum_{i=1}^{r} signJac\, G(p_i) \ .$$

We may assume that $G^{-1}(\mathbf{0})$ is contained in a ball B_R. Put $S_R = \partial B_R$ and $H = (x_1^{2N}, \ldots, x_n^{2N})$. If R is big enough then $G : S_R \longrightarrow R^n - \{\mathbf{0}\}$ is homotopic to $H : S_R \longrightarrow R^n - \{\mathbf{0}\}$ and the last mapping is homotopically trivial. Thus $deg(G \mid S_R)$, i.e., the topological degree of G restricted to S_R with respect to $\mathbf{0}$, is zero. Let d_i denote the local topological degree of G at p_i. Since $Jac\, G(p_i) \neq 0$ we have $d_i = signJac\, G(p_i)$ and then

$$0 = deg(G \mid S_R) = deg_0(G) + \sum_{i=1}^{r} d_i = deg_0(G) + \sum_{i=1}^{r} signJac\, G(p_i) \ .$$

Hence from the condition 1. we get

$$signature\, \Phi_{\ell 0} = \deg_0(G) = \deg_0(F).$$

In the remainder of the proof we follow [2]. We proved already that any linear form $\lambda : A_0 \longrightarrow \mathrm{R}$ with $\lambda(Jac_O) \neq 0$ is dualizing, that is that the bilinear symmetric form Φ_λ on $A'_0 \times A'_0$ given by $\Phi_\lambda(f,g) = \lambda(fg)$ is non degenerate.

Define a family of linear forms on A'_0 by $\lambda_t = t\,\ell_0 + (1-t)\lambda$. Let $\Phi_{\lambda t}$ be a continuous family of symmetric bilinear forms on $A'_0 \times A'_0$ given by $\Phi_{\lambda t}(f,g) =$

$\lambda_t(fg)$. We have assumed that $\lambda(Jac_O) > 0$. So, since $\ell_0(Jac_O) = \mu_0$, we have $\lambda_t(Jac_O) > 0$ for every $0 \leq t \leq 1$. Thus Φ_{λ_t} is non degenerate for every $0 \leq t \leq 1$ and

$$signature\, \Phi_\lambda = signature\, \Phi_{\lambda 0} = signature\, \Phi_{\lambda 1} = signature\, \Phi_{\ell 0} = deg_0(F).\square$$

Using similar arguments the reader may prove the well-known:

Theorem 4.5 *Let*

$$H = (h_1, \ldots, h_n) : (C^n, \mathbf{0}) \longrightarrow (C^n, \mathbf{0})$$

be a holomorphic mapping defined in some neighbourhood of $\mathbf{0}$. *If*

$$dim_C C[[x_1, \ldots, x_n]]/(h_1, \ldots, h_n) < \infty$$

then $\mathbf{0}$ *is isolated in* $H^{-1}(\mathbf{0})$ *and*

$$deg_0(H) = dim_C C[[x_1, \ldots, x_n]]/(h_1, \ldots, h_n)\ .$$

References

[1] M. E. Alonso, E. Becker, M.-F. Roy., T. Wörmann *Zero's, multiplicities and idempotents for zero dimensional systems.* in this volume.

[2] V. I. Arnold, A. N. Varchenko, S. M. Gusein-Zade: *Singularities of differentiable maps.* Vol.2. Birkhäuser, 1988.

[3] Auzinger-Stetter *An elimination theory for the computation of all zeros of a system of multivariate polynomial equations,* Numerical Mathematics, Proc. Intern. Conf. Singapore 1988. Int. Ser. Num. 8 (1988), 11–30.

[4] Becker, E. and Wörmann, T.: *On the trace formula for quadratic forms,* in Contemp. Math. 155 (1994), 271–292.

[5] Becker, T. and Weispfenning, V.: *Gröbner bases, a computational approach to commutative algebra,* Graduate Texts in Mathematics 141, Springer Verlag 1993.

[6] Cardinal J.-P.: *Dualité et algorithmes itératifs pour la solution des systèmes polynomiaux.* Thèse, Université de Rennes I. (1993).

[7] D. Eisenbud, H. I. Levine. *An algebraic formula for the degree of a* C^∞ *map germ. Annals of Mathematics,* **106** (1977), 19-44.

[8] Karpilovsky, G. *Symmetric and G-algebras, with applications to group representations,* Kluwer Academic Publishers 1990.

[9] G. M. KHIMSHIASHVILI. *On the local degree of a smooth map. Soobshch. Akad. Nauk Gruz. SSR*, **85**(1977), 309- 311 (in Russian).

[10] KREUZER, M. AND KUNZ, E.: *Traces in strict Frobenius algebras and strict complete intersections*, J. reine angew. Math. 381 (1987), 181–204.

[11] KUNZ, E.: *Kähler differentials*. Vieweg advanced lecture in Mathematics. Braunschweig, Wiesbaden 1986.

[12] KUNZ, E.: *Über den n-dimensionalen Residuensatz,* Jahresbericht der Deutschen Mathematiker-Vereinigung 94 (1972), 170–188.

[13] MAZUR, B. AND ROBERTS, L.: *Local Euler characteristics*, Invent. math. 9 (1970), 201–234.

[14] PEDERSEN, P., ROY, M.-F. AND SZPIRGLAS, A.: *Counting real zeros in the multivariate case*, in Computational Algebraic Geometry, Eysette, F. and Galligo, A. (editors), 202–223, Birkhäuser Verlag 1993.

[15] SCHEJA, G. AND STORCH, U.: *Über Spurfunktionen bei vollständigen Durchschnitten*, J. reine angew. Math. 278/279 (1975), 174–190.

[16] SCHEJA, G. AND STORCH, U.: *Quasi-Frobenius-Algebren und lokal vollständige Durchschnitte*, manuscripta math. 19 (1976), 75–104.

[17] SCHEJA, G. AND STORCH, U.: *Residuen bei vollständigen Durchschnitten*, Math. Nachr. 91 (1979), 157–170.

E. Becker (`becker@emmy.mathematik.uni-dortmund.de`)

FB Mathematik, University of Dortmund, 44221 Dortmund (Germany).

J.P. Cardinal (`cardinal@matsun1.unican.es`)

Departamento de Matemáticas, Estadística y Computación, Universidad de Cantabria, Santander 39071, Spain.

M.-F. Roy (`costeroy@univ-rennes1.fr`)

IRMAR, University of Rennes I, 35042 Rennes Cedex (France).

Z. Szafraniec

Instytut Matematyki, Uniwersytet Gdanski, Gdansk (Poland).

Progress in Mathematics, Vol. 143, © 1996 Birkhäuser Verlag Basel/Switzerland

Some effective methods in pseudo-linear algebra

M. Bronstein

1 Introduction

This paper gives an introduction to pseudo-linear algebra, which is the study of linear operatorial equations in general. We first introduce the basic objects of pseudo-linear algebra (pseudo-derivations, skew polynomials, and pseudo-linear maps), and then outline some fundamental algorithms acting on them. As applications, we are able to uncouple first-order linear systems of differential and (q-) difference equations, and in some cases to solve them.

2 Pseudo-derivations

Let k be a field of characteristic 0 and σ an endomorphism of k. A *pseudo-derivation with respect to* σ is any map $\delta : k \to k$ such that $\delta(a+b) = \delta a + \delta b$ and $\delta(ab) = \sigma(a)\delta b + b\delta a$ for any $a, b \in k$. If $\sigma = 1_k$, then k is a differential field with derivation δ. If $\sigma \neq 1_k$, then it is easy to show that $\delta = u(\sigma - 1_k)$ for some $u \in k$. The *constants of* k *(with respect to* σ *and* δ*)* are $\mathrm{Const}_{\sigma,\delta}(k) = \{a \in k$ such that $\sigma(a) = a$ and $\delta a = 0\}$ and are easily checked to form a subfield of k.

3 Skew-polynomials

Let k, σ, δ be as above and x be an indeterminate. The *skew-polynomial ring* $k[x; \sigma, \delta]$ consists of the elements of the usual polynomial ring $k[x]$, with the usual polynomial addition, and the multiplication given by $xa = \sigma(a)x + \delta a$ for any $a \in k$. This multiplication is extended uniquely to arbitrary polynomials via associativity and distributivity. We have $\deg(pq) = \deg(p) + \deg(q)$ for $p, q \in k[x; \sigma, \delta]$, which makes $k[x; \sigma, \delta]$ into a non-commutative integral domain.

Let $A, B \in k[x; \sigma, \delta] \setminus \{0\}$, ax^n and bx^m be their leading terms, and suppose that $n \geq m$. We can then perform a right Euclidean division of A by B: let

$$Q_0 = \frac{a}{\sigma^{n-m}(b)} x^{n-m}.$$

The leading monomial of $Q_0 B$ is ax^n, so we can recursively divide $A - Q_0 B$ by B on the right, obtaining $Q_1, R \in k[x; \sigma, \delta]$ such that $A - Q_0 B = Q_1 B + R$ and $\deg(R) < m$. We then have $A = QB + R$, where $Q = Q_0 + Q_1$ and $\deg(R) < \deg(B)$. R is called the *right-remainder* of A by B and is denoted rrem(A, B), while Q is called the *right-quotient* of A by B and is denoted by rquo(A, B). If σ is an automorphism of k, then there is a similar left Euclidean division, where we let

$$Q_0 = \sigma^{-m}\left(\frac{a}{b}\right) x^{n-m}$$

and obtain $Q, R \in k[x; \sigma, \delta]$ such that $A = BQ + R$ and $\deg(R) < m$. Q and R are called the left-quotient and left-remainder of A by B in that case. We can also compute the *right (resp. left) Euclidean remainder sequence* given by $R_0 = A, R_1 = B$ and $R_i = \text{rrem}(R_{i-2}, R_{i-1})$ (resp. $\text{lrem}(R_{i-2}, R_{i-1})$) for $i \geq 2$, and the *greatest common right (resp. left) divisor* of A and B which is the last non-zero element of that sequence. When $AB \neq 0$, the extended right (resp. left) Euclidean algorithm [2, 6], yields a non-zero least common left (resp. right) multiple of A and B, thereby showing that $k[x; \sigma, \delta]$ is always a left Ore ring, and a right Ore ring if σ is an automorphism of k.

Example 1 For any differential field k with derivation δ, $k[D; 1_k, \delta]$ is a ring of linear ordinary differential operators. If $k = \mathbb{C}(n)$ and σ is the automorphism of k over $\mathbb{C}$ that takes n to $n+1$, then $k[E; \sigma, 0]$ is the ring of linear ordinary recurrence operators, while $k[E; \sigma, \Delta]$ is the ring of linear ordinary difference operators where $\Delta = \sigma - 1_k$. If $k = \mathbb{C}(t)$ and σ is the automorphism of k over $\mathbb{C}$ that takes t to qt for a given $q \in \mathbb{C}^*$, then $k[B; \sigma, \Delta]$ is the ring of linear ordinary q-difference operators where $\Delta = (\sigma - 1_k)/(t(q-1))$.

4 Pseudo-linear maps

Let k, σ, δ be as above and V be a vector space over k. A map $\theta : V \to V$ is called *k-pseudo-linear (with respect to σ and δ)* if $\theta(u+v) = \theta u + \theta v$ and $\theta(au) = \sigma(a)\,\theta u + \delta a\; u$ for any $u, v \in V$ and $a \in k$. A k-pseudo linear map is clearly linear with respect to the constant subfield of k. Suppose now that $\dim_k(V) = n$ is finite, and let $\mathcal{B} = (b_1, \dots, b_n)$ be a given basis for V over k. Then the *matrix of θ with respect to $\mathcal{B}$* is the matrix $M_{\mathcal{B}}(\theta) = (m_{ij})$ with entries in k given by $\theta b_i = \sum_{j=1}^{n} m_{ji} b_j$ for all i's. The action of θ on the coordinates with respect to $\mathcal{B}$ is then given by

$$\theta \begin{bmatrix} v_1 \\ \vdots \\ v_n \end{bmatrix} = M_{\mathcal{B}}(\theta)\, \sigma \begin{bmatrix} v_1 \\ \vdots \\ v_n \end{bmatrix} + \delta \begin{bmatrix} v_1 \\ \vdots \\ v_n \end{bmatrix}.$$

Conversely, for any $n \times n$ matrix M with entries in k, the map defined on V by the above formula with $M_{\mathcal{B}}(\theta)$ replaced by M is k-pseudo linear, and its matrix with respect to $\mathcal{B}$ is M.

Let now $k[x;\sigma,\delta]$ be a skew polynomial ring. Any k-pseudo-linear map θ of V induces an action $*_\theta : k[x;\sigma,\delta] \times V \to V$ given by

$$\left(\sum_{i=0}^{n} a_i x^i\right) *_\theta u = \sum_{i=0}^{n} a_i \theta^i u$$

for any $u \in V$. This action is linear with respect to the constants of k, so the elements of $k[x;\sigma,\delta]$ can be viewed as linear operators acting on V. Furthermore, $(pq) *_\theta u = p *_\theta (q *_\theta u)$ for any $p, q \in k[x;\sigma,\delta]$ and $u \in V$, which means that the multiplication in $k[x;\sigma,\delta]$ corresponds to the composition of linear operators. Given a pseudo-linear map θ and $p \in k[x;\sigma,\delta]$, we say that $\alpha \in V$ is a *zero of p (with respect to θ)* if $p *_\theta \alpha = 0$.

Let K be a field extension of k, $\theta : K \to K$ be k-pseudo linear, $\alpha \in K^*$ be such that $u = \theta(\alpha)/\alpha \in k$, and $p \in k[x;\sigma,\delta]$. Doing a right Euclidean division of p by $x - u$ we obtain $p = q(x-u) + r$ where $q \in k[x;\sigma,\delta]$ and $r \in k$. We then have

$$p *_\theta \alpha = q *_\theta (x-u) *_\theta \alpha + r\alpha = q *_\theta (\theta\alpha - u\alpha) + r\alpha = q *_\theta 0 + r\alpha = r\alpha$$

which implies that α is a zero of p with respect to θ if and only if $x - \theta(\alpha)/\alpha$ divides p exactly on the right. We make use of this remark in the factorisation algorithm which we present next.

5 Factorisation of skew-polynomials

We say that $p \in k[x;\sigma,\delta] \setminus k$ is *irreducible* if $p = ab$ for $a, b \in k[x;\sigma,\delta]$ implies that $a \in k$ or $b \in k$. We say that $r \in k[x;\sigma,\delta]$ is *similar* to $s \in k[x;\sigma,\delta]$ if $\operatorname{lclm}(s,t) = rt$ for some $t \in k[x;\sigma,\delta]$ such that $\operatorname{gcrd}(s,t) \in k^*$. It is clear that any monic $p \in k[x;\sigma,\delta]$ can be written as a product of irreducibles. Such a factorisation is however not unique, but a fundamental theorem of Ore [6] states that any two factorisations of p into irreducibles must have the same number of factors, and that the factors are similar in pairs (which implies in particular that the multisets of degrees are the same).

We outline now an algorithm (developped jointly with M. Petkovšek) that reduces the problem of factoring in $k[x;\sigma,\delta]$ to the problem of finding all the irreducible right factors of degree 1. See [2, 3] for proofs and detailed descriptions.

Suppose first that $\sigma \neq 1_k$, and let $u \in k$ be such that $u \neq \sigma(u)$. It can then be shown that the map $\phi_u : k[x;\sigma,\delta] \to k[y;\sigma,0]$ given by

$$\phi_u\left(\sum_i a_i x^i\right) = \sum_i a_i \left(\frac{y + \delta u}{u - \sigma(u)}\right)^i$$

is an isomorphism of skew polynomial rings, which means that factoring in $k[x;\sigma,\delta]$ is equivalent to factoring in $k[y;\sigma,0]$ in this case. We can thus assume in the rest of this section that either $\sigma = 1_k$ or that $\delta = 0_k$. We let $\theta : k \to k$ be σ if $\delta = 0_k$, δ otherwise.

The basic idea behind the factoring algorithm is trial division: let $p = x^n - \sum_{i=0}^{n-1} a_i x^i \in k[x;\sigma,\delta]$, and suppose that $q = x^m - \sum_{i=0}^{m-1} b_i x^i$ is a right factor of p. If we can determine the b_i's up to some undetermined constants c_{ij}'s, then equating the right-remainder of p by q to 0 yields a system of algebraic equations with coefficients in $C = \mathrm{Const}_{\sigma,\delta}(k)$ for the $c'_{ij}s$. If that system has no solution in C then p does not have a right factor of degree m in $k[x;\sigma,\delta]$, otherwise any solution gives rises to such a factor (the same applies if C is replaced by an algebraic extension of C). Therefore we proceed to determine the b_i's. Since $\sigma = 1_k$ or $\delta = 0_k$, we know from the theory of linear differential and difference equations that there exists a field extension K of k such that θ can be extended to either an automorphism or derivation of K, and $y_1, \dots, y_m \in K$, linearly independent over $\mathrm{Const}_{\sigma,\delta}(K)$, such that $q *_\theta y_j = 0$ for each j. Note then that $p *_\theta y_j = 0$ for each j since q is a right factor of p. Let $\mathcal{M}$ be the $n \times m$ matrix

$$\mathcal{M} = \begin{pmatrix} y_1 & y_2 & \cdots & y_m \\ \theta y_1 & \theta y_2 & \cdots & \theta y_m \\ \theta^{n-1} y_1 & \theta^{n-1} y_2 & \cdots & \theta^{n-1} y_m \end{pmatrix}$$

and for any set $S = \{s_1, \dots, s_m\}$ of m integers with $1 \le s_1 < \cdots < s_m \le n$, let $[S] = [s_1, \dots, s_m]$ be the minor obtained from the rows $s_1, \dots, s_m$ of $\mathcal{M}$. It can be shown that

(i) $[1, \dots, m] \neq 0$,

(ii) $[S] \neq 0 \Rightarrow \theta([S])/[S] \in k$

(iii) $b_i = (-1)^{m-i+1} [\overline{i+1}]/[1, \dots, m]$ where $\overline{j} = \{1, \dots, m+1\} \setminus \{j\}$.

Point (iii) reduces the problem of determining the b_i's to determining $[\overline{i}]$ for $1 \le i \le m+1$. Let w be the vector composed of all the N minors of size m of $\mathcal{M}$, where $N = \binom{n}{m}$. Using only combinatorial operations on sets of m integers in $\{1, \dots, n\}$, we can compute a matrix $M_m(p)$ with entries in k such that $\theta w = M_m(p) \cdot w$ [3]. From $M_m(p)$, we compute matrices $A_S(p)$ with entries in k such that

$$A_S(p) \cdot w = \left(\theta[S], \theta^2[S], \dots, \theta^N[S]\right)^T$$

for any set S of m integers in $\{1, \dots, n\}$. Suppose first that $A_S(p)$ is invertible for some particular set S. Then the row corresponding to $[S]$ of

$$w = A_S(p)^{-1} \cdot \left(\theta[S], \theta^2[S], \dots, \theta^N[S]\right)^T$$

yields an element p_S of $k[x;\sigma,\delta]$ such that $p_S *_\theta [S] = 0$, while all the other rows yield elements $p_{S,T}$ of $k[x;\sigma,\delta]$ such that $[T] = p_{S,T} *_\theta [S]$. In particular, $[1, \dots, m] = p_{S,\{1,\dots,m\}} *_\theta [S]$ is non-zero by (i), which implies that $[S] \neq 0$. Therefore $u_S = \theta([S])/[S] \in k$ by (ii) which implies that $x - u_S$ is a right factor of p_S in $k[x;\sigma,\delta]$. If we have an algorithm that returns all the u's in k (up to undetermined constants) such that $x-u$ is a right factor of p_S, then this yields all the possible candidates for $[S]$, hence for the $[\bar{i}]$'s, hence all the possible candidate right factors of p of degree m.

If all the $A_S(p)$'s are singular, then any non-zero element of the kernel of $A_S(p)^T$ yields a linear dependence between its rows, hence a linear dependence between $\theta[S],\ldots,\theta^N[S]$, hence an element p_S of $k[x;\sigma,\delta]$ such that $p_S *_\theta [S] = 0$. As before we obtain that for each i, either $[\bar{i}] = 0$ or $x - u_i$ is a right factor of $p_{[\bar{i}]}$ in $k[x;\sigma,\delta]$ where $u_i = \theta([\bar{i}])/[\bar{i}]$, so an algorithm that returns all the right factors of degree 1 in $k[x;\sigma,\delta]$ yields all the possible candidate right factors of p of degree m.

Thus, we have reduced factoring in $k[x;\sigma,\delta]$ to finding right factors of degree 1. There exist known algorithms for finding those factors in at least 2 important cases: if $\sigma = 1_k$ and k is a Liouvillian extension of its constant subfield [8], and if k is a rational function field, σ is the shift operator and $\delta = 0$ [7]. Our algorithm is therefore applicable in those cases.

6 A normal form for pseudo-linear maps

Let V be a finite-dimensional vector space over k, $\mathcal{B}$ a given basis for V over k, and $\theta : V \to V$ a k-pseudo linear map with associated matrix $M_{\mathcal{B}}(\theta)$. Let A be an invertible $n \times n$ matrix with entries in k. Then, $\mathcal{C} = A\mathcal{B}$ is a basis for V over k and we have

$$M_{\mathcal{C}}(\theta) = A^{-1} M_{\mathcal{B}}(\theta)\sigma(A) + A^{-1}\delta(A)$$

where σ and δ are applied to all the entries of A. A natural question is then to ask whether there exists invertible matrices for which $M_{\mathcal{C}}(\theta)$ has a given form. B. Zürcher has recently generalized Danilevski's weak Frobenius algorithm [4] to show that any pseudo-linear map can be brought via a change of basis to a block-diagonal form, where each block is a companion matrix of the form

$$\begin{pmatrix} 0 & 1 & 0 & \ldots & 0 \\ 0 & 0 & 1 & & \vdots \\ \vdots & & \ddots & \ddots & \\ 0 & 0 & \ldots & 0 & 1 \\ c_0 & c_1 & \ldots & c_{N-2} & c_{N-1} \end{pmatrix} \tag{1}$$

Theorem 1 (Zürcher [9]) *For any pseudo-linear map $\theta : V \to V$, we can compute a basis for V over k such that the matrix M of θ with respect to this basis is of the form $M = diag(C_1, C_2, \ldots, C_m)$ where the C_i's are all companion matrices.*

His algorithm can be used to uncouple linear systems of differential or difference equations, producing skew polynomials which can then be given to the factorisation algorithm. We outline in the next subsections how to apply this algorithm to systems of differential and difference equations.

6.1 Differential equations

Let $y' = My + v$ be a first-order differential system where M is an $m \times m$ matrix with entries in a differential field k with derivation $'$, and $v \in k^m$. Let K be any differential extension of k, $\sigma = 1_K$, $\delta : K \to K$ be given by $\delta a = -a'$, and $\theta : K^m \to K^m$ be the pseudo-linear map whose matrix with respect to the canonical basis is M. Zürcher's algorithm produces an invertible matrix A and companion matrices $C_1, \dots, C_q$ such that for $z = A^{-1}y$, $\theta z = \mathrm{diag}(C_1, C_2, \dots, C_q) \cdot z - z'$. Assuming without loss of generality that there is only one companion block C of the form (1), we get that $y \in K^m$ is a solution of $y' = My + v$ if and only if $z = A^{-1}y$ is a solution of $z' = Cz + w$ where $w = A^{-1}v \in k^m$. That last system is of the form $z_{i+1} = z_i' - w_i = z_1^{(i)} - \sum_{j=1}^{i} w_j^{(i-j)}$ for $1 < i < m$, and the last equation is $z_m' = \sum_{i=0}^{m-1} c_i z_{i+1} + w_m$, which becomes

$$z_1^{(m)} - \sum_{j=1}^{m} w_j^{(m-j)} = \sum_{i=0}^{m-1} c_i z_1^{(i)} - \sum_{i=0}^{m-1} c_i \sum_{j=1}^{i} w_j^{(i-j)} + w_m \qquad (2)$$

which is an uncoupled differential equation for z_1.

Example 2 ([1]) Consider the differential system

$$t^2 dy/dt = My = \begin{pmatrix} 4t+1 & -5t & 7t & -8t & 8t & -6t \\ -10t & 9t+1 & -14t & 16t & -16t & 12t \\ -5t & 5t & -8t+1 & 8t & -8t & 6t \\ 10t & -10t & 14t & -17t+1 & 16t & -12t \\ 5t & -5t & 7t & -8t & 7t+1 & -6t \\ -5t & 5t & -7t & 8t & -8t & 5t+1 \end{pmatrix} \cdot y$$

Taking $' = t^2 dy/dt$, $\sigma = 1$, $\delta = -t^2 d/dt$ and applying Zürcher's algorithm to the pseudo-linear map whose matrix is M, we get the invertible matrix

$$A = \begin{pmatrix} 1 & 0 & 0 & 0 & 0 & 0 \\ \frac{-5t+1}{5t} & -\frac{1}{5t} & \frac{7}{5} & -\frac{8}{5} & \frac{8}{5} & -\frac{6}{5} \\ -1 & 0 & 1 & 0 & 0 & 0 \\ 2 & 0 & 0 & 1 & 0 & 0 \\ 1 & 0 & 0 & 0 & 1 & 0 \\ -1 & 0 & 0 & 0 & 0 & 1 \end{pmatrix}$$

and the companion matrices

$$C_1 = \begin{pmatrix} 0 & 1 \\ 5t^2 - 5t - 1 & 5t + 2 \end{pmatrix}, \qquad C_2 = C_3 = C_4 = C_5 = (1 - t)$$

such that y is a solution of $t^2 dy/dt = My$ if and only if $z = A^{-1}y$ is a solution of $t^2 dz/dt = Tz$ where $T = \mathrm{diag}(C_1, C_2, C_3, C_4, C_5)$. Using (2) we get the uncoupled equations $z_2 = t^2 dz_1/dt$, $t^2 dz_i/dt = (1-t)z_i$ for $i \in \{3, 4, 5, 6\}$. and

$$t^2 \frac{d}{dt}\left(t^2 \frac{dz_1}{dt}\right) = \left(5t^2 - 5t - 1\right) z_1 + (5t+2)t^2 \frac{dz_1}{dt}.$$

The corresponding skew polynomial can be factored as a product of operators of degree 1 [8], yielding the general solution

$$z_1 = \left(c_1 t^5 + \frac{c_2}{t}\right) e^{-1/t}$$

while for $i \in \{3, 4, 5, 6\}$, the equations $t^2 dz_i/dt = (1-t)z_i$ have general solutions $z_i = c_i e^{-1/t}/t$. Using $y = Az$ yields the general solution of the original system.

6.2 Difference equations

Let $Ey = My + v$ be a first-order recurrence system where M is an $m \times m$ matrix with entries in a difference field k with transform E (for example, which sends n to $n+1$), and $v \in k^m$. Let K be any difference extension of k, $\sigma = E^{-1}$, $\delta = \sigma - 1_K$, and $\theta : K^m \to K^m$ be the pseudo-linear map whose matrix with respect to the canonical basis is $\sigma(M) - I$. Zürcher's algorithm produces an invertible matrix A and companion matrices $C_1, \dots, C_q$ such that for $z = A^{-1}y$, $\theta z = \mathrm{diag}(C_1, C_2, \dots, C_q) \cdot \sigma(z) + \delta z$. Assuming without loss of generality that there is only one companion block C of the form (1), we get that $y \in K^m$ is a solution of $Ey = My + v$ if and only if $z = A^{-1}y$ is a solution of $C\sigma(z) + \delta z = A^{-1}v$. Applying E on both sides of that system, we get $E(C)z + z - Ez = w$ where $w = E(A^{-1}v)$, hence $Ez = Tz - w$ where $T = E(C) + I$. That last system is of the form $z_{i+1} = Ez_i - z_i + w_i = \Delta^i z_1 + \sum_{j=1}^{i} E^{i-j} w_j$ for $1 < i < m$, where $\Delta = E - 1_K$ is the associated difference operator. The last equation is $Ez_m = \sum_{i=0}^{m-1} E(c_i) z_{i+1} + z_m + w_m$ which becomes

$$\Delta^n z_1 + \sum_{j=1}^{i} \Delta E^{i-j} w_j = \sum_{i=0}^{m-1} E(c_i) \Delta^i z_1 + \sum_{i=0}^{m-1} E(c_i) \sum_{j=1}^{i} E^{i-j} w_j + w_m \qquad (3)$$

which is an uncoupled difference equation for z_1.

Example 3 Consider the recurrence system $y(n+1) = M(n)y(n)$ where $M(n)$ is

$$\begin{pmatrix} n & 1 & 0 & n+3 \\ s_n & 4n^2+36n+80 & 3n+13 & 4n^3+48n^2+186n+233 \\ -2n^2-10n & -2n-10 & -2 & -2n^2-16n-29 \\ -4n^2-18n+2 & -4n-20 & -3-\frac{1}{n+4} & -4n^2-32n-58-\frac{1}{n+4} \end{pmatrix}$$

with $s_n = 4n^3 + 34n^2 + 71n - 7$. Taking $En = n + 1$, $\sigma n = n - 1$, $\delta = \sigma - 1$ and applying Zürcher's algorithm to the pseudo-linear map whose matrix is $M(n-1) - I$, we get the invertible matrix

$$A = \begin{pmatrix} 1 & 0 & 0 & 0 \\ 4n^2 + 27n + 49 & 1 & -3n - 9 & -n - 3 \\ -2n - 8 & 0 & 1 & 0 \\ -4n - 16 & 0 & 3 & 1 \end{pmatrix}$$

and the companion matrices

$$C_1 = \begin{pmatrix} 0 & 1 \\ 2n - 1 & n - 2 \end{pmatrix}, \qquad C_2 = \begin{pmatrix} 0 & 1 \\ \frac{-4}{n+3} & -\frac{2n+7}{n+3} \end{pmatrix}$$

such that y is a solution of $y(n+1) = M(n)y(n)$ if and only if $z = A^{-1}y$ is a solution of $z(n+1) = Tz(n)$ where $T = \operatorname{diag}(C_1(n+1), C_2(n+1)) + 1$. Using (3) we get the uncoupled equations $z_2 = \Delta z_1$, $z_4 = \Delta z_3$,

$$\Delta^2 z_1 = (2n+1)z_1 + (n-1)\Delta z_1, \quad \text{and} \quad \Delta^2 z_3 = -\frac{4}{n+4} z_3 - \frac{2n+9}{n+4} \Delta z_3 \,.$$

The corresponding skew polynomials can be factored as products of operators of degree 1 [7], yielding the general solutions

$$z_1 = n! \left(c_1 + c_2 \sum_{i=0}^{n} \frac{(-1)^i}{i!} \right), \qquad z_3 = \frac{c_3 + c_4(-1)^n(2n+3)}{(n+1)(n+2)}$$

and $y = Az$ yields the general solution of the original system.

References

[1] M. A. Barkatou (1993): An algorithm for computing a companion block diagonal form for a system of linear differential equations, *Applicable Algebra in Engineering, Communication and Computing* **4,** 185–195.

[2] M. Bronstein & M. Petkovšek (1994): On Ore rings, linear operators and factorisation, *Programmirovanie* **1,** 27–45. Also Research Report 200, Informatik, ETH Zürich.

[3] M. Bronstein & M. Petkovšek (1994): On the factorisation of skew polynomials, submitted to the Journal of Symbolic Computation.

[4] A. Danilewski (1937): The numerical solution of the secular equation (in russian), *Mat. Sbornik* **2,** 169–171.

[5] N. Jacobson (1937): Pseudo-linear transformations, *Annals of Mathematics* **38,** 484–507.

[6] O. Ore (1933): Theory of non commutative polynomials, *Annals of Mathematics* **34,** 480–508.

[7] M. Petkovšek (1992): Hypergeometric solutions of linear difference equations with polynomial coefficients, *Journal of Symbolic Computation* **14,** 243–264.

[8] M. Singer (1991): Liouvillian solutions of linear differential equations with liouvillian coefficients, *Journal of Symbolic Computation* **11,** 251–273.

[9] B. Zürcher (1994): *Rationale Normalform von pseudo-linearen Abbildungen*, Diplomarbeit Bericht, Mathematik, ETH Zürich.

M. Bronstein (`bronstein@inf.ethz.ch`)

Inst. for Scientific Computation, ETH Zürich, 8092 Zürich (Switzerland).

Progress in Mathematics, Vol. 143, © 1996 Birkhäuser Verlag Basel/Switzerland

Gröbner basis in the classification of characteristically nilpotent filiform Lie algebras of dimension 10

F. J. Castro-Jiménez[1], J. Núñez-Valdés[2]

1 Introduction

The aim of this paper is to study complex filiform characteristically nilpotent Lie algebras of dimension 10. Filiform Lie algebras were defined by M. Vergne in her Thèse de 3e cycle [10]. Characteristically nilpotent Lie algebras were introduced by Dixmier and Lister [4].

According to the classification of nilpotent Lie algebras of dimension at most 6, due to Morozov [8], there are not characteristically nilpotent Lie algebras of dimension ≤ 6 over $\mathbb{C}$. In [2], R. Carles proves that the set of characteristically nilpotent Lie algebras of dimension n is a (non-empty) Zariski constructible subset of the variety of laws of nilpotent Lie algebras over $\mathbb{C}^n$, $n \geq 7$. Morever, in [6] Godfrey proves that there exist three (up to isomorphism) filiform characteristically nilpotent Lie algebras of dimension 7. Related to dimension 8, Echarte and Núñez exhibt, in [5], seven (up to isomorphism) filiform Lie algebras and five uniparametric families of filiform Lie algebras which are characteristically nilpotent. Finally, in [3], we study the case of dimension 9 and we prove that the set of filiform characteristically nilpotent Lie algebras is a Zariski constructible set in the variety of filiform Lie algebras and we give a expression of this set as a finite union of locally closed sets.

The structure of this paper is the following. In sections 1 and 2 we introduce some notations and we give a proof of a result of Hakimjanov, Ancochea and Goze (see [7]). In fact, we prove that the set of laws (up to isomorphism) of filiform Lie algebras over $\mathbb{C}^{10}$ is an affine algebraic subset V_{10} of $\mathbb{C}^{13}$ with Krull dimension 10. This set has three irreducible components, each of which is defined by hyperquadrics.

In section 3 we give the system $\mathcal{S}$ of 13 linear homogeneous equations whose solutions are the derivations of a filiform Lie algebra of dimension 10. This system has coefficients in the quotient ring of the affine algebraic set V_{10}.

In the last three sections we study the solutions of the system $\mathcal{S}$ on each irreducible component of V_{10}. On such components the set of points corresponding to characteristically nilpotent Lie algebras is given as a finite union of Zariski locally closed subsets. We apply the Buchberger's Gröbner basis theory and some of computations were made by using the *Mathematica* package.

---

(1) Partially supported by DGICYT PB90-0883

(2) Partially supported by Junta de Andalucía, Ayuda a Grupos

2 Notation

2.1 Filiform Lie algebras

In this paper, all the Lie algebras which appear will be considered over the complex number $\mathbb{C}$.

Let $\mathbf{g} = (\mathbb{C}^n, \mu)$ be a Lie algebra of dimension n, with μ the associated law. We consider the lower central series of $\mathbf{g}$ defined by $C^1\mathbf{g} = \mathbf{g}, \quad C^i\mathbf{g} = \mu(\mathbf{g}, C^{i-1}\mathbf{g})$. The Lie algebra $\mathbf{g}$ is *filiform* if $\dim_{\mathbb{C}} C^i\mathbf{g} = n - i$ for $2 \leq i \leq n$. If $x \in \mathbf{g}$ we denote by $ad(x)$ the adjoint mapping associated to x (i.e. the map $y \mapsto \mu(x, y)$).

Let $\mathbf{g}$ be a filiform Lie algebra of dimension n. Then there exists a basis $\mathcal{B} = \{e_1, \ldots, e_n\}$ of $\mathbf{g}$ such that $e_1 \in \mathbf{g} \setminus C^2\mathbf{g}$, the matrix of $ad(e_1)$ with respect to $\mathcal{B}$ has a Jordan block of order $n-1$ and $C^i\mathbf{g}$ is the vector space generated by $\{e_2, \ldots, e_{n-(i-1)}\}$ with $2 \leq i \leq n-1$. A such basis is called an *adapted basis*. Sometimes, we will use $[x, y]$ instead of $\mu(x, y)$ for the Lie bracket in a Lie algebra.

2.2 Affine algebraic sets

We denote by $\mathbb{C}[\underline{x}]$ the ring of polynomials in n variables $\underline{x} = (x_1, \ldots, x_n)$ over $\mathbb{C}$. Let I be an ideal of $\mathbb{C}[\underline{x}]$. We denote by $\mathcal{V}(I)$ the affine algebraic set defined by I (i.e., $\mathcal{V}(I) = \{\underline{a} \in \mathbb{C}^n \mid f(\underline{a}) = 0, \forall f \in I\}$). If $\{f_1, \ldots, f_m\}$ is a set of generators for I, we denote $\mathcal{V}(f_1, \ldots, f_m) = \mathcal{V}(I)$. The *Zariski topology* on the affine space $\mathbb{C}^n$ is the one whose closed sets are $\mathcal{V}(I)$, where I is an ideal. For each $f \in \mathbb{C}[\underline{x}]$, let $D(f)$ denote the complement of $\mathcal{V}(f)$ in $\mathbb{C}^n$. If $\{f_1, \ldots, f_m\} \subset \mathbb{C}[\underline{x}]$, let $< f_1, \ldots, f_m >$ denote the ideal generated by the f_i.

2.3 Gröbner bases

Let $\prec$ be a total ordering on $\mathbb{N}^n$ such that $1 \prec \alpha$ and $(\alpha \prec \beta \Rightarrow \alpha + \gamma \prec \beta + \gamma)$ for all $\alpha, \beta, \gamma \in \mathbb{N}^n$. For each

$$f = \sum_{\alpha \in \mathbb{N}^n} f_\alpha \underline{x}^\alpha \in \mathbb{C}[\underline{x}]$$

let $\mathcal{N}(f)$ denote the *Newton diagram* of f (i.e., $\mathcal{N}(f) = \{\alpha \in \mathbb{N}^n \mid f_\alpha \neq 0\}$). We call the *leading monomial* of f – and we denote it by $LM(f)$ – the maximal element, *with respect to* $\prec$, of the set $\mathcal{N}(f)$. Obviously, $LM(f.g) = LM(f) + LM(g)$. For each ideal I of $\mathbb{C}[\underline{x}]$, we denote by $LM(I)$ the set $\{LM(f) \mid f \in I\}$. Note that $LM(I) + \mathbb{N}^n = LM(I)$.

A set of generators $\{f_1, \ldots, f_m\}$ for an ideal I is called a Gröbner basis for I (*with respect to* $\prec$) if

$$LM(I) = \bigcup_{i=1}^{m} (LM(f_i) + \mathbb{N}^n).$$

All the ideals of $\mathbb{C}[\underline{x}]$ have a Gröbner basis (*with respect to* $\prec$), which can be computed by a suitable algorithm starting from a set of generators. Let I be an ideal of $\mathbb{C}[\underline{x}]$ and let $\mathcal{F} = \{f_1, \ldots, f_m\}$ a Gröbner basis of I. The main result related to this theory is the following: Any $f \in \mathbb{C}[\underline{x}]$ has an unique *normal form* $r(f; \mathcal{F})$ with respect to $\mathcal{F}$ and f belongs to I if and only if $r(f; \mathcal{F}) = 0$.

For a general overviews of the theory of Gröbner bases the reader can consult [1].

3 The set of filiform Lie algebras of dimension 10

3.1 The equations

Proposition 3.1.1 ([7]) *The set of filiform Lie algebras laws over* $\mathbb{C}^{10}$ *can be parametrized (up to isomorphism) by the points of an affine algebraic set* $V_{10} \subset \mathbb{C}^{13}$ *of Krull dimension* 10. *Furthermore, if* $\mathbf{g} = (\mathbb{C}^{10}, \mu)$ *is a filiform Lie algebra, then there exists a basis* $\{e_1, \ldots, e_{10}\}$ *of* $\mathbf{g}$ *such that:*

1. $\mu(e_1, e_i) = e_{i-1},\ i \geq 3$
2. $\mu(e_4, e_9) = a_1 e_2$
3. $\mu(e_4, e_{10}) = a_2 e_2 + a_1 e_3$
4. $\mu(e_5, e_8) = -a_1 e_2$
5. $\mu(e_5, e_9) = b_2 e_2$
6. $\mu(e_5, e_{10}) = b_3 e_2 + (a_2 + b_2) e_3 + a_1 e_4$
7. $\mu(e_6, e_7) = a_1 e_2$
8. $\mu(e_6, e_8) = c_2 e_2$
9. $\mu(e_6, e_9) = c_3 e_2 + (b_2 + c_2) e_3$
10. $\mu(e_6, e_{10}) = c_4 e_2 + (b_3 + c_3) e_3 + (a_2 + 2b_2 + c_2) e_4 + a_1 e_5$
11. $\mu(e_7, e_8) = d_1 e_2 + c_2 e_3$
12. $\mu(e_7, e_9) = d_2 e_2 + (c_3 + d_1) e_3 + (b_2 + 2c_2) e_4$
13. $\mu(e_7, e_{10}) = d_3 e_2 + (d_2 + c_4) e_3 + (b_3 + 2c_3 + d_1) e_4 + (a_2 + 3b_2 + 3c_2) e_5 + a_1 e_6$
14. $\mu(e_8, e_9) = f_1 e_2 + d_2 e_3 + (c_3 + d_1) e_4 + (b_2 + 2c_2) e_5$
15. $\mu(e_8, e_{10}) = f_2 e_2 + (d_3 + f_1) e_3 + (c_4 + 2d_2) e_4 + (b_3 + 3c_3 + 2d_1) e_5 + (a_2 + 4b_2 + 5c_2) e_6 + a_1 e_7$
16. $\mu(e_9, e_{10}) = g_1 e_2 + f_2 e_3 + (d_3 + f_1) e_4 + (c_4 + 2d_2) e_5 + (b_3 + 3c_3 + 2d_1) e_6 + (a_2 + 4b_2 + 5c_2) e_7 + a_1 e_8$

with $a_i, b_i, c_i, d_i, f_i, g_i \in \mathbb{C}$ *verifying the equations:* $P_1 = P_2 = P_3 = 0$ *where*

$$P_1 := 3b_2^2 + 3b_2c_2 - 2a_2c_2$$

$$P_2 := a_1(2a_2 + 7b_2 + 7c_2)$$

$$P_3 := -3b_3c_2 + 7b_2c_3 + 2c_2c_3 + 2a_1c_4 - 2a_2d_1 - 2b_2d_1 - 7c_2d_1 + 5a_1d_2$$

Proof. Let $\mathcal{B} = \{e_1, \dots, e_{10}\}$ be an adapted basis of $\mathbf{g}$ (see 2.1). So, we have

$$C^i\mathbf{g} = \mathbb{C} < e_2, \dots, e_{10-(i-1)} >, \ i = 2, \dots, 9 \qquad (3.1.1)$$
$$C^{10}\mathbf{g} = 0$$

We denote by $c_{i,j}^k$ the constants of structure of $\mathbf{g}$, i.e.

$$\mu(e_i, e_j) = \sum_{k=1}^{10} c_{i,j}^k e_k.$$

We have $\mu(e_1, e_i) = e_{i-1}$, $i = 3, \dots, 10$. Since $C^{10}\mathbf{g} = 0$ then $\mu(e_2, e_i) = 0$, $i = 1, \dots, 10$.

From Jacobi identities related to (e_1, e_3, e_j) for $j \geq 4$ we can obtain, by using 3.1.1, that

1. $\mu(e_3, e_j) = 0, \ \ j = 4, \dots, 9$
2. $\mu(e_3, e_{10}) = c_{3,10}^2 e_2$

However, by using the base change given by $e'_{10} = e_{10} + c_{3,10}^2 e_1$, $e'_k = e_k$, $k = 1, \dots, 9$, we can get that e_3 commutes with all of the vectors of the new basis of $\mathbf{g}$ except with the vector e_1. So, from now on, we will suppose that $\mu(e_3, e_j) = 0, \ \ j = 4, \dots, 10$. From Jacobi identities related to (e_1, e_i, e_j) for $i \geq 4, j \geq i+1$, we obtain

1. $\mu(e_4, e_j) = 0, \ j = 5, \dots, 8; \quad \mu(e_4, e_9) = c_{4,9}^2 e_2; \quad \mu(e_4, e_{10}) = c_{4,10}^2 e_2 + c_{4,9}^2 e_3$
2. $\mu(e_5, e_j) = 0, \ j = 6, 7; \quad \mu(e_5, e_8) = c_{5,8}^2 e_2; \quad \mu(e_5, e_9) = c_{5,9}^2 e_2 + (c_{5,8}^2 + c_{4,9}^2) e_3$
3. $\mu(e_5, e_{10}) = c_{5,10}^2 e_2 + (c_{5,9}^2 + c_{4,10}^2) e_3 + (c_{5,8}^2 e_2 + 2c_{4,9}^2) e_4$
4. $\mu(e_6, e_7) = c_{6,7}^2 e_2; \quad \mu(e_6, e_8) = c_{6,8}^2 e_2 + (c_{5,8}^2 + c_{6,7}^2) e_3$
5. $\mu(e_6, e_9) = c_{6,9}^2 e_2 + (c_{5,9}^2 + c_{6,8}^2) e_3 + (c_{4,9}^2 + 2c_{5,8}^2 + c_{6,7}^2) e_4$
6. $\mu(e_6, e_{10}) = c_{6,10}^2 e_2 + (c_{5,10}^2 + c_{6,9}^2) e_3 + (c_{4,10}^2 + 2c_{5,9}^2 + c_{6,8}^2) e_4 + (3c_{4,9}^2 + 3c_{5,8}^2 + c_{6,7}^2) e_5$

7. $\mu(e_7, e_8) = c_{7,8}^2 e_2 + c_{6,8}^2 e_3 + (c_{5,8}^2 + c_{6,7}^2) e_4$

8. $\mu(e_7, e_9) = c_{7,9}^2 e_2 + (c_{6,9}^2 + c_{7,8}^2) e_3 + (c_{5,9}^2 + 2c_{6,8}^2) e_4 + (c_{4,9}^2 + 3c_{5,8}^2 + 2c_{6,7}^2) e_5$

9. $\mu(e_7, e_{10}) = c_{7,10}^2 e_2 + (c_{6,10}^2 + c_{7,9}^2) e_3 + (c_{5,10}^2 + 2c_{6,9}^2 + c_{7,8}^2) e_4 + (c_{4,10}^2 + 3c_{5,9}^2 + 3c_{6,8}^2) e_5 + (4c_{4,9}^2 + 6c_{5,8}^2 + 3c_{6,7}^2) e_6$

10. $\mu(e_8, e_9) = c_{8,9}^2 e_2 + c_{7,9}^2 e_3 + (c_{6,9}^2 + c_{7,8}^2) e_4 + (c_{5,9}^2 + 2c_{6,8}^2) e_5 + (c_{4,9}^2 + 3c_{5,8}^2 + 2c_{6,7}^2) e_6$

11. $\mu(e_8, e_{10}) = c_{8,10}^2 e_2 + (c_{7,10}^2 + c_{8,9}^2) e_3 + (c_{6,10}^2 + 2c_{7,9}^2) e_4 + (c_{5,10}^2 + 3c_{6,9}^2 + 2c_{7,8}^2) e_5 + (c_{4,10}^2 + 4c_{5,9}^2 + 5c_{6,8}^2) e_6 + (5c_{4,9}^2 + 9c_{5,8}^2 + 5c_{6,7}^2) e_7$

12. $\mu(e_9, e_{10}) = c_{9,10}^2 e_2 + c_{8,10}^2 e_3 + (c_{7,10}^2 + c_{8,9}^2) e_4 + (c_{6,10}^2 + 2c_{7,9}^2) e_5 + (c_{5,10}^2 + 3c_{6,9}^2 + 2c_{7,8}^2) e_6 + (c_{4,10}^2 + 4c_{5,9}^2 + 5c_{6,8}^2) e_7 + (5c_{4,9}^2 + 9c_{5,8}^2 + 5c_{6,7}^2) e_8$

For the sake of simplicity we denote

$$\begin{array}{lllll} c_{4,9}^2 = a_1 & c_{4,10}^2 = a_2 & c_{5,8}^2 = b_1 & c_{5,9}^2 = b_2 & c_{5,10}^2 = b_3 \\ c_{6,7}^2 = c_1 & c_{6,8}^2 = c_2 & c_{6,9}^2 = c_3 & c_{6,10}^2 = c_4 & c_{7,8}^2 = d_1 \\ c_{7,9}^2 = d_2 & c_{7,10}^2 = d_3 & c_{8,9}^2 = f_1 & c_{8,10}^2 = f_2 & c_{9,10}^2 = g_1 \end{array}$$

If $P(i, j, k, l)$ denotes the coefficient of the vector e_l in the Jacobi identity related to (e_i, e_j, e_k), a system of generators of the ideal I' of

$$R' = \mathbb{C}[a_1, a_2, b_1, b_2, b_3, c_1, c_2, c_3, c_4, d_1, d_2, d_3, f_1, f_2, g_1]$$

generated by the set

$$\{P(5,9,10,l), P(6,8,10,l), P(6,9,10,l), P(7,8,9,l), \\ P(7,8,10,l), P(7,9,10,l), P(8,9,10,l) \mid l = 1, \dots, 10\}$$

is given by the following polynomials:

$P_4 := -3a_1b_1 - 3b_1^2 - 5a_1c_1 - 10b_1c_1 - 5c_1^2$

$P_5 := -2a_1^2 - 6a_1b_1 - 9b_1^2 - 5b_1c_1$

$P_6 := 2a_2b_1 - 5a_1b_2 - 3b_1b_2 - 5b_2c_1 - 6a_1c_2 - 9b_1c_2 - 10c_1c_2$

$P_7 := -2a_1^2 - 9a_1b_1 - 12b_1^2 - 5a_1c_1 - 15b_1c_1 - 5c_1^2$

$P_8 := -3b_1^2 + 2a_1c_1 + b_1c_1 + 2c_1^2$

$P_9 := -3b_1b_2 + 2a_2c_1 + 4b_2c_1 - 4a_1c_2 - 9b_1c_2 + 2c_1c_2$

$P_{10} := -3a_1b_1 - 6b_1^2 - 3a_1c_1 - 9b_1c_1 - 3c_1^2 = -3(b_1 + c_1)(a_1 + 2b_1 + c_1)$

$P_{11} := -3b_2^2 + 3b_1b_3 + 3b_3c_1 + 2a_2c_2 - 3b_2c_2 - 6a_1c_3 - 6b_1c_3 - 6a_1d_1 - 9b_1d_1 - 3c_1d_1$

$P_{12} := 2a_2b_1 - 5a_1b_2 - 6b_1b_2 + 2a_2c_1 - b_2c_1 - 10a_1c_2 - 18b_1c_2 - 8c_1c_2$

$P_{13} := -2a_1^2 - 12a_1b_1 - 18b_1^2 - 8a_1c_1 - 24b_1c_1 - 8c_1^2 = -2(a_1 + 3b_1 + 2c_1)^2$

$P_{14} := 3b_3c_2 - 7b_2c_3 - 2c_2c_3 + 4b_1c_4 + 2c_1c_4 + 2a_2d_1 + 2b_2d_1 + 7c_2d_1 - 7a_1d_2 - 7b_1d_2$

$P_{15} := -3b_2^2 + 3b_1b_3 + 3b_3c_1 + 2a_2c_2 - 3b_2c_2 - 6a_1c_3 - 6b_1c_3 - 6a_1d_1 - 9b_1d_1 - 3c_1d_1$

$P_{16} := 2a_2b_1 - 5a_1b_2 - 6b_1b_2 + 2a_2c_1 - b_2c_1 - 10a_1c_2 - 18b_1c_2 - 8c_1c_2$

$P_{17} := -2a_1^2 - 12a_1b_1 - 18b_1^2 - 8a_1c_1 - 24b_1c_1 - 8c_1^2$

Now, from $P_{10} = P_{13} = 0$ we obtain $b_1 = -a_1$ and $c_1 = a_1$. Let I be the ideal corresponding to $I' + (a_1 + b_1, a_1 - c_1)$ in the quotient ring

$$R = R'/(a_1 + b_1, a_1 - c_1) \simeq \mathbb{C}[a_1, a_2, b_2, b_3, c_2, c_3, c_4, d_1, d_2, d_3, f_1, f_2, g_1].$$

Thus, a minimal system of generators for the ideal I is $\{P_1, P_2, P_3\}$ with

$P_1 := 3b_2^2 + 3b_2c_2 - 2a_2c_2$

$P_2 := a_1(2a_2 + 7b_2 + 7c_2)$

$P_3 := -3b_3c_2 + 7b_2c_3 + 2c_2c_3 + 2a_1c_4 - 2a_2d_1 - 2b_2d_1 - 7c_2d_1 + 5a_1d_2$

It is clear that the Krull dimension of the algebraic subset V_{10} of $\mathbb{C}^{13}$ defined by $\{P_1 = P_2 = P_3 = 0\}$ is 10. □

3.2 The irreducible components

Lemma 3.2.1 ([7]) *With above notations, the decomposition of V_{10} as a finite union of irreducible components is $V_{10} = V_{10}^1 \bigcup V_{10}^2 \bigcup V_{10}^3$ where*

$$\begin{aligned} V_{10}^1 &= \mathcal{V}(-3b_2^2 + 2a_2c_2 - 3b_2c_2, \\ &\qquad 3b_3c_2 - 7b_2c_3 - 2c_2c_3 + 2b_2d_1 + 7c_2d_1 + 2a_2d_1, a_1) \\ V_{10}^2 &= \mathcal{V}(3b_2 + 7c_2, a_2 + 2b_2, 43b_2c_3 + 14a_1c_4 + 35b_2d_1 + 35a_1d_2 + 9b_2b_3) \\ V_{10}^3 &= \mathcal{V}(b_2 + c_2, a_2, 5b_2c_3 + 2a_1c_4 + 5b_2d_1 + 5a_1d_2 + 3b_2b_3) \end{aligned}$$

Proof. Let $H_1 := (7/4)a_1P_1 + ((3/4)b_2 - (1/2)a_2)P_2 = -a_1a_2(a_2 + 2b_2)$ and $H_2 := 7a_1P_1 + 3b_2P_2 = 2a_1a_2(3b_2 + 7c_2)$. So,

$$< P_1, P_2, P_3 >=< P_1, P_2, P_3, H_1, H_2 >,$$

which proves the decomposition. Now, to prove the irreducibility it suffices to note that V_{10}^i is generated by linear forms and hyperquadrics in $\mathbb{C}^{13}$ and that the hyperquadrics

$$\begin{gathered} \{-3b_2^2 + 2a_2c_2 - 3b_2c_2, \\ 3b_3c_2 - 7b_2c_3 - 2c_2c_3 + 2b_2d_1 + 7c_2d_1 + 2a_2d_1, \\ 5b_2c_3 + 2a_1c_4 + 5b_2d_1 + 5a_1d_2 + 3b_2b_3, \\ 43b_2c_3 + 14a_1c_4 + 35b_2d_1 + 35a_1d_2 + 9b_2b_3\} \end{gathered}$$

are irreducible in the ring R due to the dimension of the singular point set is, in each case, less than 9. □

4 Characteristically nilpotent filiform Lie algebras. Derivations of filiform Lie algebras

Let $\mathbf{g} = (\mathbb{C}^n, \mu)$ be a nilpotent Lie algebra of dimension n. We say that $\mathbf{g}$ is a derived Lie algebra if there exists a Lie algebra $\mathbf{g}' = (\mathbb{C}^{n+1}, \mu')$ such that $\mathbf{g}$ is a subalgebra of $\mathbf{g}'$ and $C^2\mathbf{g}' = \mathbf{g}$.

A Lie algebra $\mathbf{g}$ is called characteristically nilpotent if all its derivations are nilpotent. The following lemma is easy to prove:

Lemma 4.0.2 *A filiform Lie algebra* $\mathbf{g}$ *is characteristically nilpotent if and only if* $\mathbf{g}$ *is not a derived algebra.*

4.1 Derivations of a filiform Lie algebra

Let $D : \mathbf{g} \to \mathbf{g}$ be a derivation of $\mathbf{g}$. Let $A = (a_{i,j}) \in Mat(n \times n, \mathbb{C})$ be the matrix of D with respect to an adapted basis $\mathcal{B} = \{e_1, \ldots, e_n\}$ of $\mathbf{g}$. Since $D(C^k\mathbf{g}) \subset C^k\mathbf{g}$ for any k, we have

$$\begin{aligned} a_{i,j} &= 0, \qquad 2 \leq i \leq n-1; \quad j > i \\ a_{i,1} &= 0, \qquad 2 \leq i \leq n-1 \end{aligned} \tag{4.1.0}$$

Moreover, from $D[e_3, e_n] = [De_3, e_n] + [e_3, De_n]$ we deduce that $a_{n,1} = 0$. Thus, the eigenvalues of D are $\{a_{1,1}, \ldots, a_{n,n}\}$.

For each pair (i, j) with $1 \leq i < j \leq n$, we will denote by $ec(i, j, k)$ the coefficient of e_k $(1 \leq k \leq n)$ in the expression of $D[e_i, e_j] - [De_i, e_j] - [e_i, De_j]$ with respect to the basis $\mathcal{B}$. The elements of the matrix A verify the homogeneous linear system defined by

$$\mathcal{S} = \{ec(i, j, k) = 0,\ 1 \leq i < j \leq n,\ 1 \leq k \leq n\}. \tag{4.1.1}$$

Conversely, any matrix $A = (a_{i,j}) \in Mat(n \times n, \mathbb{C})$ verifying the system $\mathcal{S}$, defines a derivation in $\mathbf{g}$.

In [9] and [5] it is proved that if all the solutions of the system $\mathcal{S}$ verify $a_{1,1} = 0$, then $a_{i,i} = 0$ for all $2 \leq i \leq n$. In this case, all the derivations of $\mathbf{g}$ are nilpotent and thus the Lie algebra $\mathbf{g}$ is characteristically nilpotent. Obviously, if there exists a solution of $\mathcal{S}$ verifying $a_{1,1} \neq 0$, then the corresponding derivation is not nilpotent and the Lie algebra $\mathbf{g}$ is not characteristically nilpotent.

Note. In [9] and [5] the authors prove that, with the exception of two kinds of filiform Lie algebras (called by them, in [5], *algèbre modèle* and *algèbre especiale*), any derivation D of a filiform Lie algebra with matrix A with respect to an adapted basis verifies $a_{i,i} = \lambda_i a_{1,1}$ with $\lambda_i \in \mathbb{C} \setminus \{0\}$ and $2 \leq i \leq n$. These authors also prove that those two kinds of filiform Lie algebras are derived algebras and thus are not characteristically nilpotent.

4.2 The case $n = 10$

Let $\underline{u} = (a_1, a_2, b_2, b_3, c_2, c_3, c_4, d_1, d_2, d_3, f_1, f_2, g_1)$ be a point of V_{10} and $\{e_1, \ldots, e_{10}\}$ be a basis as in 3.1.1 of the corresponding filiform Lie algebra $\mathbf{g}_{\underline{u}} = (\mathbb{C}^{10}, \mu_{\underline{u}})$. Let $\mathcal{S}_{\underline{u}}$ the corresponding homogeneous linear system (see 4.1.1).

We can consider the linear system $\mathcal{S}_{\underline{u}}$ as a system whose coefficients are in the quotient ring R/I where I is the ideal generated by $\{P_1, P_2, P_3\}$ (see 3.1.1).

It there exists an algorithmic proceedure to decide if the filiform Lie algebra $\mathbf{g}_{\underline{u}}$ is not a characteristically nilpotent Lie algebra.

Indeed, it suffices to prove that there exists a solution of the system $\mathcal{S}_{\underline{u}}$ verifying $a_{1,1} \neq 0$. As we said above, if all the solutions of $\mathcal{S}_{\underline{u}}$ verify $a_{1,1} = 0$, then the filiform Lie algebra $\mathbf{g}_{\underline{u}}$ is characteristically nilpotent.

Fortunately, as we are using adapted base, the system $\mathcal{S}_{\underline{u}}$ has a very particular form and it can be reduced, by using elementary transformations in the quotient ring R/I (we use straightforward calculations in *Mathematica*), to a new system

$$\mathcal{S}'_{\underline{u}} = \{ec_i = 0\}_{i=1,\ldots,13}$$

with

$$ec_1 = 9a_1a_{1,1} - 7{a_1}^2a_{1,10} - a_1a_{2,2}$$

$$ec_2 = 10a_2a_{1,1} - a_2a_{2,2} + a_{1,10}(-13a_1a_2 - 14a_1b_2 - 14a_1c_2)$$

$$ec_3 = 10a_{1,1}b_2 - a_{2,2}b_2 + a_{1,10}(4a_1a_2 + 5a_1b_2 + 14a_1c_2)$$

$$ec_4 = -2{a_1}^2a_{1,8} - 2a_1a_2a_{1,9} + 2a_1a_{4,2} + 11a_{1,1}b_3 - a_{2,2}b_3 + a_{1,10}(-2{a_2}^2 - 7a_2b_2 - 14{b_2}^2 - 3a_1b_3 - 14b_2c_2 + 9a_1c_3 + 5a_1d_1)$$

$$ec_5 = 10a_{1,1}c_2 - a_{2,2}c_2 + a_{1,10}(-2a_1a_2 - 7a_1b_2 - 16a_1c_2)$$

$$ec_6 = 2{a_1}^2a_{1,8} - 2a_1a_{4,2} + a_{1,9}(-7a_1b_2 - 7a_1c_2) + 11a_{1,1}c_3 - a_{2,2}c_3 + a_{1,10}(-3a_2b_2 - 3{b_2}^2 - 7a_1b_3 - 6a_2c_2 - 15b_2c_2 - 14{c_2}^2 - 19a_1c_3 - 5a_1d_1)$$

$$ec_7 = -2a_2a_{1,9}b_2 + 2a_{4,2}b_2 + a_{1,8}(2a_1a_2 + 5a_1b_2 + 7a_1c_2) + 12a_{1,1}c_4 - a_{2,2}c_4 + a_{1,10}(-5a_2b_3 - 3b_2b_3 - 6b_3c_2 - 7a_2c_3 - 14b_2c_3 - 23c_2c_3 - 13a_1c_4 - 5c_2d_1 - 5a_1d_2)$$

$$ec_8 = -2{a_1}^2a_{1,8} + 2a_1a_{4,2} + a_{1,9}(7a_1b_2 + 7a_1c_2) + 11a_{1,1}d_1 - a_{2,2}d_1 + a_{1,10}(7a_1b_3 - 4a_2c_2 - 6b_2c_2 - 4{c_2}^2 + 9a_1c_3 - 5a_1d_1)$$

$$ec_9 = -2{a_1}^2a_{1,6} + 2a_1a_{6,2} + a_{5,2}(3b_2 + 3c_2) + a_{1,7}(-2a_1a_2 - 4a_1b_2 - 6a_1c_2) + 2a_{4,2}c_3 + a_{1,8}(-2a_2c_2 - 4a_1c_3) + a_{1,9}(-3b_2b_3 - 3b_3c_2 - 2a_2c_3 - 2a_1c_4 + 13a_{1,1}d_3 - a_{2,2}d_3) + a_{1,10}(-3{b_3}^2 - 3b_3c_3 - 6a_2c_4 - 6b_2c_4 - 4c_2c_4 - 6b_3d_1 - 9c_3d_1 - 5{d_1}^2 - 7a_2d_2 - 14b_2d_2 - 14c_2d_2 - 7a_1d_3 + 2a_1f_1)$$

$$ec_{10} = 2{a_1}^2 a_{1,6} - 2a_1 a_{6,2} + 5a_1 a_{1,7} c_2 + 3a_{5,2} c_2 + 2a_{4,2} d_1 + a_{1,8}(-3b_2 c_2 + 2a_1 c_3 - 2a_1 d_1) + a_{1,9}(-7b_2 c_3 - 2c_2 c_3 + 2b_2 d_1 + 7c_2 d_1 - 5a_1 d_2) + 13a_{1,1} f_1 - a_{2,2} f_1 + a_{1,10}(-4b_3 c_3 - 6{c_3}^2 - 3b_2 c_4 + 4c_2 c_4 + 5b_3 d_1 + 6c_3 d_1 + 5{d_1}^2 - 5a_2 d_2 - 13b_2 d_2 - 4c_2 d_2 - 5a_1 d_3 - 14a_1 f_1)$$

$$ec_{11} = -2a_1 a_{1,8} c_2 + 2a_{4,2} c_2 + a_{1,9}(-3{b_2}^2 - 3b_2 c_2) + 12a_{1,1} d_2 - a_{2,2} d_2 + a_{1,10} (-3b_2 b_3 + 3b_3 c_2 - 4a_2 c_3 - 16b_2 c_3 - 6c_2 c_3 - 4a_2 d_1 - 13b_2 d_1 - 7c_2 d_1 - 11a_1 d_2)$$

$$ec_{12} = a_{6,2}(4b_2 + 6c_2) + a_{1,6}(2a_1 a_2 + 3a_1 b_2 + a_1 c_2) + a_{5,2}(3c_3 + 3d_1) + a_{1,7} (-4b_2 c_2 - 6{c_2}^2 + 3a_1 c_3 + 3a_1 d_1) + 2a_{4,2} d_2 + a_{1,8}(-3b_3 c_2 - 4c_2 c_3 + 2a_1 c_4 - 2a_2 d_1 - 5b_2 d_1 - 7c_2 d_1 + 3a_1 d_2) + a_{1,9}(-3b_3 c_3 - 4b_2 c_4 - 6c_2 c_4 - 3b_3 d_1 - 2a_2 d_2) + 14a_{1,1} f_2 - a_{2,2} f_2 + a_{1,10}(-7b_3 c_4 - 6c_3 c_4 + c_4 d_1 - 3b_3 d_2 + 5d_1 d_2 - 7a_2 d_3 - 10b_2 d_3 - 6c_2 d_3 - 7a_2 f_1 - 14b_2 f_1 - 12c_2 f_1 - 13a_1 f_2)$$

$$ec_{13} = -2{a_1}^2 a_{1,4} + 2a_1 a_{8,2} + a_{1,5}(-2a_1 a_2 - 2a_1 b_2) + a_{7,2}(5b_2 + 5c_2) + a_{1,6}(-3{b_2}^2 + 2a_2 c_2 - 3b_2 c_2 - 4a_1 c_3) + a_{6,2}(4c_3 + 2d_1) + 3a_{5,2} d_2 + a_{1,7}(3b_3 c_2 - 7b_2 c_3 - 6c_2 c_3 - 2a_1 c_4 + 2a_2 d_1 - 3b_2 d_1 - 2a_1 d_2) + 2a_{4,2} f_1 + a_{1,8}(-4{c_3}^2 - 2c_3 d_1 - 8b_2 d_2 - 5c_2 d_2 - 4a_1 f_1) + a_{1,9}(-4c_3 c_4 - 2c_4 d_1 - 3b_3 d_2 - 5b_2 d_3 - 5c_2 d_3 - 2a_2 f_1 - 2a_1 f_2) + 15a_{1,1} g_1 - a_{2,2} g_1 + a_{1,10}(-4{c_4}^2 - c_4 d_2 + 5{d_2}^2 - 8b_3 d_3 - 6c_3 d_3 - 5d_1 d_3 + 3b_3 f_1 + 11c_3 f_1 + 5d_1 f_1 - 8a_2 f_2 - 12b_2 f_2 - 14c_2 f_2 - 11a_1 g_1)$$

Note. The system $\mathcal{S}'_{\underline{u}}$ is not equivalent to the system $\mathcal{S}_{\underline{u}}$, but if $A = (a_{i,j})$ is a solution of $\mathcal{S}_{\underline{u}}$ and $A' = (a'_{i,j})$ is a solution of $\mathcal{S}'_{\underline{u}}$ then $a_{1,1} = 0$ if and only if $a'_{1,1} = 0$.

So, to determine if the filiform Lie algebra $\mathbf{g}_{\underline{u}}$ is characteristically nilpotent, we have to check that any solution $A = (a_{i,j})$ of $\mathcal{S}'_{\underline{u}}$ verifies $a_{1,1} = 0$. To do this, we proceed, in V_{10}, component by component. We will prove that the set of points $\underline{u} \in V_{10}^i$ such that there exists a solution $A = (a_{i,j})$ of the system $\mathcal{S}'_{\underline{u}}$ verifying $a_{1,1} \neq 0$ is a Zariski constructible set and we will give an expression of this set as a finite union of Zariski locally closed subsets. The idea consists in cutting each component V_{10}^i with suitable hypersurfaces (hyperplanes in general) and studying the system induced by $\mathcal{S}'_{\underline{u}}$ in the corresponding set.

5 The irreducible component V_{10}^1

A system of generators of the ideal of the component V_{10}^1 is $\{P_1, a_1, P_3\}$, or equivalently, $\{2a_2c_2 - b_2(3b_2 + 3c_2), a_1, -3b_3c_2 + 7b_2c_3 + 2c_2c_3 - 2a_2d_1 - 2b_2d_1 - 7c_2d_1\}$.

From now on, $\mathcal{S}'_{\underline{u}}$ will denote the system $\mathcal{S}'_{\underline{u}}$ under the condition $a_1 = 0$.

5.1

We will first operate in the open set $D(a_2) \bigcap V_{10}^1 = \{\underline{u} \in V_{10}^1 \mid a_2 \neq 0\}$. In this set we have $a_{2,2} = 10a_{1,1}$. We distinguish two cases:

5.1.1 We work in $V^1_{10}\bigcap\mathcal{V}(b_2)\bigcap D(a_2)$. From $P_1 = P_3 = 0$ we have $c_2 = d_1 = 0$. Obviously, from $ec_6 = ec_{10} = ec_{11} = 0$ we deduce that in the open set $D(c_3)\bigcup D(d_2)\bigcup D(f_1)$ all the solutions of $\mathcal{S}'_{\underline{u}}$ will verify $a_{1,1} = 0$. So all the algebras in $D(c_3)\bigcup D(d_2)\bigcup D(f_1)$ are characteristically nilpotent Lie algebras.

Therefore, we will operate in the set

$$Z = V^1_{10}\bigcap D(a_2)\bigcap \mathcal{V}(b_2, c_3, d_2, f_1)$$

where the fourth equation of $\mathcal{S}'_{\underline{u}}$ reduces to $-2a_2^2a_{1,10} + a_{1,1}b_3 = 0$. Hence, $a_{1,10} = a_{1,1}b_3/(-2a_2^2)$ and the system is now reduced to

$$Q_1a_{1,1} = Q_2a_{1,1} = Q_3a_{1,1} = Q_4a_{1,1} = 0$$

with

$$Q_1 := -5b_3^2 + 4a_2c_4 \qquad Q_2 := -3b_3^3 - 6a_2b_3c_4 + 6a_2^2d_3$$
$$Q_3 := -7b_3^2c_4 + 8a_2^2f_2 - 7a_2b_3d_3 \quad Q_4 := -2b_3c_4^2 - 4b_3^2d_3 + 5a_2^2g_1 - 4a_2b_3f_2$$

Hence, in $Z\bigcap(\bigcup D(Q_i))$ all the filiform Lie algebras are characteristically nilpotent Lie algebras, whereas all the filiform Lie algebras in the set $Z\bigcap\mathcal{V}(Q_1, Q_2, Q_3, Q_4)$ are not. Note that the set $Z\bigcap\mathcal{V}(Q_1, Q_2, Q_3, Q_4)$ (which is a Zariski closed subset of $D(a_2)$) has Krull dimension 4.

5.1.2 Within $D(a_2b_2)\bigcap V^1_{10}$ we have $a_2 \neq 0$ and $b_2 \neq 0$ and thus, $c_2 \neq 0$, $b_2 + c_2 \neq 0$. By using elementary transformations in the ring of the component V^1_{10}, that is, in $R/ < a_1, P_1, P_3 >$, the system $\mathcal{S}'_{\underline{u}}$ is now reduced to the following:

$$Q_{2,i}a_{1,1} + Q'_{2,i}a_{1,10} = 0 \quad 1 \leq i \leq 5$$

with

$$Q_{2,1} := 3b_3 \quad Q'_{2,1} := -6a_2^2 - 21a_2b_2 - 28a_2c_2$$
$$Q_{2,2} := c_3 \quad Q'_{2,2} := -3a_2b_2 - 8a_2c_2 - 12b_2c_2 - 14{c_2}^2$$
$$Q_{2,3} := d_1 \quad Q'_{2,3} := -4a_2c_2 - 6b_2c_2 - 4{c_2}^2$$

$$Q_{2,4} := -4c_2c_4 + 4b_2d_2$$

$$Q'_{2,4} := 9{b_2}^2b_3 + 39b_2b_3c_2 + 81b_3{c_2}^2 - 8a_2b_2c_3 - 39{b_2}^2c_3 - 132b_2c_2c_3 + 51{b_2}^2d_1 + 129b_2c_2d_1 + 171{c_2}^2d_1$$

$$Q_{2,5} := 8c_2c_3c_4 - 8b_2c_4d_1 - 8c_2c_4d_1 - 12b_2c_2d_3 + 12{b_2}^2f_1 + 12b_2c_2f_1$$

$$Q'_{2,5} := -18b_2{b_3}^2c_2 - 243{b_3}^2{c_2}^2 + 24{b_2}^2b_3c_3 + 499b_2b_3c_2c_3 + 32{b_2}^2{c_3}^2 + 228b_2c_2{c_3}^2 + 24{b_2}^3c_4 + 64{b_2}^2c_2c_4 + 32b_2{c_2}^2c_4 - 201{b_2}^2b_3d_1 - 365b_2b_3c_2d_1 - 918b_3{c_2}^2d_1 - 44{b_2}^2c_3d_1 + 797b_2c_2c_3d_1 - 855{c_2}^2{d_1}^2 - 20a_2{b_2}^2d_2 - 40{b_2}^3d_2 + 40b_2{c_2}^2d_2$$

So, if we consider the minors $m_1, \ldots, m_{10}$ of order 2×2 of the coefficient matrix of this system, we can deduce the following two results:

1. In the open set $(\bigcup D(m_i)) \bigcap D(a_2 b_2) \bigcap V_{10}^1$ of V_{10}^1 we have $a_{1,1} = 0$. So all the algebras in this set are characteristically nilpotent Lie algebras.

2. We consider a point $\underline{u}$ in the locally closed subset

$$(\bigcap \mathcal{V}(m_i)) \bigcap D(a_2 b_2) \bigcap V_{10}^1$$

of V_{10}^1. Then:

2.1 If $\underline{u}$ is in $\bigcap_{j=1}^{5} \mathcal{V}(Q_{2,j})$ then the corresponding Lie algebra is not a characteristically nilpotent Lie algebra, because the system $S'_{\underline{u}}$ has a solution verifying $a_{1,1} \neq 0$. The Krull dimension of $V_{10}^1 \bigcap (\bigcap \mathcal{V}(Q_{2,j}))$ is 7.

2.2 If $\underline{u}$ is in $\bigcup_{j=1}^{5} D(Q_{2,j})$ then there exists j_0 such that $Q_{2,j_0}(\underline{u}) \neq 0$. Therefore if $Q'_{2,j_0}(\underline{u}) = 0$ then the corresponding filiform Lie algebra is a characteristically nilpotent Lie algebra. On the contrary, if $Q'_{2,j_0}(\underline{u}) \neq 0$ then the corresponding Lie algebra is not a characteristically nilpotent Lie algebra.

5.2

Finally, we will operate in the closed set $V_{10}^1 \bigcap \mathcal{V}(a_2)$. From $P_1 = 0$ we deduce $b_2(b_2 + c_2) = 0$. Now, we distinguish:

5.2.1 $V_{10}^1 \bigcap \mathcal{V}(a_2, b_2)$. From $ec_5 = 0$ we have $c_2(10a_{1,1} - a_{2,2}) = 0$. Besides, from $P_3 = 0$ we obtain $c_2(-3b_3 + 2c_3 - 7d_1) = 0$. So

5.2.1.1 Let suppose $c_2 = 0$. Now, we distinguish two cases depending on $b_3 \neq 0$ or $b_3 = 0$.

5.2.1.1.A Suppose $\underline{u} \in V_{10}^1 \bigcap \mathcal{V}(a_2, b_2, c_2) \bigcap D(b_3)$. From $ec_4 = 0$ we have $a_{2,2} = 11a_{1,1}$ and from $ec_7 = ec_{11} = 0$ we have $a_{1,1}c_4 = a_{1,1}d_2 = 0$. Obviously, in the subset $D(c_4) \bigcup D(d_2)$ we have $a_{1,1} = 0$. So all the algebras in this subset are characteristically nilpotent Lie algebras. On the other hand, in the subset defined by $\mathcal{V}(c_4, d_2)$ we have two new possibilities depending on $\underline{u}$ belongs to the open subset $D(c_3.(c_3 + d_1).(2c_3 + d_1))$ or $\underline{u}$ belongs to the closed subset given by $\mathcal{V}(c_3) \bigcup \mathcal{V}(c_3 + d_1) \bigcup \mathcal{V}(2c_3 + d_1)$. In the first case, the system $S'_{\underline{u}}$ is reduced to

$$Q_{2,6}a_{1,1} + Q'_{2,6}a_{1,10} = 0$$

with

$$Q_{2,6} := -2d_1d_3 + 2f_1c_3$$

$$Q'_{2,6} := -4b_3c_3^2 - 6c_3^3 + 8b_3d_1c_3 + 3b_3^2d_1 + 6c_3^2d_1 + 14d_1^2c_3 + 6b_3d_1^2 + 5d_1^3$$

Thus, if either $Q_{2,6}(\underline{u}) = 0$ or ($Q_{2,6}(\underline{u}) \neq 0$ and $Q'_{2,6}(\underline{u}) \neq 0$) then there exists a solution of the system $\mathcal{S}'_{\underline{u}}$ verifying $a_{1,1} \neq 0$ and thus, the corresponding Lie algebra is not a characteristically nilpotent Lie algebra. Otherwise, if $Q_{2,6}(\underline{u}) \neq 0$ and $Q'_{2,6}(\underline{u}) = 0$ then $a_{1,1} = 0$ and the corresponding filiform Lie algebra are characteristically nilpotent Lie algebras.

Now, to continue, we will divide the second case into the following three subcases, in all of which we will do the same reasoning as previous similar cases:

Subcase $(c_3 + d_1 = 0)$. In this case, if either $f_2 \neq 0$ or ($f_2 = 0$ and $c_3 = 0$ and $f_1 \neq 0$) then $a_{1,1} = 0$ and all the algebras are characteristically nilpotent Lie algebras. If $f_2 = 0$ and $c_3 \neq 0$, the system $\mathcal{S}'_{\underline{u}}$ is the only equation

$$-3(b_3 + c_3)^2 a_{1,10} + 2(d_3 + f_1)a_{1,1} = 0.$$

Finally, if $f_2 = 0$ and $f_1 = 0$ and $c_3 = 0$ the system $\mathcal{S}'_{\underline{u}}$ reduces to

$$\begin{aligned} 2d_3a_{1,1} - 3b_3^2a_{1,10} &= 0 \\ 4g_1a_{1,1} - 8b_3d_3a_{1,10} &= 0 \end{aligned}$$

Subcase $(c_3 = 0)$. In this case we can suppose that $d_1 \neq 0$. Then, the system $\mathcal{S}'_{\underline{u}}$ is now reduced to the only equation:

$$(-3b_3^2 - 6b_3d_1 - 5d_1^2)a_{1,10} + 2d_3a_{1,1} = 0.$$

Subcase $(2c_3 + d_1 = 0)$. In this case we can also suppose that $d_1 \neq 0$. Then, the system $\mathcal{S}'_{\underline{u}}$ is now reduced to the following:

$$\begin{aligned} (8d_3 + 4f_1)a_{1,1} &- (12b_3^2 + 4b_3d_1 + d_1^2)a_{1,10} = 0 \\ (-2f_1^2 + 4d_1g_1)a_{1,1} &- (8b_3d_3d_1 + 2d_1^2d_3 + 4b_3d_1f_1 + d_1^2f_1)a_{1,10} = 0 \end{aligned}$$

We denote by Δ the determinant of the coefficients matrix.

5.2.1.1.B Now, suppose $\underline{u} \in V_{10}^1 \cap \mathcal{V}(a_2, b_2, c_2, b_3)$. From $ec_6 = ec_7 = ec_8 = ec_{11} = 0$ we deduce that, if $\underline{u} \in (D(c_4) \cup D(d_2)) \cap (D(c_3) \cup D(d_1))$, all the algebras are characteristically nilpotent Lie algebras. So, we can suppose that $\underline{u}$ belongs to $\mathcal{V}(c_4, d_2) \cup \mathcal{V}(c_3, d_1)$. Let consider first $\underline{u} \in \mathcal{V}(c_3, d_1)$. By doing the same reasoning as previous cases we find that the corresponding algebra is a characteristically nilpotent Lie algebra if and only if $\underline{u}$ belongs to some of the following locally closed subset of $\mathcal{V}(c_3, d_1)$, whose union is denoted by Γ :

$$\begin{gathered} D(d_2d_3) \cup D(d_2f_1) \\ \mathcal{V}(d_2) \cap (D(f_1c_4) \cup D(f_1f_2)) \\ \mathcal{V}(d_2, f_1) \cap (D(c_4d_3) \cup D(c_4f_2)) \\ \mathcal{V}(d_2, f_1, c_4) \cap (D(d_3) \cup D(f_2) \cup D(g_1)) \end{gathered}$$

Now, we can suppose that $\underline{u} \in \mathcal{V}(c_4, d_2) \cap (D(c_3) \cup D(d_1))$. We distinguish the two following subcases:

Subcase ($c_3 + d_1 = 0$). In this subset, the only characteristically nilpotent Lie algebra corresponds to the points $\underline{u}$ with $f_2 \neq 0$.

Subcase ($c_3 + d_1 \neq 0$). If $\underline{u}$ is in $\mathcal{V}(2c_3 + d_1)$ then the system $\mathcal{S}'_{\underline{u}}$ reduces to the equation

$$Q_{2,7}a_{1,1} = (8d_3^2 + 8d_3f_1 + 3f_1^2 + 4c_3g_1)a_{1,1} = 0.$$

So, the only characteristically nilpotent Lie algebra corresponds to points in $D(Q_{2,7})$.

If $\underline{u}$ is in $D(2c_3 + d_1)$, then the system reduces to the equation:

$$(-6c_3^3 + 6c_3^2d_1 + 14c_3d_1^2 + 5d_1^3)a_{1,10} + 2a_{1,1}(c_3f_1 - d_1d_3) = 0,$$

and we can obtain similar conclusions as before.

5.2.1.2 Now, we operate in the set $V_{10}^1 \cap \mathcal{V}(a_2, b_2) \cap D(c_2)$. From $P_3 = 0$ we can deduce that $3b_3 = 2c_3 - 7d_1$ and $a_{2,2} = 10a_{1,1}$. Then, from the fourth equation of the system $\mathcal{S}'_{\underline{u}}$, we can deduce that on the complementary of the hyperplane $b_3 = 0$ (i.e., on the subset $V_{10}^1 \cap \mathcal{V}(a_2, b_2) \cap D(c_2b_3)$) all the algebras are characteristically nilpotent Lie algebras. If $b_3 = 0$, the system $\mathcal{S}'_{\underline{u}}$ can be reduced to the following

$$Q_{2,8}a_{1,1} = Q_{2,9}a_{1,1} = 0$$

with

$$Q_{2,8} := -16c_4c_2 + 171d_1^2 \quad Q_{2,9} := 9d_1^3 + 60c_2d_1d_2 - 24c_2^2d_3 + 24c_2^2f_1$$

and so, in the set $D(Q_{2,8}) \cup D(Q_{2,9})$ all the algebras are characteristically nilpotent Lie algebras, and, on the other hand, in the set $\mathcal{V}(Q_{2,8}) \cap \mathcal{V}(Q_{2,9})$ the corresponding filiform Lie algebra are not.

5.2.2 Consider now the set $V_{10}^1 \cap \mathcal{V}(a_2) \cap D(b_2)$. From $P_1 = 0$ we deduce $b_2 + c_2 = 0$ and from $P_3 = 0$ we have $3b_3 + 5c_3 + 5d_1 = 0$. Thus, from $ec_3 = 0$ we obtain $a_{2,2} = 10a_{1,1}$. The system $\mathcal{S}'_{\underline{u}}$ is now reduced to the following:

$$Q_{2,10}a_{1,1} = Q_{2,11}a_{1,1} = Q_{2,12}a_{1,1} = Q_{2,13}a_{1,1} = 0$$

with

$$\begin{aligned} Q_{2,10} &:= c_3 + d_1, & Q_{2,11} &:= c_4 + d_2, \\ Q_{2,12} &:= d_3c_2 - d_1c_4, & Q_{2,13} &:= -d_2f_1 + d_1f_2 + c_2g_1. \end{aligned}$$

So, in the set

$$\bigcup_{i=10}^{13} D(Q_{2,i})$$

we have $a_{1,1} = 0$ and thus all the algebras are characteristically nilpotent Lie algebras, whereas in the set $\mathcal{V}(Q_{2,10}, Q_{2,11}, Q_{2,12}, Q_{2,13})$ the corresponding filiform Lie algebras are not. Observe that the Krull dimension of this last subset is 5 because of it is the intersection of two hyperquadrics, in generic position, in an affine space of dimension 7.

So, in this section we have proved the following theorem: let consider the complex affine space $\mathbb{C}^{13}$ with coordinates

$$(a_1, a_2, b_2, b_3, c_2, c_3, c_4, d_1, d_2, d_3, f_1, f_2, g_1).$$

Theorem 5.2.3 *Let $\mathbf{V}_{10}^1 \subset \mathbb{C}^{13}$ be the first irreducible component of $\mathbf{V}_{10}$ (see 3.2). The Zariski constructible subset of non characteristically nilpotent Lie algebras of $\mathbf{V}_{10}^1$ is defined as the union of the following subsets:*

$$D(a_2) \cap \begin{cases} \mathcal{V}(b_2, c_3, d_2, f_1, Q_1, Q_2, Q_3, Q_4) \\ D(b_2) \cap (\cap_{i=1}^{10} \mathcal{V}(m_i)) \cap \begin{cases} \bigcap_{j=1}^{5} \mathcal{V}(Q_{2,j}) \\ \bigcup_{i=1}^{5} D(Q_{2,j} Q'_{2,j}) \end{cases} \end{cases}$$

$$\mathcal{V}(a_2, b_2, c_2, c_4, d_2) \cap D(b_3) \cap \begin{cases} D(c_3(c_3+d_1)(2c_3+d_1)) \cap (\mathcal{V}(Q_{2,6}) \cup D(Q_{2,6} Q'_{2,6})) \\ \mathcal{V}(c_3+d_1, f_2) \cap D(c_3) \cap (\mathcal{V}(d_3+f_1) \cup D((d_3+f_1)(b_3+c_3)) \\ \mathcal{V}(c_3, f_1, f_2, d_1, 4d_3^2 - 3b_3 g_1) \\ \mathcal{V}(c_3) \cap D(d_1) \cap (\mathcal{V}(d_3) \cup D(d_3(3b_3^2 + 6b_3 d_1 + 5d_1^2)) \\ \mathcal{V}(2c_3+d_1) \cap D(d_1) \cap \begin{cases} \mathcal{V}(2d_3+f_1, -2d_3^2 + d_1 g_1) \\ \mathcal{V}(\Delta) \cap D((2d_3+f_1)(12b_3^2 + 4b_3 d_1 + d_1^2)) \end{cases} \end{cases}$$

$$\mathcal{V}(a_2, b_2, b_3, c_2) \cap \begin{cases} \mathcal{V}(c_3, d_1) \setminus \Gamma \\ \mathcal{V}(c_4, d_2) \cap (D(c_3) \cup D(d_1)) \begin{cases} \mathcal{V}(c_3+d_1, f_2) \\ D(c_3+d_1) \cap \begin{cases} \mathcal{V}(2c_3+d_1, Q_{2,7}) \\ D(2c_3+d_1) \cap \mathcal{V}(c_3 f_1 - d_1 d_3) \\ D((2c_3+d_1)(c_3 f_1 - d_1 d_3) \\ \quad (6c_3^3 - 6c_3^2 d_1 - 14 c_3 d_1^2 - 5d_1^3)) \end{cases} \end{cases} \end{cases}$$

$$\mathcal{V}(a_2, b_2, b_3, Q_{2,8}, Q_{2,9}) \cap D(c_2)$$

$$\mathcal{V}(a_2, Q_{2,10}, Q_{2,11}, Q_{2,12}, Q_{2,13}) \cap D(b_2))$$

6 The irreducible component V_{10}^2

A Gröbner basis of the ideal of the component V_{10}^2 is

$$\{a_2 + 2b_2, 3b_2 + 7c_2, P_{32}\}$$

with $P_{32} = 9b_2 b_3 + 43 b_2 c_3 + 14 a_1 c_4 + 35 b_2 d_1 + 35 a_1 d_2$. Besides, as $V_{10}^2 \bigcap \mathcal{V}(a_1) \subset V_{10}^1$, we will only need to consider points in $V_{10}^2 \bigcap D(a_1)$. So we distinguish two cases:

6.1

Firstly, we consider points in $V_{10}^2 \bigcap D(a_1) \bigcap \mathcal{V}(b_2)$. This implies $a_2 = c_2 = 2c_4 + 5d_2 = 0$. Now we distinguish two new cases according to $c_3 + d_1$ is or not equal to 0.

6.1.1 Now, we operate in $V_{10}^2 \bigcap D(a_1) \bigcap \mathcal{V}(b_2, (c_3 + d_1)$. In this subcase, by using elementary transformations in the ring

$$R / < a_2 + 2b_2, 3b_2 + 7c_2, P_{32} >.$$

we can reduce the system $\mathcal{S}'_{\underline{u}}$ to the following one:

$$Q_{3,i} a_{1,1} + Q'_{3,i} a_{1,10} = 0 \quad 1 \le i \le 4$$

with

$$\begin{array}{lclllcl} Q_{3,1} & := & 2b_3 - 2d_1, & Q'_{3,1} & := & -3a_1 b_3 + 3a_1 d_1, \\ Q_{3,2} & := & 4d_3 + 4f_1, & Q'_{3,2} & := & -3b_3^2 - 3d_1^2 + 6b_3 d_1 - 5a_1 d_3 - 5a_1 f_1, \\ Q_{3,3} & := & 3d_2, & Q'_{3,3} & := & -4a_1 d_2, \\ Q_{3,4} & := & 30 f_2, & Q'_{3,4} & := & 47b_3 d_2 - 51 d_1 d_2 - 36 a_1 f_2. \end{array}$$

So, if we consider the minors $m_1, \ldots, m_6$ of order 2×2 of the coefficient matrix of this system, we can deduce the following two results:

6.1.1.1 In the open set of $V_{10}^2 \bigcap D(a_1) \bigcap \mathcal{V}(b_2, (c_3 + d_1)$

$$\bigcup_{i=1}^{6} D(m_i)$$

we have $a_{1,1} = 0$. So all the algebras in this open set are characteristically nilpotent Lie algebras.

6.1.1.2 We consider a point $\underline{u}$ in the closed set

$$(\bigcap \mathcal{V}(m_i)) \bigcap V_{10}^2 \bigcap D(a_1) \bigcap \mathcal{V}(b_2, c_3 + d_1).$$

Then we distinguish:

6.1.1.2.A If $\underline{u}$ is in $\bigcap_{j=1}^4 \mathcal{V}(Q_{3,j})$ then the corresponding Lie algebra is not a characteristically nilpotent Lie algebra. The Krull dimension of $V_{10}^2 \bigcap \mathcal{V}(b_2, c_3 + d_1) \bigcap (\bigcap \mathcal{V}(Q_{3,j}))$ is 5.

6.1.1.2.B If $\underline{u}$ is in the open set $\bigcup_{j=1}^4 D(Q_{3,j})$ then there exists j_0 such that $Q_{3,j_0}(\underline{u}) \neq 0$. Therefore if $Q'_{3,j_0}(\underline{u}) = 0$ then the corresponding Lie algebra is a characteristically nilpotent Lie algebra. On the other hand, if $Q'_{3,j_0}(\underline{u}) \neq 0$ then the corresponding Lie algebra is not.

6.1.2 Now, we operate in $V_{10}^2 \bigcap D(a_1) \bigcap \mathcal{V}(b_2) \bigcap D(c_3 + d_1)$.

In this set, by using elementary transformations in the corresponding localized ring, the system $\mathcal{S}'_{\underline{u}}$ is now reduced to the following:

$$d_2 a_{1,10} = Q_{4,1} a_{1,10} = 0$$

where $Q_{4,1} = 3{b_3}^2 + 14 b_3 c_3 + 15{c_3}^2 + 8 b_3 d_1 + 17 c_3 d_1 + 5{d_1}^2 - a_1 d_3 - a_1 f_1$.

6.2

Consider the set $V_{10}^2 \bigcap D(a_1 b_2)$, that is, $a_1 \neq 0$ and $b_2 \neq 0$.

In this case we have $a_{1,1} = 2 a_1 a_{1,10}$. Thus, by using elementary transformations in the ring

$$R / < a_2 + 2b_2, 3b_2 + 7c_2, P_{32} >$$

the system $\mathcal{S}'_{\underline{u}}$ is now reduced to the following system

$$Q_{5,j}\; a_{1,10} = 0, \;\; j = 1, \ldots, 5$$

with

$Q_{5,1} := -2b_2^2 + 7a_1 b_3 + 7a_1 c_3$, $\quad Q_{5,2} := 14 b_2^3 + 11 a_1 b_2 b_3 + 6 a_1 b_2 c_3 - 35 a_1^2 d_2 - 20 a_1 b_2 d_1$,

$Q_{5,3} := 20 b_2^2 - 49 a_1 b_3 + 49 a_1 d_1$, $\quad Q_{5,4} := -6 b_2^3 - 6 a_1 b_2 b_3 - 11 a_1 b_2 d_1 + 14 a_1^2 + a_1 b_2 d_1$,

$$\begin{aligned} Q_{5,5} := {} & -315 a_1 {b_2}^2 {b_3}^2 + 210 {b_2}^4 c_3 - 1715 a_1 {b_2}^2 b_3 c_3 + 1029 {a_1}^2 {b_3}^2 c_3 \\ & - 2471 a_1 {b_2}^2 {c_3}^2 + 5145 {a_1}^2 b_3 {c_3}^2 + 5145 {a_1}^2 {c_3}^3 + 150 a_1 {b_2}^3 c_4 - \\ & 343 {a_1}^2 b_2 b_3 c_4 - 1323 {a_1}^2 b_2 c_3 c_4 + 280 {b_2}^4 d_1 - 1050 a_1 {b_2}^2 b_3 d_1 + 1029 {a_1}^2 {b_3}^2 d_1 \\ & - 3262 a_1 {b_2}^2 c_3 d_1 + 8232 {a_1}^2 b_3 c_3 d_1 + 10976 {a_1}^2 {c_3}^2 d_1 - 980 {a_1}^2 b_2 c_4 d_1 - \\ & 1211 a_1 {b_2}^2 {d_1}^2 + 3087 {a_1}^2 b_3 {d_1}^2 + 7546 {a_1}^2 c_3 {d_1}^2 + 1715 {a_1}^2 {d_1}^3 + 548 a_1 {b_2}^3 d_2 \\ & - 539 {a_1}^2 b_2 b_3 d_2 - 2058 {a_1}^2 b_2 c_3 d_2 - 1617 {a_1}^2 b_2 d_1 d_2 + 462 {a_1}^2 {b_2}^2 d_3 - \\ & 1029 {a_1}^3 c_3 d_3 - 1029 {a_1}^3 d_1 d_3 + 602 {a_1}^2 {b_2}^2 f_1 - 1029 {a_1}^3 c_3 f_1 \\ & - 1029 {a_1}^3 d_1 f_1 + 196 {a_1}^3 b_2 f_2. \end{aligned}$$

However, as it can be checked by computing a Gröbner basis of this ideal, $Q_{5,4}$ belongs to the ideal generated by

$$\{a_2 + 2b_2, 3b_2 + 7c_2, P_{32}, Q_{5,1}, Q_{5,3}\}.$$

In the same way the system $(\mathcal{S}'_{\underline{u}})$ is now reduced to the system

$$Q_{6,j}\; a_{1,10} = 0, \; j = 1, 2, 3, 4$$

with

$$Q_{6,1} := Q_{5,1}, \quad Q_{6,2} := 58c_2^3 - 9a_1c_2d_1 + 9a_1^2d_2, \quad Q_{6,3} := Q_{5,3},$$

$$Q_{6,4} := c_2(85c_2^3c_3 + 186c_2^3d_1 - 15a_1c_2d_1^2 - 75a_1c_2^2d_2 + 9a_1^2d_1d_2 + 126a_1^2c_2d_3 + 156a_1^2c_2f_1 - 18c_1^3f_2).$$

So by using the same reasoning as in previous cases we deduce that if $\underline{u}$ is in the open set $\bigcup D(Q_{6,j})$, then the corresponding Lie algebra is a characteristically nilpotent Lie algebra. Otherwise, if $\underline{u}$ is in the closed set $\bigcap \mathcal{V}(Q_{6,j})$ then the corresponding Lie algebra is not. The Krull dimension of $V_{10}^2 \bigcap(\bigcap \mathcal{V}(Q_{6,j}))$ is 9.

Remember that $\mathbf{V}_{10}^2 \subset \mathbb{C}^{13}$ is the second irreducible component of $\mathbf{V}_{10}$ (see 3.2). We have $\mathbf{V}_{10}^2 \cap \mathcal{V}(a_1) \subset \mathbf{V}_{10}^1$. According to theorem 5.2.3, we will only consider points in the set $\mathbf{V}_{10}^2 \cap D(a_1)$. So, in this section we have proved the following theorem:

Theorem 6.2.1 *The Zariski constructible subset of non characteristically nilpotent Lie algebras of* $\mathbf{V}_{10}^2 \cap D(a_1)$ *is defined as the union of the following subsets:*

$$\mathcal{V}(b_2, c_3 + d_1) \cap (\cap_{i=1}^6 \mathcal{V}(m_i)) \cap \begin{cases} \bigcap_{j=1}^4 \mathcal{V}(Q_{3,j}) \\ \bigcup_{j=1}^4 (D(Q_{3,j}) \cap D(Q'_{3,j})) \end{cases}$$

$$D(c_3 + d_1) \cap \mathcal{V}(b_2, d_2, 3b_3^2 + 14b_3c_3 + 15c_3^2 + 8b_3d_1 + 17c_3d_1 + 5d_1^2 - a_1d_3 - a_1f_1)$$

$$D(b_2) \cap (\bigcap_{j=1}^4 \mathcal{V}(Q_{6,j}))$$

7 The irreducible component V_{10}^3

A Gröbner basis of the ideal of the component V_{10}^3 is $\{a_2, b_2 + c_2, P_{33}\}$, with $P_{33} = 5b_2c_3 + 2a_1c_4 + 5b_2d_1 + 5a_1d_2 + 3b_2b_3$. Furthermore, as $V_{10}^3 \bigcap \mathcal{V}(a_1) \subset V_{10}^1$, it suffices to consider points in $V_{10}^3 \bigcap D(a_1)$. Moreover, since $V_{10}^3 \bigcap \mathcal{V}(b_2) \subset V_{10}^2$, it will be sufficient to consider points in the open set $V_{10}^3 \bigcap D(a_1b_2)$.

So, in this component the system $S'_{\underline{u}}$ is reduced, by using elementary transformations in the corresponding localized ring, to the system:

$$Q_{7,j}\, a_{1,10} = 0, \; j = 1, \ldots, 5$$

with

$$Q_{7,1} := -2b_2^2 + a_1b_3 + a_1c_3, \quad Q_{7,2} := b_2b_3 + a_1d_2,$$

$$Q_{7,3} := 2b_2^2 - a_1b_3 + a_1d_1, \quad Q_{7,4} := 2b_2b_3 - b_2c_3 - b_2d_1 + 2a_1d_2,$$

$Q_{7,5} := -9a_1{b_3}^2c_3 + 3{b_2}^2{c_3}^2 - 45a_1b_3{c_3}^2 - 45a_1{c_3}^3 - 6{b_2}^3c_4 - 8a_1b_2c_3c_4 + 10{a_1}^2{c_4}^2 - 9a_1{b_3}^2d_1 - 7{b_2}^2c_3d_1 - 72a_1b_3c_3d_1 - 96a_1{c_3}^2d_1 + 3a_1b_2c_4d_1 - 10{b_2}^2{d_1}^2 - 27a_1b_3{d_1}^2 - 66a_1c_3{d_1}^2 - 15a_1{d_1}^3 + 6a_1b_2c_3d_2 + 37{a_1}^2c_4d_2 + 8a_1b_2d_1d_2 + 30{a_1}^2{d_2}^2 + 12a_1{b_2}^2d_3 + 9{a_1}^2c_3d_3 + 9{a_1}^2d_1d_3 + 9{a_1}^2c_3f_1 + 9{a_1}^2d_1f_1 + 12{a_1}^2b_2f_2.$

However, we find that in the open set $V_{10}^3 \bigcap D(a_1b_2)$ it is verified that

$$Q_{7,4} = 2Q_{7,2} - (b_2/a_1)(Q_{7,1} + Q_{7,3}).$$

Hence, the system is now reduced to the following:

$$Q_{7,1}a_{1,10} = Q_{7,2}a_{1,10} = Q'_{7,3}a_{1,10} = Q_{7,5}a_{1,10} = 0$$

with $Q'_{7,3} = c_3 + d_1$.

Now, to continue reducing the system we compute a Gröbner basis of the ideal generated by $\{a_2, b_2 + c_2, P_{33}, c_3 + d_1, Q_{7,1}, Q_{7,2}\}$ in R. Let $Q'_{7,5} = c_2(b_3c_2d_1 + a_1c_2d_3 - a_1^2f_2)$. $Q'_{7,5}$ is a representative of the congruence class of $Q_{7,5}$ modulo this ideal.

So, according to the reasoning made in previous cases, in the set $\mathcal{V}(Q_{7,1}, Q_{7,2}, Q'_{7,3}, Q'_{7,5})$ all the algebras are not characteristically nilpotent Lie algebras. The Krull dimension of $V_{10}^3 \bigcap \mathcal{V}(Q_{7,1}, Q_{7,2}, Q'_{7,3}, Q'_{7,5})$ is 8.

On the other hand, in the set $D(Q_{7,1}) \bigcup D(Q_{7,2}) \bigcup D(Q'_{7,3}) \bigcup D(Q'_{7,5})$, the Lie algebras are characteristically nilpotent Lie algebras.

Remember that $\mathbf{V}_{10}^3 \subset \mathbb{C}^{13}$ is the third irreducible component of $\mathbf{V}_{10}$ (see 3.2). We have $\mathbf{V}_{10}^3 \cap \mathcal{V}(a_1) \subset \mathbf{V}_{10}^1$ and $\mathbf{V}_{10}^3 \cap \mathcal{V}(b_2) \subset \mathbf{V}_{10}^2$. According to theorems 5.2.3 and.6.2.1 we will only consider points in the set $\mathbf{W}_{10}^3 := \mathbf{V}_{10}^3 \cap D(a_1b_2)$. So, in this section we have proved the following theorem:

Theorem 7.0.2 *The Zariski constructible subset of non characteristically nilpotent Lie algebras of* $\mathbf{W}_{10}^3$ *is* $\mathbf{W}_{10}^3 \cap \mathcal{V}(Q_{7,1}, Q_{7,2}, Q'_{7,3}, Q'_{7,5})$

References

[1] B. Buchberger. Gröbner Bases: An algorithmic method in polynomial ideal theory In *Multidimensional Systems Theory (N.K. Bose (ed.))*, 184–232, 1985, Reidel Pub. Co.

[2] R. Carles. Sur les algèbres de Lie caractéristiquement nilpotentes. *Prépublication Université de Poitiers*, 1984.

[3] F. Castro y J. Núñez. On filiform Lie algebras of dimension 9. *Prepub. de la Facultad de Matemáticas de la Uni. de Sevilla* n° 8, 1993.

[4] J. Dixmier and W. G. Lister. Derivations of nilpotent Lie algebras. *Proc. Amer. Math. Soc.* 8:155–158, 1957.

[5] F. J. Echarte, J. R. Gómez, J. Núñez. Les algèbres de Lie filiformes complexes dérivées d'autres algèbres de Lie. *Collection Travaux en Cours. Hermann, Paris*, to appear.

[6] C. Godfrey. Tables of coadjoint orbits for nilpotent Lie algebras. *University of Massachusset at Boston*, Boston, Mass. 021125, U.S.A.

[7] Y. B. Hakimjanov, J. M. Ancochea Bermúdez et M. Goze. Sur la réductibilité de la variété des algèbres de Lie nilpotentes complexes. *C.R. Acad. Sci. Paris 313, série I, 1991, p. 59-62.*

[8] V. V. Morozov. Classification des algèbres de Lie nilpotentes de dimension 6. *Isv. Vyss. Ucheb. Zav. Math.* 190:161-171, 1958.

[9] J. Núñez. Las algebras de Lie filiformes complejas según sean o no derivadas de otras. Tesis Doctoral Universidad de Sevilla, 1991.

[10] M. Vergne. Sur la variété des algèbres de Lie nilpotente. Thèse de 3e cycle, Paris, 1966.

F. J. Castro-Jiménez (`castro@algebra.us.es`)

J. Núñez-Valdés

Facultad de Matemáticas. Universidad de Sevilla. Apartado 1160. 41080 Sevilla (Spain).

Progress in Mathematics, Vol. 143,

Computing multidimensional residues

E. Cattani, A. Dickenstein, B. Sturmfels

Introduction

Given n polynomials in n variables with a finite number of complex roots, for any of their roots there is a local residue operator assigning a complex number to any polynomial. This is an algebraic, but generally not rational, function of the coefficients. On the other hand, the global residue, which is defined as the sum of the local residues over all roots, has invariance properties which guarantee its rational dependence on the coefficients [9],[27]. In this paper we present symbolic algorithms for evaluating that rational function.

Under the assumption that the deformation to the initial forms is flat, for some choice of weights on the variables, we express the global residue as a single residue integral with respect to the initial forms. When the input equations are a Gröbner basis with respect to a term order, this leads to an efficient series expansion algorithm for global residues, and to a vanishing theorem with respect to the corresponding cone in the Gröbner fan.

The global residue of a polynomial is shown to equal the highest coefficient of its (Gröbner basis) normal form, and, conversely, the entire normal form is expressed in terms of global residues. This yields a new method for evaluating traces over zero-dimensional complete intersections. Applications to be discussed include the counting of real roots (as in [4],[22]), the computation of the degree of a polynomial map (cf. [12]), and the evaluation of multivariate symmetric functions (cf. [16],[21]). All results and algorithms are illustrated for an explicit system in three variables.

0 Basic properties of multidimensional residues

Multidimensional residues play a fundamental role in complex analysis and geometry. Recent applications of residues in computational algebra include explicit division formulae [5],[6], the evaluation of symmetric functions [21], the membership problem for polynomial ideals [9],[10], the effective Nullstellensatz [13], and numerical algorithms for solving polynomial systems [7]. In most of these articles the emphasis lies on degree estimates and complexity results. Our goal here is to develop practical tools for computing global residues. We thus refrain from using "univariate projections" or "linear changes of coordinates"; instead we seek algorithms involving Gröbner bases and sparsity-preserving series expansions. While initially our discussion follows a path similar to [25],[27], it then proceeds to systematically develop the interplay between residues and Gröbner bases.

This paper is organized as follows. In §1 we consider n polynomials in n variables which form a Gröbner basis (or H-basis) with respect to some choice of positive weights. This hypothesis has natural geometric (1.3'), algebraic (1.5) and analytic (1.7) interpretations. In (1.17) we express the global residue as a single residue integral with respect to the initial forms. In §2 we specialize to the case of a Gröbner basis in the usual sense, with respect to a term order. In (2.3) we express the residue as a coefficient of a certain polynomial. This yields a polyhedral vanishing theorem (2.5), and a bound on the degree of the residue as a polynomial in the trailing coefficients (2.7). The constructions of §1 and §2 lead to algorithms, which will be presented in §3. In §4 we relate global residues to the coefficients in the (Gröbner basis) normal form. The global residue of a polynomial is shown to equal the highest coefficient of its normal form (4.2). This results in fast procedures for computing residues and traces (4.8). In §5 we present applications to symmetric functions, to computing the degree of a polynomial map, and to counting real roots. Finally, we study in §6 an explicit system in three variables.

In this section (§0) we review the complex-analytic definition and basic properties of multidimensional residues. Details and proofs can be found in [1],[15],[25]. For the algebraic counterpart to the analytic theory, see, e.g., [3],[19],[20],[23]. We shall discuss the equivalence of the algebraic and the analytic approach briefly at the end of §0.

Given n holomorphic functions $g_1, \ldots, g_n$ in an open set $U \subset \mathbb{C}^n$ with a single common zero p in U, one can associate to any holomorphic function $h \in \mathcal{O}(U)$ the local residue at p of the meromorphic n-form

$$\omega = \frac{h(\mathbf{x})\, d\mathbf{x}}{g_1(\mathbf{x}) \cdots g_n(\mathbf{x})}\,, \qquad d\mathbf{x} = dx_1 \wedge \cdots \wedge dx_n\,.$$

This defines the $\mathbb{C}$-linear operator

$$(0.1) \qquad \mathcal{O}(U) \to \mathbb{C}, \quad h \mapsto \operatorname{Res}_p\left(\frac{h(\mathbf{x})\, d\mathbf{x}}{g_1(\mathbf{x}) \ldots g_n(\mathbf{x})}\right) := \frac{1}{(2\pi i)^n} \int_{\Gamma_{\mathbf{g}}(\epsilon)} \omega\,,$$

where $\Gamma_{\mathbf{g}}(\epsilon)$ is the real n-dimensional cycle

$$\Gamma_{\mathbf{g}}(\epsilon) = \{\mathbf{x} \in U \;:\; |g_i(\mathbf{x})| = \epsilon_i \,, i = 1, \ldots, n\}$$

with orientation defined by the n-form $d\arg(g_1) \wedge \cdots \wedge d\arg(g_n)$, and where $\epsilon = (\epsilon_1, \ldots, \epsilon_n)$ is any n-tuple of sufficiently small, positive real numbers.

The Cauchy formula in n complex variables provides the simplest example of a residue operator: If $g_i(\mathbf{x}) = (x_i - p_i)^{a_i+1}$, $a_i \in \mathbb{N}$, then

$$(0.2) \qquad \operatorname{Res}_p\left(\frac{h(\mathbf{x})\, d\mathbf{x}}{\prod_{i=1}^n (x_i - p_i)^{a_i+1}}\right) = \frac{1}{a_1! \cdots a_n!}\left(\frac{\partial^{a_1+\cdots+a_n} h}{\partial x_1^{a_1} \cdots \partial x_n^{a_n}}\right)(p)\,.$$

Let $J_{\mathbf{g}} := \det\left(\frac{\partial g_i}{\partial x_j}\right)$ denote the Jacobian of $\mathbf{g} = (g_1, \ldots, g_n)$. Then,

$$\text{(0.3)} \qquad \operatorname{Res}_p\left(\frac{h\, J_{\mathbf{g}}\, d\mathbf{x}}{g_1 \cdots g_n}\right) = \mu_{\mathbf{g}}(p)\, h(p)\,.$$

where $\mu_{\mathbf{g}}(p)$ denotes the intersection multiplicity of $\mathbf{g} = (g_1, \ldots, g_n)$ at p [15, p. 662ff.]. If p is a simple root, hence $J_{\mathbf{g}}(p) \neq 0$, then

$$\text{(0.4)} \qquad \operatorname{Res}_p\left(\frac{h\, d\mathbf{x}}{g_1 \cdots g_n}\right) = \frac{h(p)}{J_{\mathbf{g}}(p)}\,.$$

Suppose now that $g_1, \ldots, g_n \in \mathbb{C}[\mathbf{x}]$ are n-variate polynomials whose zero set $Z(\mathbf{g})$ is a non-empty finite subset of $\mathbb{C}^n$. We can consider the *global residue operator* ([9],[15],[25]) which assigns to a polynomial $h \in \mathbb{C}[\mathbf{x}]$ the complex number

$$\text{(0.5)} \qquad \operatorname{Res}_{\mathbf{g}}(h) = \operatorname{Res}\left(\frac{h\, d\mathbf{x}}{g_1 \cdots g_n}\right) := \sum_{p \in Z(\mathbf{g})} \operatorname{Res}_p\left(\frac{h\, d\mathbf{x}}{g_1 \cdots g_n}\right)\,.$$

The main functorial properties of the global residue are encapsulated in the following two results, whose proof may be found, for example, in [25, II.8.3-4].

(0.6) Transformation law. *Suppose that $f_1, \ldots, f_n \in \mathbb{C}[\mathbf{x}]$ have finitely many common roots and that we can write*

$$f_i = \sum_{j=1}^{n} A_{ij} g_j\,; \quad A_{ij} \in \mathbb{C}[\mathbf{x}]\,, \quad i = 1, \ldots, n\,.$$

Then, for $h \in \mathbb{C}[\mathbf{x}]$,

$$\operatorname{Res}\left(\frac{h\, d\mathbf{x}}{g_1 \cdots g_n}\right) = \operatorname{Res}\left(\frac{h\, \det(A_{ij})\, d\mathbf{x}}{f_1 \cdots f_n}\right)\,.$$

If $g_1, \ldots, g_n \in \mathbb{C}[\mathbf{x}]$ are as above, the ideal I generated by them is zero-dimensional, and therefore $V = \mathbb{C}[\mathbf{x}]/I$ is a finite-dimensional $\mathbb{C}$-vector space. Since the global residue $\operatorname{Res}_{\mathbf{g}}(h)$ vanishes for $h \in I$ (see [15, p. 650]), it defines a $\mathbb{C}$-linear map:

$$\operatorname{Res}_{\mathbf{g}}\colon V \to \mathbb{C}\,, \quad h \mapsto \operatorname{Res}_{\mathbf{g}}(h)\,.$$

(0.7) Duality. *A polynomial $h \in \mathbb{C}[\mathbf{x}]$ lies in I if and only if $\operatorname{Res}_{\mathbf{g}}(fh) = 0$ for all $f \in \mathbb{C}[\mathbf{x}]$.*

This duality law may be interpreted as follows: Let $V^* = \mathrm{Hom}_{\mathbb{C}}(V, \mathbb{C})$. Then the pairing

$$V \times V^* \to V^*, \quad (b, \phi) \to (b.\phi),$$

where $(b.\phi)(b') = \phi(b\,b')$, makes V^* into a V-module. Statement (0.7) is equivalent to the assertion that the residue operator $\mathrm{Res}_{\mathbf{g}}$ is a generator of V^* as a V-module.

We recall that $\mathrm{Res}_{\mathbf{g}}(h)$ is a rational function, with integral coefficients, in the coefficients of $g_1, \ldots, g_n$. It may be computed in simply exponential time with respect to n, the number of variables; see [9],[21],[27]. A general procedure, suggested in these articles, is to find univariate polynomials $f_1(x_1), \ldots, f_n(x_n)$ in the ideal generated by $g_1, \ldots, g_n$ and to transform the global residue applying (0.6). The global residues with respect to $f_1, \ldots, f_n$ are then computed as a sum of products of univariate global residues. This general procedure is often too slow for practical computations. We seek more efficient algorithms for "nice" situations, such as the case when $g_1, \ldots, g_n$ are a Gröbner basis for a term order. The theory for such nice situations is to be developed in the next section.

In closing let us mention the relationship between this analytic definition of global residue and the algebraic definitions: of course, they coincide. Write for each $i = 1, \ldots, n$,

$$g_i(\mathbf{y}) - g_i(\mathbf{x}) \quad = \quad \sum_{j=1}^{n} g_{ij}(\mathbf{y}, \mathbf{x})(y_j - x_j)\,.$$

and denote $\Delta := \det(g_{ij})$. Let $U \subset \mathbb{C}^n$ be the union of relatively compact open neighborhoods isolating each of the points in $Z(\mathbf{g})$, let ϵ be any n-tuple of small positive real numbers, and define $\Pi_\epsilon := \{\mathbf{x} \in U : |g_i(\mathbf{x})| < \epsilon_i\,, \forall i = 1, \ldots, n\}$. For any holomorphic function h on U, one can deduce from (0.2) and (0.6) (cf.[25, §17]) the following integral representation known as Weil's formula [26]:

$$h(\mathbf{x}) = \frac{1}{(2\pi i)^n} \int_{\Gamma_{\mathbf{g}}(\epsilon)} \frac{h(\mathbf{y})\Delta(\mathbf{y}, \mathbf{x})d\mathbf{y}}{\prod_{i=1}^{n}(g_i(\mathbf{y}) - g_i(\mathbf{x}))}, \qquad \mathbf{x} \in \Pi_\epsilon.$$

As $|g_i(\mathbf{x})| < |g_i(\mathbf{y})|$ for any $i, \mathbf{x} \in \Pi_\epsilon$ and $\mathbf{y} \in \Gamma_{\mathbf{g}}(\epsilon)$, the integrand may be expanded as a multiple geometric series

$$\frac{h(\mathbf{y})\Delta(\mathbf{y}, \mathbf{x})}{\prod_{i=1}^{n}(g_i(\mathbf{y}) - g_i(\mathbf{x}))} \;=\; h(\mathbf{y})\Delta(\mathbf{y}, \mathbf{x}) \sum_{\alpha_i \geq 0} \left(\prod_{i=1}^{n} \frac{g_i(\mathbf{x})^{\alpha_i}}{g_i(\mathbf{y})^{\alpha_i+1}} \right),$$

which converges uniformly on compact subsets of $\Gamma_{\mathbf{g}}(\epsilon) \times \Pi_\epsilon$. As a result of term by term integration, we deduce that (cf.[25,§5],[5],[6])

(0.8)

$$h(\mathbf{x}) \;=\; \mathrm{Res}_{\mathbf{g}}(h(\,\cdot\,)\;\Delta(\,\cdot\,, \mathbf{x})) + \sum_{|\alpha| \geq 1} \mathrm{Res}\left(\frac{h(\mathbf{y})\Delta(\mathbf{y}, \mathbf{x})\,d\mathbf{y}}{\prod_{i=1}^{n} g_i(\mathbf{y})^{\alpha_i+1}} \right) \mathbf{g}^\alpha(\mathbf{x}), \quad \forall \mathbf{x} \in \Pi_\epsilon.$$

Note that the second summand on the right is in the ideal generated by $g_1, \ldots, g_n$ in $\mathcal{O}(\Pi_\epsilon)$ and that $\mathrm{Res}_{\mathbf{g}}(h(\cdot)\,\Delta(\cdot,\mathbf{x}))$ depends polynomially on $\mathbf{x}$. If, in addition, h is a polynomial, then the fact that $Z(\mathbf{g})$ is contained in Π_ϵ plus the fact that local analytic membership is equivalent to local algebraic membership ([24]), imply that

$$(0.9) \qquad h(\mathbf{x}) = \mathrm{Res}_{\mathbf{g}}(h(\cdot)\,\Delta(\cdot,\mathbf{x})) \quad \text{on the quotient ring } V\,.$$

In general, (0.8) does not provide a representation of their difference as a polynomial linear combination of $g_1, \ldots, g_n$. Under the hypothesis (1.3) below, it follows from the vanishing statement in (1.18) that the series becomes a finite sum, giving an effective division formula with remainder which involves computing only finitely many global residues associated to powers of $g_1, \ldots, g_n$. In summary, the formula (0.9) is the algebraic version of the integral representation. It proves that the global residue we are considering coincides with the "trace" associated to Δ as in [19, Appendix F] and [13], and with the Kronecker symbol (i.e., the dualizing linear form associated to Δ) as in [3].

1 Gröbner bases for a weight partial order

Let $\mathbf{K}$ be any subfield of the complex numbers $\mathbb{C}$ and let $g_i(\mathbf{x})$, $i = 1, \ldots, n$, be polynomials in $S = \mathbf{K}[\mathbf{x}]$, $\mathbf{x} = (x_1, \ldots, x_n)$. Let $\mathbf{w} = (w_1, \ldots, w_n) \in \mathbb{N}^n$ be a positive *weight vector*. The *weighted degree* of a monomial $\mathbf{x}^{\mathbf{a}} = x_1^{a_1} \ldots x_n^{a_n}$ is

$$\deg_{\mathbf{w}}(\mathbf{x}^{\mathbf{a}}) = \langle \mathbf{w}, \mathbf{a} \rangle = \sum_{i=1}^{n} w_i\, a_i\,.$$

We extend the notion of weighted degree to arbitrary polynomials in S in the usual manner. Write each polynomial $g_i(\mathbf{x})$ as

$$(1.1) \qquad g_i(\mathbf{x}) = p_i(\mathbf{x}) + q_i(\mathbf{x})$$

where p_i is $\mathbf{w}$-homogeneous, and

$$(1.2) \qquad d_i = \deg_{\mathbf{w}}(p_i) = \deg_{\mathbf{w}}(g_i)\,; \quad \deg_{\mathbf{w}}(q_i) < \deg_{\mathbf{w}}(g_i)\,.$$

Throughout this section we make the following assumption:

$$(1.3) \qquad p_1(\mathbf{x}) = \cdots = p_n(\mathbf{x}) = 0 \quad \text{if and only if} \quad \mathbf{x} = 0.$$

In what follows we will interpret this condition geometrically (1.3′), algebraically (1.5), and analytically (1.7). Let

$$(1.4) \qquad \tilde{g}_i(t;\mathbf{x}) = t^{d_i} g_i(t^{-w_1}x_1, \ldots, t^{-w_n}x_n).$$

This is a homogeneous polynomial of degree d_i in $(t; x_1, \ldots, x_n)$ relative to the weights $(1; w_1, \ldots, w_n)$.

Let $\mathbb{P}^n_{\mathbf{w}}$ denote the weighted projective space with homogeneous coordinates $(t; x_1, \ldots, x_n)$ and weights $(1; w_1, \ldots, w_n)$. The image of the hyperplane $\{t = 1\} \subset \mathbb{C}^{n+1} \backslash \{0\}$ in $\mathbb{P}^n_{\mathbf{w}}$ is identified with $\mathbb{C}^n$. If

$$\tilde{D}_i = \{(t; \mathbf{x}) \in \mathbb{P}^n_{\mathbf{w}} \; : \; \tilde{g}_i(t; \mathbf{x}) = 0\}$$

then (1.3) is equivalent to the geometric condition

$$\tilde{D}_1 \cap \cdots \cap \tilde{D}_n \subset \mathbb{C}^n \qquad (1.3')$$

The algebraic meaning of (1.3) is best expressed using the following notion of a Gröbner basis: Given a polynomial $f \in S$, we denote by $\mathrm{in}_{\mathbf{w}}(f)$ its form of highest weighted degree. For any ideal $I \subset S$ we define the initial ideal $\mathrm{in}_{\mathbf{w}}(I)$ to be the ideal generated by $\mathrm{in}_{\mathbf{w}}(f)$ where f runs over I. A finite subset $\mathcal{G} \subset I$ is said to be a *Gröbner basis* for I, relative to the weight $\mathbf{w}$, provided:

$$\mathrm{in}_{\mathbf{w}}(I) \quad = \quad \langle \mathrm{in}_{\mathbf{w}}(g) \; : \; g \in \mathcal{G} \rangle \, .$$

We emphasize that $\mathrm{in}_{\mathbf{w}}(I)$ need not be a monomial ideal. Some authors prefer to call $\mathcal{G}$ an *H-basis*, a term which goes back to Macaulay in the classical case $\mathbf{w} = (1, 1, \ldots, 1)$.

(1.5) Lemma. *Suppose $\mathcal{G} = \{g_1, \ldots, g_n\} \subset S$ satisfy (1.3). Then $\mathcal{G}$ is a Gröbner basis for the ideal it generates. Conversely, suppose $\mathcal{G} = \{g_1, \ldots, g_n\} \subset S$ is a Gröbner basis, with respect to $\mathbf{w}$, for a zero-dimensional ideal I. Then $\{g_1, \ldots, g_n\}$ satisfy (1.3).*

Proof: With the same notation as above, we have $p_i = \mathrm{in}_{\mathbf{w}}(g_i)$ and $q_i = g_i - \mathrm{in}_{\mathbf{w}}(g_i)$. Since $p_1, \ldots, p_n$ define a complete intersection, the Koszul complex on these forms in exact. This implies that every syzygy $\sum h_i \cdot e_i$ on $(p_1, \ldots, p_n)$ can be written as a linear combination of the basic syzygies $p_j \cdot e_k - p_k \cdot e_j$.

Suppose that $\mathcal{G}$ is not a Gröbner basis. Then, there exists a polynomial

$$f \; = \; \sum h_i g_i \; = \; \sum h_i p_i + \sum h_i q_i$$

whose initial form does not lie in $\langle p_1, \ldots, p_n \rangle$. Hence $\sum h_i p_i \; = \; 0$. By the remark above, we can write $\sum h_i \cdot e_i \; = \; \sum_{j,k} b_{jk} \cdot (p_j \cdot e_k \; - \; p_k \cdot e_j)$, and the leading term of

$$\sum h_i g_i \; = \; \sum h_i q_i \; = \; \sum_{j,k} b_{jk} \cdot (p_j q_k \; - \; p_k q_j)$$

must lie in $\langle p_1, \ldots, p_n \rangle$. This is a contradiction, completing the proof of the first statement. To prove the converse, it suffices to note $\dim(I) = \dim(\mathrm{in}_{\mathbf{w}}(I)) = 0$ (see e.g. [17]). □

(1.6) Remarks: (i) The first part of lemma (1.5) remains true for any set of polynomials $\mathcal{G} = \{g_1, \ldots, g_m\} \subset S$ whose initial forms define a complete intersection.

(ii) The initial ideal $\mathrm{in}_{\mathbf{w}}(I)$ is a flat deformation of the given ideal I (see e.g. [11, Ch. 6]).

(iii) The results in this section can be extended to fields other than the complex numbers using the deformation techniques in [20].

For each $t \in \mathbb{C}$, consider the map $\mathbf{g}_t : \mathbb{C}^n \to \mathbb{C}^n$ defined by

$$\mathbf{g}_t(\mathbf{x}) = (\tilde{g}_1(t; \mathbf{x}), \ldots, \tilde{g}_n(t; \mathbf{x})).$$

We have the following analytic interpretation of (1.3). Recall that a map $F : \mathbb{C}^n \to \mathbb{C}^n$ is said to be *proper* if the inverse image of any compact set is compact.

(1.7) Lemma. *The polynomials $g_1, \ldots, g_n$ satisfy condition (1.3) if and only if the map $\mathbf{g}_t$ is proper for every $t \in \mathbb{C}$.*

Proof: At $t = 0$ we have $\mathbf{g}_0 = (p_1, \cdots, p_n)$. Since the polynomials p_i are weighted homogeneous, the inverse image $\mathbf{g}_0^{-1}(0)$ is compact if and only if $\mathbf{g}_0^{-1}(0) = \{0\}$. Thus, if $\mathbf{g}_0$ is proper, then condition (1.3) is satisfied.

For the converse it is enough to show that the map $\mathbf{g}$ is proper, since (1.3) is a condition on just the initial form of the polynomials. Let $\tilde{\mathbf{g}} : \mathbb{C}^{n+1} \setminus \{0\} \longrightarrow \mathbb{C}^{n+1} \setminus \{0\}$ be defined by

$$\tilde{\mathbf{g}}(t, x_1, \ldots, x_n) = (t, \tilde{g}_1(t; \mathbf{x}), \ldots, \tilde{g}_n(t; \mathbf{x})).$$

The fact that $\mathbf{g}(\mathbf{x})$ satisfies (1.3) guarantees that $\tilde{\mathbf{g}}$ takes values in $\mathbb{C}^{n+1} \setminus \{0\}$. Since

$$\tilde{\mathbf{g}}(\lambda t, \lambda^{w_1} x_1, \ldots, \lambda^{w_n} x_n) = (\lambda t, \lambda^{d_1} \tilde{g}_1(t; \mathbf{x}), \ldots, \lambda^{d_n} \tilde{g}_n(t; \mathbf{x})),$$

$\tilde{\mathbf{g}}$ defines a map from $\mathbb{P}^n_{\mathbf{w}}$ to weighted projective space $\mathbb{P}^n_{\mathbf{d}}$ with weights $(1; d_1, \ldots, d_n)$. We may now consider the embedding of $\mathbb{C}^n$ in $\mathbb{C}^{n+1} \setminus \{0\}$ as the hyperplane $\{t = 1\}$. Since t has weight one in both $\mathbb{P}^n_{\mathbf{w}}$ and $\mathbb{P}^n_{\mathbf{d}}$, the natural projection from $\mathbb{C}^{n+1} \setminus \{0\}$ to $\mathbb{P}^n_{\mathbf{w}}$ or $\mathbb{P}^n_{\mathbf{d}}$ is a homeomorphism of the hyperplane $\{t = 1\}$ to its image. Thus, $\tilde{\mathbf{g}}$ is a continuous extension of $\mathbf{g}$ to appropriate compactifications of $\mathbb{C}^n$. If $K \subset \mathbb{C}^n$ is compact then $\mathbf{g}^{-1}(K)$ is compact since it coincides with $\tilde{\mathbf{g}}^{-1}(K)$. □

(1.8) Examples: An important special case, to be investigated in detail in §2, is that of n polynomials $g_1, \ldots, g_n \in S$ with finitely many common roots in $\mathbb{C}^n$ and such that they are a Gröbner basis with respect to some term order $\prec$. We can choose a weight vector $\mathbf{w} \in \mathbb{N}^n$ such that $\mathrm{in}_\prec(g_i) = \mathrm{in}_\mathbf{w}(g_i)$. Since the ideal generated by the initial monomials $\mathrm{in}_\mathbf{w}(g_i)$ is zero-dimensional, we may assume without loss of generality that:

$$\mathrm{in}_\mathbf{w}(g_i) = \alpha_i\, x_i^{r_i+1}$$

for some $\alpha_i \in \mathbf{K} \setminus \{0\}$, and therefore they satisfy (1.3). Particular examples are:

(i) The term order $\prec$ is lexicographic order: then $g_1, \ldots, g_n$ satisfy $g_i(\mathbf{x}) = g_i(x_i, \ldots, x_n)$ and g_i is monic in x_i. This is the case studied in [9]; see also [21] for the subcase when $g_i(\mathbf{x}) = g_i(x_i)$ are univariate polynomials.

(ii) The term order is defined as total degree with ties broken by lexicographic order with $x_n > \cdots > x_1$. This is the case studied in [1, (21.3)] and [25, II.8.2]. A weighted variant, due to Aĭzenberg and Tsikh, is studied in [1, (21.5)].

In this section we are interested in studying the global residue $\mathrm{Res}_\mathbf{g}(h)$, for a polynomial $h \in \mathbb{C}[\mathbf{x}]$, under the hypothesis that $g_1(\mathbf{x}), \ldots, g_n(\mathbf{x})$ satisfy (1.3). This hypothesis makes it possible to reduce the computation of $\mathrm{Res}_\mathbf{g}(h)$ to that of residues involving only certain powers of the initial forms $p_1(\mathbf{x}), \ldots, p_n(\mathbf{x})$. This is the content of (1.20) below. In fact, considering t as a parameter, the idea that one can recover the information from the deformation to the initial forms is the core of the geometric interpretation of Gröbner bases (see e.g. [2]).

Since the map $\mathbf{g}(\mathbf{x}) = (g_1(\mathbf{x}), \ldots, g_n(\mathbf{x}))$ is proper, we can replace (0.5) by a single integral

$$\mathrm{Res}\left(\frac{h\, d\mathbf{x}}{g_1 \cdots g_n}\right) = \frac{1}{(2\pi i)^n} \int_{\Gamma(\mathbf{r})} \frac{h\, d\mathbf{x}}{g_1 \cdots g_n}$$

for any $\mathbf{r} = (r_1, \ldots, r_n) \in (\mathbb{R}_{>0})^n$, where $\Gamma(\mathbf{r})$ is the compact, real n-cycle

$$\Gamma(\mathbf{r}) = \{\mathbf{x} \in \mathbb{C}^n \;:\; |g_i(\mathbf{x})| = r_i \;;\; i = 1, \ldots n\}.$$

Similarly, for each fixed $t \in \mathbb{C}$,

$$\begin{aligned} R_h(t) := \mathrm{Res}\left(\frac{h(\mathbf{x})\, d\mathbf{x}}{\tilde{g}_1(t;\mathbf{x}) \ldots \tilde{g}_n(t;\mathbf{x})}\right) &= \\ = \frac{1}{(2\pi i)^n} \int_{\tilde{\Gamma}_t(\mathbf{r})} \frac{h(\mathbf{x})\, d\mathbf{x}}{\tilde{g}_1(t;\mathbf{x}) \ldots \tilde{g}_n(t;\mathbf{x})}, \end{aligned} \tag{1.9}$$

where the global residue is taken relative to the divisors

$$\{\mathbf{x} \in \mathbb{C}^n \ : \ \tilde{g}_i(t;\mathbf{x}) = 0\},$$

$i = 1, \ldots, n$ and $\tilde{\Gamma}_t(\mathbf{r}) = \{\mathbf{x} \in \mathbb{C}^n \ : \ |\tilde{g}_i(t;\mathbf{x})| = r_i \ ; \ i = 1, \ldots n\}$.

The family $\{\tilde{g}_1(t,\mathbf{x}), \ldots, \tilde{g}_n(t,\mathbf{x})\}$ is a Gröbner basis with respect to $\mathbf{w}' = (1, 2w_1, \ldots, 2w_n)$ for the ideal $\tilde{I}$ generated by $\{\tilde{f}, f \in I\}$ and $\mathrm{in}_{\mathbf{w}'}(\tilde{f}) = \mathrm{in}_{\mathbf{w}}(f)$, $\forall f \in S$. It then follows that the coordinates $(t, \mathbf{x})$ are in *Noether position* [8] for $\tilde{I}$ and, consequently, we may apply theorem 3.3 in [9] to deduce that $R_h(t)$ is a polynomial in t. We will reprove this and in fact obtain the stronger result (1.18).

We set

$$\tilde{G}(t;\mathbf{x}) \ := \ \prod_{i=1}^{n} \tilde{g}_i(t;\mathbf{x}) \ = \ \sum_{j=0}^{d_{\mathbf{w}}} A_j(\mathbf{x})\, t^j \tag{1.10}$$

where $d_{\mathbf{w}} = d_1 + \cdots + d_n$. Inverting the polynomial $\tilde{G}(t;\mathbf{x})$ as a rational formal power series in t, we write

$$\tilde{G}^{-1}(t;\mathbf{x}) \ = \ \sum_{j \geq 0} B_j(\mathbf{x})\, t^j \,. \tag{1.11}$$

Given positive real numbers $k_1, \ldots, k_n$, let

$$T(\mathbf{k}) := \{\mathbf{x} \in \mathbb{C}^n \ : \ |p_i(\mathbf{x})| = k_i\}.$$

(1.12) Lemma. *Given $\delta > 0$, there exist positive constants $k_1, \ldots, k_n$ so that, for $|t| \leq \delta$,*

$$R_h(t) = \frac{1}{(2\pi i)^n} \sum_{m \geq 0} \left(\int_{T(\mathbf{k})} h(\mathbf{x})\, B_m(\mathbf{x})\, d\mathbf{x} \right) t^m \,.$$

and this series is uniformly convergent for $|t| \leq \delta$.

Proof: Suppose $g_1, \ldots, g_n$ satisfy (1.3) and let

$$\tilde{g}_i(t;\mathbf{x}) = p_i(\mathbf{x}) + t\, \hat{q}_i(t;\mathbf{x}).$$

Then, given $\delta > 0$, there exist positive constants $k_1, \ldots, k_n$ such that, for $\mathbf{x} \in T(\mathbf{k})$,

$$\frac{|p_i(\mathbf{x})|}{2} > |t\, \hat{q}_i(t;\mathbf{x})| \tag{1.13}$$

for all $i = 1, \ldots, n$ and $|t| \leq \delta$. Indeed, because of the weighted homogeneity property of p_i, it suffices to take $k_i = \lambda^{d_i}$ with λ sufficiently large.

The estimate (1.13) allows us to apply *Rouché's principle for residues* [25, II.8.1] and replace, for $|t| \leq \delta$, the integration cycles $\tilde{\Gamma}_t(\mathbf{r})$ in (1.9) by the fixed cycle $T(\mathbf{k})$. Thus,

$$R_h(t) = \frac{1}{(2\pi i)^n} \int_{T(\mathbf{k})} \frac{h(\mathbf{x})\, d\mathbf{x}}{\tilde{g}_1(t;\mathbf{x}) \cdots \tilde{g}_n(t;\mathbf{x})} \qquad \text{for all } |t| \leq \delta. \tag{1.14}$$

In view of (1.13), it follows that the series

$$\sum_{j \geq 0} B_j(\mathbf{x})\, t^j \quad = \quad \frac{1}{\prod_{i=1}^{n} p_i(\mathbf{x}) \left(1 + \dfrac{t\, \hat{q}_i(t;\mathbf{x})}{p_i(\mathbf{x})}\right)}$$

is uniformly convergent for $\mathbf{x} \in T(\mathbf{k})$ and $|t| \leq \delta$. Since we can now integrate (1.14) term by term, the result follows. □

(1.15) Lemma. *Let $P(\mathbf{x}) = p_1(\mathbf{x}) \ldots p_n(\mathbf{x})$. Then $P^{m+1}(\mathbf{x})\, B_m(\mathbf{x})$ is a weighted homogeneous polynomial of degree $m\,(d_{\mathbf{w}} - 1)$ with respect to $\mathbf{w}$.*

Proof: Since $\tilde{G}(t;\mathbf{x})$ is weighted homogeneous of degree $d_{\mathbf{w}}$ and t has weight 1, the coefficients $A_j(\mathbf{x})$ in (1.10) are weighted homogeneous of degree $d_{\mathbf{w}} - j$. On the other hand, the series (1.11) inverts (1.10). This implies the following recursion relations:

$$\sum_{j=0}^{m} A_j\, B_{m-j} = 0\,, \quad m \geq 1 \tag{1.16}$$

with initial conditions $A_0\, B_0 = 1$ and $A_0(\mathbf{x}) = P(\mathbf{x})$. In particular,

$$P\, B_m \;=\; -\sum_{j=1}^{m} A_j\, B_{m-j}\,,$$

and

$$P^{m+1}\, B_m \;=\; \sum_{j=1}^{m} A_j\, P^{j-1}\, (P^{m-j+1}\, B_{m-j})\,.$$

Assuming that (1.14) holds inductively with respect to m, we obtain

$$\deg_{\mathbf{w}}(P^{m+1}\, B_m) \;=\; (d_{\mathbf{w}} - j) + (j-1)\, d_{\mathbf{w}} + (m-j)\,(d_{\mathbf{w}} - 1) \;=\; m\,(d_{\mathbf{w}} - 1)\,. \quad \square$$

The following is the main result in this section.

(1.17) Theorem. *For any monomial* $\mathbf{x}^{\mathbf{a}} = x_1^{a_1} \cdots x_n^{a_n}$, *set*

$$s(\mathbf{a}) = \langle \mathbf{w}, \mathbf{a} \rangle - d_{\mathbf{w}} + \sum_{i=1}^{n} w_i .$$

Then

$$R_{\mathbf{x}^{\mathbf{a}}}(t) \;=\; \mathrm{Res}_{\mathbf{g}}(\mathbf{x}^{\mathbf{a}}) \cdot t^{s(\mathbf{a})} ,$$

and

$$\mathrm{Res}_{\mathbf{g}}(\mathbf{x}^{\mathbf{a}}) \;=\; \frac{1}{(2\pi i)^n} \int_{T(\mathbf{k})} \mathbf{x}^{\mathbf{a}}\, B_{s(\mathbf{a})}(\mathbf{x})\, d\mathbf{x} .$$

Before proving (1.17), we note the following weighted version of the Euler-Jacobi theorem [15, p. 671]. A more general toric version was given by Khovanskii in [18].

(1.18) Corollary. $R_h(t)$ *is a polynomial in* t *of degree at most* $\deg_{\mathbf{w}}(h) - d_{\mathbf{w}} + \sum w_i$, *and*

$$\mathrm{Res}\left(\frac{h\, d\mathbf{x}}{g_1 \cdots g_n}\right) = 0 \qquad \textit{whenever} \quad \deg_{\mathbf{w}}(h) < d_{\mathbf{w}} - \sum w_i .$$

We observe also that, under the current hypothesis (1.3), this corollary implies that the terms in the series in (0.8) will vanish for $\sum_{i=1}^{n} \alpha_i d_i > \deg_{\mathbf{w}}(h)$.

Proof of (1.17): We begin by noting that, as in the case with unit weights [25, IV.20.1]:

(1.19) *If* P *and* Q *are weighted homogeneous polynomials in* $\mathbb{C}[\mathbf{x}]$, *and* $\deg_{\mathbf{w}}(P) - \deg_{\mathbf{w}}(Q) + \sum w_i \neq 0$, *then the form* $\omega = \dfrac{P(\mathbf{x})\, d\mathbf{x}}{Q(\mathbf{x})}$ *is exact.*

Indeed, we find that $\omega = (\deg_{\mathbf{w}}(P) - \deg_{\mathbf{w}}(Q) + \sum w_i)^{-1}\, d\sigma$, where

$$\sigma \;=\; \frac{P(\mathbf{x})}{Q(\mathbf{x})} \sum_{j=1}^{n} (-1)^{j-1} w_j\, x_j\, dx_1 \wedge \cdots \wedge \widehat{dx_j} \wedge \cdots \wedge dx_n .$$

The verification of this equality is a straightforward consequence of Euler's formula for weighted homogeneous polynomials:

$$\sum_{j=1}^{n} w_j\, x_j \frac{\partial P}{\partial x_j} \;=\; \deg_{\mathbf{w}}(P)\, P .$$

As in lemma (1.12), we write

$$R_{\mathbf{x}^{\mathbf{a}}}(t) = \sum_{m\geq 0} \left(\int_{T(\mathbf{k})} \mathbf{x}^{\mathbf{a}}\, B_m(\mathbf{x})\, d\mathbf{x} \right) t^m .$$

Since, by lemma (1.15), $B_m(\mathbf{x})$ is a quotient of weighted homogeneous polynomials, we can apply (1.19) to conclude that

$$\int_{T(\mathbf{k})} \mathbf{x}^{\mathbf{a}}\, B_m(\mathbf{x})\, d\mathbf{x} \; = 0$$

whenever

$$\deg_{\mathbf{w}}(\mathbf{x}^{\mathbf{a}}) + \deg_{\mathbf{w}}(B_m) + \sum w_i \quad \neq \quad 0 .$$

This inequation is equivalent to $m \neq s(\mathbf{a})$. Hence all integrals in (1.12) vanish, except for the one with $m = s(\mathbf{a})$. This was precisely the claim of (1.17). □

The second assertion of (1.17) says that we may write

(1.20) $$\operatorname{Res}\left(\frac{\mathbf{x}^{\mathbf{a}}\, d\mathbf{x}}{g_1 \cdots g_n} \right) = \frac{1}{(2\pi i)^n} \int_{T(\mathbf{k})} \frac{\mathbf{x}^{\mathbf{a}}\left(P^{s(\mathbf{a})+1}(\mathbf{x})\, B_{s(\mathbf{a})}(\mathbf{x})\right)}{P^{s(\mathbf{a})+1}(\mathbf{x})}\, d\mathbf{x} .$$

The numerator is a weighted homogeneous polynomial, by lemma (1.15). Therefore (1.20) is a residue with respect to the $(s(\mathbf{a})+1)$ power of the initial forms $p_1(\mathbf{x}), \ldots, p_n(\mathbf{x})$.

We conclude this section by observing that as a direct consequence of (1.17) and the duality theorem (0.7) we obtain (see [25, IV.20.1] for the case of unit weights):

(1.21) Macaulay's theorem. *Let $p_1(\mathbf{x}), \ldots, p_n(\mathbf{x})$ be weighted homogeneous polynomials whose only common zero is the origin. Then, any weighted homogeneous polynomial $h(\mathbf{x})$ satisfying*

$$\deg_{\mathbf{w}}(h) > d_{\mathbf{w}} - \sum_{i=1}^{n} w_i$$

is in the ideal generated by $p_1(\mathbf{x}), \ldots, p_n(\mathbf{x})$.

2 Gröbner bases for a term order

In this section we specialize to the case of n polynomials $g_1, \ldots, g_n \in S$ with finitely many roots in $\mathbb{C}^n$, which are a Gröbner basis with respect to a term order $\prec$. As in (1.8), we choose a positive weight $\mathbf{w} \in \mathbb{N}^n$ such that $\mathrm{in}_{\prec}(g_i) = \mathrm{in}_{\mathbf{w}}(g_i)$. We may assume that

$$\mathrm{in}_{\mathbf{w}}(g_i) = x_i^{r_i+1}, \quad i = 1, \ldots, n. \tag{2.1}$$

Let $\mathbf{r} = (r_1, \ldots, r_n)$ and $\mathbf{r} + \mathbf{1} = (r_1 + 1, \ldots, r_n + 1)$. Then, with notation as in §1, $d_{\mathbf{w}} = \langle \mathbf{w}, \mathbf{r} + \mathbf{1} \rangle$, and

$$P(\mathbf{x}) = x_1^{r_1+1} \ldots x_n^{r_n+1} = \mathbf{x}^{\mathbf{r}+\mathbf{1}},$$

$$s(\mathbf{a}) = \langle \mathbf{w}, \mathbf{a} \rangle - d_{\mathbf{w}} + \sum w_i = \langle \mathbf{w}, \mathbf{a} - \mathbf{r} \rangle,$$

and (1.15) implies the following homogeneity of the coefficients of (1.11):

(2.2) *$B_m(\mathbf{x})$ is a $\mathbf{w}$-homogeneous Laurent polynomial of weighted degree $-(m + d_{\mathbf{w}})$.*

Consequently, the integrand in (1.17) is a Laurent polynomial. Since

$$\int_{T(\mathbf{k})} \mathbf{x}^{\mathbf{b}}\, d\mathbf{x} = 0 \qquad \text{for } \mathbf{b} \neq (-1, \ldots, -1),$$

we obtain the following result.

(2.3) Theorem. *The residue $\mathrm{Res}_{\mathbf{g}}(\mathbf{x}^{\mathbf{a}})$ equals the $x_1^{-1} \cdots x_n^{-1}$-coefficient of $\mathbf{x}^{\mathbf{a}}\, B_{s(\mathbf{a})}(\mathbf{x})$.*

Theorem (2.3) is essentially a restatement of a formula for computing global residues due to Aĭzenberg and Tsikh [1, (21.5)], which, in a simpler version pointed out to us by J. Petean, says that $\mathrm{Res}_{\mathbf{g}}(\mathbf{x}^{\mathbf{a}})$ equals the $x_1^{-1} \cdots x_n^{-1}$-coefficient in the expression

$$\sum_{|\alpha| \leq \langle \mathbf{w}, \mathbf{a} - \mathbf{r} \rangle} \left((-1)^{|\alpha|}\, \mathbf{x}^{\mathbf{a} - (\mathbf{r}+\mathbf{1})} \prod_{i=1}^{n} (q_i(\mathbf{x})/x_i^{r_i+1})^{\alpha_i} \right).$$

The introduction of the homogenizing parameter t organizes the computation of this Laurent series and the search for the desired coefficient, as evidenced in algorithm (3.1) below. Keeping track of the homogeneity properties of the coefficients $B_j(\mathbf{x})$, also allows us to get more precise information about the global residues, such as theorems (2.5) and (2.7).

(2.4) Remark. Note that (2.2) implies that $\mathbf{x}^{\mathbf{a}}\, B_m(\mathbf{x})$ may contain a term of the form $\alpha\, x_1^{-1} \cdots x_n^{-1}$ only if

$$\langle \mathbf{w}, \mathbf{a} \rangle - (m + d_{\mathbf{w}}) = -\sum w_i.$$

that is, only if

$$m = \langle \mathbf{w}, \mathbf{a} - \mathbf{r} \rangle.$$

Combined with (1.12), this gives a simpler proof of the first statement in (1.17) in the case when $g_1, \ldots, g_n$ are a Gröbner basis with respect to some term order.

A similar argument combined with the recursive relations (1.16) makes it possible to improve on the vanishing statement in Corollary (1.18). Let $\mathcal{W}^*$ denote the polyhedral cone in $\mathbb{R}^n$ which is positively spanned by all lattice points of the form $\mathbf{r} + \mathbf{1} - \mathbf{b}$, where $\mathbf{x}^{\mathbf{b}}$ runs over all monomials appearing in the expansion of $g_1(\mathbf{x}) g_2(\mathbf{x}) \cdots g_n(\mathbf{x})$.

(2.5) Theorem. *The residue* $\mathrm{Res}_{\mathbf{g}}(\mathbf{x}^{\mathbf{a}})$ *vanishes if* $\mathbf{a} - \mathbf{r}$ *lies outside the cone* $\mathcal{W}^*$.

Proof: The condition on $\mathbf{b}$ in the definition of $\mathcal{W}^*$ is equivalent to saying that $\mathbf{x}^{\mathbf{b}}$ appears in one of the coefficients $A_j(\mathbf{x})$, $j = 1, \ldots, d_{\mathbf{w}}$, in the expansion (1.10) of $\tilde{G}(t; \mathbf{x})$ as a polynomial in t. The recursion relations (1.16) imply that $B_m(\mathbf{x})$ consists of terms $k_\alpha \mathbf{x}^\alpha$ where the n-tuples $\alpha \in \mathbb{Z}^n$ are in the translated cone $-((\mathbf{r} + \mathbf{1}) + \mathcal{W}^*)$. Indeed, if $k_\alpha \mathbf{x}^\alpha$ is a term in $B_m(\mathbf{x})$, then (1.16) implies that for some $j = 1, \ldots, m$, the Laurent polynomial $A_j(\mathbf{x})\, B_{m-j}(\mathbf{x})$ contains a term of the form $c_\alpha \mathbf{x}^{\alpha + \mathbf{r} + \mathbf{1}}$ and therefore,

$$\alpha + \mathbf{r} + \mathbf{1} \quad = \quad \mathbf{b} + \beta,$$

where $\mathbf{x}^{\mathbf{b}}$ is a monomial in $A_j(\mathbf{x})$ and $\mathbf{x}^\beta$ is a monomial in $B_{m-j}(\mathbf{x})$. Then

$$\alpha + \mathbf{r} + \mathbf{1} \quad = \quad (\mathbf{b} - (\mathbf{r} + \mathbf{1})) + (\beta + \mathbf{r} + \mathbf{1})$$

and the assertion follows by induction on m.

Now, theorem (2.3) implies that $\mathrm{Res}_{\mathbf{g}}(\mathbf{x}^{\mathbf{a}}) = 0$ unless the coefficient $B_{s(\mathbf{a})}(\mathbf{x})$ contains a term which is a non-zero multiple of $x_1^{-(a_1+1)} \cdots x_n^{-(a_n+1)}$. But this is possible only if $-(\mathbf{a} + \mathbf{1}) \in -((\mathbf{r} + \mathbf{1}) + \mathcal{W}^*)$, or equivalently, if $\mathbf{a} - \mathbf{r} \in \mathcal{W}^*$. □

Theorem (2.3) may also be used to study the dependence of $\mathrm{Res}_{\mathbf{g}}(\mathbf{x}^{\mathbf{a}})$ on the coefficients of the polynomials $g_1, \ldots, g_n$. We write

$$g_i(\mathbf{x}) = x_i^{r_i+1} - \sum_{j=1}^{\nu_i} c_{ij}\, \mathbf{x}^{\mathbf{a}_{ij}}, \tag{2.6}$$

and let $\mathcal{W} \subset \mathbf{R}^n$ denote the closed convex cone of all vectors $\mathbf{w} \in \mathbf{R}^n$ such that

$$\langle \mathbf{w}, \mathbf{a}_{ij} \rangle \leq \langle \mathbf{w}, (r_i + 1)\, \mathbf{e}_i \rangle \ ; \quad \text{for all } i = 1, \ldots, n;\ j = 1, \ldots, \nu_i.$$

This cone is the polar dual of the cone $\mathcal{W}^*$ defined above. By assumption, $\mathcal{W}$ has non-empty interior. Note that $\mathcal{W}$ is the cone in the *Gröbner fan* of I corresponding to the given term order (see e.g. [14, §3.1]). We have the following result.

(2.7) Theorem. *The residue* $\mathrm{Res}_{\mathbf{g}}(\mathbf{x}^{\mathbf{a}})$ *is a polynomial function in the coefficients* c_{ij}. *Its degree in the variable* c_{ij} *is bounded above by*

$$(2.8) \qquad \min_{\mathbf{w}\in\mathcal{W}} \frac{\langle \mathbf{w}, \mathbf{a}-\mathbf{r}\rangle}{\langle \mathbf{w}, (r_i+1)\,\mathbf{e}_i - \mathbf{a}_{ij}\rangle}$$

and the total degree in the variables $\mathbf{c} = (c_{ij})$ *is bounded by*

$$(2.9) \qquad \min_{\mathbf{w}\in int(\mathcal{W})\cap\mathbb{Z}^n} \langle \mathbf{w}, \mathbf{a}-\mathbf{r}\rangle .$$

Proof: Given $g_i(\mathbf{x})$ as in (2.6), its weighted homogenization with respect to a weight $\mathbf{w} = (w_1,\dots,w_n) \in \mathbb{Z}^n \cap \mathcal{W}$, is given by

$$\tilde{g}_i(t;\mathbf{x}) = x_i^{r_i+1} - \sum_{j=1}^{\nu_i} c_{ij}\, t^{\langle \mathbf{w}, (r_i+1)\,\mathbf{e}_i - \mathbf{a}_{ij}\rangle}\, \mathbf{x}^{\mathbf{a}_{ij}}$$

where $\mathbf{e}_i$ denotes the i-th unit vector. Set $\rho_{ij} = (r_i+1)\,\mathbf{e}_i - \mathbf{a}_{ij}$.

According to (2.3), $\mathrm{Res}_{\mathbf{g}}(\mathbf{x}^{\mathbf{a}})$ equals the $x_1^{-1}\cdots x_n^{-1}$-coefficient of $\mathbf{x}^{\mathbf{a}}\, B_{s(\mathbf{a})}(\mathbf{x})$, which is equal to the coefficient of $t^{\langle \mathbf{w},\mathbf{a}-\mathbf{r}\rangle}\, x_1^{-1}\cdots x_n^{-1}$ in the expansion of
(2.10)

$$\frac{\mathbf{x}^{\mathbf{a}}}{\mathbf{x}^{\mathbf{r}+1}} \sum_{i_1,\dots,i_n\ge 0} \left(\sum_{j=1}^{\nu_1} c_{1j}\, t^{\langle \mathbf{w},\rho_{1j}\rangle}\, \mathbf{x}^{-\rho_{1j}}\right)^{i_1} \cdots \left(\sum_{j=1}^{\nu_n} c_{nj}\, t^{\langle \mathbf{w},\rho_{nj}\rangle}\, \mathbf{x}^{-\rho_{nj}}\right)^{i_n} .$$

It is now clear that the residue depends polynomially on the coefficients c_{ij} and that, for a given choice of $\mathbf{w} \in \mathcal{W}$, its degree in c_{ij} is bounded above by $\langle \mathbf{w}, \mathbf{a}-\mathbf{r}\rangle / \langle \mathbf{w}, (r_i+1)\,\mathbf{e}_i - \mathbf{a}_{ij}\rangle$.

To prove the bound in (2.9), it suffices to observe that if a monomial $\mathbf{c}^{\mathbf{k}}$ appears in $\mathrm{Res}_{\mathbf{g}}(\mathbf{x}^{\mathbf{a}})$, then by (2.10)

$$\sum_{i,j} k_{ij} \langle \mathbf{w}, \rho_{ij}\rangle \;=\; \langle \mathbf{w}, \mathbf{a}-\mathbf{r}\rangle .$$

As $\langle \mathbf{w}, \rho_{ij}\rangle \ge 1$ for $\mathbf{w} \in int(\mathcal{W})$ integral and all i, j, the claim follows. □

(2.11) Remarks: i) By the Cauchy-Schwarz inequality, the bound (2.8) is minimized by vectors $\mathbf{w} \in \mathcal{W}$ which are as far as possible from $\mathbf{a}-\mathbf{r}$ and as close as possible to $(r_i+1)\,\mathbf{e}_i - \mathbf{a}_{ij}$.

ii) If we are given a system of polynomials $f_1,\dots,f_n$ which satisfy (2.1), and supposing that only the constant coefficients are perturbed, say

$$g_i(\mathbf{x}) \;=\; f_i(\mathbf{x}) - c_i \qquad i = 1,\dots,n .$$

then the bound in (2.9) implies the following: If a monomial $c_1^{k_1} \cdots c_n^{k_n}$ appears in $\mathrm{Res}_{\mathbf{g}}(\mathbf{x}^{\mathbf{a}})$, then $\langle \mathbf{w}, \mathbf{k} \rangle \leq \langle \mathbf{w}, \mathbf{a} - \mathbf{r} \rangle$, that is, the weighted degree with respect to $\mathbf{w}$ of $\mathrm{Res}_{\mathbf{g}}(\mathbf{x}^{\mathbf{a}})$ in $(c_1, \ldots, c_n)$ is bounded by $\langle \mathbf{w}, \mathbf{a} - \mathbf{r} \rangle$ for all integral vectors $\mathbf{w} \in int(\mathcal{W})$.

3 Deformation algorithms for global residues

Let $\{g_1, \ldots, g_n\} \subset S$ be a Gröbner basis as in §2, and let $\mathbf{w} \in \mathbb{N}^n$ such that (2.1) holds. We have shown that for any polynomial

$$h = \sum_{\langle \mathbf{w}, \mathbf{a} \rangle \leq d} c_{\mathbf{a}} \, \mathbf{x}^{\mathbf{a}} \quad \in \quad S,$$

the computation of the global residue

$$\mathrm{Res}\left(\frac{h \, d\mathbf{x}}{g_1 \cdots g_n} \right) = \sum_{\langle \mathbf{w}, \mathbf{a} \rangle \leq d} c_{\mathbf{a}} \, \mathrm{Res}\left(\frac{\mathbf{x}^{\mathbf{a}} \, d\mathbf{x}}{g_1 \cdots g_n} \right)$$

can be reduced to a sum of residues (at the origin) with respect to the family of monomials $\{ x_1^{r_1+1}, \ldots, x_n^{r_n+1} \}$ and some of their powers. Theorem (2.3) gives the following algorithm for computing all global residues up to a given weighted degree. Note that algorithm (3.1) respects possible sparsity of the input polynomials.

(3.1) Algorithm.

Input: $\mathbf{w} \in \mathbb{N}^n$, $g_1, \ldots, g_n \in S$ *satisfying (2.1), and* $d \in \mathbb{N}$.
Output: *The global residues* $\mathrm{Res}_{\mathbf{g}}(\mathbf{x}^{\mathbf{a}})$, *for all* $\mathbf{a} \in \mathbb{N}^n$ *such that* $\langle \mathbf{w}, \mathbf{a} \rangle \leq d$.

Step 1: Define the weighted homogenizations

$$\tilde{g}_i(t; \mathbf{x}) = t^{w_i(r_i+1)} \, g_i(t^{-w_1} x_1, \ldots, t^{-w_n} x_n).$$

Step 2: If $\langle \mathbf{w}, \mathbf{a} - \mathbf{r} \rangle \leq 0$, then $\mathrm{Res}_{\mathbf{g}}(\mathbf{x}^{\mathbf{a}}) = 0$.
Step 3: Set $d' = d - \langle \mathbf{w}, \mathbf{r} \rangle$. Compute the Taylor polynomial $\sum_{j=0}^{d'} B_j(\mathbf{x}) \, t^j$ of degree d' for $(\prod \tilde{g}_i(t; \mathbf{x}))^{-1}$.
Step 4: For each $\mathbf{a}$ such that $\langle \mathbf{w}, \mathbf{r} \rangle \leq \langle \mathbf{w}, \mathbf{a} \rangle \leq d$, find the coefficient of $\mathbf{x}^{-(\mathbf{a}+\mathbf{1})}$ in the Laurent polynomial $B_{\langle \mathbf{w}, \mathbf{a} - \mathbf{r} \rangle}(\mathbf{x})$. It equals $\mathrm{Res}_{\mathbf{g}}(\mathbf{x}^{\mathbf{a}})$.

The verification that $\mathcal{G} = \{g_1, \ldots, g_n\}$ is a Gröbner basis and, if so, the choice of a compatible weight $\mathbf{w} \in \mathbb{N}^n$ may be accomplished in at most

$$O(n^{n+2} \, m^{2n-1} \, d^{(2n+1)n})$$

arithmetic operations, where m is a bound for the number of monomials in each g_i and d is a bound for their degrees. This was shown in [14, §3.2].

Naturally, algorithm (3.1) will be most efficient when it is known *a priori* that $\mathcal{G}$ is a Gröbner basis and a "small" compatible weight is given. However, even if the given equations $\mathcal{G}$ are not a Gröbner basis for any $\mathbf{w}$, then algorithm (3.1) still serves as a useful subroutine. To illustrate this, we describe a general procedure for computing the global residue associated to *any* complete intersection zero-dimensional ideal:

(3.2) Algorithm.

Input: *Polynomials* $g_1, \ldots, g_n$ *in* S *whose ideal* I *is zero-dimensional.*
Output: *The global residue* $\mathrm{Res}_{\mathbf{g}}(h)$, *for any specified polynomial* $h \in S$.

Step 1: Choose a "good" term order "$\prec$".
Step 2: Starting with $\{g_1, \ldots, g_n\}$, run the Buchberger algorithm towards a Gröbner basis, until the current basis of I contains polynomials $f_1, \ldots, f_n$ with $\mathrm{in}_\prec(f_i) = x_i^{r_i+1}$.
Step 3: By keeping track of coefficients during Step 2, we obtain an $n \times n$-matrix $A = (A_{ij})$ of polynomials such that $f_i = \sum_{i=1}^n A_{ij}\, g_j$, for $i = 1, \ldots, n$.
Step 4: Compute the desired residue via the following formula:

$$\mathrm{Res}\left(\frac{h\, d\mathbf{x}}{g_1 \cdots g_n}\right) = \mathrm{Res}\left(\frac{h\, \det(A)\, d\mathbf{x}}{f_1 \cdots f_n}\right). \tag{3.3}$$

(3.4) Remarks. (i) In Step 1 we may take "$\prec$" to be *optimal* in the precise sense of [14, §3.3]. Such a choice is possible at almost no extra computational cost if the *Gröbner basis detection* procedure of [14, §3.2] had been run beforehand to test applicability of (3.1).

(ii) The termination and correctness of Step 2 follows from $\dim(\mathrm{in}_\prec(I)) = \dim(I) = 0$.

(iii) The correctness of (3.3) is just the transformation law (0.6). In order the evaluate the right-hand side of (3.3), we may use either algorithm (3.1), in case many residues are desired, or the formula to be presented in (4.2) below, in case only one residue is desired.

(iv) As shown in [9, theorem 3.3], it is possible to find polynomials $f_1, \ldots, f_n$ and A_{ij}, $1 \le i, j \le n$ with degrees bounded by $nd^{2n} + d^n + d$, where $d = \max(3, \max\{\deg(g_i)\})$.

(v) If the polynomials $g_1, \ldots, g_n$ satisfy (1.3) relative to some weight $\mathbf{w}$, but not necessarily (2.1), then the weighted version of the following argument due to Tsikh [25, II.8.3] describes how to find polynomials $f_1, \ldots, f_n$ in the ideal $I = \langle g_1, \ldots, g_n \rangle$ such that

$$\text{(3.5)} \qquad \mathrm{in}_{\mathbf{w}}(f_i) \;=\; x_i^{\rho} \quad \text{for some } \rho.$$

Indeed, Macaulay's theorem (1.21) implies that, if $\rho > d_{\mathbf{w}} - (w_1 + \cdots + w_n)$, then

$$x_i^{\rho} \,\in\, \mathrm{in}_{\mathbf{w}}(I) \,=\, \langle p_1, \ldots, p_n \rangle.$$

This degree bound allows the use of linear algebra (over $\mathbf{K}$) to determine polynomials A_{ij}, $1 \leq i, j \leq n$, such that $x_i^{\rho} = \sum_{i=1}^{n} A_{ij}\, p_j$. Moreover, $\deg_{\mathbf{w}}(A_{ij}) = \rho\, w_i - \deg_{\mathbf{w}}(g_j)$. Let

$$f_i \;:=\; \sum_{i=1}^{n} A_{ij}\, g_j \;=\; x_i^{\rho} + \sum_{i=1}^{n} A_{ij}\, q_j.$$

Since $\deg_{\mathbf{w}}\left(\sum A_{ij}\, q_j\right) < \rho\, w_i$, (3.5) holds, and (3.1) or (4.2) are applicable. Naturally, we may use the Buchberger algorithm to organize the computation of the A_{ij}, which leads to a version of algorithm (3.2) which is applied to the initial forms $p_1, \ldots, p_n$ rather than the entire equations $g_1, \ldots, g_n$. In summary, there is plenty of room for experimentation!

The fact that the coefficients $B_j(\mathbf{x})$ are weighted homogeneous implies that we can fix the value of one of the variables, say $x_n = 1$, and obtain a non homogeneous version of algorithm (3.1): Set $x_n = 1$ everywhere, and, for each $\mathbf{a}$ with $\langle \mathbf{w}, \mathbf{a} \rangle \leq d$, find the coefficient of $x_1^{-(a_1+1)} \cdots x_{n-1}^{-(a_{n-1}+1)}$ in

$$B_{\langle \mathbf{w}, \mathbf{a}-\mathbf{r} \rangle}(x_1, \ldots, x_{n-1}, 1).$$

We note that other terms of the form $k\, x_1^{-(a_1+1)} \cdots x_{n-1}^{-(a_{n-1}+1)}$ may appear in different coefficients $B_j(x_1, \ldots, x_{n-1}, 1)$, but the homogenizing parameter t keeps track of the only one contributing to the residue.

In the classical case $\mathbf{w} = (1, \ldots, 1)$, we can apply an argument of Yuzhakov [27] to give the following geometric interpretation of (3.1). As in §1, we imbed $\mathbb{C}^n$ in $\mathbb{P}^n$; the n-form $\dfrac{\mathbf{x}^{\mathbf{a}}\, d\mathbf{x}}{g_1(\mathbf{x}) \cdots g_n(\mathbf{x})}$ may be extended to a meromorphic form Φ in $\mathbb{P}^n$, which has $\{t = 0\}$ as a polar divisor of order $\langle \mathbf{w}, \mathbf{a} - \mathbf{r} \rangle + 1$ if and only if $\langle \mathbf{w}, \mathbf{a} - \mathbf{r} \rangle \geq 0$. The global residue in $\mathbb{C}^n$ may now be expressed as a single residue at a point at ∞: if $P = (0, \ldots, 0, 1)$, then

$$\mathrm{Res}_{\mathbf{g}}(\mathbf{x}^{\mathbf{a}}) \;=\; (-1)^n\, \mathrm{Res}_P\, \Phi \;=\; \frac{1}{\langle \mathbf{w}, \mathbf{a}-\mathbf{r} \rangle!}$$

$$\mathrm{Res}_0\left(x_1^{a_1} \cdots x_{n-1}^{a_{n-1}} \frac{\partial^{\langle \mathbf{w},\mathbf{a}-\mathbf{r}\rangle}}{\partial t^{\langle \mathbf{w},\mathbf{a}-\mathbf{r}\rangle}} \left(\frac{1}{\prod_{i=1}^n \tilde{g}_i(x_1,\ldots,x_{n-1},1;t)} \right)(0)\, dx_1 \wedge \cdots \wedge dx_{n-1} \right)$$

which, in turn, may be seen to equal the coefficient of $x_1^{-(a_1+1)} \cdots x_{n-1}^{-(a_{n-1}+1)}$ in $B_{\langle \mathbf{w},\mathbf{a}-\mathbf{r}\rangle}(x_1,\ldots,x_{n-1},1)$. With suitable modifications, the same interpretation holds for arbitrary weights.

4 Normal forms

In this section we give a formula expressing the coefficients of the (Gröbner basis) normal form in terms of global residues. In particular, the global residue of a polynomial equals the highest coefficient of its normal form (4.2). This leads to a fast algorithm for computing residues as well as traces over a zero-dimensional complete intersection.

Suppose $g_1,\ldots,g_n \in \mathbf{K}[\mathbf{x}]$ satisfy (2.1) with $\mathrm{in}_{\mathbf{w}}(g_i) = x_i^{r_i+1}$. Then $V = \mathbf{K}[\mathbf{x}]/I$ is an Artinian ring of $\mathbf{K}$-dimension $(r_1+1)(r_2+1)\cdots(r_n+1)$. Abbreviating

$$\mathcal{I} \quad := \quad \{ \mathbf{i} = (i_1,\ldots,i_n) \in \mathbb{Z}^n \ : \ 0 \le i_j \le r_j,\ j = 1,\ldots,n \},$$

the set of monomials $\{ \mathbf{x}^{\mathbf{i}} \ : \ \mathbf{i} \in \mathcal{I} \}$ is a $\mathbf{K}$-vectorspace basis of V. Every polynomial $h \in \mathbf{K}[\mathbf{x}]$ has a unique *normal form*

$$(4.1) \qquad \mathcal{NF}(h) \quad = \quad \sum_{\mathbf{i}\in\mathcal{I}} c_{\mathbf{i}}(h)\, \mathbf{x}^{\mathbf{i}}.$$

The scalars $c_{\mathbf{i}}(h) \in \mathbf{K}$ are uniquely defined by the property that $h \equiv \mathcal{NF}(h)$ (mod I). They are computed using the division algorithm modulo the Gröbner basis $\{g_1,\ldots,g_n\}$.

(4.2) Lemma. *With the notation as above, every polynomial $h \in \mathbf{K}[\mathbf{x}]$ satisfies*

$$\mathrm{Res}\left(\frac{h\, d\mathbf{x}}{g_1 \cdots g_n} \right) \quad = \quad c_{r_1,\ldots,r_n}(h)$$

Proof: By linearity of the residue operator, $\mathrm{Res}_{\mathbf{g}}(h) = \sum_{\mathbf{i}\in\mathcal{I}} c_{\mathbf{i}} \mathrm{Res}_{\mathbf{g}}(\mathbf{x}^{\mathbf{i}})$. However, for $\mathbf{i} \in \mathcal{I}$, $\mathbf{i} \neq \mathbf{r}$, we have $\langle \mathbf{w}, \mathbf{i}-\mathbf{r}\rangle < 0$ and, consequently, $\mathrm{Res}_{\mathbf{g}}(\mathbf{x}^{\mathbf{i}}) = 0$ by Corollary (1.18). On the other hand, it follows from theorem (1.17) that

$$\mathrm{Res}\left(\frac{\mathbf{x}^{\mathbf{r}}\, d\mathbf{x}}{g_1 \cdots g_n} \right) \quad = \quad \mathrm{Res}\left(\frac{\mathbf{x}^{\mathbf{r}}\, d\mathbf{x}}{x_1^{r_1+1} \cdots x_n^{r_n+1}} \right) \quad = \quad 1.$$

This proves lemma (4.2). □

We now show that all coefficients of the normal form may be computed using residues:

(4.3) Theorem. *Fix an order on the index set $\mathcal{I}$, and let M be the symmetric $|\mathcal{I}| \times |\mathcal{I}|$-matrix defined by*

$$M_{\mathbf{ij}} \quad := \quad \operatorname{Res}\left(\frac{\mathbf{x}^{\mathbf{i}}\,\mathbf{x}^{\mathbf{j}}\,d\mathbf{x}}{g_1\cdots g_n}\right), \quad \mathbf{i},\mathbf{j} \in \mathcal{I}\,.$$

Then M is invertible and for any $h \in \mathbf{K}[\mathbf{x}]$,

$$\big(c_{\mathbf{i}}(h)\big)_{\mathbf{i}\in\mathcal{I}} \quad = \quad M^{-1}\cdot\Big(\operatorname{Res}\big(\frac{h\,\mathbf{x}^{\mathbf{j}}\,d\mathbf{x}}{g_1\cdots g_n}\big)\Big)_{\mathbf{j}\in\mathcal{I}}$$

Proof: The duality theorem (0.7) implies that the symmetric bilinear form

$$\text{(4.4)} \qquad V\times V \to \mathbf{K}\,, \qquad (h_1,h_2) \mapsto \operatorname{Res}\left(\frac{h_1\,h_2\,d\mathbf{x}}{g_1\cdots g_n}\right)$$

is non degenerate. The symmetric matrix M represents (4.4) relative to the basis $\{\mathbf{x}^{\mathbf{i}} \,:\, \mathbf{i}\in\mathcal{I}\}$. Therefore M is non singular. The second claim follows from the fact that

$$\operatorname{Res}\left(\frac{\mathbf{x}^{\mathbf{i}}\,h\,d\mathbf{x}}{g_1\cdots g_n}\right) \;=\; \sum_{\mathbf{j}\in\mathcal{I}} c_{\mathbf{j}}(h)\operatorname{Res}\left(\frac{\mathbf{x}^{\mathbf{i}}\,\mathbf{x}^{\mathbf{j}}\,d\mathbf{x}}{g_1\cdots g_n}\right) \;=\; \sum_{\mathbf{j}\in\mathcal{I}} M_{\mathbf{ij}}\,c_{\mathbf{j}}(h)\,. \qquad \square$$

(4.5) Remark. As observed in [3], if we choose the lexicographical order in $\mathcal{I}$, then the matrix M has the triangular form

$$M \quad = \quad \begin{pmatrix} & & & 1 \\ & 0 & \cdot & \\ & \cdot & & \\ 1 & & * & \end{pmatrix}.$$

Consequently, $\det M = \pm 1$, and it is easy to compute the inverse of M.

We have seen in §0 that the duality law may also be interpreted as saying that the global residue operator $\operatorname{Res}_{\mathbf{g}}$ is a generator of V^* as a V-module. There is another element in this V-module which is of special interest in computational algebra, namely, the morphism $\operatorname{tr} \in V^*$, which assigns to a polynomial h the trace of the endomorphism of V given by multiplication by h. The *trace* can be computed by normal form reduction as follows:

$$(4.6) \qquad \mathrm{tr}(h) \quad = \quad \sum_{\mathrm{i}\in\mathcal{I}} c_{\mathrm{i}}(h\cdot\mathbf{x}^{\mathrm{i}}).$$

On the other hand, it is known (see e.g. [23, Satz (4.2)]) that the trace may be expressed in terms of the global residue:

$$(4.7) \qquad \mathrm{tr}(h) \quad = \quad \sum_{p\in Z(\mathbf{g})} \mu_{\mathbf{g}}(p)h(p) \quad = \quad \mathrm{Res}\left(\frac{h\, J_{\mathbf{g}}\, d\mathbf{x}}{g_1\cdots g_n}\right).$$

This expression, together with (4.2), gives the following formula for computing the trace:

(4.8) Algorithm. *Compute the trace of an element* $h \in V$ *by* $\mathrm{tr}(h) = c_{r_1\cdots r_n}(h\cdot J_{\mathbf{g}})$.

Thus, to find the trace of an element of V over $\mathbf{K}$, it suffices to run a single normal form reduction. We found algorithm (4.8) to be quite efficient in practice. Additional speed can be gained by simple tricks, such as replacing $J_{\mathbf{g}}$ by $\mathcal{NF}(J_{\mathbf{g}})$ in algorithm (4.8), and by storing previously computed normal forms of monomials.

From theorem 2.7 we can derive bounds on the degree of the trace for $g_1,\dots,g_n$ as in (2.6) which satisfy (2.1). Note that the corresponding cone $\mathcal{W}$ in the Gröbner fan has non-empty interior. For each value of the parameters c_{ij}, let $Z_{\mathbf{c}}$ denote the zero set of $g_1,\dots,g_n$.

(4.9) Theorem. *For any* $h \in K[\mathbf{x}]$, *the parametric trace*

$$\mathrm{tr}(h)(\mathbf{c}) \quad = \quad \sum_{p\in Z_{\mathbf{c}}} \mu_{\mathbf{g}}(p)\, h(p)$$

is a polynomial function of $\mathbf{c} = (c_{ij})$ *with degree bounded above by*

$$\min_{\mathbf{w}\in int(\mathcal{W})\cap\mathbb{Z}^n} \deg_{\mathbf{w}}(h).$$

Proof: Let $J_{\mathbf{g}}(\mathbf{x})$ be the Jacobian of $g_1, \ldots, g_n$ with respect to the variables $x_1, \ldots, x_n$. Given any polynomial h and $\mathbf{w} \in int(\mathcal{W}) \cap \mathbb{Z}^n$, we know by the second bound in theorem 2.7 that the degree of $\mathrm{tr}(h)(\mathbf{c}) = \mathrm{Res}_{\mathbf{g}}(h \cdot J_{\mathbf{g}})$ in the $\mathbf{c}$ variables is bounded by $\deg_{\mathbf{w}}(h \cdot J_{\mathbf{g}}) - \langle \mathbf{w}, \mathbf{r} \rangle$. As $\deg_{\mathbf{w}}(J_{\mathbf{g}}) = \langle \mathbf{w}, \mathbf{r} \rangle$, the claim follows. □

5 Real roots, degree, and symmetric polynomials

We present three applications of the computation of global residues: counting real roots using the trace form (following Pedersen-Roy-Szpirglas [22], Becker-Wörmann [4]), computing the degree of a polynomial map (following Eisenbud-Levine [12]), and evaluating elementary symmetric polynomials in a multivariate setting (following classical work of Junker [16]).

We assume as above that $g_1, \ldots, g_n \in \mathbf{K}[\mathbf{x}]$ satisfy (2.1) with $\mathrm{in}_{\mathbf{w}}(g_i) = x_i^{r_i+1}$, and again let $\mathcal{I} := \{\mathbf{i} = (i_1, \ldots, i_n) \in \mathbb{Z}^n \ : \ 0 \le i_j \le r_j\,,\ j = 1, \ldots, n\}$. Suppose that $\mathbf{K}$ is a subfield of the real numbers $\mathbb{R}$ and let T be the symmetric $|\mathcal{I}| \times |\mathcal{I}|$-matrix T defined over $\mathbf{K}$ by

$$T_{\mathbf{ij}} \quad := \quad \mathrm{tr}\left(\mathbf{x}^{\mathbf{i}}\mathbf{x}^{\mathbf{j}}\right), \quad \mathbf{i}, \mathbf{j} \in \mathcal{I}. \tag{5.1}$$

The following result is due to Becker-Wörmann and Pedersen-Roy-Szpirglas.

(5.2) Theorem. ([4],[22]) *The rank of T equals the number of distinct complex roots in $Z(\mathbf{g})$. The signature of T equals the number of distinct real roots in $Z(\mathbf{g}) \cap \mathbb{R}^n$.*

Recall that the *signature* of T equals the number of positive eigenvalues minus the number of negative eigenvalues (all eigenvalues of a real symmetric matrix are real). As is pointed out in [22, Prop. 2.8], the signature can be read off directly from (the number of sign variations in) the characteristic polynomial of T. A straightforward generalization of (5.2) states that, for any $h \in V$, the signature of the matrix $T^h = \left(\mathrm{tr}(\mathbf{x}^{\mathbf{i}}\mathbf{x}^{\mathbf{j}} \cdot h)\right)_{\mathbf{i},\mathbf{j} \in \mathcal{I}}$ equals the number of distinct real roots with $h > 0$ minus the number of distinct real roots with $h < 0$. algorithms (4.8) and (3.1) provide subroutines for computing T and hence for counting real zeros of zero-dimensional complete intersections.

Viewing now $(g_1, \ldots, g_n)$ as a proper map $\mathbf{g} : \mathbb{R}^n \to \mathbb{R}^n$, its *degree* is defined as

$$\deg(\mathbf{g}) := \sum_{p \in \mathbf{g}^{-1}(q)} \deg_p(\mathbf{g})$$

where q is a regular value of $\mathbf{g}$ and $\deg_p(\mathbf{g})$ is ± 1 depending on whether $J_{\mathbf{g}}(p)$ is positive or negative. The degree is a topological invariant of $\mathbf{g}$.

Let M be the non singular, symmetric matrix M defined, as in (4.3), by

$$M_{\mathbf{ij}} \quad := \quad \operatorname{Res}\left(\frac{\mathbf{x}^{\mathbf{i}}\,\mathbf{x}^{\mathbf{j}}\,d\mathbf{x}}{g_1\cdots g_n}\right), \quad \mathbf{i},\mathbf{j}\in\mathcal{I}.$$

The following result is essentially contained in [12]; although the results there are local, the passage to the global situation may be done as in [22].

(5.3) Theorem. *The degree of* $\mathbf{g}$ *equals the signature of* M.

We may apply algorithm (3.1) or lemma (4.2) to compute the matrix M and, consequently, the degree of $\mathbf{g}$.

For our third application we need to review the concept of symmetric polynomials in a multivariate setting. This theory is classical (see Junker [16], who refers to even earlier work of MacMahon and Schläfli). It reappeared in the recent computer algebra literature in [21]. Let $A = (\alpha_{ij})$ be an $N \times n$-matrix of indeterminates over $\mathbf{K}$. The symmetric group S_N acts on the polynomial ring $\mathbf{K}[\alpha_{ij}]$ by permuting rows of A. We are interested in the invariant subring $\mathbf{K}[\alpha_{ij}]^{S_N}$, whose elements are called *symmetric polynomials*. It is known that $\mathbf{K}[\alpha_{ij}]^{S_N}$ is generated by symmetric polynomials of total degree at most N, but, in contrast to the familiar $n = 1$ case, this $\mathbf{K}$-algebra $\mathbf{K}[\alpha_{ij}]^{S_N}$ is not free for $n \geq 2$. An important set of generators are the *elementary symmetric polynomials* $e_{\mathbf{j}}(A)$, which are defined as the coefficients of the following auxiliary polynomial in $u_1, u_2, \ldots u_n$:

(5.4)

$$\prod_{i=1}^{N}(1+\alpha_{i1}u_1+\alpha_{i2}u_2+\cdots+\alpha_{in}u_n) \quad = \quad \sum_{j_1+\ldots+j_n\leq N} e_{j_1,\ldots,j_n}(A)\cdot u_1^{j_1}u_2^{j_2}\cdots u_n^{j_n}.$$

Another set of generators is given by the *power sums:*

$$(5.5)\quad h_{\mathbf{j}} := \sum_{i=1}^{N}\alpha_{i1}^{j_1}\alpha_{i2}^{j_2}\cdots\alpha_{in}^{j_n}, \quad \text{for } \mathbf{j}=(j_1,j_2,\ldots,j_n)\in\mathbb{N}^n,\ j_1+\cdots+j_n\leq N.$$

algorithms and formulas for writing the $e_{\mathbf{j}}$ in terms of the $h_{\mathbf{j}}$ and conversely are studied in detail by Junker [16]. One of his methods will be presented in (5.8)–(5.9) below.

Returning to our zero-dimensional complete intersection, let $N = \dim_{\mathbf{K}}(V)$ be the cardinality of the multiset $Z(\mathbf{g}) \subset \mathbb{C}^n$ (counting multiplicities). We fix any bijection between the rows of $A = (\alpha_{ij})$ and $Z(\mathbf{g})$. This defines a natural $\mathbf{K}$-algebra homomorphism

$$(5.6)\qquad \phi\ :\ \mathbf{K}[\alpha_{ij}]^{S_N} \quad\to\quad \mathbf{K},$$

where the indeterminate α_{ij} gets mapped to the j-th coordinate of the i-th point in $Z(\mathbf{g})$. Our objective is to evaluate the map ϕ using only operations in $\mathbf{K}$. In particular, we are interested in the problem of evaluating the elementary symmetric polynomials e_{j} under ϕ.

The punch line of our discussion is that it is easy to evaluate the power sums via the trace:

$$(5.7) \qquad \phi(h_{\mathrm{j}}) \quad = \quad \mathrm{tr}(\mathbf{x}^{\mathrm{j}}).$$

Thus to compute (5.7) we use algorithm (4.8). We then proceed using the following method due to Junker and MacMahon. Consider the image of (5.4) under ϕ,

$$(5.8) \qquad R(\mathbf{u}) \quad = \quad \prod_{p \in Z(\mathbf{g})} (1 + p_1 u_1 + \cdots + p_n u_n)^{\mu_{\mathbf{g}}(p)} \quad = \quad \sum_{\mathrm{j}} \phi(e_{\mathrm{j}}) \cdot \mathbf{u}^{\mathrm{j}}.$$

The polynomial $R(\mathbf{u})$ is the *Chow form* of the zero-dimensional scheme defined by I. In computer algebra it is known also as the *U-resultant*. Following [16, pp. 233, Eq. (4)], the formal logarithm of (5.8) equals

$$(5.9) \qquad \log\big(R(\mathbf{u})\big) \quad = \quad \sum_{d=1}^{\infty} \frac{(-1)^{d-1}}{d} \cdot \sum_{|\mathrm{j}|=d} \binom{d}{\mathrm{j}} \phi(h_{\mathrm{j}})\, \mathbf{u}^{\mathrm{j}}.$$

Here $|\mathrm{j}| = j_1 + \cdots + j_n$ and $\binom{d}{\mathrm{j}} = \frac{d!}{j_1! j_2! \cdots j_n!}$. Using (5.7) and (4.8), we can compute the formal power series (5.9) up to any desired degree d'. We then formally exponentiate this truncated series (using operations only in $\mathbf{K}$) to get the Chow form (5.8) up to the same degree d'. In order to determine (5.8) completely, which means to evaluate all elementary symmetric polynomials, it suffices to expand (5.9) up to degree $d' = N = \dim_{\mathbf{K}}(V)$.

6 An example

In this section we apply our results and algorithms to the specific trivariate system:

$$(6.1) \; g_1 \;=\; \underline{x_1^5} + x_2^3 + x_3^2 - 1, \quad g_2 \;=\; x_1^2 + \underline{x_2^2} + x_3 - 1, \quad g_3 \;=\; x_1^6 + x_2^5 + \underline{x_3^3} - 1.$$

This example is taken from [14, example 3.1.2], where it served to illustrate the problem of *Gröbner basis detection*. Indeed, the polynomials g_1, g_2, g_3 are a Gröbner basis, namely, for the weight vector $\mathbf{w} = (3, 4, 7)$. With respect to these weights, the initial monomials are the pure powers underlined above. We see that, counting possible multiplicities, the set $Z(\mathbf{g})$ consists of 30 points in $\mathbb{C}^3$. Our basic problem is to evaluate the global residue

$$(6.2) \qquad \mathrm{Res}_{\mathbf{g}}(\mathbf{x}^{\mathbf{a}}) \quad = \quad \mathrm{Res}\left(\frac{x_1^{a_1}x_2^{a_2}x_3^{a_3}\,d\mathbf{x}}{g_1(\mathbf{x})g_2(\mathbf{x})g_3(\mathbf{x})}\right),$$

for any nonnegative integer vector $\mathbf{a} = (a_1, a_2, a_3)$.

It is interesting to compare the relative efficiency of algorithm (3.1) and the Gröbner basis reduction method deduced from lemma (4.2). In step 1 of algorithm (3.1) we compute the weighted homogenizations

$$\tilde{g}_1 = x_1^5+t^3x_2^3+t^7x_3^2-t^{15}, \quad \tilde{g}_2 = x_2^2+tx_3+t^2x_1^2-t^8, \quad \tilde{g}_3 = x_3^3+tx_2^5+t^3x_1^6-t^{21}.$$

We then consider the expression

$$(6.3) \qquad \frac{1}{\tilde{g}_1(t;\mathbf{x})\cdot\tilde{g}_2(t;\mathbf{x})\cdot\tilde{g}_3(t;\mathbf{x})}$$

as a rational function in t, and we compute its Taylor expansion $\sum_{j=0}^{d} B_j(\mathbf{x})\,t^j$ up to some degree d which exceeds $3(a_1-4)+4(a_2-1)+7(a_3-2)$. Here the coefficients $B_j(\mathbf{x})$ are $\mathbf{w}$-homogeneous Laurent polynomials in x_1, x_2, x_3; for instance,

$$B_2(\mathbf{x}) \quad = \quad \frac{1}{x_1^{10}x_2^4} - \frac{1}{x_1^3x_2^4x_3^3} + \frac{x_2}{x_1^5x_3^5} + \frac{x_2^3}{x_1^{10}x_3^4} + \frac{x_3}{x_1^{15}x_2^2} + \frac{x_2^8}{x_1^5x_3^9} + \frac{1}{x_1^5x_2^6x_3}.$$

Now set $j = 3(a_1-4)+4(a_2-1)+7(a_3-2)$. The desired residue (6.2) equals the coefficient of $x_1^{-a_1-1}x_2^{-a_2-1}x_3^{-a_3-1}$ in the Laurent polynomial $B_j \in \mathbb{Z}[x_1, x_1^{-1}, x_2, x_2^{-1}, x_3, x_3^{-1}]$.

This Taylor expansion is a fairly space consuming process since the polynomials $B_j(\mathbf{x})$ grow quite large. This is witnessed by the following table, which shows the number of terms of $B_j(\mathbf{x})$ for some values of j between 2 and 40:

j:	2	5	10	15	20	25	30	35	40
#:	7	41	216	569	1102	1803	2682	3744	4964

On the other hand, the normal form method of lemma (4.2) is quite efficient for evaluating individual residues. Let I denote the ideal in $\mathbb{Q}[x_1, x_2, x_3]$ generated by (6.1). The quotient ring $V = \mathbb{Q}[x_1, x_2, x_3]/I$ is a 30-dimensional $\mathbb{Q}$-vector space. Every element $h \in V$ is uniquely represented by its normal form $\mathcal{NF}(h)$ modulo the reduction relations:

$$(6.4) \quad x_1^5 \longrightarrow -x_2^3-x_3^2+1, \qquad x_2^2 \longrightarrow -x_1^2-x_3+1, \qquad x_3^3 \longrightarrow -x_1^6-x_2^5+1.$$

By lemma (4.2), the residue (6.2) is equal to the coefficient of $x_1^4 x_2 x_3^2$ in $\mathcal{NF}(h)$. For instance, for the Jacobian

$$\begin{aligned} J(\mathbf{x}) &= \det\left(\frac{\partial f_i}{\partial x_j}\right) = \\ &= 18x_1^5x_2^2 - 24x_1^5x_2x_3 - 25x_1^4x_2^4 + 30x_1^4x_2x_3^2 + 20x_1x_2^4x_3 - 18x_1x_2^2x_3^2 \end{aligned}$$

it takes 10 reductions modulo (6.4) to reach the normal form

$$\begin{aligned} \mathcal{NF}(J) = \; & \underline{30x_1^4x_2x_3^2} - 25x_1^4x_3^2 - 152x_1^4x_2 + 146x_1^4x_3 - 251x_1^3x_2x_3 \\ & + 83x_1^3x_3^2 + 16x_1^4 + 229x_1^3x_2 + 8x_1^3x_3 - 196x_1^2x_2x_3 + 226x_1^2x_3^2 \\ & - 114x_1x_2x_3^2 - 73x_1^3 + 240x_1^2x_2 + 34x_1^2x_3 + 254x_1x_2x_3 - 62x_1x_3^2 \\ & + 69x_2x_3^2 - 260x_1^2 - 140x_1x_2 - 78x_1x_3 + 108x_2x_3 - 49x_3^2 \\ & + 140x_1 - 177x_2 - 128x_3 + 177. \end{aligned}$$

Indeed, we see that the coefficient of $x_1^4x_2x_3^2$ equals $\mathrm{Res}_{\mathbf{g}}(J) = \mathrm{tr}(1) = \dim(V) = 30$. Here is a slightly more serious example: It takes 62 reductions modulo (6.4), running less than two minutes in MAPLE on a Sparc 2, in order to find the global residue

$$\mathrm{Res}_{\mathbf{g}}(x_1^{15}x_2^{15}x_3^{15}) = -258,756,707,658,424,020,014,953,731,203.$$

We made the observation that the efficiency of the two methods is comparable when computing all residues of the form $\mathrm{Res}_{\mathbf{g}}(\mathbf{x}^{\mathbf{a}})$ with $\langle \mathbf{w}, \mathbf{a}\rangle \leq d$ for some fixed d. This is the case, for example, in the computation of the matrix M defined in §4. This is a symmetric, 30×30 matrix whose computation using algorithm (3.1) requires the knowledge of $B_j(\mathbf{x})$ for $j \leq 30$. Using MAPLE on a Sparc 2 these may be obtained in 321 seconds. It takes an additional 247 seconds to read off the desired 465 coefficients. On the other hand, it takes 324 seconds to build up the matrix M using lemma (4.2). The signature of M is zero, and hence so is the degree of the map $\mathbf{g} : \mathbb{R}^3 \to \mathbb{R}^3$ by theorem (5.3).

For further combinatorial analysis we may wish to compute the two polyhedral cones in Section 2. We first obtain the 3-dimensional quadrangular cone

$$\begin{aligned} \mathcal{W} &= \{ (w_1, w_2, w_3) \in \mathbb{R}^3 \;:\; 5w_1 \geq 2w_3,\, 2w_2 \geq w_3,\, w_3 \geq 2w_1,\, 3w_3 \geq 5w_2 \} \\ &= \mathrm{pos}\{(4,5,10),\, (1,1,2),\, (5,6,10),\, (2,3,5)\}. \end{aligned}$$

The interior of $\mathcal{W}$ consists of all weight vectors which select the underlined monomials in (6.1) to be initial. The cone polar to $\mathcal{W}$ equals

$$\begin{aligned} \mathcal{W}^* &= \mathrm{pos}\{(5,0,-2),\, (0,2,-1),\, (-2,0,1),\, (0,-5,3)\} \\ &= \{ (a_1,a_2,a_3) \in R^3 \;:\; 4a_1 + 5a_2 + 10a_3 \geq 0,\, a_1 + a_2 + 2a_3 \geq 0, \\ &\qquad 5a_1 + 6a_2 + 10a_3 \geq 0,\, 2a_1 + 3a_2 + 5a_3 \geq 0 \}. \end{aligned}$$

By theorem (2.5), the residue (6.2) vanishes whenever

$$(a_1, a_2, a_3) \quad \notin \quad (4,1,2) + \mathcal{W}^*, \qquad \text{or equivalently,}$$

$4a_1 + 5a_2 + 10a_3 < 41$ or $a_1 + a_2 + 2a_3 < 9$ or $5a_1 + 6a_2 + 10a_3 < 46$ or $2a_1 + 3a_2 + 5a_3 < 21$. For instance, $(6,1,1)$ satisfies the first inequality and therefore $\mathrm{Res}_{\mathbf{g}}(x_1^6 x_2 x_3) = 0$.

In Section 4 we have shown that the trace $\mathrm{tr}(h)$ of an element h in $V = \mathbb{Q}[\mathbf{x}]/I$ can be computed easily as the coefficient of $x_1^4 x_2 x_3^2$ in $\mathcal{NF}(h \cdot J)$. Using this technique, let us now analyze the zero set $Z(\mathbf{g})$ with respect to multiple roots, real roots, etc. We compute the symmetric, integer 30×30 matrix representing the trace form T as in (4.8). The largest entry in T appears in the lower-right corner:

$$\mathrm{tr}(x_1^4 x_2 x_3^2 \cdot x_1^4 x_2 x_3^2) \quad = \quad \mathrm{Res}_{\mathbf{g}}\big(x_1^8 x_2^2 x_3^4 \cdot J(\mathbf{x})\big) \quad = \quad 16,049,138,278.$$

The rank of the matrix T equals 20. By theorem (5.2), this is the number of *distinct* roots of $\mathbf{g}$. The characteristic polynomial of T has 13 positive real roots and 7 negative real roots. Therefore the signature of T equals 6, and this is the number of distinct *real* roots of $\mathbf{g}$. It turns out that there are four rational roots, and they account for all multiplicities: the root $(1,0,0)$ has multiplicity 3, the root $(0,1,0)$ has multiplicity 4, the root $(0,0,1)$ has multiplicity 6, while the root $(-1,1,-1)$ is simple. The remaining 16 roots, two real and 14 imaginary, are all simple and they are conjugates over $\mathbb{Q}$.

We finally come to the problem of computing the Chow form

$$R(u_1, u_2, u_3) \quad = \quad \prod_{(\alpha_1,\alpha_2,\alpha_3)\in Z(\mathbf{g})} (1 + \alpha_1 u_1 + \alpha_2 u_2 + \alpha_3 u_3)^{\mu_{\mathbf{g}}(\alpha)}.$$

Note that each of the three non simple roots appears with its multiplicity in this product. The $\binom{33}{3} = 5,456$ rational coefficients of $R(u_1, u_2, u_3)$ are the values of the elementary symmetric polynomials at the roots of $\mathbf{g} = (g_1, g_2, g_3)$. Following (5.9), (5.7) and using algorithm (4.8), we compute the following formal power series up to a chosen degree:

$$\begin{aligned}
\log\big(R(u_1, u_2, u_3)\big) \quad &= \quad \mathrm{tr}(x_1)u_1 + \mathrm{tr}(x_2)u_2 + \mathrm{tr}(x_3)u_3 - \frac{1}{2}\cdot\big(\mathrm{tr}(x_1^2)u_1^2 \\
&+ 2\mathrm{tr}(x_1x_2)u_1u_2 + 2\mathrm{tr}(x_1x_3)u_1u_3 + \mathrm{tr}(x_2^2)u_2^2 + 2\mathrm{tr}(x_2x_3)u_2u_3 + \mathrm{tr}(x_3^2)u_3^2\big) \\
&+ \frac{1}{3}\cdot\big(\mathrm{tr}(x_1^3)u_1^3 + \ldots \\
&= \quad 5u_2 - 5u_3 + 37u_1u_2 - 121u_1u_3 - \frac{35}{2}u_2^2 + 106u_2u_3 - \frac{485}{2}u_3^2 + 17u_1^3 \\
&- 74u_1^2u_2 + 177u_1^2u_3 - 172u_1u_2^2 + 536u_1u_2u_3 - 686u_1u_3^2 + \frac{185}{3}u_2^3 \\
&- 667u_2^2u_3 + 1084u_2u_3^2 + \ldots
\end{aligned}$$

By formally exponentiating this series, we obtain the Chow form

$$\begin{aligned} R(u_1,u_2,u_3) &= 1+5u_2-5u_3+37u_1u_2-121u_1u_3-5u_2^2+81u_2u_3 \\ &- 230u_3^2+17u_1^3-74u_1^2u_2+177u_1^2u_3+13u_1u_2^2-254u_1u_2u_3-81u_1u_3^2 \\ &- 5u_2^3-112u_2^2u_3-596u_2u_3^2+\ldots \end{aligned}$$

and hence all elementary symmetric polynomials. For instance, we see that $\sum \alpha_1\beta_2\gamma_3 = -254$, where the sum is taken over all triples of roots $(\alpha_1,\alpha_2,\alpha_3)$, $(\beta_1,\beta_2,\beta_3)$ and $(\gamma_1,\gamma_2,\gamma_3)$ in $Z(\mathbf{g})$.

Acknowledgments. This project began during the 1992 NSF Regional Geometry Institute at Amherst College. We thank its organizers and, most particularly, its Research Director, David Cox, for their hospitality. We express our gratitude to Adrian Paenza and Paul Pedersen for their help and support, and to the Center for Applied Mathematics of Cornell University for its hospitality during the preparation of this paper. E. Cattani was partially supported by NSF Grant DMS-9107323, A. Dickenstein was partially supported by UBA-CYT and CONICET, and B. Sturmfels was partially supported by NSF grants DMS-9201453, DMS-9258547 (NYI) and a David and Lucile Packard Fellowship.

Note added in proof: During the MEGA 94 meeting we became aware of the paper: [M. Kreuzer and E. Kunz: Traces in strict Frobenius algebras and strict complete intersections. *J. reine angew. Math.* **381** (1987), 181–204]. Our assumption (1.3) is equivalent, by their proposition (4.2), to the statement that the K-algebra $V = K[\mathbf{x}]/I$ is a strict complete intersection. Consequently, theorem (1.17), in the case $s(\mathbf{a}) \leq 0$ (in particular the Euler-Jacobi theorem (Corollary (1.18)) is contained in their Corollary (4.6) and theorem (4.8).

References

[1] I. A. Aĭzenberg and A. P. Yuzhakov: Integral representations and residues in multidimensional complex analysis. Translations of Mathematical Monographs **58**. American Mathematical Society, 1983.

[2] D. Bayer and D. Mumford: What can be computed in algebraic geometry?, In: *"Computational Algebraic Geometry and Commutative Algebra"* (eds. D. Eisenbud, L. Robbiano), Proceedings Cortona 1991, Cambridge University Press, 1993, pp. 1–48.

[3] E. Becker, J.-P. Cardinal, M.-F. Roy, and Z. Szafraniec: Multivariate Bezoutians, Kronecker symbol and Eisenbud-Levine formula, this volume.

[4] E. Becker and T. Wörmann: On the trace formula for quadratic forms, Recent Advances in Real Algebraic Geometry and Quadratic Forms. Proceedings of the RAGSQUAD Year, Berkeley 1990–1991. W. B. Jacob, T.-Y. Lam, R. O. Robson (editors), *Contemporary Mathematics*. 155. pp. 271–291.

[5] C. Berenstein and A. Yger: Une formule de Jacobi et ses conséquences. *Ann. scient. Ec. Norm. Sup.* 4e série, **24** (1991) 369–377.

[6] C. Berenstein and A. Yger: Effective Bezout identities in $\mathbb{Q}[z_1, \ldots, z_n]$. *Acta Math.* **166** (1991) 69–120.

[7] J.-P. Cardinal: Dualité et algorithmes itératifs pour la résolution de systèmes polynomiaux, Thèse Univ. Rennes I, Janvier 1993.

[8] A. Dickenstein, N. Fitchas, M. Giusti, and C. Sessa: The membership problem for unmixed polynomial ideals is solvable in single exponential time. *Discrete Applied Math.* **33** (1991) 73–94.

[9] A. Dickenstein and C. Sessa: An effective residual criterion for the membership problem in $\mathbb{C}[z_1, \ldots, z_n]$. *Journal of Pure and Applied Algebra* **74** (1991) 149–158.

[10] A. Dickenstein and C. Sessa: Duality methods for the membership problem, In: *"Effective Methods in Algebraic Geometry"* (eds. T. Mora, C. Traverso), Proceedings MEGA-90, Progress in Math. 94. Birkhäuser. 1991, pp. 89–103.

[11] D. Eisenbud: Commutative algebra with a view toward algebraic geometry. To Appear.

[12] D. Eisenbud and H. Levine: An algebraic formula for the degree of a C^∞ map germ. *Annals of Mathematics* **106** (1977) 19–44.

[13] N. Fitchas, M. Giusti and F. Smietanski: Sur la complexité du théorème des zéros, Preprint, 1993.

[14] P. Gritzmann and B. Sturmfels: Minkowski addition of polytopes: Computational complexity and applications to Gröbner bases. *SIAM J. Discr. Math.* **6** (1993) 246–269.

[15] P. A. Griffiths and J. Harris: Principles of algebraic geometry. Wiley-Interscience, New York, 1978.

[16] F. Junker: Über symmetrische Funktionen von mehreren Reihen von Veränderlichen, *Mathematische Annalen*, **43** (1893) 225–270.

[17] M. Kalkbrener and B. Sturmfels: Initial complexes of prime ideals. *Advances in Math.*, to appear.

[18] A. G. Khovanskii: Newton's polyhedron and the Euler-Jacobi formula. *Uspekhi Mat. Nauk* **33**, no. 6 (1978) 245–246; English transl. in *Russian Math. Surveys* **33** (1978).

[19] E. Kunz: Kähler differentials, Advanced Lectures in Mathematics, Vieweg Verlag, 1986.

[20] E. Kunz and R. Waldi: Deformations of zero-dimensional intersection schemes and residues, *Note di Matematica* **11** (1991) 247–259.

[21] P. Pedersen: Calculating multidimensional symmetric functions using Jacobi's formula, Proceedings AAECC 9, (eds. H.F. Mattson, T. Mora, T.R.N. Rao), Springer Lecture Notes in Computer Science, **539**, 1991, pp. 304–317.

[22] P. Pedersen, M.-F. Roy, and A. Szpirglas: Counting real zeros in the multivariate case, In: *"Computational Algebraic Geometry"* (eds. F. Eyssette, A. Galligo), Proceedings MEGA-92, Progress in Math. 109, Birkhäuser, 1993, pp. 203–223.

[23] G. Scheja and U. Storch: Über Spurfunktionen bei vollständigen Durchschnitten, *J. Reine u. Angewandte Mathematik* **278/9** (1975) 174–190.

[24] J.-P. Serre: G.A.G.A, *Annales de l'Institut Fourier VI* (1956),1–42.

[25] A. K. Tsikh: Multidimensional residues and their applications. Translations of Mathematical Monographs **103**. American Mathematical Society, 1992.

[26] A. Weil: L'intégrale de Cauchy et les fonctions de plusieurs variables. *Math. Ann.* **111** (1935), 178–182.

[27] A. P. Yuzhakov: On the computation of the complete sum of residues relative to a polynomial mapping in $\mathbb{C}^n$. *Dokl. Akad. Nauk. SSSR* **275** (1984), 817–820; English transl. in *Soviet. Math. Dokl.* **29(2)** (1984) .

E. Cattani

Department of Mathematics, University of Massachusetts, Amherst, MA 01003 (USA).

A. Dickenstein (alidick@mate.dm.uba.edu.ar)

Departamento de Matemática, Universidad de Buenos Aires, 1428 Buenos Aires (Argentina).

B. Sturmfels (bernd@math.cornell.edu)

Department of Mathematics, Cornell University, Ithaca NY 14853 (USA).

Progress in Mathematics, Vol. 143, © 1996 Birkhäuser Verlag Basel/Switzerland

The arithmetic of hyperelliptic curves

E. V. Flynn*

Introduction

The constructive theory of hyperelliptic curves has been advanced significantly during the last year. It is intended to give here an indication of the current level of progress, and an outline of the main methods employed. The emphasis in Sections 1 to 4 will be on the group of rational points on the Jacobian of a hyperelliptic curve. Section 5 will concern itself with the use of the Jacobian to help to determine the rational points on the curve itself.

1 The group law and formal group on the Jacobian

Let $\mathcal{C}$ be a hyperelliptic curve of genus g:

$$\mathcal{C} : Y^2 = F(X), \tag{1}$$

where $deg(F) = 2g+1$ or $2g+2$ and F has non-zero discriminant.

We shall assume that $\mathcal{C}$ is *defined over* $\mathbb{Q}$; that is, the coefficients of F are in $\mathbb{Q}$. By a *divisor* of $\mathcal{C}$ we shall mean (with slight abuse of notation and uniqueness) an unordered set of g points on the curve, where multiplicities are permitted. When $deg(F) = 2g+1$, we include ∞ as a point on $\mathcal{C}$, and denote $\{\infty, \ldots, \infty\} = \{g \cdot \infty\}$ by $\mathcal{O}$. When $deg(F) = 2g+2$ and g is even, we must include ∞^+ and ∞^- (the branches of the singularity at infinity) as separate points on $\mathcal{C}$, and we take $\mathcal{O} = \{g/2 \cdot \infty^+, g/2 \cdot \infty^-\}$. When $deg(F) = 2g+2$ and g is odd, then such an $\mathcal{O}$ is not defined over $\mathbb{Q}$ - however this technicality need not concern us here, as our examples will avoid that situation. Given a point $P = (x, y)$ on $\mathcal{C}$, its *flip* $\overline{P} = (x, -y)$. The inverse of a divisor $\{P_1, \ldots, P_g\}$ will then be $\{\overline{P_1}, \ldots, \overline{P_g}\}$.

We shall say that three such divisors D_1, D_2, D_3 sum to $\mathcal{O}$ if there exists a function of the form

$$R(X) \cdot Y - S(X), \text{ where } deg(R) \leq g/2 - 1, \text{ and } deg(S) \leq 3g/2 \tag{2}$$

which is satisfied by the $3g$ points contained in the sets D_1, D_2, D_3. We let $\mathcal{J} = \mathcal{J}(\mathcal{C})$, the *Jacobian of* $\mathcal{C}$, denote all such sets of g points: then the above laws give $\mathcal{J}$ the structure of an abelian group, with identity $\mathcal{O}$, which generalises the usual group law on an elliptic curve (the case $g = 1$). A divisor $D =$

(*) The author thanks SERC for financial support.

$\{P_1, \ldots, P_g\} = \{(x_1, y_1), \ldots, (x_g, y_g)\}$ in $\mathcal{J}$ is *rational* if there exist polynomials ϕ and ψ of degree g, with coefficients in $\mathbb{Q}$, such that:

$$\phi(X) = \prod_{i=1}^{n} (X - x_i) \text{ and } y_i = \psi(x_i), \text{ for all } i. \tag{3}$$

The rational divisors form a subgroup of $\mathcal{J}$, denoted $\mathcal{J}(\mathbb{Q})$. A divisor D in $\mathcal{J}$ is of *finite order* (or *torsion*) if there exists a positive integer N such that $ND = \mathcal{O}$; the smallest such N is the *order* of D. It is well-known that the subgroup of rational torsion divisors, $\mathcal{J}_{tors}(\mathbb{Q})$, is finite. It is also well-known that $\mathcal{J}(\mathbb{Q})$ is finitely generated, and so there exists a nonnegative integer r such that $\mathcal{J}(\mathbb{Q}) \cong \mathcal{J}_{tors}(\mathbb{Q}) \times \mathbb{Z}^r$. This integer r is the *rank* of the Jacobian. One of main aims of techniques developed during the last five years, has been – given a hyperelliptic curve – to find a set of generators for $\mathcal{J}(\mathbb{Q})$.

As well as being an abelian group, the Jacobian can also be given the structure of a smooth projective variety. By a theorem of Lefschetz ([15], p.105), we can find an embedding into $\mathbb{P}^{4^g-1}$.

Theorem 1.1. *Let C be a hyperelliptic curve of genus g, with coefficients in $\mathbb{Q}$. Then there is an embedding of $\mathcal{J}$ into $\mathbb{P}^{4^g-1}$ (which maps $\mathcal{J}(\mathbb{Q})$ into $\mathbb{P}^{4^g-1}(\mathbb{Q})$) as a smooth variety of dimension g, with defining equations given by quadratic forms, and the group law given by a biquadratic map. Further, the Kummer variety, obtained by taking the quotient of the Jacobian by $\pm$, may be embedded into $\mathbb{P}^{2^g-1}$, with the duplication law given by quartic forms on both the Jacobian and Kummer varieties.* □

In practice, it is difficult to compute a set of defining equations for the Jacobian, due to the sheer size of the expressions involved. In the case of genus 2, however, the equations have been derived explicitly with the help of the computer algebra package Maple (see [3],[6],[8]).

Theorem 1.2. *Let $\mathbf{a} = (a_0, \ldots, a_{15})$ be the 16 functions given in [8]. Then these provide an embedding of the Jacobian into $\mathbb{P}^{15}$, with defining equations given by the 72 quadratic forms given in Appendix A of [6].* □

When considering local properties of the Jacobian (viewed over $\mathbb{Q}_p$) "near" $\mathcal{O}$, it is convenient to work with a power-series description of the group law. It is necessary to find a basis of local parameters $\mathbf{s} = (s_1, \ldots, s_g)$, which are expressed in terms of the coordinate functions of the projective embedding of the Jacobian. A set of local parameters must have the property that they uniquely determine any $D \in \mathcal{J}$ which is sufficiently close to $\mathcal{O}$. There exists an associated vector $\mathcal{F} = (\mathcal{F}_1, \ldots, \mathcal{F}_g)$, where each $\mathcal{F}_i = \mathcal{F}_i(s_1, \ldots, s_g, t_1, \ldots, t_g)$ is a power-series in $2g$ variables. Let $\mathbf{a}, \mathbf{b}, \mathbf{c} \in \mathcal{J}(\mathbb{Q}_p)$ have local parameters $\mathbf{s}, \mathbf{t}, \mathbf{u}$, respectively; then, in a neighbourhood of $\mathcal{O}$, we have $\mathbf{u} = \mathcal{F}(\mathbf{s}, \mathbf{t})$. For the genus 2 case, a pair of local parameters is given by $s_1 = a_1/a_0, s_2 = a_2/a_0$, where $\mathbf{a}$

is as in theorem 1.2. In this case, a method for deriving terms of the formal group is described in [6], [8].

In the case when a curve of genus 2 can be written over $\mathbb{Q}$ in the form $Y^2 = \textit{quintic in } X$, then the variety describing the Jacobian may be embedded into $\mathbb{P}^8$ rather than $\mathbb{P}^{15}$. The resulting algebra is considerably simpler, both for computing the defining equations of the Jacobian, and the terms of the formal group. This situation has been considered in detail in [13].

2 Rational torsion sequences

In the case of an elliptic curve $\mathcal{E}$ over $\mathbb{Q}$, the possible torsion groups $\mathcal{E}_{tors}(\mathbb{Q})$ which can occur have been completely determined by Mazur in [18].

Theorem 2.1. *Let $\mathcal{E}$ be an elliptic curve defined over $\mathbb{Q}$. Then the torsion subgroup $\mathcal{E}_{tors}(\mathbb{Q})$ is one of the fifteen groups: $\mathbb{Z}/N\mathbb{Z}$ for $N = 1, \ldots, 10, 12$, or $\mathbb{Z}/2\mathbb{Z} \times \mathbb{Z}/2N\mathbb{Z}$ for $N = 1, \ldots, 4$.* □

A result which applies to elliptic curves over a number field K has recently been found by Merel in [21], where it is shown that any prime torsion order p must be bounded by d^{3d^2}, where d is the degree of K over $\mathbb{Q}$. No result along these lines has been found for Jacobians of curves of higher genus, and it is a natural question to ask what new torsion orders can occur in $\mathcal{J}_{tors}(\mathbb{Q})$ as the genus increases.

In order to derive hyperelliptic curves for which the torsion orders in $\mathcal{J}(\mathbb{Q})$ increase quickly with respect to the genus, the strategy is to choose sequences of curves of genus g with rational points $P_1, \ldots, P_n$, so that n different functions meet the curve only at these points. If these functions induce n $\mathbb{Z}[g]$-linear conditions given by:

$$A \begin{pmatrix} P_1 \\ \vdots \\ P_n \end{pmatrix} = \begin{pmatrix} \mathcal{O} \\ \vdots \\ \mathcal{O} \end{pmatrix} \tag{4}$$

where $A \in M_n[\mathbb{Z}[g]]$, then it is immediate (on multiplying both sides on the left by $\det(A) \cdot A^{-1} \in M_n[\mathbb{Z}[g]]$) that

$$\det(A) \begin{pmatrix} P_1 \\ \vdots \\ P_n \end{pmatrix} = \begin{pmatrix} \mathcal{O} \\ \vdots \\ \mathcal{O} \end{pmatrix} \tag{5}$$

so that, for $i = 1, \ldots, n$, $\det(A) \cdot P_i = \mathcal{O}$ (where, as always, everything is up to linear equivalence). This provides a divisor of order dividing $\det(A)$, which can often be shown to have order exactly $\det(A)$.

For the purpose of deriving quadratic sequences, we require only two such points and two such functions. We have used this technique in [7] to find the following sequences.

Result 2.2. *The 1-parameter space of curves of genus g ($t \neq 0$):*

$$\mathcal{C} : Y^2 = -tX^{g-r}(X-1)^{g+r+1} + \psi(X)^2$$

where $0 \leq r \leq g-1$, and $\psi(X) = X^{g+1} - t(X-1)^g - X^{g-r}(X-1)^{r+1}$ (degree g in X), has a divisor of positive torsion order dividing: $2g^2+2g+r+1$. In particular, when $r=0$, the divisor $D = \{(1,1), (g-1)\cdot\infty\}$ has exact order $2g^2+2g+1$. □

Result 2.3. *In even genus g, there exists $\mathbb{Q}$ rational torsion divisors of all orders in the interval $[g^2+2g+1, g^2+3g+1]$. Explicitly, the 1-parameter space of curves of genus g (g even, $t \neq 0$):*

$$\mathcal{C} : Y^2 = (\psi(X))^2 - 2t(X^{g+2} + X^{r+1}) + t^2(X-1)^2$$

where $0 \leq r \leq g$, and $\psi(X) = \sum_{i=1}^{g-r+1} X^{r+i} = (X^{g+2} - X^{r+1})/(X-1)$, has a divisor of exact order $g^2+3g+1-r$. □

More recently, Leprévost in [16],[17] has improved Result 2.2 to find sequences of the form: $2g^2+kg+1$, for $k=2,3,4$.

3 Complete 2-descent and descent via isogeny

An intermediary step towards resolving $\mathcal{J}(\mathbb{Q})$ is to find $\mathcal{J}(\mathbb{Q})/2\mathcal{J}(\mathbb{Q})$, which is known to be finite. For a hyperelliptic curve $Y^2 = F(X)$ defined over $\mathbb{Q}$, let $F(X) = F_1(X)\cdots F_n(X)$ be the irreducible factorisation of $F(X)$ over $\mathbb{Q}$ and, for each i, let $K_i = \mathbb{Q}(\theta_i)$, where θ_i is a root of $F_i(X)$. Then, there is a well-known [2] finite group M, which can be given as a subgroup of $K_1^*/(K_1^*)^2 \times \cdots \times K_n^*/(K_n^*)^2$, and an injection $\psi : \mathcal{J}(\mathbb{Q})/2\mathcal{J}(\mathbb{Q}) \longrightarrow M$. The construction of M and ψ guarantees that $\mathcal{J}(\mathbb{Q})/2\mathcal{J}(\mathbb{Q})$ is finite and provides an upper-bound for its size, but does not guarantee that $\mathcal{J}(\mathbb{Q})/2\mathcal{J}(\mathbb{Q})$ can be found completely. The standard technique is to make use of the commutative diagram:

$$\begin{array}{ccc} \mathcal{J}(\mathbb{Q})/2\mathcal{J}(\mathbb{Q}) & \xrightarrow{\psi} & M \\ \downarrow{\scriptstyle i_p} & & \downarrow{\scriptstyle j_p} \\ \mathcal{J}(\mathbb{Q}_p)/2\mathcal{J}(\mathbb{Q}_p) & \xrightarrow{\psi_p} & M_p \end{array} \qquad (6)$$

where the bottom row is constructed in the same way as the top row, but with respect to $\mathbb{Q}_p$, the p-adic numbers. The maps i_p and j_p are natural maps on the quotient induced by the inclusion map from $\mathbb{Q}$ into $\mathbb{Q}_p$ (note that i_p and j_p are not injective in general). It turns out that, for any p, it is straightforward to compute $\mathcal{J}(\mathbb{Q}_p)$, ψ_p and M_p completely. The preimage of M_p under j_p can then be used to bound the image of ψ. We define the *Selmer group*, S, by:

$$S = \bigcap_p j_p^{-1}(im(\psi_p)).$$

The group S may be viewed as those members of M which cannot be discarded as potential members of $im\,\psi$ merely by "congruence" arguments. Clearly $im\,\psi \le S$. It may turn out that $im\,\psi = S$, in which case the above method determines $\mathcal{J}(\mathbb{Q})/2\mathcal{J}(\mathbb{Q})$, and hence the rank of $\mathcal{J}(\mathbb{Q})$, completely. The extent to which S fails to determine $im\,\psi$ completely is described by a portion of a structure called the *Tate-Shafarevich group*. The method is not an algorithm, since there is no known effective procedure for determining the Tate-Shafarevich group.

The above methodology has long been employed to find ranks of elliptic curves (the case $g = 1$). It is only very recently that non-trivial examples have been computed on Jacobians of curves of higher genus. The first successful method (complete 2-descent) was due to Gordon and Grant [12], which applies to curves of genus 2 which split completely over $\mathbb{Q}$:

$$Y^2 = (X - a_1)(X - a_2)(X - a_3)(X - a_4)(X - a_5),\ a_i \in \mathbb{Q}.$$

Note that, for a general curve of genus 2, given by $Y^2 = F(X)$, the 2-torsion subgroup of $\mathcal{J}$ is given by $\mathcal{O}$ and divisors of the form $\{(x_1,0),(x_2,0)\}$, where x_1, x_2 are distinct roots of $F(X)$. This gives a 2-torsion group of size 16 in $\mathcal{J}$, when viewed over $\mathbb{C}$. The above condition imposed by Gordon and Grant guarantees that all points of order 2 in $\mathcal{J}$ lie in $\mathcal{J}(\mathbb{Q})$, which simplifies the construction of the finite group M above (which lies inside products of $\mathbb{Q}^*/(\mathbb{Q}^*)^2$). For each $d \in M$, they construct homogeneous spaces $\mathcal{J}_d$, which have the property that $d \in im\,\psi \iff \mathcal{J}_d(\mathbb{Q}) \ne \emptyset$. For each $\mathcal{J}_d$, one then tries either to find a rational point, or to find a contradiction in some $\mathbb{Q}_p$. Two examples were computed by this method in [12].

Example 3.1. *Let C be the curve $y^2 = x(x-1)(x-2)(x-5)(x-6)$. Then $\mathcal{J}(\mathbb{Q})/2\mathcal{J}(\mathbb{Q})$ is generated by the 2-torsion divisors, and the divisor $\{(3,6),\infty\}$, which has infinite order. It follows that $\mathcal{J}(\mathbb{Q})$ has rank 1.* □

Example 3.2. *Let C be the curve $y^2 = x(x-3)(x-4)(x-6)(x-7)$. Then $\mathcal{J}(\mathbb{Q})/2\mathcal{J}(\mathbb{Q})$ is generated by the 2-torsion divisors, and $\mathcal{J}(\mathbb{Q})$ has rank 0.* □

Note that, in Example 3.1, the rank of $\mathcal{J}(\mathbb{Q})$ was deduced from the size of $\mathcal{J}(\mathbb{Q})/2\mathcal{J}(\mathbb{Q})$, by first finding the image of $\mathcal{J}_{tors}(\mathbb{Q})$ on $\mathcal{J}(\mathbb{Q})/2\mathcal{J}(\mathbb{Q})$, and then using the fact that each independent divisor of infinite order on $\mathcal{J}(\mathbb{Q})$ contributes precisely one to the number generators of $\mathcal{J}(\mathbb{Q})/2\mathcal{J}(\mathbb{Q})$. Therefore, the rank has been determined, without any guarantee that the divisors found are actual generators for $\mathcal{J}(\mathbb{Q})$. We will return to this point in Section 4.

A second method was developed by the author in [9], which applies to the more general class of curves of genus 2 which may be written in the form: $\mathcal{C} : Y^2 = q_1(X)q_2(X)q_3(X)$, where each $q_i(X)$ is a quadratic defined over $\mathbb{Q}$. In this case, the 2-torsion group of $\mathcal{J}(\mathbb{Q})$ must have size at least 4, including $\mathcal{O}$ and the 3 rational divisors of order 2, each given by a conjugate pair of roots to

$q_i(X)$. This group of size 4 can be taken to be the kernel of a homomorphism (an *isogeny* of degree 4) to the Jacobian of an associated curve [1].

Definition 3.3. Let $\mathcal{C}$ be the curve of genus 2 defined over $\mathbb{Q}$ as:

$$\mathcal{C} : Y^2 = q_1(X)q_2(X)q_3(X) = (f_1X^2 + g_1X + h_1)(f_2X^2 + g_2X + h_2)(f_3X^2 + g_3X + h_3). \tag{7}$$

For any two polynomials $p(X)$, $q(X)$, let $[p, q]$ denote $p'q - pq'$. Define $\widehat{\mathcal{C}}$ by:

$$\widehat{\mathcal{C}} : \Delta Y^2 = \hat{q}_1(X)\hat{q}_2(X)\hat{q}_3(X) = [q_2, q_3][q_3, q_1][q_1, q_2]$$

where

$$\Delta = \begin{vmatrix} h_1 & g_1 & f_1 \\ h_2 & g_2 & f_2 \\ h_3 & g_3 & f_3 \end{vmatrix}.$$

Denote $b_{ij} = \text{resultant}(q_i, q_j)$, $b_i = b_{ij}b_{ik}$, $\hat{b}_{ij} = \text{resultant}(\hat{q}_i, \hat{q}_j)$, $\hat{b}_i = \hat{b}_{ij}\hat{b}_{ik}$. Let $\mathcal{J}$, $\widehat{\mathcal{J}}$ be the Jacobians of $\mathcal{C}$ and $\widehat{\mathcal{C}}$, and let α_i denote the point of order 2 in $\mathcal{J}(\mathbb{Q})$ corresponding to $q_i(X)$; similarly for $\hat{\alpha}_i$.

It has been shown in [1] that $\mathcal{J}, \widehat{\mathcal{J}}$ are isogenous. There exists isogenies: $\phi : \mathcal{J} \to \widehat{\mathcal{J}}$ with kernel $\{\mathcal{O}, \alpha_1, \alpha_2, \alpha_3\}$ and $\hat{\phi} : \widehat{\mathcal{J}} \to \mathcal{J}$ with kernel $\{\widehat{\mathcal{O}}, \hat{\alpha}_1, \hat{\alpha}_2, \hat{\alpha}_3\}$, such that $\hat{\phi} \circ \phi = [2]$, the duplication map on $\mathcal{J}$. As with elliptic curves, there is a natural injection [9] from $\widehat{\mathcal{J}}(\mathbb{Q})/\phi(\mathcal{J}(\mathbb{Q}))$ into a known finite group which provides the foundation for descent via isogeny.

Theorem 3.4. *Let $\mathcal{C}, \widehat{\mathcal{C}}$ be as in definition 1.1, and let $\mathbf{w} \in \widehat{\mathcal{J}}(\mathbb{Q})$. Then there exists a unique pair $(d_1, d_2) \in \mathbb{Q}^*/(\mathbb{Q}^*)^2 \times \mathbb{Q}^*/(\mathbb{Q}^*)^2$ such that for every $\mathbf{v} \in \phi^{-1}(\mathbf{w})$, the sets $\{\mathbf{v}\}$, $\{\mathbf{v}, \mathbf{v} + \alpha_i\}$, $\{\mathbf{v}, \mathbf{v} + \alpha_1, \mathbf{v} + \alpha_2, \mathbf{v} + \alpha_3\}$ are defined over $\mathbb{Q}(\sqrt{d_1}, \sqrt{d_2})$, $\mathbb{Q}(\sqrt{d_i})$, $\mathbb{Q}$, respectively ($i = 1, 2, 3$, $d_3 = d_1d_2$). Let $\psi^\phi : \widehat{\mathcal{J}}(\mathbb{Q})/\phi(\mathcal{J}(\mathbb{Q})) \mapsto \mathbb{Q}^*/(\mathbb{Q}^*)^2 \times \mathbb{Q}^*/(\mathbb{Q}^*)^2 : \mathbf{w} \mapsto (d_1, d_2)$. Then ψ^ϕ is a well defined injective homomorphism. Let $\mathcal{S} = \{p : p \mid \Delta b_1b_2b_3\hat{b}_1\hat{b}_2\hat{b}_3\} \cup \{2\} = \{p_1 \cdots p_r\}$, and $\mathbb{Q}(\mathcal{S}^\phi) = \{\pm p_1^{e_1} \cdots p_r^{e_r}\} \leq \mathbb{Q}^*/(\mathbb{Q}^*)^2$. Then $im\,\psi^\phi \leq \mathbb{Q}(\mathcal{S}) \times \mathbb{Q}(\mathcal{S})$.* □

The problem of finding $\widehat{\mathcal{J}}(\mathbb{Q})/\phi(\mathcal{J}(\mathbb{Q}))$ is therefore reduced to that of determining, for each member of $\mathbb{Q}(\mathcal{S}) \times \mathbb{Q}(\mathcal{S})$, whether a preimage exists under ψ^ϕ. As before, we use a commutative diagram similar to equation (6), except with $\mathcal{J}(\mathbb{Q})/2\mathcal{J}(\mathbb{Q})$ replaced by $\widehat{\mathcal{J}}(\mathbb{Q})/\phi(\mathcal{J}(\mathbb{Q}))$, and with $\mathbb{Q}(\mathcal{S}) \times \mathbb{Q}(\mathcal{S})$ performing the role of the finite group M. We again hope that the image of ψ^ϕ is completely determined by p-adic considerations. Having found $\widehat{\mathcal{J}}(\mathbb{Q})/\phi(\mathcal{J}(\mathbb{Q}))$, we then perform the same process with respect to the dual isogeny to find $\mathcal{J}(\mathbb{Q})/\hat{\phi}(\mathcal{J}(\mathbb{Q}))$. Then the exact sequence ($\mathcal{U} = \{\widehat{\mathcal{O}}, \hat{\alpha}_1, \hat{\alpha}_2, \hat{\alpha}_3\}$)

$$0 \to \mathcal{U} \to \widehat{\mathcal{J}}(\mathbb{Q})/\phi(\mathcal{J}(\mathbb{Q})) \xrightarrow{\hat{\phi}} \mathcal{J}(\mathbb{Q})/2\mathcal{J}(\mathbb{Q}) \to \mathcal{J}(\mathbb{Q})/\hat{\phi}(\widehat{\mathcal{J}}(\mathbb{Q})) \to 0$$

may be used to give generators for $\mathcal{J}(\mathbb{Q})/2\mathcal{J}(\mathbb{Q})$, and hence the rank of $\mathcal{J}(\mathbb{Q})$ as before.

Descent via isogeny can be viewed as breaking the work of finding $\mathcal{J}(\mathbb{Q})/2\mathcal{J}(\mathbb{Q})$ into 2 easier pieces. It has the considerable advantage that the computations are performed over number fields of smaller degree than if a complete 2-descent were attempted. In [9], 12 worked examples were given of the following type.

Example 3.5. *Let $\mathcal{C}$, $\widehat{\mathcal{C}}$ be as follows:*

$$\mathcal{C} : Y^2 = (X^2 + 6X + 7)(X^2 + 4X + 1)(X^2 + 2X + 3).$$
$$\widehat{\mathcal{C}} : Y^2 = (X^2 - 2X - 5)(X^2 + 2X - 1)(X^2 + 6X + 11).$$

Then $\mathcal{J}(\mathbb{Q})$ and $\widehat{\mathcal{J}}(\mathbb{Q})$ have rank 2. □

Recent improvements have significantly improved the speed of both methods. For example, Schaefer [22] has computed the following genus 3 example.

Example 3.6. *Let $\mathcal{C}$ be the curve:*

$$Y^2 = X(X-2)(X-3)(X-4)(X-5)(X-7)(X-10).$$

Then $\mathcal{J}(\mathbb{Q})$ has rank 2. □

Improvements to the technique of descent via isogeny are described in [11], in which further ranks are computed. So far, the various techniques have computed over 100 ranks, and it hoped that rank tables will soon be made available by anonymous ftp.

4 Height functions on the Kummer variety

It was observed in Section 3 that the methods for finding $\mathcal{J}(\mathbb{Q})/2\mathcal{J}(\mathbb{Q})$ found the rank of $\mathcal{J}(\mathbb{Q})$, but did not provide a way of showing that the divisors generating $\mathcal{J}(\mathbb{Q})/2\mathcal{J}(\mathbb{Q})$ also generate $\mathcal{J}(\mathbb{Q})$. A possible route from $\mathcal{J}(\mathbb{Q})/2\mathcal{J}(\mathbb{Q})$ to generators of $\mathcal{J}(\mathbb{Q})$ is via a height function defined on $\mathcal{J}(\mathbb{Q})$. We first define a height function on a general abelian group.

Definition 4.1. Let G be an abelian group. A *height function* H is a map $H : G \mapsto \mathbb{R}^+$ satisfying:

(1) There exists a constant C_1 such that, for all $P, Q \in G$, $H(P+Q)H(P-Q) \leq C_1 H(P)^2 H(Q)^2$.
(2) There exists a constant C_2 such that, for all $P \in G$, $H(2P) \geq H(P)^4/C_2$.
(3) For any constant C_3, the set $\{P \in G : H(P) \leq C_3\}$ is finite.

The constants C_1, C_2 depend only on the group G and the height function H, and we shall refer to them as the *height constants*.

The following property of abelian groups with a height function is proved in [23], p.199.

Lemma 4.2. *Let G be an abelian group with height function H, such that $G/2G$ is a finite set: $\{Q_1, \ldots, Q_n\}$, say. Then G is finitely generated. Explicitly, if $\epsilon = \min\{H(P) : P \in G\}$, and $C_1' = \max\{H(Q_i)^2 : 1 \leq i \leq n\} \cdot C_1/\epsilon$, then G is generated by the finite set: $\{P \in G : H(P) \leq \sqrt{C_1' C_2}\} \cup \{Q_1, \ldots, Q_n\}$.* □

In general, if $G/2G$ has already been computed, and if there is a height function on G, then the above lemma reduces the task of finding generators for G to a finite computation. This is precisely the situation we have after the techniques for Section 3 have been successfully applied. In principle, therefore, it is sufficient to define a height function on $\mathcal{J}(\mathbb{Q})$. Such a function may be found by first embedding the Jacobian variety into $\mathbb{P}^{4^g-1}$, and the Kummer variety into $\mathbb{P}^{2^g-1}$, as in theorem 1.1., and then taking the standard height of the resulting point in $\mathbb{P}^{2^g-1}(\mathbb{Q})$.

Definition 4.3. Let $\kappa : \mathcal{J}(\mathbb{Q}) \longrightarrow \mathbb{P}^{2^g-1}(\mathbb{Q})$ be an embedding of the Kummer surface. For any $D \in \mathcal{J}(\mathbb{Q})$, let $\kappa(D) = (v_0, \ldots, v_{2^g-1}) \in \mathbb{P}^{2^g-1}(\mathbb{Q})$. We may choose $v_0, \ldots, v_n$ to be integers with no common factor. Now define, $H_\kappa(D) = \max_i |v_i|$.

This is the natural generalization of the usual x-coordinate height function on an elliptic curve, for which $H_\kappa(\{(x,y)\}) = \max(v_0, v_1)$, where $x = v_1/v_0$, with v_0, v_1 coprime integers.

In any genus, it is straightforward to show that H_κ is a height function. In genus 2, the height constant C_1 is easy to compute. However, the constant C_2 is more difficult. In principle, C_2 can be found by applying Hilbert's Nullstellensatz to the non-degenerate quartics which define the duplication map. In practice, this is not computationally viable even for curves with small coefficients. An improvement has been found [10], in which the duplication law on the Kummer variety is factored as:

$$\kappa(2D) = W_1 \tau W_2 \tau W_3 \kappa(D),$$

where $\tau : (v_0, \ldots, v_{2^g-1}) \mapsto (v_0^2, \ldots, v_{2^g-1}^2)$ and where W_1, W_2, W_3 are linear maps. The derivation of the equations describing W_1, W_2, W_3 makes use of the isogeny of definition 3.3. A vastly smaller value of the constant C_2 may then be expressed in terms of the entries of the matrices for W_1, W_2, W_3.

Example 4.4. *Let C be the curve $y^2 = x(x-1)(x-2)(x-5)(x-6)$, as in Example 3.1. Then $\mathcal{J}(\mathbb{Q})$ is generated by the 2-torsion divisors, and the divisor $\{(3,6), \infty\}$.* □

Several other examples have been computed in [10]. However, it should be emphasised that this approach will become too slow as the size of the coefficients of C increases, and considerable work needs to be done before there is a viable, widely applicable method for finding generators for $\mathcal{J}(\mathbb{Q})$.

5 Implementing theorems of Coleman

The following classical result of Chabauty [4] gives a way of deducing information about a curve from its Jacobian.

Proposition 5.1. *Let C be a curve of genus g defined over a number field K, whose Jacobian has Mordell-Weil rank $\leq g-1$. Then C has only finitely many K-rational points.*

This is a strictly weaker result than Falting's theorem (which gives the same result unconditionally); however it has been shown by Coleman [5] that Chabauty's method – when applicable – can be used in many situations to give good bounds for the number of points on a curve. In particular, there are two potential genus 2 applications [5], [14].

Proposition 5.2. *Let $\mathcal{C}$ be a curve of genus 2 defined over $\mathbb{Q}$, and $p \geq 4$ be a prime of good reduction. If the Jacobian of $\mathcal{C}$ has rank at most 1 and $\widetilde{\mathcal{C}}$ is the reduction of $\mathcal{C}$ mod p then $\#\mathcal{C}(\mathbb{Q}) \leq \#\widetilde{\mathcal{C}}(\mathbb{F}_p) + 2$.* □

Proposition 5.3. *Let $\mathcal{C}$ be a curve of genus 2 defined over $\mathbb{Q}$ with 4 rational branch points and good reduction at 3, whose Jacobian has rank at most 1. Then $\#\mathcal{C}(\mathbb{Q}) \leq 6$.* □

If the rational branch points of the curve in Proposition 0.3 are mapped to $(0,0), (1,0), (-1,0), (1/\lambda, 0)$, then there is the following situation for which Coleman's method is guaranteed to determine $\mathcal{C}(\mathbb{Q})$ completely.

Proposition 5.4. *Let $\mathcal{C}$ be the curve of genus 2:*

$$\mathcal{C} : Y^2 = X(X^2 - 1)(X - 1/\lambda)(X^2 + aX + b)$$

with $\lambda, a, b \in \mathbb{Z}$. Suppose $3^{2r} \| \lambda$, for some $r > 0$, and 3 does not divide $b(1 - a + b)(1 + a + b)$, and that the Jacobian of $\mathcal{C}$ has rank at most 1. Then $\mathcal{C}(\mathbb{Q})$ contains precisely the points $(0,0), (1,0), (-1,0), (1/\lambda, 0)$ and the 2 rational points at infinity. □

There is only one non-trivial application of Proposition 5.2 in the literature, which is the curve already given as Example 3.1, due to Gordon and Grant [14].

Example 5.5. *Let $\mathcal{C}$ be the curve $Y^2 = X(X-1)(X-2)(X-5)(X-6)$ defined over $\mathbb{Q}$. Then $\#\mathcal{C}(\mathbb{Q}) = \#\widetilde{\mathcal{C}}(\mathbb{F}_7) + 2 = 10$.* □

It seems unlikely that there will be many direct applications of Proposition 5.2, which will resolve $\#\mathcal{C}(\mathbb{Q})$ completely, since one has to be fortunate for the bound $\#\widetilde{\mathcal{C}}(\mathbb{F}_p) + 2$ to be attained. However, there have recently been applications of Proposition 5.4 in [11], such as the following example.

Example 5.6. *The Jacobian of the curve: $Y^2 = X(X^2-1)(X-\frac{1}{9})(X^2-18X+1)$ has rank 1 over $\mathbb{Q}$. Hence, by Proposition 5.4, there are no $\mathbb{Q}$-rational points on the curve apart from the points $(0,0), (1,0), (-1,0), (1/9, 0)$ and the 2 rational points at infinity.* □

We also refer the reader to the work of McCallum [19],[20], who makes use of Coleman's version of Chabauty's theorem to obtain conditional bounds on the number of rational points on the Fermat curves.

6 Work in progress

Work currently in progress emphasises enhancements of the techniques described in Section 3 (computing the rank of $\mathcal{J}(\mathbb{Q})$) and Section 5 (applying the theorems of Chabauty and Coleman).

The main impediment to a fast and widely applicable implementation of the descent procedures of Section 3 is the difficulty in explicitly describing generators of the finite group M into which $\mathcal{J}(\mathbb{Q})/2\mathcal{J}(\mathbb{Q})$ injects. This is the main step which requires genuine work in a number field. An example of a key slow step is the following: *Given α in the ring of integers of a number field K, find all irreducibles which divide α.* This type of problem is straightforward when K has class number 1, but otherwise can quickly become time consuming. Any progress with this slow step would have a dramatic effect on the speed at which ranks of Jacobians could be computed.

Techniques for applying Chabauty's theorem are rapidly being made more flexible and widely applicable beyond the special cases indicated in Section 5. The formal group of the Jacobian (mentioned at the end of Section 1) is being used to construct formal power-series, defined over $\mathbb{Q}_p$ (for some choice of p), which must be satisfied by n, where n is the number of $\mathbb{Q}$-rational points on the original curve. This power-series induces a bound on n which experimentally appears very sharp. In the 20 examples computed so far, the bound was attained in 17 cases (finding $C(\mathbb{Q})$ completely), and in the 3 remaining cases, the bound was only 1 greater than the number of known rational points on the curve.

References

[1] Bost, J. B. and Mestre, J.-F. *Moyenne arithmético-géometrique et périodes des courbes de genre 1 et 2.* Gaz. Math. Soc. France, **38** (1988), 36–64.

[2] Cassels, J. W. S. *The Mordell-Weil Group of Curves of Genus 2.* Arithmetic and Geometry papers dedicated to I. R. Shafarevich on the occasion of his sixtieth birthday, Vol. **1.** Arithmetic, 29–60, Birkhäuser, Boston (1983).

[3] Cassels, J. W. S. *Arithmetic of curves of genus 2.* Number theory and applications (ed. R.A. Mollin), 27–35. NATO ASI Series C,265. Kluwer Academic Publishers, 1989.

[4] Chabauty C. *Sur les points rationels des variétés algébriques dont l'irrégularité et supérieur à la dimension.* Comptes Rendus, Paris **212** (1941), 882–885.

[5] Coleman, R. F. *Effective Chabauty.* Duke Math. J. **52** (1985), 765–780.

[6] Flynn, E. V. *The Jacobian and Formal Group of a Curve of Genus 2 over an Arbitrary Ground Field.* Math. Proc. Camb. Phil. Soc. **107** (1990), 425–441.

[7] Flynn, E. V. *Sequences of rational torsions on abelian varieties.* Inventiones Math. **106** (1991), 433–442.

[8] Flynn, E. V. *The group law on the Jacobian of a curve of genus 2.* J. Reine Angew. Math. **439** (1993), 45–69.

[9] Flynn, E. V. *Descent via isogeny on the Jacobian of a curve of genus 2.* Acta Arithmetica LXVII.1 (1994), 23–43.

[10] Flynn, E. V. *An explicit theory of heights in dimension 2.* Preprint, February 1994.

[11] Flynn, E. V. *On a theorem of Coleman.* Preprint, April 1994.

[12] Gordon, D.M. and Grant, D. *Computing the Mordell-Weil rank of Jacobians of curves of genus 2.* Trans. A.M.S., **337**, Number 2, (1993), 807–824.

[13] Grant, D. *Formal Groups in Genus 2.* J. Reine Angew. Math. **411** (1990), 96–121.

[14] Grant, D. *A curve for which Coleman's Chabauty bound is sharp.* Preprint, 1991.

[15] Lang, S. *Introduction to Algebraic and Abelian Functinos.* 2nd edition. Graduate Texts in Math. no. 89 (Springer-Verlag, 1982).

[16] Leprévost, F. *Torsion sur des familles de courbes de genre g.* Manuscripta math. **75** (1992), 303–326.

[17] Leprévost, F. *Famille de Courbes Hyperelliptiques de Genre g munies d'une Classe de Diviseurs Rationnels d'Ordre* $2g^2+4g+1$. Preprint, 1993.

[18] Mazur, B. *Rational points of modular curves*, Modular Functions of One Variable, V. Lecture Notes in Math. **601** (1977), 107–148.

[19] McCallum, W.G. *On the Shafarevich-Tate group of the Jacobian of a quotient of the Fermat curve.* Invent. Math. **93** (1988), 637–666.

[20] McCallum, W.G. *The Arithmetic of Fermat Curves.* Math. Ann. **294** (1992), 503–511.

[21] Merel, L. *Bornes pours la torsion des courbes elliptiques sur les corps de nombres.* Preprint, 1994.

[22] Schaefer, E.F. *2-descent on the Jacobians of hyperelliptic curves.* J. Number Theory (to appear).

[23] Silverman, J. H. *The Arithmetic of Elliptic Curves.* Springer-Verlag, New York (1986).

E. V. Flynn (`evflynn@liverpool.ac.uk`)
D.P.M.M.S., University of Cambridge, 16 Mill Lane, Cambridge CB2 1SB (England).

Progress in Mathematics, Vol. 143, © 1996 Birkhäuser Verlag Basel/Switzerland

Viro's method and T-curves

I. Itenberg

1 Introduction

Let A be a real algebraic plane projective curve of degree m, i.e., a real homogeneous polynomial in three variables of degree m considered up to multiplication by a non-zero real number. We suppose the curve to be non singular, which means that the polynomial does not have singular points in $\mathbf{R}^3 \setminus 0$.

This polynomial has a well-defined zero locus $\mathbf{R}A$ in the real projective plane $\mathbf{R}P^2$. The set $\mathbf{R}A$ is a union of non intersecting circles embedded in $\mathbf{R}P^2$. The topological type of the pair $(\mathbf{R}P^2, \mathbf{R}A)$ is defined by the scheme of disposition of the components of $\mathbf{R}A$. This scheme is called *the real scheme of curve* A.

In 1900, D. Hilbert [Hi] included the following question in the 16th problem of his famous list: What types of real schemes can be realized by non singular curves of a given degree? The complete answer is known now only for curves of degree not greater than 7.

To solve the problem it is necessary to work in two main directions: First, to find the restrictions for the topological types of pairs $(\mathbf{R}P^2, \mathbf{R}A)$, and, second, to give the constructions of curves for realizable real schemes. Many deep and important results were obtained in the first direction using the modern machinery of algebraic and differential topology (see, for example, the survey papers [Vi 5], [Wi]). However, the methods of constructions had not been seriously changed since the XIX century until 1980, when O. Viro proposed a principally new method to construct curves (see [Vi 3], [Vi 4], [Vi 6], [Ri]).

In the present paper we discuss a special case of Viro's method, which is proved to be useful and has some fruitful applications. In this case Viro's method gives a possibility to construct curves using a simple combinatorial procedure. Slight modifications of this method allow to construct different objects like, for example, real polynomials in two variables with prescribed collections of critical points (see [Sh]) and real polynomial vector fields in $\mathbf{R}^2$ with prescribed collections of non degenerated singular points (see [It-Sh]).

I would like to thank V. Kharlamov and O. Viro for the useful comments and discussions.

2 T-curves

2.1 Construction

Let m be a positive integer number and T be the triangle in $\mathbf{R}^2$

$$\{x \geq 0, \ \ y \geq 0, \ \ x + y \leq m\}.$$

Suppose that T is triangulated in such a way that the vertices of the triangles are integer, and that some distribution of signs, $a_{i,j} = \pm$ at the vertices of the triangulation, is given. Then there arises a naturally associated piecewise-linear curve L in $\mathbf{R}P^2$.

The construction of L is the following.

Take copies

$$T_x = s_x(T), \ \ T_y = s_y(T), \ \ T_{xy} = s(T)$$

of T, where $s = s_x \circ s_y$ and s_x, s_y are reflections with respect to the coordinate axes. Extend the triangulation of T to a symmetric triangulation of $T \cup T_x \cup T_y \cup T_{xy}$ and extend the distribution of signs to a distribution on the vertices of the extended triangulation which verifies the modular property: $g^*(a_{i,j}x^i y^j) = a_{g(i,j)}x^i y^j$ for $g = s_x$, s_y and s (other words, the sign at a vertex is the sign of the corresponding monomial in the quadrant containing the vertex).

If a triangle of the triangulation has vertices of different signs, select a midline separating them. Denote by L' the union of the selected midlines (see, for example, Figure 1). It is contained in $T \cup T_x \cup T_y \cup T_{xy}$. Glue by s the sides of $T \cup T_x \cup T_y \cup T_{xy}$. The resulting space T_* is homeomorphic to $\mathbf{R}P^2$. Take the curve L to be the image of L' in T_*.

A pair (T_*, L) is called *a chart* of a real algebraic plane projective curve A, if there exists a homeomorphism of pairs $(T_*, L) \longrightarrow (\mathbf{R}P^2, \mathbf{R}A)$.

Let us introduce two additional assumptions: The considered triangulation of T is *primitive* and *convex*. The first condition means that all triangles are of area $1/2$ (or, equivalently, that all integer points of T are vertices of the triangulation). Primitiveness of the triangulation is not really needed for the following theorem. We put this assumption just to make the combinatorics easier in applications. The second condition means that there exists a convex piecewise-linear function $T \longrightarrow \mathbf{R}$ which is linear on each triangle of the triangulation and not linear on the union of two triangles. We can suppose that this function takes integer values in integer points of T.

Theorem 2.1 (O. Viro) *Under the assumptions made above on the triangulation of the triangle T, there exists a non singular real algebraic plane projective curve A of degree m with the chart (T_*, L).*

This statement is the special case of Viro's theorem [Vi 4, Th. 1.4]. We will not discuss a proof in the present paper. Let us just mention that the main

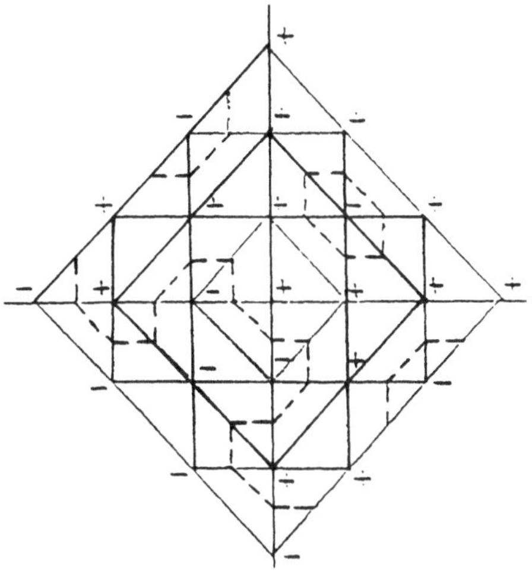

Figure 1

idea is to consider a polynomial

$$Q_t(x,y) = \sum_{(i,j)\in T} a_{i,j} x^i y^j t^{\nu(i,j)}$$

(where i,j are integer numbers, t is a parameter, $a_{i,j}$ is the sign of the integer point (i,j), and ν is a convex function defining the triangulation of T) and to remark that the projectivisation of the polynomial $Q_t(x,y)$ for sufficiently small positive values of t defines a curve with the required chart.

A curve having the chart (T_*, L) is called *a T-curve*. This notion was introduced by S. Orevkov [Or].

Theorem 2.1 gives a combinatorial way to construct curves. One should choose a primitive convex triangulation of the triangle T and signs at the vertices of the triangulation, and then, using the procedure described above, draw the curve L.

Example The construction of a T-curve of degree 3 with two connected components of the real point set is shown in Figure 1.

Let us write down a polynomial defining this curve. Take the convex piecewise-linear function $T \longrightarrow \mathbf{R}$ with the values at the integer points presented in Figure 2. Then the polynomial is as follows:

$$t^3 + tx + ty + tx^2 - xy + ty^2 + t^3x^3 + tx^2y + txy^2 + t^3y^3.$$

One can verify in a simple way that for $0 < t < 1/7$ the projectivisation of this polynomial defines the curve required. In general, we are not able to precise an acceptable value of t.

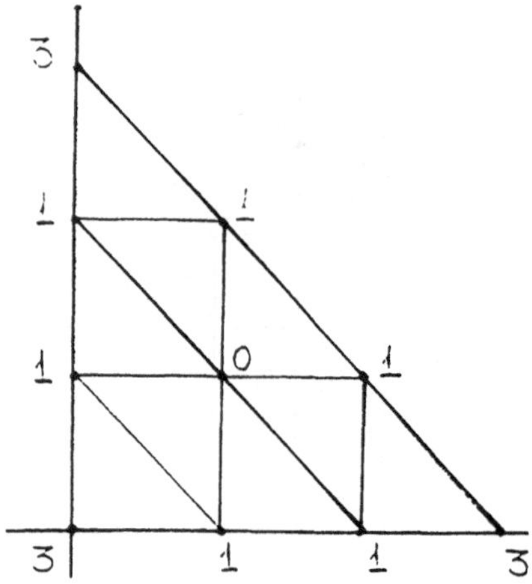

Figure 2

2.2 T-curves among all curves

It is natural to pose the following question: Can the real scheme of an arbitrary non singular real algebraic plane projective curve be realized by a T-curve of the same degree?

One can immediately find a trivial restriction: Evidently, the empty real scheme of a curve of an even degree cannot be realized by T-curves. We will formulate another, more serious restriction.

Let us, first, give the necessary definitions. *An M-curve* is a curve having the maximal possible number of connected components of the real point set for a given degree. It was proved by Harnack [Har] that this maximal number is equal to $\frac{(m-1)(m-2)}{2} + 1$ for the degree m.

Each connected component of the real point set $\mathbf{R}A$ of a curve of even degree is called *an oval.* It divides $\mathbf{R}P^2$ in two parts. We call the part homeomorphic to a disk *the interior* of the oval.

A pair of ovals is called *injective* if one oval of this pair lies inside of the other one. Let us denote by J the number of ovals of a curve containing inside of them at least one injective pair.

Proposition 2.2 *For any T-curve of degree* m *which is an M-curve the following inequality holds*

$$J \leq 3m/2.$$

This proposition has a purely combinatorial proof.

Remark that Proposition 2.2 gives a strong restriction on the topology of T-curves being M-curves. One can easily construct a family of M-curves of increasing degrees such that the numbers J of the curves of this family depend quadratically in the degree.

There are many open questions in the subject under discussion. For example,

i) how large is the class of T-curves (is it true that in a sense almost all curves are T-curves)?

ii) is it true that a T-curve of degree m being an M-curve has no more than $O(m)$ non-empty ovals?

A statement similar to the statement of Theorem 2.1 can be formulated and be proved in any dimension. Shustin [Sh] proved that the number of connected components of T-surfaces of degree m is not greater than $m^3/6 + O(m^2)$. However, Viro [Vi 1] constructed surfaces of degree m with $(7m^3 - 24m^2 + 32m)/24$ connected components for any $m = 4l + 2$ (l is a positive integer number). That means, in particular, that these surfaces are not T-surfaces, if we take m large enough.

We will give some constructions of curves using Theorem 2.1 in the following section. These examples show that the class of T-curves is sufficiently rich. Subsection 3.2 is devoted to the counterexamples to Ragsdale conjecture. subsection 3.3 – to the classification of M-curves of degree $4l + 2$ with one non-empty oval.

3 Examples of T-curves

3.1 Construction of Harnack curves

In this subsection we will describe, using Theorem 2.1, the construction of some M-curves (a special case of Harnack curves). This construction will play an important role in subsections 3.2 and 3.3.

Let $m = 2k$ be a positive even number, and T again be the triangle in $\mathbf{R}^2$

$$\{x \geq 0,\ \ y \geq 0,\ \ x + y \leq m\}.$$

An integer point of T is called *even*, if i, j are both even, and *odd* if not.

Let us consider the following distribution of signs at the integer points of the triangle T:

A point has sign "$-$", if it is even, and has sign "$+$", if it is odd.

We will call this rule *Harnack distribution of signs*.

We use the system of notations for the real schemes of non singular curves suggested by Viro [Vi 2]. The scheme consisting of a single oval is denoted by the symbol $< 1 >$, the empty scheme – by the symbol $< 0 >$. If a symbol $< A >$ stands for some set of ovals, then the set of ovals obtained by addition of an oval surrounding all old ovals is denoted by $< 1 < A >>$. If a scheme is the union of two non intersecting sets of ovals denoted by $< A >$ and $< B >$ respectively with no oval of one set surrounding an oval of the other set, then this scheme is denoted by the symbol $< A \cup B >$. Besides, if A is the notation for some set of ovals then a part $A \cup \cdots \cup A$ of another notation where A repeats n times is denoted by $n \times A$; a part $n \times 1$ is denoted by n.

Proposition 3.1 *An arbitrary primitive convex triangulation of T with the Harnack distribution of signs at the vertices produces a T-curve of degree $m = 2k$ with the real scheme*

$$< \frac{3k^2 - 3k}{2} \cup 1 < \frac{(k-1)(k-2)}{2} >> .$$

Remark A curve with this real scheme has $\frac{(m-1)(m-2)}{2} + 1$ connected components of the real point set. So, it is an M-curve.

Proof Let us, first, notice that the number of interior (i.e., lying strongly inside of the triangle T) integer points is equal to $\frac{(m-1)(m-2)}{2}$, the number of even interior points is equal to $\frac{(k-1)(k-2)}{2}$, and the number of odd interior points is equal to $\frac{3k^2-3k}{2}$.

Take an arbitrary even interior vertex of a triangulation of the triangle T. It has the sign "$-$". All neighbouring vertices (i.e., the vertices connected with the taken vertex by edges of the triangulation) are odd, and thus they all have the sign "$+$". It means that the star of an even interior vertex contains an oval of the curve L (*the star* of a vertex of the triangulation is the union of all triangles of the triangulation containing this vertex). The number of such ovals is equal to $\frac{(k-1)(k-2)}{2}$.

Take now an odd interior vertex of the triangulation. It has the sign "$+$". There are two vertices with "$-$" and one vertex with "$+$" among the three symmetric images of the taken vertex under $s = s_x \circ s_y$ and s_x, s_y (where s_x, s_y are the reflections with respect to the coordinate axes). Consider the symmetric copy of the taken vertex with the sign "$+$". It is easy to verify, that all its neighbouring vertices have the sign "$-$". It means again that the star of this copy contains an oval of the curve L. The number of such ovals is equal to $\frac{3k^2-3k}{2}$.

Remark that

$$\frac{(k-1)(k-2)}{2} + \frac{3k^2 - 3k}{2} = \frac{(m-1)(m-2)}{2}$$

and, thus, we can have only one oval more.

This oval exists because, for example, the curve L intersects the coordinate axes.

To finish the proof it remains to notice that the union of the segments

$$\{x - y = -m,\ x \leq 0,\ y \geq 0\} \cup$$

$$\{y = 0,\ -m \leq x \leq 0\} \cup \{x = 0,\ -m \leq y \leq 0\}$$

is not contractible in T_* and contains only the signs "$-$". It means that $\frac{3k^2-3k}{2}$ ovals corresponding to odd interior points and containing the sign "$+$" inside of them are situated outside of the non-empty oval. •

3.2 Counterexamples to Ragsdale conjecture

Let us consider a non singular real algebraic plane projective curve of even degree $m = 2k$. The real point set $\mathbf{R}A$ of this curve divides the real projective plane $\mathbf{R}P^2$ in two parts with a common boundary $\mathbf{R}A$ (these parts are the subsets of $\mathbf{R}P^2$ where a polynomial defining the curve has positive or, respectively, negative values). One of these parts is non orientable, we will denote it by $\mathbf{R}P^2_-$. The other one will be denoted by $\mathbf{R}P^2_+$.

The topology of $\mathbf{R}P^2_-$ and $\mathbf{R}P^2_+$ is closely connected with the topological type of the pair $(\mathbf{R}P^2, \mathbf{R}A)$. Let p be the number of connected components of $\mathbf{R}P^2_+$, and $n+1$ be the number of connected components of $\mathbf{R}P^2_-$ (exactly one component of $\mathbf{R}P^2_-$ is non orientable).

The numbers p and n can be described in another way. An oval of a curve is called *even* (resp. *odd*) if it lies inside of an even (resp. odd) number of other ovals of this curve.

It is easy to see that p is the number of even ovals of a curve, and n is the number of odd ovals.

In 1906, V. Ragsdale [Ra] studying the results of Harnack's and Hilbert's constructions proposed two conjectures:

$$p \leq \frac{3k^2 - 3k + 2}{2}, \quad n + 1 \leq \frac{3k^2 - 3k + 2}{2}$$

and

$$p - n \leq \frac{3k^2 - 3k + 2}{2}, \quad n - p + 1 \leq \frac{3k^2 - 3k + 2}{2}.$$

In 1938, I. Petrovsky [Pe] proved the second Ragsdale conjecture and also proposed a conjecture similar to the first one:

$$p \leq \frac{3k^2 - 3k + 2}{2}, \quad n \leq \frac{3k^2 - 3k + 2}{2}.$$

In 1980, O. Viro [Vi 2] constructed curves of degree $2k$ with $n = \frac{3k^2-3k+2}{2}$ for any even $k \geq 4$. These curves are counterexamples to the original Ragsdale conjecture, but not to the conjecture of Petrovsky.

The following theorem gives counterexamples to the "corrected" Ragsdale conjecture (or to the conjecture of Petrovsky) (see also [It]).

Theorem 3.2 *For any integer number $k \geq 1$*
a) there exists a non singular real algebraic plane projective curve of degree $2k$ with

$$p = \frac{3k^2 - 3k + 2}{2} + \left[\frac{(k-3)^2 + 4}{8}\right]$$

(where $[x]$ denotes the maximal integer not greater than x),
b) there exists a non singular real algebraic plane projective curve of degree $2k$ with

$$n = \frac{3k^2 - 3k + 2}{2} + \left[\frac{(k-3)^2 + 4}{8}\right] - 1.$$

Proof We will construct T-curves with the stated properties. Let us show, first, how to construct a curve of degree $m = 2k$ with $p = \frac{3k^2-3k+2}{2} + 1$.

Suppose that the hexagon S shown in Figure 3 is placed inside of the triangle $T = \{x \geq 0,\ \ y \geq 0,\ \ x + y \leq m\}$ in such a way that the center of S has both coordinates odd. Any convex primitive triangulation of a convex part of a convex polygon is extendable to a convex primitive triangulation of the polygon. Inside of the hexagon S, let us take the convex primitive triangulation shown in Figure 3 and extend it to T.

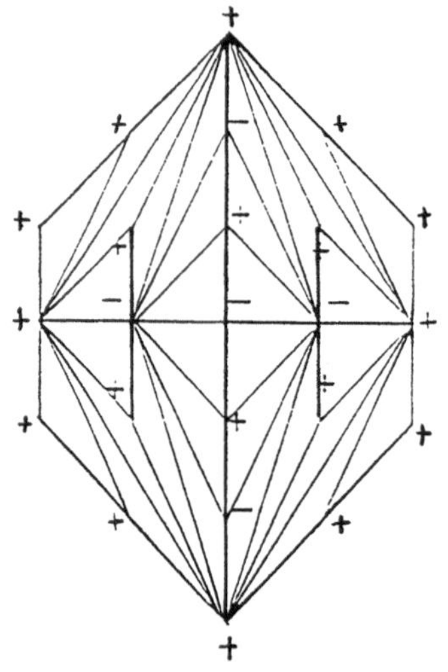

Figure 3

To apply Theorem 2.1 we need to choose signs at the vertices in T. Inside of S put signs according to Figure 3, outside, use the Harnack rule of distribution of signs (see subsection 3.1): A vertex (i, j) gets sign "$-$", if i, j are even, and sign "$+$" otherwise.

It is easy to calculate that the corresponding piecewise-linear curve L has exactly one even oval more than the M-curve constructed in subsection 3.1 (i.e., now $p = \frac{3k^2-3k+2}{2} + 1$). One can verify that the curve obtained has the real scheme

$$< \frac{3k^2 - 3k - 2}{2} \cup 1 < \frac{(k-1)(k-2) - 8}{2} \cup 1 < 2 >>>$$

This curve is an (M $-$ 2)-curve (it means that the number of connected components of the real point set is equal to $\frac{(m-1)(m-2)}{2} - 1$).

Now, consider the partition of the triangle T shown in Figure 4. Let us take in each marked hexagon the triangulation and the signs of S. The triangulation of the union of the marked hexagons can be extended to the primitive convex triangulation of T. Let us fix such an extension. Outside of the union of the marked hexagons again choose the signs at the vertices of the triangulation using the Harnack rule.

One can calculate that for the corresponding piecewise-linear curve L we have

$$p = \frac{3k^2 - 3k + 2}{2} + a$$

where a is the number of the marked hexagons, and

$$a = \left[\frac{(k-3)^2 + 4}{8}\right].$$

The curve constructed has the following real scheme

$$< \frac{3k^2 - 3k - 2a}{2} \cup 1 < \frac{(k-1)(k-2) - 8a}{2} \cup a \times 1 < 2 >>>.$$

To prove the part b) of the statement of the theorem, let us take again the partition of the triangle T shown in Figure 4 with the triangulation and the signs of each marked hexagon coinciding with the triangulation and the signs of S. Fix, in addition, a triangulation of some part P of a neighbourhood of the axis OY and the signs at the vertices of this triangulation as it is shown in Figure 5 (more precisely, only the case $k \equiv 1 \; mod \; 4$ is presented in this figure, if $k \not\equiv 1 \; mod \; 4$ one should change a little the triangulation near the point $(0, m)$). The chosen primitive convex triangulation of the union of the marked hexagons and of the part P can be extended to a primitive convex triangulation of the triangle T. Outside of the union of the marked hexagons and of the part P, let us choose again the signs at the vertices of the triangulation using the Harnack rule.

For the corresponding piecewise-linear curve L

$$n = \frac{3k^2 - 3k + 2}{2} + \left[\frac{(k-3)^2 + 4}{8}\right]$$

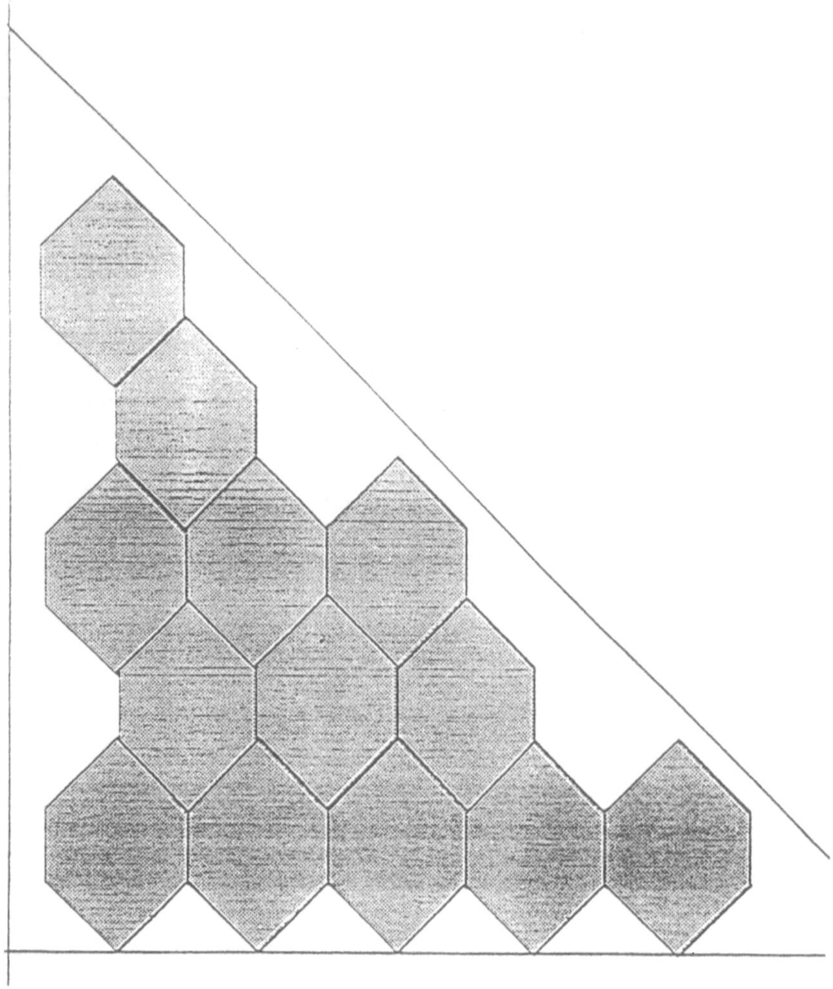

Figure 4

(the case $k \equiv 1 \; mod \; 4$) or

$$n = \frac{3k^2 - 3k + 2}{2} + \left[\frac{(k-3)^2 + 4}{8}\right] - 1$$

(the case $k \not\equiv 1 \; mod \; 4$). •

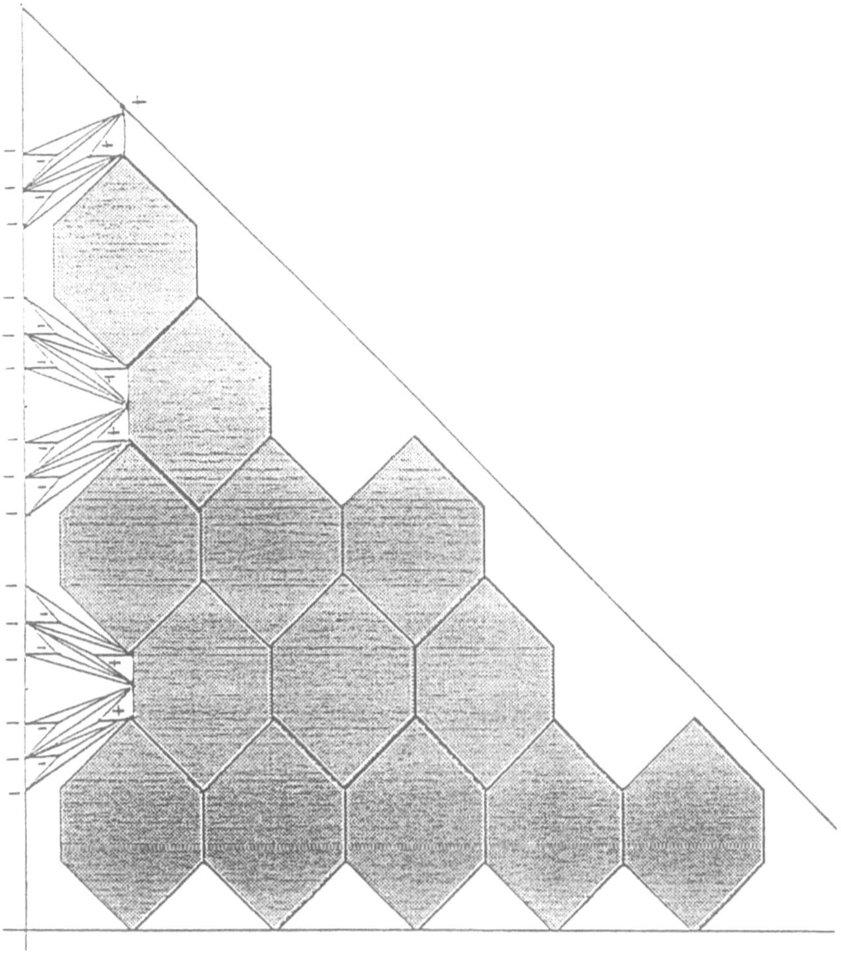

Figure 5

Recently, B. Haas [Has] constructed examples of T-curves of degree $2k$ with

$$p = \frac{3k^2 - 3k + 2}{2} + \left[\frac{k^2 - 7k - 10}{6}\right].$$

3.3 M-curves with one non-empty oval

Recall that an M-curve is a curve with the maximal possible number of connected components of the real point set for a given degree. This maximal number is equal to $\frac{(m-1)(m-2)}{2} + 1$ for the degree m.

In this subsection we discuss a classification of the real schemes of M-curves of the degree $2k = 4l + 2$ with one non-empty oval.

Each non singular curve of even degree with one non-empty oval has the real scheme

$$< (p-1) \cup 1 < n >> .$$

Restrictions

We need to use here two well-known restrictions for the topology of real plane projective curves (see, for example, the survey articles [Vi 5], [Wi]).

Gudkov-Rokhlin congruence

$p - n \equiv k^2$ *mod* 8 *for an M-curve of degree* $2k$.

The Euler characteristic of a connected component of $\mathbf{R}P^2 \setminus \mathbf{R}A$ is called *the characteristic* of the outer bounding oval of the component. Denote by p_- (resp. by n_-) the number of even (resp. odd) ovals of $\mathbf{R}A$ with negative characteristics.

Improved Petrovsky inequalities

$$p - n_- \leq \frac{3k^2 - 3k + 2}{2}, \quad n - p_- + 1 \leq \frac{3k^2 - 3k + 2}{2}$$

for a curve of degree $2k$.

Remark that $p_- \leq 1, \ \ n_- = 0$ in the case of curves with one non-empty oval, and the improved Petrovsky inequalities give the following ones:

$$p \leq \frac{3k^2 - 3k + 2}{2}, \quad n \leq \frac{3k^2 - 3k + 2}{2}$$

It is easy to see, using the Gudkov-Rokhlin congruence, that for M-curves of degree $4l + 2$ with one non-empty oval, the second inequality can be improved by 1:

$$n \leq \frac{3k^2 - 3k}{2}.$$

Construction

The following theorem states that there are no other restrictions (except the Gudkov-Rokhlin congruence and the improved Petrovsky inequalities) for the topology of M-curves of degree $m = 4l + 2$ with one non-empty oval.

Theorem 3.3 *Suppose that* $m = 2k = 4l + 2$, *where* l *is a positive integer number. Then for any positive integer numbers* p, n *such that*

$$p + n = \frac{(m-1)(m-2)}{2} + 1$$

satisfying the Gudkov-Rokhlin congruence and the improved Petrovsky inequalities there exists a real algebraic plane projective M-curve of degree m *with the real scheme*

$$< (p-1) \cup 1 < n >> .$$

Proof Recall that the Harnack distribution of signs in the vertices of a triangulation is the rule:

A vertex (i, j) gets sign "$-$", if the numbers i, j are both even, and it gets sign "+" in the opposite case.

Let us call *inverse Harnack distributions of signs* the following ones:

(1) a vertex (i, j) has sign "$-$", if i, j are both odd, and sign "+" otherwise,

(2) a vertex (i, j) has sign "$-$", if i is even and j is odd. and it has sign "+" otherwise,

(3) a vertex (i, j) has sign "$-$", if j is even and i is odd. and it has sign "+" otherwise.

Remark that each inverse Harnack distribution of signs can be formulated as the Harnack one for the appropriate quadrant of the plane exchanging "+" and "$-$".

Thus, Proposition 3.1 is also true for the inverse Harnack distributions.

Let us divide the triangle T in two polygons T_1 and T_2 (T_1 is a quadrangle, T_2 is a triangle) by a segment with the following properties:

(i) the ends of the segment lie on the boundary of T and are odd integer points,

(ii) the segment contains no integer points except the ends.

Consider an arbitrary convex primitive triangulation in each polygon T_1, T_2 (the union of these triangulations is a convex primitive triangulation of T, because the chosen segment does not contain vertices of the triangulations except the ends). Let us choose in T_1 the Harnack distribution of signs, and take in T_2 the only inverse Harnack one compatible in the common boundary of the polygons with the chosen distribution in T_1 (it is possible due to the assumptions on the segment).

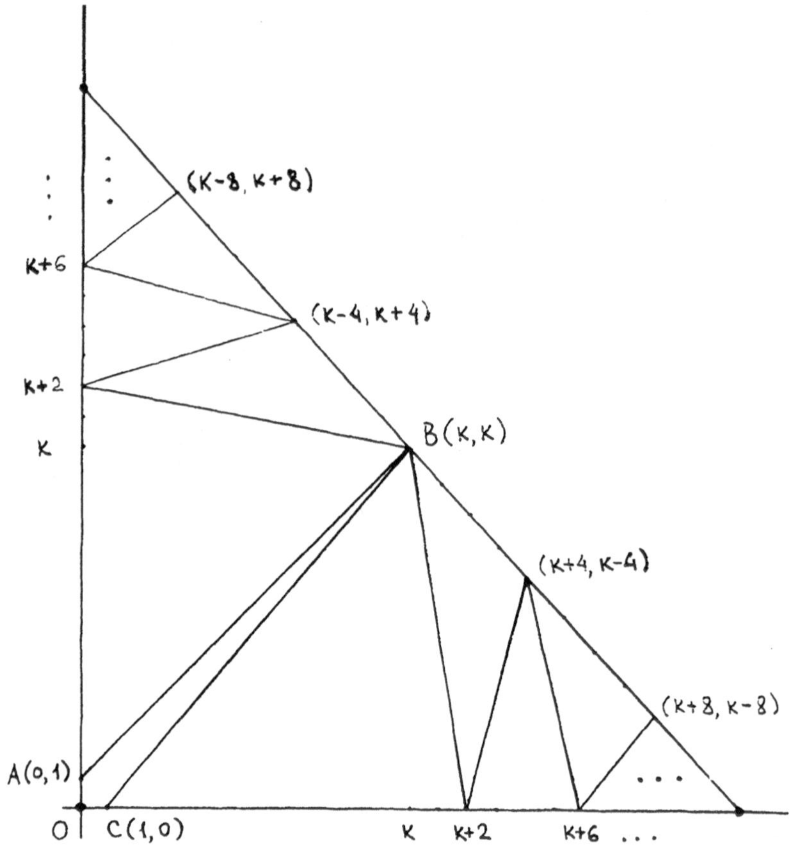

Figure 6

The arguments of the proof of proposition 3.1 show again that the triangulation and the distribution of signs described above give an M-curve with one non-empty oval.

Compute the number of even ovals of this curve. Let P_1, P_2 be the numbers of interior even points of T_1 and T_2, and N_1, N_2 be the numbers of interior odd points of these polygons.

One can easily see that for the curve obtained

$$p = N_1 + P_2 + 1, \quad n = P_1 + N_2.$$

Remark The Gudkov-Rokhlin congruence has a nice corollary:

The numbers P_2, N_2 of even and odd interior points of the triangle T_2 are congruent modulo 4.

To prove Theorem 3.3, let us divide the triangle T by segments with the properties (i), (ii) as it shown in Figure 6 (here we use the fact that $m = 2k = 4l + 2$, because we need the point (k, k) to be odd; the vertices on the axis OX have the coordinates $(k + (4i + 2),\ 0)$ the vertices on the axis OY - the coordinates $(0,\ k + (4i + 2))$, and the vertices on the line $x + y = m$ - the coordinates $(k \pm 4i,\ k \mp 4i)$ with appropriate values of a non negative integer i). Now we take in the quadrangle $OABC$ the Harnack distribution of signs, and extend it to the distribution of signs in T choosing at each dividing line what type of distribution (the Harnack one or the appropriate inverse Harnack one) should be used in the next triangle. Finally, we choose an arbitrary primitive convex triangulation in each part of the described subdivision of T (actually, the real scheme of the resulting T-curve does not depend of this choice of primitive triangulations). It is easy to verify that all possible (in the sense of the statement of Theorem 3.3) pairs p, n can be realized using this procedure. For example, to realize two extremal cases one can take the Harnack distribution of signs in the whole triangle T (the case $p = \frac{3k^2-3k+2}{2}$, $n = \frac{(k-1)(k-2)}{2}$ - a Harnack curve) or the Harnack distribution in $OABC$ and inverse Harnack ones in $T \setminus OABC$ (the opposite extremal case $p = \frac{(k-1)(k-2)}{2} + 1$, $n = \frac{3k^2-3k}{2}$).

References

[Has] B. Haas, *Private communication.*

[Har] A. Harnack, *Über Vieltheiligkeit der ebenen algebraischen Curven,* Math. Ann. 10 (1876), 189–199.

[Hi] D. Hilbert, *Mathematische Probleme,* Arch. Math. Phys. 1 (1901), 213 237.

[It] I. Itenberg, *Contre-exemples à la conjecture de Ragsdale*. C. R. Acad. Sci. Paris, t. 317, Série I (1993), 277–282.

[Or] S. Yu. Orevkov, *Private communication*

[Pe] I. G. Petrovsky, *On the topology of real plane algebraic curves*, An. Math. 39 (1938), no. 1, 187–209.

[Ra] V. Ragsdale, *On the arrangement of the real branches of plane algebraic curves,* Amer. J. Math. 28 (1906), 377–404.

[Ri] J.-J. Risler, *Construction d'hypersurfaces réelles (d'après Viro)*, Séminaire N. Bourbaki (1992), no. 763.

[Sh] E. Shustin, *Critical points of real polynomials, subdivisions of Newton polyhedra and topology of real algebraic hypersurfaces*. Preprint, Tel Aviv University (1993).

[Sh-It] E. Shustin, I. Itenberg, *Singular points of planar polynomial vector fields,* (to appear).

[Vi 1] O. Ya. Viro, *Construction of multicomponent real algebraic surfaces,* Dokl. Akad. Nauk SSSR. 248 (1979), no. 2 (in Russian, English translation in Soviet Math. Dokl. 20 (1979), no. 5, 991–995).

[Vi 2] O. Ya. Viro, *Curves of degree 7, curves of degree 8, and Ragsdale conjecture,* Soviet Math. Dokl. 22 (1980), 566–569.

[Vi 3] O. Ya. Viro, *Gluing of algebraic hypersurfaces, smoothing of singularities and construction of curves* (in Russian), Intern. Topol. Conf., Leningrad, 1983, 149–197.

[Vi 4] O. Ya. Viro, *Gluing of plane real algebraic curves and construction of curves of degree 6 and 7,* Lecture Notes in Mathematics, 1060 (1984), 187–200.

[Vi 5] O. Ya. Viro, *Progress in the topology of real algebraic varieties over the last six years,* Rus. Math. Surv. 41 (1986), no. 3, 55–82.

[Vi 6] O. Ya. Viro, *Real algebraic plane curves: constructions with controlled topology,* Alg. i Analiz. 1 (1989), no. 5, 1–73 (in Russian, English translation in Leningrad Math. J. 1 (1990) 1059–1134).

[Wi] G.Wilson, *Hilbert's sixteenth problem,* Topology. 17 (1978), no. 1, 53–73.

I. Itenberg (`itenberg@univ-rennes1.fr`)

IRMAR, University of Rennes I, 35042 Rennes Cedex (France).

Progress in Mathematics, Vol. 143, © 1996 Birkhäuser Verlag Basel/Switzerland

A computational method for diophantine approximation

T. Krick[1], L. M. Pardo[2]

1 Introduction

The procedures to solve algebraic geometry elimination problems have usually been designed from the point of view of commutative algebra. For instance, let us consider the problem of deciding whether a given system of polynomial equalities has a solution. This means that we have to eliminate a single block of quantifiers in a formula with polynomial equations. More precisely, let $\mathbb{Z}$ be the ring of integer numbers, $\mathbb{Q}$ the field of rationals and $\mathbb{C}$ the field of complex numbers. Assume your are given (input) polynomials $f_1, \dots, f_s \in \mathbb{Z}[X_1, \dots, X_n]$ and you want to decide whether the following formula holds:

$$\exists x_1 \in \mathbb{C}, \dots, \exists x_n \in \mathbb{C} : \qquad f_1(x_1, \dots, x_n) = 0, \dots, f_s(x_1, \dots, x_s) = 0. \quad (1)$$

By means of Hilbert's Nullstellensatz this question can be reduced to the following problem in terms of ideals: Decide whether the integer 1 belongs to the ideal generated by $f_1, \dots, f_s$ in $\mathbb{Q}[X_1, \dots, X_n]$, i.e., decide whether there exist polynomials $g_1, \dots, g_s \in \mathbb{Q}[X_1, \dots, X_n]$ such that the following identity holds:

$$1 = g_1 f_1 + \cdots + g_s f_s. \quad (2)$$

Thus, an *effective Nullstellensatz* solves the following two problems:

- *Decision problem:* Decide whether there exist polynomials g_i's such that equality (2) holds.
- *Representation problem:* In the affirmative case, compute polynomials g_i's such that equality (2) holds.

In this form, the previous elimination problem is transformed in a particular case of the general *membership problem to polynomial ideals* of ideal theory (or commutative algebra). This more general problem can be stated in the following terms: given polynomial inputs $f_1, \dots, f_s$ and g in $\mathbb{Z}[X_1, \dots, X_n]$ the problem ask for the following two issues:

- *Decision problem:* Decide whether g belongs to the ideal generated by the polynomials $f_1, \dots, f_s$ in the ring $\mathbb{Q}[X_1, \dots, X_n]$.

[1]Partially supported by UBACYT and CONICET (Argentina)

[2]Partially supported by DGICyT PB 92-0498–C02–01, PB93-0472-C02-02, TIC 1026–CE and "POSSO", ESPRIT-BRA 6846.

- *Representation problem:* In the affirmative case, find polynomials $g_1, \dots, g_s \in \mathbb{Q}[X_1, \dots, X_n]$ such that

$$g = g_1 f_1 + \cdots + g_s f_s.$$

The complexity of the membership problem was characterized by E. Mayr and A. Meyer in [42] and [43]. They showed that this problem is exponential space complete (EXPSPACE-complete). As an immediate consequence the time to solve it is at least exponential (and probably doubly exponential) and this lower bound cannot be improved because of the time and space hierarchy theorems. Therefore, the general problem of ideal membership is placed out of the scope of any tractable (theoretical or practical) complexity class.

Nevertheless, the effective Nullstellensatz has a particular feature ($g = 1$ and its geometrical meaning). This suggests that an adapted specific procedure might solve this problem within a much lower complexity. In these pages we'll show that polynomial upper bounds (for time and space) can be obtained for this problem by means of probabilistic algorithms.

Clearing denominators in the identity (2), one can reformulate the problem as the computation of a non-zero integer $a \in \mathbb{Z} - \{0\}$ and integer polynomials $g_1, \dots, g_s \in \mathbb{Z}[X_1, \dots, X_n]$ such that the following equality holds:

$$a = g_1 f_1 + \cdots + g_s f_s. \tag{3}$$

After some preprocessing (see section 6), the effective Nullstellensatz can be reinterpreted as a particular case of the *membership problem in a complete intersection ideal*, i.e., decide whether a polynomial g belongs to the ideal generated by $f_1, \dots, f_s$ in $\mathbb{Q}[X_1, \dots, X_n]$ (decision problem) and if the answer is affirmative compute an integer $a \in \mathbb{Z} - \{0\}$ and integer polynomials $g_1, \dots, g_s$ such that $a\, g = g_1 f_1 + \cdots + g_s f_s$ (representation problem) in the particular case where $f_1, \dots, f_s, g \in \mathbb{Z}[X_1, \dots, X_n]$ $(s \le n+1)$ verify the conditions:

- $(f_1, \dots, f_{s-1})$ is a proper ideal of dimension $n - s + 1$.
- The polynomial f_s is not a zero divisor in $\mathbb{Q}[X_1 \dots, X_n]/(f_1, \dots, f_{s-1})$.

Both problems (effective Nullstellensatz and membership in complete intersections) have a similar treatment as we shall see in the forthcoming pages.

From the point of view of Turing machine complexity, it is commonly accepted that the space in a Turing machine is measured as the space used by the working tapes. In some sense, this is CPU space, while the space required by input and output tapes is an "external memory" requirement. Nevertheless, this last amount of space gives an immediate lower bound for the sequential time of the procedure. In particular, exponential space encoding of outputs imply exponential time lower bounds. Hence, any improvement of the complexity of the geometric problems demands an encoding of the output polynomials in, at least, polynomial space.

Most of the complexity analyses of the effective Nullstellensatz naïvely consider dense representations both for the input and the output polynomials. This simply means that one writes all the coefficients (integer or rational numbers) of all monomials of the f_i's and the g_i's. Let us define the (absolute) height of an integer polynomial as the maximum of the absolute values of its coefficients (then the binary size of an integer number is the logarithm of its height). Assume that d is an upper bound for the degree of the polynomial inputs f_i's and that η is an upper bound for the height of their coefficients. Finally, recall that the number of monomials of an n-variate polynomial of degree d is asymptotically d^n. Hence, the input length is of order $sd^n \log_2 \eta$.

The studies of complexity of the effective Nullstellensatz were focused on the estimation of upper bounds for the degrees and heights of the output polynomials $a \in \mathbb{Z}$ and $g_1, \ldots, g_s \in \mathbb{Z}[X_1, \ldots, X_n]$ in terms of s, d and η.

A first bound for the degree of the g_i's was obtained in 1926 as a consequence of the work of G. Hermann ([31]). Later on, D. W. Masser and G. Wüstholz ([47]) extended transcendency proofs of A. Gel'fond ([21] and [49]). Using the methods of [31], they obtained upper bounds for the height of a and the polynomials $g_1, \ldots, g_s$. Both results lead to bounds of order d^{2^n} for the degrees and $d^{2^n} \log_2 \eta$ for the binary size of a and for the coefficients of the g_i's, which is far from being tractable.

The bounds for the degrees of the polynomials $g_1, \ldots, g_s$ were successively improved in [10], [12] and [34] to $\deg f_i g_i \leq \max\{3, d\}^n$. A well-known example of Mora-Lazard-Masser-Philippon shows that this bound for the degree is essentially optimal.

These estimations reduce the effective Nullstellensatz to the resolution of a linear system of equations by comparision of coefficients in the identity (3). The linear system contains as many variables as monomials of degree at most $\max\{3, d\}^n$ (while a is simply the determinant of a maximal minor of the system matrix). This leads to an exponential number of coefficients in the dense representation (sd^{n^2}) of binary size $d^{O(n^2)} \log_2 \eta^{O(1)}$. As observed in [12], by transforming Berkowitz' algorithm for the determinant in Boolean circuits and transforming the parallel time in sequential space (cf. [6]), this gives poly-logarithmic bounds $(n \log_2 d)^{O(1)}$ for the working space. Nevertheless, this procedure still requires dense representation of polynomial outputs, i.e., an amount of exponential space in "external memory" is required ($d^{O(n^2)}$) and, in particular, this yields to an exponential lower bound ($d^{O(n^2)}$) for the sequential time with respect to the input size.

In the realm of transcendental number theory and diophantine approximation, it was still interesting to get more precise bounds for the height of the output polynomials. A typical application of these bounds are Liouville estimations (cf. [39]): Lower bounds for the absolute value of multivariate integer polynomials when evaluated at algebraic points). In [2] and [3], C. Berenstein and A. Yger obtained a division theorem in terms of complex integrals which implies the following precise bound for the heights in identity (3):

$$\max\{\log_2 ht(a), \log_2 ht(g_i)\} \leq \kappa(n) d^{8n+3} (\log_2 \eta + d \log_2 d)$$

(where $\kappa(n)$ is an effective constant depending only on n).

As in the case of the degree bounds, these bounds for the height are optimal as shows the following simple example for $\eta \in \mathbb{Z} - \{0\}$:

$$f_1 := X_1 - \eta\,,\ f_2 := X_2 - X_1^d\,, \dots,\ f_n := X_n - X_{n-1}^d\,,\ f_{n+1} := X_n.$$

Since $f_1, \dots, f_s$ don't share any complex zero they satisfy identity (3). Specializing this equation in the integer point $(\eta, \eta^d, \dots, \eta^{d^{n-1}}) \in \mathbb{Z}^n$ one immediately remarks that $\log_2 ht(a) \geq d^{n-1} \log_2 |\eta|$.

This optimal bound suggests an improvement: One still has an exponential number of coefficients, but the size of each one of them is polynomial in the input length. Nevertheless, it does not solve all the difficulties and some other alternatives have to be considered when dealing with the effective Nullstellensatz since no advance in terms of complexity can be expected while keeping the underlying dense encoding of polynomials.

One of the possible alternatives might be the sparse representation of polynomials: One encodes only the non-zero coefficients both of the input and of the output polynomials. However, this approach presents many difficulties since it is still not known whether the output polynomials in the Nullstellesatz have few monomials of small height coefficients. For instance, if we consider the complete intersection ideal generated by $(X - 1)$ in $\mathbb{Q}[X]$ and the polynomial $X^n - 1$ which is a sparse polynomial in this ideal, the output polynomial is $X^{n-1} + X^{n-2} + \cdots + X + 1$ which is no more sparse. On the other hand, "sparsity" is very unstable when dealing with geometric problems (for instance, a linear change of coordinates transforms sparse polynomials into polynomials with many non-zero coefficients).

Another alternative to improve the algorithmic resolution of the Nullstellensatz is described in [23], [24], and [18]. These papers introduce the encoding of polynomials by means of straight-line programs (a natural idea coming from algebraic complexity theory and semi-numerical algorithms). A straight-line program consists of a description of the evaluation scheme one chooses in order to evaluate a given polynomial at a concrete point, using the arithmetic operations $+, -, \times$, and eventually $/$ and some fixed integer constants (that we call parameters). For instance, to "write" the determinant polynomial of an indeterminate matrix of order n would require $n!$ places, while to evaluate its specialization can be performed in $O(n^3)$ arithmetic steps.

By a dimension argument, one can easily show that most polynomials of degree D in n variables require D^n arithmetic operations to be evaluated.

However, in their work the authors show that even if the output and the intermediate polynomials in the effective Nullstellensatz are of degree $d^{O(n)}$, their evaluation schemes are also of order $d^{O(n)}$ (instead of the expectable $d^{O(n^2)}$). The authors also present an algorithm to obtain these polynomial outputs.

These papers are a first step to improvements, since the exponential lower bound does not hold in these conditions. Nevertheless, there were still some problems when a Turing machine has to simulate the procedure the authors describe.

In order to understand these problems, we need to introduce here the representation of a straight-line program by means of a (directed acyclic) graph, and the complexity parameters we consider will be essentially the size L of the graph (corresponding to the sequential time of the evaluation), its depth ℓ (the parallel time) and the maximum η of the heights of the integer parameters introduced along the graph.

Proposition 15 below shows then that the growing of the intermediate values obtained when applying a straight-line program to an integer point depends only polynomially on the size L of the graph and on the height η of the parameters while it is doubly exponential on the depth ℓ.

In [23] and [18] the authors encode intermediate and output polynomials by straight-line programs of size $d^{O(n)}$ and depth $O(n^2 \log_2^2 d)$. This last bound could imply that they are manipulating integer numbers of exponential binary size $d^{O(n^2)}$. Since arithmetic operations between integer numbers of exponential length require exponential time, these procedures risk to be still exponential (even if they are polynomial in the algebraic complexity model). However, if one succeeds to control the depth of the graphs (together with the height of the parameters), it would be possible to encode the output polynomials in polynomial space $d^{O(n)}$. Moreover, controlled bounds for the heights of the outputs are also interesting because of their applications to transcendental number theory and diophantine approximation.

In the present paper, we reconsider these questions introducing the encoding of polynomials by means of (division free) non-scalar straight-line programs. This scheme has the following features:

- Linear operations are taken for free.

- No division is allowed.

Our main goal here is to control the non-scalar depth (below $O(n \log_2 d)$) and the height of the parameters of non-scalar straight-line programs arising in both problems. We are able to show the following result in terms of non-scalar straight-line programs:

Theorem 1 (Effective Nullstellensatz) *Let $f_1, \dots, f_s$ be polynomials in $\mathbb{Z}[X_1, \dots, X_n]$ of degree and (absolute) height bounded by $d \geq n$ and η respectively.*

which don't share any zero in $\mathbb{C}^n$. Then there exist a non-scalar straight-line program of size $sd^{O(n)}$, depth $O(n \log_2 d)$ and integer parameters of absolute height $\max\{d^{O(n)}, \eta\}$ *which evaluates an integer number $a \in \mathbb{Z} - \{0\}$ and polynomials $g_1, \dots, g_s \in \mathbb{Z}[X_1, \dots, X_n]$ such that:*

$$a = g_1 f_1 + \cdots + g_s f_s.$$

Similarly, we deal with the membership problem in complete intersection ideals. We obtain the following result (which proof is similar to the previous one):

Theorem 2 (Membership problem in a complete intersection)
Let $f_1, \dots, f_r$ ($r \le n$) and g be polynomials in $\mathbb{Z}[X_1, \dots, X_n]$ of degree and height bounded by $d \ge n$ and η respectively. Assume that $(f_1, \dots, f_r)$ is a proper ideal in $\mathbb{Q}[X_1, \dots, X_n]$ of dimension $n-r$ and that $g \in (f_1, \dots, f_r)$ holds. Then, there exists a non-scalar straight-line program of size $d^{O(n)}$, depth $O(r \log_2 d)$ and integer parameters of height $\max\{d^{O(n)}, \eta\}$ *which evaluates an integer $a \in \mathbb{Z} - \{0\}$ and polynomials $g_1, \dots, g_r \in \mathbb{Z}[X_1, \dots, X_n]$ such that:*

$$a\, g = g_1 f_1 + \cdots + g_r f_r.$$

As we mentioned before, proposition 15 shows that the growing of the intermediate values in a non-scalar straight-line program is polynomial in the size of the graph and the height of the parameters and doubly exponential on the depth. For the particular non-scalar straight-line programs we are considering in these two statements, the evaluation at an integer point can be simulated by a Turing machine using time and space bounded by $d^{O(n)}$. This is due to the controlled depth of the graph and heights of the parameters. Hence, using our analysis of correct test sequences for the non-scalar model (section 3 below) and checking the intermediate polynomials we can design in subsection 7.2 (following the notations of [1, chapter 6]) a probabilistic Turing machine which shows that:

Corollary 3 *The effective Nullstellensatz and the membership problem to complete intersection ideals are in the BPP-class of bounded error probabilistic polynomial time machines, i.e., there is a probablistic Turing machine that solves both problems within time polynomial in the dense representation of the input polynomials. In particular, both problems can be solved in polynomial space counting also "external memory" requirements.*

Theorems 1 and 2 have also consequences of a more mathematical nature. First of all, combining theorem 1 with the estimations of proposition 15 we recover for the effective Nullstellensatz the bounds for the degrees of [10],[12], and [34] and the bounds for the heights of [2] and [3] (simultaneously with the property of "quick" evaluation of [18]):

Corollary 4 *Let $f_1, \dots, f_s$ be polynomials in $\mathbb{Z}[X_1, \dots, X_n]$ of degree and (absolute) height bounded by $d \geq n$ and η respectively, which don't share any zero in $\mathbb{C}^n$. Then there exist an integer $a \in \mathbb{Z} - \{0\}$ and polynomials $g_1, \dots, g_s \in \mathbb{Z}[X_1, \dots, X_n]$ such that:*

- $\deg g_i = d^{O(n)} \quad (1 \leq i \leq s)$
- $\max\{\log_2 ht(a), \log_2 ht(g_1), \dots, \log_2 ht(g_s)\} = d^{O(n)} \max\{\log_2 \eta, \log_2 s\}$.

In [16], M. Elkadi obtained bounds for the degrees and the heights of the polynomials arising in the membership problem to a complete intersections. This author uses a slight modification of the arguments of [2] and [3]. We independently obtain similar results as a consequence of the two theorems above:

Corollary 5 *Let $f_1, \dots, f_r$ $(r \leq n)$ and g be polynomials in $\mathbb{Z}[X_1, \dots, X_n]$ of degree and height bounded by $d \geq n$ and η respectively. Assume that $(f_1, \dots, f_r)$ is a proper ideal in $\mathbb{Q}[X_1, \dots, X_n]$ of dimension $n - r$ and that $g \in (f_1, \dots, f_r)$ holds. Then there exist an integer $a \in \mathbb{Z} - \{0\}$ and polynomials $g_1, \dots, g_r \in \mathbb{Z}[X_1, \dots, X_n]$ such that:*

- $\deg g_i = d^{O(r)} \quad (1 \leq i \leq r)$
- $\max\{\log_2 ht(a), \log_2 ht(g_1), \dots, \log_2 ht(g_r)\} = d^{O(r)} n \log_2(\eta)$.

One can also apply the bounds arising in the straight-line programs to the analysis of the height of varieties. In particular, the results of section 4 on the isolated points of affine algebraic varieties lead to a bound for the heights which is essentially the same that follows from [7].

Corollary 6 *Let $f_1, \dots, f_r$ $(r \leq n)$ and g be polynomials in $\mathbb{Z}[X_1, \dots, X_n]$ of degree and height bounded by $d \geq n$ and η, respectively, and let V denote the algebraic affine variety of $\mathbb{C}^n$ defined by:*

$$V := \{f_1 = 0, \dots, f_s = 0\}.$$

Then V has at most d^n isolated points whose heights verify:

$$\log_2 ht(P) = d^{O(n)}(\log_2 s + \log_2 \eta).$$

In particular, applying straightforward the bounds for the complex solutions of an integer polynomial we obtain the following corollary:

Corollary 7 *With the same notation, the coordinates of every isolated point $(x_1, \dots, x_n) \in \mathbb{C}^n$ of the algebraic set $V = \{f_1 = 0, \dots, f_s = 0\}$ verify:*

$$\frac{1}{d^{\Omega(n)}(\log_2 s + \log_2 \eta)} \leq \log_2 \mid x_i \mid \leq d^{O(n)}(\log_2 s + \log_2 \eta).$$

We conclude this introduction observing that the results suggest the existence of some trade-off between complexity analysis, encoding of polynomials as straight-line programs and bounds in diophantine approximation and number theory. We hope that this trade-off can be improved in forthcoming works on the subject.

2 Non-scalar straight-line programs

2.1 Basic notations

This section is devoted to introduce a geometric model for polynomial evaluation.

Each polynomial $P(X_1, \ldots, X_n) \in \mathbb{Z}[X_1, \ldots, X_n]$ has a dense representation of the form:

$$P(X_1 \ldots, X_n) = \sum_{|\mu| \leq deg(P)} P^\mu X_1^{\mu_1} \cdots X_n^{\mu_n}$$

where $\deg(P)$ denotes the total degree of P, $\mu := (\mu_1, \ldots, \mu_n) \in \mathbb{N}_0^n$ is a multi-index, $|\mu| := \mu_1 + \cdots + \mu_n$ is its length and P^μ is an integer.
We define the (absolute) *height of* P as the maximal absolute value of all its coefficients:

$$ht(P) := max\{| P^\mu | : | \mu | \leq deg(P)\}.$$

Similarly, for a finite family $\mathcal{F} \subset \mathbb{Z}$, we define $ht(\mathcal{F})$ as the maximal absolute value of all its elements.

In the sequel we shall work with the complexity model of non-scalar straight-line programs (see for instance [27], [55], or [46]): a non-scalar straight-line program is a device which evaluates (and hence represents) a given polynomial of $\mathbb{Z}[X_1, \ldots, X_n]$ taking $\mathbb{Z}$-linear operations for free. We shall tacitly assume that our straight-line programs do not contain any division.
We shall represent a straight-line program for the evaluation of a polynomial $P \in \mathbb{Z}[X_1, \ldots, X_n]$ by a *directed acyclic graph* $\mathcal{G}$ whose nodes are labelled gates which perform arithmetical operations. Therefore we identify the nodes of $\mathcal{G}$ with the corresponding gates. The graph $\mathcal{G}$ disposes of $n+1$ particular nodes labelled by the variables $X_1, \ldots, X_n$ and the constant 1. These nodes are called the input gates of $\mathcal{G}$.
We define the depth of a gate ν of our graph as the length of the largest path which joins ν with some input gate. Let us denote the gates of the directed acyclic graph by pairs of integer numbers (i, j), where i represents the depth of the gate and j is the corresponding value of an arbitrary numbering imposed to the set of gates of depth i (this encoding can be seen, for instance, in [45]).

Definition 8 *A division-free non-scalar straight-line program with inputs* X_1, $\ldots, X_n$ *is a pair* $\Gamma := (\mathcal{G}, Q)$, *where* $\mathcal{G}$ *is a directed acyclic graph, with* $n+1$

input gates, unbounded fan-in, and Q is a function that assigns to every gate (i,j) one of the following instructions:

$$i=0: \qquad Q_{0,1}:=1\ ,\ Q_{0,2}:=X_1\ ,\ \ldots\ ,\ Q_{0,n+1}:=X_n$$
$$1\leq i\leq \ell:$$

$$Q_{i,j}:=\Big(\sum_{r\leq i-1,1\leq s\leq L_r} A_{i,j}^{r,s}\,Q_{r,s}\Big)\cdot\Big(\sum_{r'\leq i-1,1\leq s'\leq L_{r'}} B_{i,j}^{r',s'}\,Q_{r',s'}\Big).$$

Here, $A_{i,j}^{r,s}$, $B_{i,j}^{r',s'}$ are indeterminates over $\mathbb{Z}$ called the parameters introduced in Γ. The size of the straight-line program Γ is $L(\Gamma)=L_0+\cdots+L_\ell$ (where $L_0:=n+1$) and its depth $\ell(\Gamma)=\ell$ (these notions coincide with the notions of size and depth of the underlying computation graph).

Let us mention that in our notation the subindices of the parameters $A_{i,j}^{r,s}$ and $B_{i,j}^{r',s'}$ represent the gate of the multiplication they are assigned and the superindices correspond to the previous result they involve in the multiplication. We abbreviate $\underline{A}=(A_{i,j}^{r,s})$ and $\underline{B}=(B_{i,j}^{r',s'})$.
Semantically speaking, the straight-line program Γ defines an evaluation algorithm of the polynomials (intermediate results):

$$Q_{i,j}=\sum_{|\mu|\leq 2^i} Q_{i,j}^{\mu}(\underline{A},\underline{B})\,X_1^{\mu_1}\cdots X_n^{\mu_n} \tag{4}$$

Here, each coefficient $Q_{i,j}^{\mu}(\underline{A},\underline{B})$ belongs to the polynomial ring $\mathbb{Z}[\underline{A},\underline{B}]$. The result $Q_{i,j}$ has degree at most 2^i with respect to the variables $X_1,\ldots,X_n$.
We obtain a *specialization* in $\mathbb{Z}$ of the non-scalar straight-line program Γ substituting the parameter sets $\underline{A}$ and $\underline{B}$ by integers $\underline{\alpha}=(\alpha_{i,j}^{r,s})$ and $\underline{\beta}=(\beta_{i,j}^{r',s'})$ (we insist on the fact that $\alpha_{i,j}^{r,s},\beta_{i,j}^{r',s'}$ belong to $\mathbb{Z}$).
A specialization $\underline{A}\to\underline{\alpha}$, $\underline{B}\to\underline{\beta}$ of the parameters of Γ induces a straight-line program (computation) in $\mathbb{Z}[X_1,\ldots,X_n]$ in the most obvious way. The intermediate results of this specialized straight-line program γ are the polynomials of the form $Q_{i,j}(\underline{\alpha},\underline{\beta},X_1,\ldots,X_n)$.
In this sense we shall say that a given polynomial $P\in\mathbb{Z}[X_1,\ldots,X_n]$ is evaluable, or computable, by (a specialization of) the straight-line program Γ if there exists a specialization $\underline{A}\longrightarrow\underline{\alpha}$, $\underline{B}\longrightarrow\underline{\beta}$ of the parameters of Γ such that for some gate (i,j) the following equality holds:

$$P(X_1,\ldots,X_n)=Q_{i,j}(\underline{\alpha},\underline{\beta},X_1,\ldots,X_n). \tag{5}$$

Taking into account the representation of (4) we can rewrite identity (5) as:

$$P^{\mu}=Q_{i,j}^{\mu}(\underline{\alpha},\underline{\beta})$$

for all μ with $|\mu|\leq 2^i$ and $P^{\mu}=0$ for $|\mu|\ 2^i$.

Finally, for η a positive number we say that a specialization $\underline{A} \longrightarrow \underline{\alpha}$, $\underline{B} \longrightarrow \underline{\beta}$ has *height at most* η or equivalently that $P \in \mathbb{Z}[X_1, \ldots, X_n]$ is computable by a straight-line program Γ with parameters of height η if $ht(\underline{\alpha}) \leq \eta$ and $ht(\underline{\beta}) \leq \eta$ hold.
We will also say that an integer $\alpha \in \mathbb{Z}$ is computable by a non-scalar straight-line program Γ when it is considered as an element of $\mathbb{Z}[X]$.

2.2 Some technical lemmas

Most basic examples of straight-line programs belong to the field of linear algebra. We overview some of them under the scope of non-scalar measure. For instance, the next proposition describes the bounds naturally arising from Berkowitz' computation of determinant and characteristic polynomial (cf. [4]):

Proposition 9 (Berkowitz, 1986) *Let R be a ring. There is a non-scalar straight-line program of size $O(n^5)$, depth $O(\log_2 n)$, and parameters in $\{-1, 0, 1\}$ which computes from the entries of every matrix $A \in \mathcal{M}_n(R)$ the coefficients of its characteristic polynomial and, in particular, its determinant $det(A)$.*

Proof. We only need to follow the proof of [4]. Assume you are given a matrix in $\mathcal{M}_n(R)$:

$$A = \begin{pmatrix} a_{1\,1} & \ldots & a_{1\,n} \\ \vdots & \ddots & \vdots \\ a_{n\,1} & \ldots & a_{n\,n} \end{pmatrix}.$$

For every $m \in \{1, \ldots, n-1\}$ let us define the matrices:

$$\begin{aligned} M_m : &= \begin{pmatrix} a_{m+1\,m+1} & \ldots & a_{m+1\,n} \\ \vdots & \ddots & \vdots \\ a_{n\,m+1} & \ldots & a_{n\,n} \end{pmatrix} \\ R_m : &= \begin{pmatrix} a_{m\,m+1} & \ldots & a_{m\,n} \end{pmatrix} \\ S_m : &= {}^t\begin{pmatrix} a_{m+1\,m} & \ldots & a_{n\,m} \end{pmatrix}. \end{aligned}$$

Now we define the lower $(n-m+1) \times (n-m)$ triangular Toeplitz matrix C_m:

$$(C_m)_{i\,j} := \begin{cases} -1 & \text{if } i = j \\ a_{m\,m} & \text{if } i = j+1 \\ -R_m\, M_m^{k-1}\, S_m & \text{if } i = j+k \quad (k1) \\ 0 & \text{otherwise.} \end{cases}$$

Finally, let us denote the characteristic polynomial of A by:

$$\mathcal{X}_A(T) := \det(A - T.Id) = \sum_{k=0}^{n} p_k\, T^k.$$

In [4] it is shown that:

$$\begin{pmatrix} p_0 \\ \vdots \\ p_n \end{pmatrix} = \prod_{m=1}^{n} C_m.$$

Therefore we can compute the coefficients of $\mathcal{X}_A$ by means of the following straight-line program:

- For every $m \in \{1, \dots, n-1\}$, we compute in parallel gates the coordinates of all the powers M_m^i for $i = 0, \dots, n$. For each m, this computation can be done straight-forward in size $O(n^4)$ and non-scalar depth $\log_2 n + 2$, with parameters in $\{0, 1\}$. This gives a total size of $O(n^5)$ and the same non-scalar depth $\log_2 n + 2$.
- Now, for each m and each i, we compute $-R_m\, M_m^i\, S_m$. We can perform this computation in additional size $O(n^3)$ and non-scalar depth 3. Thus, to compute all $-R_m\, M_m^i\, S_m$ requires additional size $O(n^5)$ and non-scalar depth 3.
- Finally we compute the product $\prod_{m=1}^{n} C_m$. This can be done in additional size $O(n^4)$ and non-scalar depth $\log_2 n + 2$.

The announced bounds are obtained by adding all these amounts. □

As a consequence of this proposition, we can show the following estimations for non-scalar straight-line programs computing gcd's and square-free decompositions.

Lemma 10 *Let R be an integral domain and $f, g \in R[X]$ be two univariate polynomials of degree bounded by D. Then there is a non-scalar straight-line program Γ of length $O(D^6)$, depth $O(\log_2 D)$, and parameters in $\{-1, 0, 1\}$ which, from the coefficients of f and g, computes the coefficients of a greatest common divisor in $R[X]$ of these polynomials. These coefficients are integer polynomials in the coefficients of f and g of degree bounded by $2D$.*

Proof. We follow here a scheme of [5] or [20]. Set $f := f_d X^d + \cdots + f_0$ and $g := g_e X_e + \cdots + g_0$, where $0 \le e \le d \le D$ $(f_d\, g_e \neq 0)$.
For $0 \le i \le e$ consider the $(d + e - 2i)$-square submatrix P_i of the Sylvester matrix of (f, g) which consists on the first $(e - i)$ columns of f's, the first $(d - i)$ columns of g's , and the first $(d + e - 2i)$ rows.
Then, for $0 \le i \le e$, compute $\det(P_i)$ and choose the minimal value k such that $\det(P_k) \neq 0$. This will be the degree of the desired $gcd(f, g)$.
Now compute the last column $(y_{e-k-1}, \dots, y_0,\ z_{d-k-1}, \dots, z_0)$ of the adjoint matrix of P_k. The polynomial $(y_{e-k-1} X^{e-k-1} + \cdots + y_0) f + (z_{d-k-1} X^{d-k-1} + \cdots + z_0) g$ is the desired (non-null scalar multiple of) $gcd(f, g)$. □

Lemma 11 *Let R be an integral domain and $f_1, \ldots, f_s \in R[X]$ be univariate polynomials of degree bounded by D. Then there is a non-scalar straight-line program Γ of length $O(s^5 D^6)$, depth $O(\log_2(sD))$ and parameters in $\{-1, 0, 1\}$ which computes from the coefficients of $f_1, \ldots, f_s$ the coefficients of a greatest common divisor $gcd(f_1, \ldots, f_s) \in R[X]$ of these polynomials; these coefficients are integer polynomials in the coefficients of $f_1, \ldots, f_s$ of degree bounded by $2D + 1$.*

Proof. We also follow [20]: we need to determine $h \in R[X] \setminus \{0\}$ of minimal degree such that:

$$h = r_1 f_1 + \cdots + r_s f_s, \qquad \deg(s_i) \le D \ \ (1 \le i \le s)$$

Setting $f_i := \sum f_{ij} X^j$ and $r_i := \sum r_{ij} X^j$, it is enough to consider the consistency of the linear systems describing the condition "$\sum r_i f_i$ is monic of degree k ($\le D$)", which are given by the equations:

$$\sum_{\substack{1 \le i \le D \\ 0 \le j \le m}} r_{ij} f_{i\,m-j} = \begin{cases} 0 & \text{for } k < m < 2D \\ 1 & \text{for } m = k \end{cases}$$

Thus there are $D + 1$ systems of $2D - k$ equations in $s(D + 1)$ variables, whose consistency depends on analysis of ranks. Once detected the degree d of $gcd(f_1, \ldots, f_s)$, we solve it in R using the adjoint and the determinant of an invertible square submatrix of maximal order. □

Lemma 12 *Let R be an integral domain and $f \in R[X]$ be an univariate polynomial of degree bounded by D. Then there is a non-scalar straight-line program Γ verifying:*

i) Γ computes from the coefficients of f an element $\theta \in R$ and the coefficients of a square-free polynomial $f^* \in R[X]$ associated to f, i.e.,:

$$f^* = \theta \frac{f}{gcd(f, f')}$$

(where f' denotes the derivative of f).

ii) The element θ and the coefficients of f^* are integer polynomials in the coefficients of f of degree bounded by $2D(D + 1)$.

iii) The size of Γ is of order $O(D^6)$, the depth is $O(\log_2 D)$, and the parameters introduced in Γ are $\{-1, 0, 1, \ldots, D\}$.

Proof. First compute the coefficients of f', then apply lemma 10 to compute the (scalar multiple of) $gcd(f, f')$, and finally solve the linear system determined by the relation:

$$f = gcd(f, f') \cdot f^*$$

(the value θ is simply a maximal minor of the coefficient matrix). □

Another useful technique is to know how to evaluate the homogeneous components of a polynomial given by a non-scalar straight-line program:

Lemma 13 *Suppose that a polynomial $P := \sum P^\mu X_1^{\mu_1} \cdots X_n^{\mu_n} \in \mathbb{Z}[X_1, \ldots, X_n]$ is computable by a non-scalar straight-line program Γ with inputs $X_1, \ldots, X_n$, of size L, depth ℓ and parameters in $\mathcal{F} \subset \mathbb{Z}$, and fix $\delta \in \mathbb{N}$. Then there is a straight-line program Γ' which computes all the homogeneous components of P of degree at most δ, such that $L(\Gamma') = (\delta + 1)^2 L$, $\ell(\Gamma') = 2\ell$, and $\mathcal{F}(\Gamma') = \mathcal{F}$.*

Proof. The straight-line program Γ' computes all the homogeneous components of the intermediate polynomials $Q_{i,j}$ in Γ. We will replace the original straight-line program Γ by a new one which is obtained "decomposing" each step of Γ as follows:

- For $i = 1$ the homogeneous components of each $Q_{1,j}$ can be computed in size 5 and depth 2, using the parameters in $\mathcal{F}$ (for instance if $Q_{11} = (A_0 + A_1 X_1 + \cdots + A_n X_n)\,(B_0 + B_1 X_1 + \cdots + B_n X_n)$, one computes in a first step $A_0 B_0$, $(A_1 X_1 + \cdots + A_n X_n)\,(B_1 X_1 + \cdots + B_n X_n)$, $A_0\,(B_1 X_1 + \cdots + B_n X_n)$ and $B_0\,(A_1 X_1 + \cdots + A_n X_n)$ and in the next one $A_0\,(B_1 X_1 + \cdots + B_n X_n) + B_0\,(A_1 X_1 + \cdots + A_n X_n))$.
- For $i = 0$: Suppose that the homogeneous components of all $Q_{r,s}$, $r < i$, of degree $\leq \delta$ are computed. We use the following simple observation to compute the homogeneous components $Q_{i,j}^{(t)}$ of degree $t \leq \delta$ of $Q_{i,j}$: If in Γ, $Q_{i,j}$ was obtained by means of the formula

$$Q_{i,j} := \Big(\sum_{r,s} A_{i,j}^{r,s} Q_{r,s}\Big) \cdot \Big(\sum_{r',s'} B_{i,j}^{r',s'} Q_{r',s'}\Big)$$

$$\text{then } Q_{i,j}^{(t)} := \sum_{t_1 + t_2 = t} \left(\sum_{r,s} A_{i,j}^{r,s} Q_{r,s}^{(t_1)}\right) \cdot \left(\sum_{r',s'} B_{i,j}^{r',s'} Q_{r',s'}^{(t_2)}\right)$$

  for the same parameters $(\underline{A}, \underline{B})$. So, in order to get $\{Q_{(i,j)}^{(t)},\ t < \delta\}$, we compute for each t at most $(\delta + 1)$ products

$$\left(\sum_{r,s} A_{i,j}^{r,s} Q_{r,s}^{(t_1)}\right) \cdot \left(\sum_{r',s'} B_{i,j}^{r',s'} Q_{r',s'}^{(t_2)}\right)$$

  and add them in a new step. Thus there are in total $(\delta + 1)^2$ nodes for each $Q_{i,j}$, and we obtain: $L(\Gamma') = (\delta + 1)^2 L$, $\ell(\Gamma') = 2\ell$ and $\mathcal{F}(\Gamma') = \mathcal{F}$. □

2.3 Some useful bounds

We are in conditions to show some bounds arising from the encoding of polynomials by non-scalar straight-line programs. First of all, we can easily bound by $2L(L-(n+1))$ the number of parameters used by a non-scalar straight-line program Γ of size L in n variables. We can also bound the degrees of the polynomials $Q^{\mu}_{i,j}$ (of formula 4) as elements in $\mathbb{Z}[\underline{A},\underline{B}]$:

Lemma 14 *Given a non-scalar straight-line program Γ, the degree of all polynomials $Q^{\mu}_{i,j} \in \mathbb{Z}[\underline{A},\underline{B}]$ is $2^{i+1}-2$ (independently from μ and j).*

Proof. First of all, it is clear that the degree of $Q^{\mu}_{i,j}$ does not depend on j in a given step i.
Denote by $d(i,\mu)$ the degree of such polynomials. Then definition 8 implies that:

$$Q^{\mu}_{i,j} = \sum_{\mu_1+\mu_2=\mu} \Big(\sum_{0\le r\le i-1, 1\le s\le L_r} A^{r,s}_{i,j}\, Q^{\mu_1}_{r,s}(\underline{A},\underline{B})\Big) \cdot \Big(\sum_{0\le r'\le i-1, 1\le s'\le L_{r'}} A^{r',s'}_{i,j}\, Q^{\mu_2}_{r',s'}(\underline{A},\underline{B})\Big). \tag{6}$$

Thus we get the following relation:

$$d(i,\mu) = \max_{\mu_1+\mu_2=\mu} \left\{ \max_{1\le r\le i-1} (d(r,\mu_1)+1) + \max_{1\le r'\le i-1} (d(r',\mu_2)+1) \right\}.$$

Now we proceed by induction on i:

- For $i=1$, $d(i,\mu)=2$ and it is independent from μ.
- For $i>1$, by hypothesis $d(r,\mu_1)=2^{r+1}-2$ and $d(r',\mu_2)=2^{r'+1}-2$ for $r,r'\le i-1$ and $|\mu_1|\le 2^r$, $|\mu_2|\le 2^{r'}$. Therefore

$$d(i,\mu) = \max_{\mu_1+\mu_2=\mu} \left\{ \max_{1\le r\le i-1} (2^{r+1}-1) + \max_{1\le r'\le i-1} (2^{r'+1}-1) \right\} = 2\,(2^i-1)$$

and the proof is complete. □

The next proposition estimates the height of a polynomial given by a non-scalar straight-line program. Observe that this bound depends doubly exponentially only on the depth of the graph, while it is polynomial on its size and on the height of the parameters.

Proposition 15 *A specialization $\underline{A} \longrightarrow \underline{\alpha}$, $\underline{B} \longrightarrow \underline{\beta}$ of height η on a straight-line program Γ of size L and depth ℓ produces the following bound for the height of the evaluated polynomials:*

$$ht(Q_{i,j}(\underline{\alpha},\underline{\beta})) \le \sum_{0\le|\mu|\le 2^i} |\, Q^{\mu}_{i,j}(\underline{\alpha},\underline{\beta}) \,| \le (L\eta)^{2^{i+1}-2}.$$

Proof. Let us consider the polynomials $Q_{i,j}(|\,\underline{\alpha}\,|, |\,\underline{\beta}\,|, X_1, \ldots, X_n)$, obtained specializing the parameters in Γ according to:

$$A_{i,j}^{r,s} \longmapsto |\,\alpha_{i,j}^{r,s}\,|$$

$$B_{i,j}^{r',s'} \longmapsto |\,\beta_{i,j}^{r',s'}\,|$$

As the polynomials $Q_{i,j}^{\mu}(\underline{A}, \underline{B})$ have positive coefficients, we have

$$|\,Q_{i,j}^{\mu}(\underline{\alpha}, \underline{\beta})\,| \leq Q_{i,j}^{\mu}(|\underline{\alpha}|, |\underline{\beta}|)$$

and

$$\sum_{0\leq\mu\leq 2^i} |\,Q_{i,j}^{\mu}(\underline{\alpha}, \underline{\beta})\,| \leq \sum_{0\leq\mu\leq 2^i} Q_{i,j}^{\mu}(|\underline{\alpha}|, |\underline{\beta}|) = Q_{i,j}(|\underline{\alpha}|, |\underline{\beta}|, 1, \ldots, 1).$$

Now let us define:

$$w(r) := max\{Q_{r,s}(|\,\underline{\alpha}\,|, |\,\underline{\beta}\,|, 1, \ldots, 1) : 1 \leq s \leq L_r\}.$$

Hence, we follow the relations stated from the straight-line program Γ to have:

$$Q_{i,j}(|\,\underline{\alpha}\,|, |\,\underline{\beta}\,|, \underline{1}) \leq$$

$$\leq \left(\sum_{r,s} |\,\alpha_{i,j}^{r,s}\,|\, Q_{r,s}(|\,\underline{\alpha}\,|, |\,\underline{\beta}\,|, \underline{1})\right)\left(\sum_{r',s'} |\,\beta_{i,j}^{r',s'}\,|\, Q_{r',s'}(|\,\underline{\alpha}\,|, |\,\underline{\beta}\,|, \underline{1})\right).$$

which yields to the following recursion rule:

$$w(i) \leq L^2\eta^2\,(w(i-1))^2.$$

Finally, since $w(1) \leq (L\eta)^2$, one easily infers the desired bound. □

Remark Let $P \in \mathbb{Z}[X_1, \ldots, X_n]$ be a polynomial which can be evaluated by a non-scalar straight-line program Γ of size L, depth ℓ, and parameters of height η. Let $x := (x_1, \ldots, x_n) \in \mathbb{Z}^n$ be a point of height bounded by h. Then, there exists a deterministic Turing machine that outputs $P(x) \in \mathbb{Z}$ in time $L^2 2^{O(\ell)}(\log_2 L + \log_2 \eta + \log_2 h)$. (We have to make $2L$ sums and one product for every gate of Γ, each operation corresponds to integer numbers of binary size bounded by $2^{2\ell}(\log_2 L + \log_2 \eta + \log_2 h)$.)

Remark The previous upper bound for the height of the polynomials is also useful to study lower bounds for polynomial evaluation. An example are Pochhammer (also called Wilkinson) polynomials, which are defined by the following rule:

$$P_d(T) := \binom{T-1}{d} d! = (T-1)\cdots(T-d).$$

These polynomials arise in a natural way from the effective Nullstellensatz (cf. also [52] or [28]). Consider the following system of polynomial equations in the variables $X_1, \ldots, X_n, T$:

$$X_1^2 - X_1 = 0 , \ldots, X_n^2 - X_n = 0 , \sum_{i=1}^{n} 2^{i-1} X_i = T.$$

The system has a solution if and only if T is a root of the Pochhammer polynomial for $d = 2^n - 1$:

$$P_{2^n-1}(T) := \prod_{i=0}^{2^n-1} (T - i).$$

Our bounds for the height of P_{2^n-1} and the Stirling formula for the factorial imply that this polynomial is "hard to compute" in the following sense: if there exists a straight-line program which evaluates P_{2^n-1} in optimal depth n using parameters in $\mathbb{C}$ bounded by $\frac{1}{2}2^{\frac{n}{3}}$, its non-scalar size will be greater than $2^{\frac{n}{3}}$.

3 Questors and Vermeidung von Divisionen

3.1 Questors or correct test sequences

One of the resources used in [18], [23] and [24] are the questors (correct test sequences) as points determining simpler probabilistic or non-uniform algorithms for elimination in commutative algebra. They keep inside all the gestions and distribution of resources required by this kind of algorithms. This section is essentially devoted to state that the maximal height of a questor set in characteristic zero depends only on the depth under the non-scalar model. We will later apply this property to analyze the points, in the avoiding division method of Strassen in [54] (cf. also [18]).

For the analysis of the height of questors we follow similar steps to those in [29]. First, we associate to a fixed generic non-scalar straight-line program Γ with exactly one output node $Q_\ell := Q_{\ell,1}$ a regular morphism Φ that describes from the parameters $(\underline{A}, \underline{B})$ the polynomial it evaluates by specialization, i.e.,

$$\Phi \ : \ (\underline{A}, \underline{B}) \longmapsto (Q_\ell^\mu(\underline{A}, \underline{B}))_{|\mu| \le 2^\ell} \cdot$$

We analyze this morphism following similar arguments of those in [30]. As always, L is the size of Γ, ℓ is its depth, and n is the number of variables.

Then, the image set $Im\,\Phi$ has irreducible Zariski closure that we shall denote by $W(\Gamma)$. Hence, the dimension of $W(\Gamma)$ is at most $2L(L-(n+1))$ (i.e., at most the number of parameters in Γ). Recall that the degree of an irreducible algebraic set V of dimension d is defined as the maximal cardinal of finite intersections of V with d generic linear hyperplanes. We can state now the following lemma, which is essentially [30, lemma 1]:

Lemma 16 *The irreducible algebraic set $W(\Gamma)$ is $\mathbb{Q}$-definible and:*

$$\deg W(\Gamma) \leq (2^{\ell+1} - 2)^{2L(L-(n+1))}.$$

Proof. We follow the same argument as in [30, lemma 1]: Assume $D := \dim W(\Gamma)$; since $W(\Gamma)$ is irreducible, there exist hyperplanes $H_1, \ldots, H_D$ which intersect $W(\Gamma)$ in exactly $\deg W(\Gamma)$ points. Moreover, we can assume that they intersect $W(\Gamma)$ in $Im\,\Phi$ (cf. [51] or [26]).
Now, we observe that from the bounds on $\deg Q_\ell^\mu$ (cf. lemma 14) it follows that $\Phi^{-1}(H_1), \ldots, \Phi^{-1}(H_D)$ are hypersurfaces of degree bounded by $(2^{\ell+1} - 2)$. Define

$$\mathcal{C} := \{C :\ C \text{ is an irreducible component of } \Phi^{-1}(H_1) \cap \ldots \cap \Phi^{-1}(H_D)\}$$

The Bezout inequality of [26] implies that:

$$\sharp\mathcal{C} \leq \sum_{C \in \mathcal{C}} \deg C \leq \prod_{i=1}^{D} \deg \Phi^{-1}(H_i) \leq (2^{\ell+1} - 2)^D.$$

Finally, Φ maps the elements of $\mathcal{C}$ onto the points of $W(\Gamma) \cap H_1 \cap \cdots \cap H_D$. Therefore, $\deg W(\Gamma) \leq \sharp\mathcal{C}$ as wanted. □

Now we are in conditions to deal with questors (correct test sequences) for polynomials.

Definition 17 *Given a set $\mathcal{F} \subset \mathbb{Z}[X_1, \ldots, X_n]$ (which contains the null polynomial) we say that a finite set $\mathcal{Q} \subset \mathbb{Z}^n$ is a questor (or a correct test sequence) for $\mathcal{F}$ iff for all $P \in \mathcal{F}$ the following holds:*

$$P|_{\mathcal{Q}} = 0 \Longrightarrow P \equiv 0\ .$$

The following proposition, translated to our context from [29] is crucial for us:

Proposition 18 *Under the above notation, let $u := (2^{\ell+1} - 2)(2^\ell + 1)^2$ and $t := 6L(L - (n+1))$. Then, in $\{1, \ldots, u\}^n$ there are at least $u^{nt}(1 - u^{-\frac{t}{6}})$ correct test sequences of length t for $W(\Gamma)$.*

In the proof, we will make use of some results of [26] which rely on a notion of the degree of an arbitrary (not necessarily irreducible) variety introduced there, useful for our complexity purposes: The degree of a variety V, $\deg V$, is defined as the sum of the degrees of its irreducible components.

Proof. The proof follows the arguments of [29]. We define the algebraic subset $E(\Gamma, t) \in \mathbb{C}^{nt+N}$ by the conditions:

$$(\underline{a}_1, \ldots, \underline{a}_t, P) \in E(\Gamma, t) \iff P \in W(\Gamma) \quad \text{and} \quad P(\underline{a}_i) = 0 \quad (1 \leq i \leq t)$$

(here each $\underline{a}_t$ is a point of $\mathbb{C}^n$ and we identify P with its vector of N coefficients in $W(\Gamma)$). Therefore the algebraic set $E(\Gamma,t)$ is defined by the equations defining $W(\Gamma)$ and the following equations of degree $2^\ell+1$:

$$\sum_{|\mu|\leq 2^\ell} Z_\mu \cdot X_{1j}^{\mu_1}\cdots X_{nj}^{\mu_n}=0 \qquad (1\leq j\leq t)$$

From Bezout's inequality (cf. [26]) and lemma 16:

$$\deg\ E(\Gamma,t)\leq \deg\ W(\Gamma)\,(2^\ell+1)^t\leq (2^{\ell+1}-2)^{\frac{t}{3}}(2^\ell+1)^t$$

Now we have the two canonical projections:

$$\pi_1: E(\Gamma,t)\longrightarrow \mathbb{C}^{nt} \quad \text{and} \quad \pi_2: E(\Gamma,t)\longrightarrow \mathbb{C}^N$$

Let C_j, $j\in J$, be all those irreducible components of $E(\Gamma,t)$ such that $\pi_2(C_j)$ contains some point representing a polynomial $P\not\equiv 0$. Thus, $\pi_1(\cup_{j\in J}C_j)\subset\mathbb{C}^{nt}$ is the set of all incorrect questors for $E(\Gamma,t)$.
We can bound the cardinality of $\pi_1(\cup_{j\in J}C_j)\cap\{1,\ldots,u\}^{nt}$ as in [29] but with the bounds of lemma 16. First of all, the Krull dimension of $\pi_2^{-1}(P)$ is $(n-1)t$ and the theorem of dimension of fibers implies that:

$$\dim\ \pi_2^{-1}(P)\geq \dim\ C_j-\dim\ W(\Gamma)$$

Therefore, $\dim\ C_j\leq (n-\frac{2}{3})t$.
Now applying [29, proposition 2.3] we have:

$$\begin{aligned}\sharp\left(\pi_1(\cup C_j)\cap\{1,\ldots,u\}^{nt}\right) &\leq \deg\left((\cup C_j)\cap\{1,\ldots,u\}^{nt}\right)\\ &\leq \deg(\cup C_j)u^{(n-\frac{2}{3})t}\\ &\leq (2^{\ell+1}-2)^{\frac{t}{3}}(2^\ell+1)^t\,u^{(n-\frac{2}{3})t}\\ &\leq u^{nt}\,u^{-\frac{t}{6}}\end{aligned}$$

since $u:=(2^{\ell+1}-2)\,(2^\ell+1)^2$. Thus, there are $u^{nt}-u^{nt}\,u^{-\frac{t}{6}}=u^{nt}(1-u^{-\frac{t}{6}})$ correct test sequences of length t in $\{1,\ldots,u\}^n$, which ends the proof of the statement. □

Let us consider now the algebraic set $W(n,\ell,L)$ defined as the union of all $W(\Gamma)$, where Γ is a non-scalar straight-line program of size bounded by L, depth bounded by ℓ, which computes n-variate polynomials. Then, proposition 18 above allows us to obtain the following bounds on questors for $W(n,\ell,L)$:

Corollary 19 *For $n,\ell,L\in\mathbb{N}$, $L\geq n+1$, let:*

$$u:=(2^{\ell+1}-2)\,(2^\ell+1)^2 \qquad \text{and} \qquad t:=6\,(\ell L)^2.$$

Then the set $\{1,\ldots,u\}^n\subset\mathbb{Z}^n$ contains at least $u^{nt}\,(1-u^{-\frac{t}{6}})$ correct test sequences of length t for $W(n,\ell,L)$. (Therefore it contains at least one correct test sequence of this length.)

Proof. Given n, ℓ, L we can design a non-scalar straight-line program Γ_0 that has L gates at every level, excepting for the input gates (depth 0) and the output gate (depth ℓ) . Then, every non-scalar straight-line program of size L, depth ℓ can be modelled as a subgraph of Γ_0. Then, $W(n, \ell, L)$ is a subvariety of $W(\Gamma_0)$. We conclude observing that the depth of Γ_0 is ℓ and its size is at most ℓL. □

The previous bounds combined with those of proposition 15 will allow us to determine a probabilistic polynomial time algorithm that tests whether a polynomial given by a non-scalar straight-line program evaluates the zero polynomial. First of all, observe that a non-scalar straight-line program of size L, depth ℓ, and parameters of height η can be encoded in binary size $L^{O(1)} \log_2 \eta$. Then, we can design the following algorithm:

- Input: A non-scalar straight-line program Γ.
- Guess a questor set $\mathcal{Q}$ of $\{1, \ldots, u\}^{nt}$, for u and t as in the corollary above.
- For every point $P \in \mathcal{Q}$, eval Γ at P.
- Accept if and only if $\Gamma(P) = 0$, for all P.

This algorithm verifies the following:

Corollary 20 *Let c be a fixed constant. There exists a probabilistic algorithm running in time $L^{O(1)} \log_2 \eta^{O(1)}$ which decides whether a polynomial in $\mathbb{Z}[X_1, \ldots, X_n]$ described by means of a non-scalar straight-line program of size L, depth $\ell \leq c \log_2 L$, and parameters η is null. The error probability of the algorithm is bounded by $\frac{1}{u^t}$, where $u := (2^{\ell+1} - 2)(2^\ell + 1)^2$ and $t := 6(\ell L)^2$. Hence the problem of deciding whether a polynomial is null or not belongs to the bounded probability polynomial time class of complexity BPP* (cf. [1, chapter 6] for notations).

Proof. Observe that the algorithm uses only sequences of t points in $\{1, \ldots, u\}^n$. The probability that one of such guesses is a correct test sequence is at least $(1 - u^{\frac{t}{6}}) \geq \frac{1}{2}$. This simply says that the previous step describes a bounded probability Turing machine.
The encoding of the input is given by the list of gates, every gate being given as a list $(i, j, \underline{\alpha}_{i,j}, \underline{\beta}_{i,j})$, where the lists $\underline{\alpha}_{i,j}, \underline{\beta}_{i,j}$ are the parameters introduced at the gate (i, j). Hence, simply reading the input one decides whether non-scalar straight-line program has depth bounded by $c \log_2 L$.
To evaluate the straight-line program at a point $P \in \mathcal{Q}$ requires $O(L^2)$ multiplications by a parameter and additions, and one multiplication by an integer number for each gate. The height of the parameters is η. The bounds of proposition 15 imply that the precomputed integers can be encoded in binary size $2^{2\ell}(\log_2 L + \log_2 \eta + \log_2 u)$. Now since $\log_2 u = O(\ell)$, we can simulate these

additions and products in sequential time $2^{O(\ell)}(\log_2 L + \log_2 \eta)^{O(1)}$, i.e., in polynomial time in the input size $L^{O(1)} \log_2 \eta$ (since $\ell = O(\log_2 L)$). Hence, the evaluation time is polynomial in the input size. Finally, the whole procedure takes polynomial time and space since the number of iterations of the previous step is bounded by $t := 6\,(\ell L)^2$. □

3.2 Vermeidung von Divisionen

Corollary 19 allows to apply, within our complexity bounds requirements, Strassen's procedure of *Vermeidung von Divisionen* (cf. [54]):

Proposition 21 (Strassen, 1973) *Let Γ be a non-scalar straight-line program of size L, depth ℓ, and parameters of height η that computes $\{f_0, \dots, f_m\} \subseteq \mathbb{Z}[X_1, \dots, X_n]$. Assume that $f_0 \neq 0$ and that $f_0 | f_i$ in $\mathbb{Z}[X_1, \dots, X_n]$ for all i, $1 \leq i \leq m$. Then, there is a non-scalar straight-line program Γ' verifying:*

i) Γ' computes polynomials $\{P_1, \dots, P_m\}$ in $\mathbb{Z}[X_1, \dots, X_n]$ and an integer $\theta \in \mathbb{Z} - \{0\}$ such that

$$P_i := \theta \frac{f_i}{f_0} \qquad (1 \leq i \leq m).$$

ii) Γ' has size of order $O(2^{2\ell}(L + n + 2^\ell + m))$, depth of order $O(\ell)$ and parameters of height $\max\{\eta, u\}$, *where* $u := (2^{\ell+1} - 2)(2^\ell + 1)^2$.

In particular, we can bound the height of θ by:

$$|\,\theta\,| \leq (L \max\{\eta, u\})^{(2^{\ell+1}-2)(2^\ell+1)}\,.$$

Proof. Since the polynomial f_0 can be evaluated by means of Γ of size L and depth ℓ, if we set $u := (2^{\ell+1} - 2)\,(2^\ell + 1)^2$, there exists $(\gamma_1, \dots, \gamma_n) \in \mathbb{Z}^n$ of height bounded by u such that $\rho := f_0(\gamma_1, \dots, \gamma_n) \neq 0$ (cf. corollary 19).
We define the following polynomials

$$F_i(X_1, \dots, X_n) := f_i(X_1 + \gamma_1, \dots, X_n + \gamma_n)$$

(such that $F_0(0, \dots, 0) := f_0(\gamma_1, \dots, \gamma_n) \neq 0$), and $Q := \rho - F_0$.
The polynomials $F_0, \dots, F_m$, $Q \in \mathbb{Z}[X_1, \dots, X_n]$, and $\rho \in \mathbb{Z}$ can be computed by a straight-line program of size $2L + n + 1$ and depth $\ell + 2$, using integer parameters of height at most $\max\{\eta, u\}$.
Finally set $\theta := \rho^{2^\ell + 1}$. Thus:

$$\frac{\theta\, F_i}{F_0} = \frac{\theta\, F_i}{\rho - Q} = \frac{\rho^{2^\ell}\, F_i}{1 - \frac{Q}{\rho}} = \left(\sum_{k=0}^{2^\ell} \rho^{2^\ell - k} Q^k + \sum_{k=2^\ell+1}^{\infty} \rho^{k-2^\ell} Q^k \right) F_i.$$

Since $\frac{\theta F_i}{F_0}$ is a polynomial in $\mathbb{Z}[X_1, \ldots, X_n]$ of degree bounded by 2^ℓ (for f_i is evaluable by Γ of depth ℓ), and Q^i is of order at least i in 0, we can conclude that $\frac{\theta F_i}{F_0}$ depends only on the homogeneous components of degree $\leq 2^\ell$ of $(\sum_{k=0}^{2^\ell} \rho^{2^\ell - k} Q^k)\, F_i$ (which itself belongs to $\mathbb{Z}[X_1, \ldots, X_n]$).
As in the proof of proposition 9 we compute in additional size $2^{\ell+1} + 2$, depth $\ell + 1$, and the same set of parameters, the powers $\rho, \ldots, \rho^{2^\ell}$ and $\{Q, \ldots, Q^{2^\ell}\}$. Afterwards we compute the sum $\sum_{k=0}^{2^\ell} \rho^{2^\ell - k} Q^k$ in additional size 2^ℓ and depth 1, and we finish multiplying the result by F_i. Therefore we can compute all the polynomials $(\sum_{k=0}^{2^\ell} \rho^{2^\ell - k} Q^k)\, F_i$ in total size $O(L + n + m + 2^{\ell+1})$ and depth $O(\ell)$, with same bounds on integer parameters. Finally, applying lemma 13, we compute and sum all the homogeneous components of degree bounded by 2^ℓ of this polynomial: This can be done in size $O(2^{2\ell}\,(L + n + m + 2^\ell))$ and depth $O(\ell)$.
We conclude the proof computing the polynomials

$$P_i := \theta \frac{f_i}{f_0} = \theta \frac{F_i(X_1 - \gamma_1, \ldots, X_n - \gamma_n)}{F_0(X_1 - \gamma_1, \ldots, X_n - \gamma_n)}$$

in additional size n and depth 1.
The bound for $|\theta|$ is only a consequence of the bound for $\rho := F_0(0)$ of lemma 15. □

4 Isolated points

In [23] the authors developped a procedure to determine the isolated points of an affine algebraic set given by a system of equations. This method is central for our purposes for two main reasons: It is primordial for the determination of a basis of a reduced complete intersection in terms of a primitive element, and it also provides sufficient polynomial conditions of good degree to check the triviality (and dimension) of an ideal. Moreover the analysis of the procedure under the scope of non-scalar straight-line programs allows to obtain upper bounds for the heights of the isolated points of a variety.

For the whole section we set $R := \mathbb{Z}[Y_1, \ldots, Y_r]$ a polynomial ring, and $\mathbb{K}$ denotes an algebraic closure of the fraction field K of R.

We want to describe the isolated points of an algebraic set $V \subseteq \mathbb{K}^n$ given by polynomial equations $f_1 = 0, \ldots, f_s = 0$, where $f_i \in R[X_1, \ldots, X_n]$ are polynomials of degree at most d.
The analysis, based on [23] we develop in this section will provide a primitive element, i.e., a linear form γ that distinguishes the isolated points of V (i.e., $\gamma(P) \neq \gamma(Q)$ for every two different isolated points $P, Q \in V$). In the case $s = n$ and the ideal $(f_1, \ldots, f_n)$ is radical of dimension r in $\mathbb{Q}[Z_1, \ldots, Z_r, X_1, \ldots, X_n]$,

this linear form γ and its powers will generate the K-vector space $K[X_1, \ldots, X_n]/(f_1, \ldots, f_n)$.
More precisely, the main result of this section is the following:

Theorem 22 *Let $(f_1, \ldots, f_n) \subset R[X_1, \ldots, X_n]$ be a reduced complete intersection ideal generated by polynomials of degree (in $\underline{X}$) bounded by $d \geq n$. Let V be the variety in $\mathbb{K}^n$ defined by these polynomials.*
Then, there exists a linear form $\gamma := \gamma_1 X_1 + \cdots + \gamma_n X_n \in \mathbb{Z}[X_1, \ldots, X_n]$, an element $\theta \in R$, univariate polynomials v_i $(1 \leq i \leq n)$ and p in $R[Z]$ such that the following holds:

i) The linear form γ separates the zeros of V, and $p(\gamma) \in (f_1, \ldots, f_n)$.
The minimal integral equation of γ over R is $\frac{1}{\alpha} p(Z)$ (where α is the leading coefficient of p).

ii) $\theta X_i - v_i(\gamma) \in (f_1, \ldots, f_n) \qquad (1 \leq i \leq n)$.

iii) $\deg p \leq d^n, \quad \deg v_i \leq d^n \qquad (1 \leq i \leq n)$.

iv) $ht(\gamma_i) = d^{O(n)} \qquad (1 \leq i \leq n)$.

v) Both θ and the coefficients of the v_i's and p are integer polynomials in the coefficients of $\{f_1, \ldots, f_n\}$ and they can be evaluated from them by a non-scalar straight-line program within size of order $d^{O(n)}$, depth of order $O(n \log_2 d)$ and parameters in $\mathbb{Z}$ of height bounded by $d^{O(n)}$.

The technical tools for the proof of this result are mainly divided in three steps. First, given a linear form γ, we need to determine an equation p which annihilates it on the isolated zeros of the variety. This is done by reduction to the case where the isolated points are locally complete intersection, which is itself performed by an homotopic reduction to a projective zero-dimensional case. Next, to determine a primitive element, one generically chooses a candidate for the linear form γ, and then selects concrete parameters for the generic equations, by analyzing all posible two variables cases. Once the primitive element has been chosen, we parametrize the variables as in the shape lemma. Finally the minimal equation of the primitive element allows to determine completely the isolated points and the basis of the K-vector space $K[X_1, \ldots, X_n]/(f_1, \ldots, f_n)$.

4.1 The case of projective dimension zero

In this subsection we analyze dependence equations of linear forms over zero-dimensional homogeneous ideals without zeros at infinity.

Proposition 23 *Let $g_1, \ldots, g_n \subset R[X_0, \ldots, X_n]$ be homogeneous polynomials of degree $d+1$ which generate an (homogeneous) ideal $\mathcal{J} \subset k[X_0, \ldots, X_n]$*

without zeros at infinity, i.e., at the projective hyperplane $\{X_0 = 0\}$ *in the* n*-dimensional projective space over* $\mathbb{K}$.
Then, for every linear form $\gamma := \gamma_0 X_0 + \cdots + \gamma_n X_n$ $(\neq X_0) \in R[X_0, \ldots, X_n]$, *there exists a polynomial* $\Delta(X_0, Z) \in R[X_0, Z]$, *monic in the variable* Z *(up to a constant in* R*), such that the following properties hold:*

i) $\Delta(X_0, \gamma) \in \mathcal{J}$.

ii) $\deg \Delta(X_0, Z) \leq (d+1)^n$.

iii) The coefficients of $\Delta(X_0, Z)$ *are integer polynomials in the coefficients of* $\{g_1, \ldots, g_n\}$ *of degree of order* $O(n^n\, d^{3n})$, *and in the coefficients of* γ *of degree of order* $O(d^n)$.

iv) The coefficients of $\Delta(X_0, Z)$ *can be computed from the coefficients of* g_i*'s and* γ *by a non-scalar straight-line program of size of order* $(nd)^{O(n)}$, *depth at most* $O(n\,(\log_2 n + \log_2 d))$ *and parameters in* $\{-1, 0, 1\} \subseteq \mathbb{Z}$.

v) The polynomial $\Delta(X_0, Z)$ *can be computed by a non-scalar straight-line program of the same characteristics.*

Proof. In [23] the authors apply the results of [40] or [9] in order to be in a context where Cayley-Hamilton theorem holds. Applying for example [40] we observe that the graded $\mathbb{K}$-algebra

$$A := \mathbb{K}[X_0, \ldots, X_n]/\mathcal{J} = A_0 + A_1 + \cdots + A_s + \cdots$$

satisfies the following two facts:

- Its Hilbert regularity is bounded by $N := n(d+1) + 1 - n = nd + 1$. This says that for all $s \geq N$ the $\mathbb{K}$-vector spaces A_s are isomorphic, of rank D bounded by their degree, which are themselves bounded by $(d+1)^n$.

- The fact that the ideal $\mathcal{J}$ doesn't define any point at the hyperplane $X_0 = 0$ implies that for all $s \geq N$ the homothesy induced by X_0

$$\eta_{X_0} : A_s \longrightarrow A_{s+1}$$

  is an isomorphism.

Thus, the linear mapping

$$\varphi := (\eta_{X_0})^{-1} \eta_\gamma : A_N \longrightarrow A_N$$

is an endomorphism of the $\mathbb{K}$-vector space A_N, and the Cayley-Hamilton theorem implies that its characteristic polynomial $P(T)$ is a non-zero polynomial which annihilates him, i.e., $P(\varphi) = \eta_{P(\frac{\gamma}{X_0})}$ vanishes, and so does $\eta_{X_0}^D P(\varphi)$.

As X_0 is not a zero-divisor in A, the non-zero polynomial $Q(X_0, Z) := X_0^D P(\frac{Z}{X_0})$ verifies that $Q(X_0, \gamma)$ vanishes in A (and $Q(X_0, Z)$ is monic in Z up to a constant of K).
Essentially we only need now to compute the polynomial

$$Q(X_0, Z) = \det(X_0 M_\varphi - Z\, Id)$$

by means of the matrix M_φ in a basis $\mathcal{B}$ of A_N. In fact, we will compute a slightly different polynomial $\Delta(X_0, Z)$ with the same property, but in $R[X_0, Z]$, i.e., without denominators.
For this purpose, we will compute monomial bases $\mathcal{B}$ and $\mathcal{B}'$ of A_N and A_{N+1}, respectively, and the matrices M_{X_0} and M_γ of the homothesies η_{X_0} and η_γ in these bases.
In order to get the bases $\mathcal{B}$ and $\mathcal{B}'$ let us denote by $\mathcal{F}_s := \{m_1, \dots, m_{r_s}\}$ the set of all mononomials of degree s in the variables $X_0, \dots, X_n$ (where $r_s := \binom{s+n}{n}$). We construct first a matrix $P \in \mathcal{M}_{n\, r_{(N-d-1)} \times r_N}(R)$ of relations of the monomials in $\mathcal{F}_N$ modulo $\mathcal{J}_N$ (the relations are obtained multiplying the monomials of degree $N - (d+1)$ by the polynomials g_i's, thus the entries of the matrix correspond to coefficients of the g_i's).
If we triangularize P in K, we obtain a matrix $\overline{P}$ which allows to extract a subfamily $\mathcal{B} := \{e_1, \dots, e_D\} \subset \mathcal{F}_N$ which is a monomial basis of A_N (given by the "parameter" columns in the triangularization). In fact, we observe that it is enough to compute a square submatrix of maximal rank of P: The basis will be determined by the remainder columns.
Similarly, we compute a square submatrix of maximal rank $\overline{Q}$ of the matrix Q of relations in A_{N+1}. In this way we extract a monomial basis $\mathcal{B}' := \{e'_1, \dots, e'_D\} \subset \mathcal{F}_{N+1}$ of A_{N+1}, and its remainder set $\{v'_1, \dots, v'_t\} := \mathcal{F}_{N+1} \setminus \mathcal{B}'$. Moreover, we obtain an element $\lambda := \det \overline{Q} \in R$ and a matrix $B \in \mathcal{M}_{t \times D}(R)$ (given by the remainder columns of $Adj(\overline{Q})\, Q$) such that

$$\lambda \begin{pmatrix} v'_1 \\ \vdots \\ v'_t \end{pmatrix} = B \begin{pmatrix} e'_1 \\ \vdots \\ e'_D \end{pmatrix}.$$

The bases $\mathcal{B}$ and $\mathcal{B}'$ and the matrix B are obtained by computing first the matrices of relations, and then a square submatrix of maximal rank: We choose first a maximal number of independent rows computing their rank, in nr_{N-d} steps and applications of the Berkowitz algorithm (proposition 9), using for instance that in our case rk(A)=rk($A \cdot {}^t A$) and computing its characteristic polynomial; afterwards we proceed with the columns.
We compute the element λ and the matrices $M_{\lambda X_0}$, $M_{\lambda\gamma}$ also applying proposition 9.
Now we are able to describe the matrix $M_{\lambda X_0} = \lambda M_{X_0}$: For every $e_i \in \mathcal{B}$, $\lambda X_0 e_k \in A_{N+1}$ is either a monomial of the form $\lambda e'_j$ or a monomial $\lambda v'_k$. In the

first case the corresponding column of λM_{X_0} is a vector with all coordinates equal to zero excepting one where λ occurs. In the second case, this column is simply a row of the matrix B. (Note that $M_{\lambda X_0} \in \mathcal{M}_D(R)$.)
Analogously, we obtain the matrix $M_{\lambda\gamma} = \sum_{i=0}^{n} \gamma_i \lambda M_{X_i} \in \mathcal{M}_D(R)$.
Now, instead of computing

$$Q(X_0, Z) = \det(X_0 M_\varphi - Z\, Id) = \det(X_0 (M_{X_0})^{-1} M_\gamma - Z\, Id),$$

it would be enough to compute the polynomial $\lambda^D \det(M_{\lambda X_0}) Q(X_0, Z) = \det(X_0 M_{\lambda\gamma} - Z M_{\lambda X_0}) \in R[X_0, Z]$ which verifies all what we need ($\Delta \neq 0$ since M_{X_0} is invertible, and $\Delta(X_0, \gamma) \in \mathcal{J}$ holds). However, as we are interested in computing the coefficients of this polynomial (and not only the whole polynomial) we need to view it as a characteristic polynomial. That's why we will compute a multiple of it:

$$\Delta(X_0, Z) = \det(M_{\lambda X_0})^{D-1} \det(X_0 M_{\lambda\gamma} - Z M_{\lambda X_0}) \in R[X_0, Z]$$

which has the same properties.
We shall in fact compute the characteristic polynomial $\mathcal{X}$ of $Adj(M_{\lambda X_0}) M_{\lambda\gamma}$, since if $\mathcal{X} := \sum p_j Y^j$, then $\Delta(X_0, Z) = \sum p_j \det(M_{\lambda X_0})^j Z^j {X_0}^{D-j}$. (Note that: $\Delta(X_0, Z) = \det(X_0 Adj(M_{\lambda X_0}) M_{\lambda\gamma} - \det(M_{\lambda X_0}) Z\, Id)$.)
The bound for $\deg \Delta$ is clear from the size of the matrices involved. Also λ and $\det(M_{\lambda X_0})$ are integer polynomials in the entries of Q and B, i.e., the coefficients of the g_i's, of degree $\leq r_N \leq e(N+n)^n$, and the same occurs with the entries of B. So that the degree of the coefficients of Δ in the coefficients of g_i are of order $O(D\, r_N)$. (For the bound in the coefficients of γ observe that these appear only linearly in $M_{\lambda\gamma}$.)
The bounds for the straight-line program are clear since we only apply a controlled number of repetitions of proposition 9. □

4.2 As many equations as unknowns

Proposition 23 above allows us to consider the case were isolated points are locally complete intersections, i.e., as many equations as unknowns: The method relies on reducing the given polynomials $f_1, \ldots, f_n$ to the zero-dimensional projective case by homothopy techniques. These techniques are classical in numerical analysis (cf. [19]) and were also used in computer algebra in [13], [25], or [23].

Lemma 24 *Let $f_1, \ldots, f_n \subset R[X_1, \ldots, X_n]$ be polynomials of degree $d \geq n$, and let $\gamma \in R[X_1, \ldots, X_n]$ be a (non-zero) linear form. Then, there exists a polynomial $p \in R[Z]$, monic in the variable Z (up to a constant in R), such that the following properties hold:*

i) $p(\gamma)$ vanishes on the isolated points in $\mathbb{K}^n$ of $\{f_1 = 0, \ldots, f_n = 0\}$.

ii) $\deg p(Z) \leq (d+1)^n$.

iii) *The coefficients of* $p(Z)$ *are integer polynomials in the coefficients of* $\{f_1, \ldots, f_n\}$ *of degree of order* $d^{O(n)}$, *and in the coefficients of* γ *of order* $O(d^n)$.

iv) *The coefficients of* $p(Z)$ *are evaluated from the coefficients of the* f_i*'s and* γ *by a non-scalar straight-line program* Γ *of size* $d^{O(n)}$, *depth* $O(n \log_2 d)$, *and parameters in* $\{-1, 0, 1\}$.

Proof. We follow here the arguments of [23, section 3]. Let $\{F_1, \ldots, F_n\}$ be the homogenizations of the polynomials $\{f_1, \ldots, f_n\}$ with respect to the variable X_0. We introduce a new variable ε that we further use either as parameter or as variable, and we define the following polynomials in $R[\varepsilon, X_0, X_1, \ldots, X_n]$:

$$g_i := X_0\, F_i + \varepsilon\, X_i^{1+\deg f_i} \qquad (1 \leq i \leq n).$$

These polynomials are homogeneous of degree $d+1$ as polynomials in $\mathbb{K}(\varepsilon)[X_0, \ldots, X_n]$. Let $\mathbb{K}_\varepsilon$ be the algebraic closure of $\mathbb{K}(\varepsilon)$. In these conditions, it is well-known that the ideal $(g_1, \ldots, g_n) \subset \mathbb{K}_\varepsilon[X_0, \ldots, X_n]$ is a zero-dimensional ideal without zeros at infinity in $\mathbb{P}_n(\mathbb{K}_\varepsilon)$ (see for instance [23, section 3.3] for a geometrical proof).

Applying lemma 23 above for the (homogeneous) linear form γ we can compute the coefficients $(p_0, \ldots, p_D)$ of the polynomial

$$\Delta(X_0, Z) = \sum_{j=1}^{D} p_j(\varepsilon) Z^j X_0^{D-j} \quad \in \quad R[\varepsilon][X_0, Z]$$

by means of an adequate non-scalar straight-line program (as the straight-line program contains no divisions, it is clear that $p_j(\varepsilon)$ are polynomials in ε). Reducing $\Delta(X_0, Z)$ dividing it by the greatest common power of ε of its content, we define

$$\Delta'(\varepsilon, X_0, Z) := \sum_{j=1}^{D} p'_j(\varepsilon) Z^j X_0^{D-j}.$$

In [23, proposition 3.3.4] it is shown that the polynomial

$$p(Z) := \Delta_1(0, 1, Z) := \sum_{j=1}^{D} p'_j(0) Z^j$$

is the desired polynomial, and assertion i) is proved.

Assertions ii) and iii) follow directly from the construction of the polynomial p. From a non-scalar straight-line program that evaluates a polynomial $q := a_k \varepsilon^k + \cdots + a_n \varepsilon^n \quad (n \geq k, a_k \neq 0)$ one can easily describe a (division-free) non-scalar straight-line program that evaluates the polynomial $q' := a_k + \cdots + a_n\varepsilon^{n-k}$, within similar bounds for height, depth, and height of the parameters. This shows assertion iv) and concludes the proof. □

4.3 Separating points of two univariate polynomials

We need to introduce here some notation. Let $T_1, \ldots, T_n$ be new variables. For each i, $1 \leq i \leq n$, we define the ring

$$R_i := R[T_1, \ldots, T_{i-1}, T_{i+1}, \ldots, T_n].$$

We denote by K_i its quotient field and by $\mathbb{K}_i$ an algebraic closure of it. We also denote by K_T the quotient field of the ring $R[T_1, \ldots, T_n]$ and by $\mathbb{K}_T$ an algebraic closure of it.

Lemma 25 *Let $p \in R_i[Z]$ be a non-constant polynomial of degree (in Z) bounded by D, whose coefficients in R_i are of degree bounded by δ, and let $q \in R[Z]$ be a non-zero polynomial of degree bounded also by D. For every $t := (t_1, \ldots, t_n) \in \mathbb{K}^n$ define the algebraic set*

$$V_t := \{(x, y) \in \mathbb{K}^2 : q(x) = 0, p_t(y) = 0\}$$

(where p_t denotes to specialize in the coefficients of p the variables T by t). Then there exists a non-zero polynomial $F(T_1, \ldots, T_n) \in R[T_1, \ldots, T_n]$ of degree bounded by $\delta\, D^{O(1)}$ such that $F(t_1, \ldots, t_n) \neq 0$ implies that the linear form $t_i X + Y$ separates the points of V_t.
(If the coefficients of p and q in the polynomial ring R are also of degree bounded by δ, the degrees of the coefficients of F in R are of the same order.)

Proof. We are looking for a polynomial $F \in R[T_1, \ldots, T_n]$ such that for all $t \in \mathbb{K}^n$ the conditions $F(t) \neq 0$, $q(x_1) = q(x_2) = 0$, $p_t(y_1) = p_t(y_2) = 0$ imply that $t_i x_1 + y_1 \neq t_i x_2 + y_2$.
For this purpose we introduce the new variables X_1, X_2, Y_1, Y_2 and the ideals $\mathcal{I}_1 := (q(X_1), q(X_2))$, $\mathcal{I}_2 := (p(Y_1), p(Y_2))$ in $K_T[X_1, X_2]$ and $K_T[Y_1, Y_2]$, respectively. Let $\alpha \in R$ and $\beta \in R_i \subseteq R_T$ be the respective leading coefficients of q and p. We now consider the homothesies:

$$\eta_{\alpha(X_1-X_2)} : K_T[X_1, X_2]/\mathcal{I}_1 \longrightarrow K_T[X_1, X_2]/\mathcal{I}_1$$

$$\eta_{\beta(Y_1-Y_2)} : K_T[Y_1, Y_2]/\mathcal{I}_2 \longrightarrow K_T[Y_1, Y_2]/\mathcal{I}_2$$

The matrices of the homothesies $\eta_{\alpha(X_1-X_2)}$ and $\eta_{\beta(Y_1-Y_2)}$ in the natural monomial bases of both residue rings are of order M bounded by D^2 and their coordinates are given respectively by coefficients of q and p. We compute the characteristic polynomials of both homothesies:

$$H(Z_1) := \mathcal{X}_{\alpha(X_1-X_2)} = Z_1^M + a_{M-1}Z_1^{M-1} + \cdots + a_m Z_1^m \in R[Z_1] \quad (a_m \neq 0)$$
$$G(Z_2) := \mathcal{X}_{\beta(Y_1-Y_2)} = Z_2^M + b_{M-1}Z_2^{M-1} + \cdots + b_{m'} Z_2^{m'} \in R_i[Z_2] \quad (b_{m'} \neq 0).$$

(Observe that the coefficients of G are polynomials in T of degree bounded by δD^2, and that the same bounds arise for the coefficients of H and G if R is a polynomial ring and the coefficients in R of p and q are also of degree bounded by δ.)
The case $m = M$, i.e., the homothesy $\eta_{\alpha(X_1-X_2)}$ is nilpotent, means that $q(x_1) = q(x_2) \Longrightarrow x_1 = x_2$, and we can obviously take $F := 1$ (since $q(x_1) = q(x_2) = 0, p_t(y_1) = p_t(y_2) = 0$ and $t_i x_1 + y_1 = t_i x_2 + y_2$ would imply $(x_1, y_1) = (x_2, y_2)$). Analogously, in the case $m' = M$, the polynomial $F = T_i$ is the obvious solution.
Thus, we shall assume $m, m' < M \leq D^2$ and we define:

$$H'(Z_1) := \frac{H(Z_1)}{Z_1^m} \quad , \quad G'(Z_2) := \frac{G(Z_2)}{Z_2^{m'}}$$

and the non-trivial ideal $\mathcal{J} := (H', G')$ they define in $K_T[Z_1, Z_2]$.
The matrices of the homothesies η_{Z_1} and η_{Z_2} in the natural monomial basis of the residue ring $K_T[Z_1, Z_2]/\mathcal{J}$ are of order $\leq D^4$ and their coefficients are easy to determine: For instance the matrix of η_{Z_1} is constituted of $M - m'$ diagonal blocks, each of one being the (invertible) companion matrix of H'. The matrix of η_{Z_2} is the same modulo a change of order in the basis.
Then, they belong to R_i and have degree at most δD^2. (If we are interested in the degrees of these coefficients as polynomials in R, the bound which appears is $2\delta D^2$.)
Finally we consider the matrix $M_{T_i Z_1 + Z_2}$ of the homothesy $\eta_{(T_i Z_1 + Z_2)}$ in the same basis: its coordinates are polynomials in $R[T_1, \ldots, T_n]$ of degree bounded by $\delta D^2 + 1$ ($2\delta D^2 + 1$ taking into account the degrees in R), and we define:

$$F(T_1, \ldots, T_n) := \beta T_i \, \det(M_{T_i Z_1 + Z_2})$$

which is a non-zero polynomial since the coefficient of T_i^1 corresponds essentially to the determinant of the matrix of η_{Z_2}, which is invertible.
We claim that this polynomial verifies the required conditions: The condition $\det(M_{t_i Z_1 + Z_2}) \neq 0$ implies that the linear form $t_i Z_1 + Z_2$ is not a zero divisor in the zero-dimensional residue ring $\mathbb{K}_t[Z_1, Z_2]/\mathcal{J}_t$, and thus doesn't belong to any prime (= maximal) ideal associated to $\mathcal{J}_t$.
On the other hand, if $(x_1, y_1) \neq (x_2, y_2) \in V_t$, then $q(x_1) = q(x_2) = 0$, $p_t(y_1) = p_t(y_2) = 0$, and therefore, as $H(\eta_{\alpha(X_1-X_2)}) = 0$, $G(\eta_{\beta(t)(Y_1-Y_2)}) = 0$ in the residue rings and $\alpha \neq 0$, $\beta(t) \neq 0$, we can conclude that $H(x_1 - x_2) = G(y_1 - y_2) = 0$ (with $x_1 - x_2 \neq 0$ or $y_1 - y_2 \neq 0$). If $x_1 = x_2$ or $y_1 = y_2$, as $t_i \neq 0$ it is

clear that $t_ix_1+y_1 \neq t_ix_2+y_2$. Otherwise, $H'(x_1-x_2)=0$ and $G'(y_1-y_2)=0$, which finally implies that $t_i(x_1-x_2)+(y_1-y_2)\neq 0$ (since (x_1-x_2, y_1-y_2) determines a maximal prime of $\mathcal{J}_t$). □

4.4 The case of two variables

We are going to present here an effective "straight-line program" version of the well-known "shape lemma" for a particular case in two variables we will use in the sequel. We recall here its proof only because we need to take care of the growing of the coefficients when we perform this change of basis.

Lemma 26 *Let $p(Y), q(X)$ be two univariate square-free polynomials with coefficients in R, and denote by $\mathcal{J} := (p(Y), q(X)) \subset K[X,Y]$ the ideal they generate. Set $D := \deg p \cdot \deg q$. Let $\gamma := Y + \alpha X$ ($\alpha \in \mathbb{Z}$ fixed) be a linear form which separates the zeros of $\mathcal{J}$ in $\mathbb{K}^2$.*
Then there exist two univariate polynomials $u, v \in R[Z]$ and a non-zero element $\theta \in R$ such that:

i) $\mathcal{J} = (u(\gamma), \theta X - v(\gamma)) \subset K[X,\gamma]$.

ii) $\deg(u) = D$.

iii) $\deg(v) \leq D-1$.

iv) The coefficients of u are polynomials in the coefficients of p and q of degree bounded by $4D$; θ and the coefficients of v are polynomials in the coefficients of q and p of degrees bounded by $2D^2$.

v) The coefficients of u are evaluable by a non-scalar straight-line program with inputs the coefficients of p and q of size $O(D^{12})$ and non-scalar depth $6\log_2 D + 12$, using $\{-1, 0, 1, \alpha\}$ as set of parameters.

vi) θ and the coefficients of v are evaluable by a non-scalar straight-line program with the same inputs of size $O(D^{10})$, depth $6\log_2 D + 11$ and the same set of parameters.

Proof. The ideal $\mathcal{J}$ is a radical zero-dimensional ideal which defines $D := \deg p \cdot \deg q$ isolated points. Also the K-vector space $K[X,Y]/\mathcal{J}$ admits the basis

$$\mathcal{B} := \{X^iY^j : 0 \leq i < \deg q, 0 \leq j < \deg p\}.$$

Since $\mathcal{J}$ is a radical ideal and the linear form γ separates all the zeros of $\mathcal{J}$, it is easy to verify that the minimal polynomial of γ in $K[X,Y]/\mathcal{J}$ over K coincides with the characteristic polynomial $\mathcal{X}_\gamma$ of the homothesy η_γ induced by γ, so that γ is a primitive element of the K-extension $K[X,Y]/\mathcal{J}$ (which defines the basis $1, \gamma, \ldots, \gamma^{D-1}$).

Now, in order to avoid denominators in K, we shall compute a slightly different characteristic polynomial u: let α and β denote the leading coefficients of q and p respectively, and set $\rho := \alpha\,\beta$. We define:

$$u(Z) := \rho^D \mathcal{X}_\gamma(Z) = \det(M_{\rho\gamma} - \rho Z Id) = \det(\alpha M_{\beta Y} + \alpha\beta M_{\alpha X} - \alpha\beta Z Id)$$

(where M_* denotes the matrix of the homothesy induced by $*$ in the basis $\mathcal{B}$). It is immediate that $u \in R[Z]$ is a polynomial of degree D such that $u(\gamma) \in \mathcal{J}$ (since $u(\eta_\gamma) = \eta_{u(\gamma)} = 0$, and we can apply it to the first element of the basis, 1). Moreover, the coefficients of u are polynomials in the coefficients of q and p of degree $4D$.

Finally, in order to compute $\theta \in R$ and $v \in R[\gamma]$ such that $\mathcal{J} = (u(\gamma), \theta X - v(\gamma)) \in k[X, \gamma]$, we observe that there exists an invertible matrix $P \in \mathcal{M}_D(R)$ verifying:

$$\begin{pmatrix} 1 \\ \rho\gamma \\ \vdots \\ (\rho\gamma)^{D-1} \end{pmatrix} = P \begin{pmatrix} 1 \\ X \\ \vdots \\ X^{\deg q - 1} Y^{\deg p - 1} \end{pmatrix}.$$

Therefore, $(\det P)X = v(\gamma) \pmod{\mathcal{J}}$, where v is a polynomial of degree bounded by $D-1$ which comes from a row of the matrix $Adj(P)$.

We set $\theta := \det P$. Observing that the matrix P can be obtained as the first columns of the matrices $(M_{\rho\gamma})^k$, $0 \le k < D$, it is clear that θ and the coefficients of v are polynomials in the coefficients of q and p of degree $2D^2$.

Now, since the ideal $(u(\gamma), \theta X - v(\gamma))$ is a radical ideal contained in the radical zero-dimensional ideal $\mathcal{I} \subset K[X, \gamma]$ which defines exactly the same points, we conclude that modulo the linear change of variables, $\mathcal{J} = (u(\gamma), \theta X - v(\gamma))$.

Now we are going to produce the straight-line program required in items 5 and 6.

We will compute the coefficients of u by means of the coefficients of the characteristic polynomial of the matrix $M_{\rho\gamma}$ in the basis $\mathcal{B}$ (if $\mathcal{X}_{\rho\gamma}(Z) = \sum a_k Z^k$, then $u(Z) = \sum(\rho^k a_k) Z^k$). First we compute the coordinates of the matrix $M_{\rho\gamma}$ by means of a straight-line program of length $O(D^2)$ and depth 2, using $\alpha \in \mathbb{Z}$ as a parameter. Then the Berkowitz algorithm gives a straight-line program of length $O(D^5)$ and depth $O(\log_2 D)$ to compute the coefficients of u, with the same parameter.

In order to produce a straight-line program which computes θ and the coefficients of v, we need to compute the matrix P, i.e., the powers of the matrix $M_{\rho\gamma}$. This can be done in length $O(D^4)$ and depth $O(\log_2 D)$. Afterwards we compute $Adj(P)$ (given by minors of the matrix P) and $\det P$ in length $O(D^6)$ and depth $O(\log_2 D)$. $\square$

4.5 Describing the isolated zeros of n n-variate polynomials

Now we are in conditions to show the following result, which describes the isolated zeros of a variety in the case where there are as many equations as unknowns.

Proposition 27 *Let $f_1, \dots, f_n \in R[X_1, \dots, X_n]$ be polynomials of degree (in $\underline{X}$) bounded by $d \geq n$. Let $V \in \mathbb{K}^n$ denote the variety defined by these polynomials. Then there exists a linear form $\gamma := \gamma_1 X_1 + \cdots + \gamma_n X_n \in \mathbb{Z}[X_1, \dots, X_n]$, an element $\theta \in R$ and univariate polynomials $v_i \in R[Z]$ $(1 \leq i \leq n)$ such that the following holds:*

i) The linear form γ separates the isolated zeros of V.

ii) The coordinates $x := (x_1, \dots, x_n)$ of the isolated zeros of V verify the equations $\theta x_i = v_i(\gamma(x))$ $\quad (1 \leq i \leq n)$.

iii) $\deg v_i = d^{O(n)} \quad (1 \leq i \leq n)$.

iv) $ht(\gamma_i) = d^{O(n)} \quad (1 \leq i \leq n)$.

v) Both θ and the coefficients of the v_i's are integer polynomials in the coefficients of $\{f_1, \dots, f_n\}$ of degree bounded by $d^{O(n)}$ and they can be evaluated from them by a non-scalar straight-line program within size of order $d^{O(n)}$, depth of order $O(n \log_2 d)$ and parameters in $\mathbb{Z}$ of height bounded by $d^{O(n)}$.

Proof. We consider the generic elements $Y := T_1 X_1 + \cdots + T_n X_n$ and $Y_i := Y - T_i X_i$ $(1 \leq i \leq n)$ in the ring $R[T_1, \dots, T_n][X_1, \dots, X_n]$.
Applying lemma 24 above to the linear forms Y_i, X_i $(1 \leq i \leq n)$ we can find non-constant polynomials $p_i \in R_i[Z]$, $q_i \in R[Z]$ $(1 \leq i \leq n)$ of degree $O(d^n)$ such that $p_i(Y_i)$, $q_i(X_i)$ vanish on the isolated zeros of the system $f_1 = 0, \dots, f_n = 0$ in $\mathbb{K}^n_T$. Also the coefficients of p_i are integer polynomials in $T_1, \dots, T_n$ of degree $O(d^n)$, and in the coefficients of $f_1, \dots, f_n$ of degree $d^{O(n)}$.
Fix i, $1 \leq i \leq n$: By lemma 25 applied to $p_i(Y_i), q_i(X_i)$ there exists a non-zero polynomial $F_i \in R[T_1, \dots, T_n]$ of degree bounded by $d^{O(n)}$ such that $F_i(t_1, \dots, t_n) \neq 0$ implies that the linear form $Y_i + t_i X_i := t_1 X_1 + \cdots + t_n X_n$ separates the solutions of $\{p_i(t, y_i) = 0,\ q_i(x_i) = 0\}$.
The non-zero polynomial $F := \prod_{i=1}^n F_i \in R[T_1, \dots, T_n]$ is of degree bounded by $d^{O(n)}$. An adequate specialization of the variables of the polynomial ring R allows to obtain a non-zero polynomial $F^* \in \mathbb{Z}[T_1, \dots, T_n]$, of degree of the same order, which thus can be evaluated by means of a straight-line program over $\mathbb{Z}$ of depth $O(n \log_2 d)$. Corollary 19 implies that there exists $(\gamma_1, \dots, \gamma_n) \in \mathbb{Z}^n$ of height of order $d^{O(n)}$ such that $F^*(\gamma_1, \dots, \gamma_n) \neq 0$, and therefore $F(\gamma_1, \dots, \gamma_n) \neq 0$ as wanted.
We conclude that the linear form $\gamma := \gamma_1 X_1 + \cdots + \gamma_n X_n$ separates the solutions in $\mathbb{K}^2$ of the n systems:

$$p_i(\gamma_1, \dots, \gamma_n, Y_i) = 0 \quad , \quad q_i(X_i) = 0 \qquad (1 \leq i \leq n)$$

and in particular separates the isolated zeros of the system $\{f_1 = 0, \ldots, f_n = 0\}$ in $\mathbb{K}^n$.
Recall that the polynomials $p_i(Y_i), q_i(X_i)$ are polynomials in the coefficients of $f_1, \ldots, f_n$ and in $T_1, \ldots, T_n$ of degree $d^{O(n)}$, which can be computed from them by a non-scalar straight-line program of size $d^{O(n)}$, depth $O(n \log_2 d)$, and parameters in $\{-1, 0, 1, \gamma_1, \ldots, \gamma_n\}$.
To reduce them as in lemma 12 in order to obtain the square-free polynomials p_i^*, q_i^* doesn't modify essentially size, depth and parameters.
Finally, we apply lemma 26 to $\mathcal{J} := (p_i^*(\gamma_1, \ldots, \gamma_n, Y_i), q_i^*(X_i)) \in \mathbb{K}^2$ and the linear form $Y_i + \gamma_i X_i$ to compute $\theta_i \in R$ and the coefficients of $v_i' \in R[Z]$ such that $\theta_i X_i - v_i'(\gamma) \in \mathcal{J}$, i.e., vanish on the isolated zeros of $f_1, \ldots, f_n$.
We conclude the proof setting $\theta := \prod_{i=1}^n \theta_i$ and $v_i := (\prod_{j \neq i} \theta_j) v_i'$. □

4.6 The primitive element for a reduced complete intersection

The previous proposition is the key to obtain the main result of this section, theorem 22, that we state again for the commodity of the reader:

Theorem 22 *Let $(f_1, \ldots, f_n) \subset R[X_1, \ldots, X_n]$ be a reduced complete intersection ideal generated by polynomials of degree (in $\underline{X}$) bounded by $d \geq n$. Let V be the variety in $\mathbb{K}^n$ defined by these polynomials.*
Then there exists a linear form $\gamma := \gamma_1 X_1 + \cdots + \gamma_n X_n \in \mathbb{Z}[X_1, \ldots, X_n]$, an element $\theta \in R$, univariate polynomials v_i $(1 \leq i \leq n)$ and p in $R[Z]$ such that the following holds:

i) The linear form γ separates the zeros of V, and $p(\gamma) \in (f_1, \ldots, f_n)$.
The minimal integral equation of γ over R is $\frac{1}{\alpha} p(Z)$ (where α is the leading coefficient of p).

ii) $\theta X_i - v_i(\gamma) \in (f_1, \ldots, f_n) \qquad (1 \leq i \leq n)$.

iii) $\deg p \leq d^n \quad , \quad \deg v_i \leq d^n \qquad (1 \leq i \leq n)$.

iv) $ht(\gamma_i) = d^{O(n)} \qquad (1 \leq i \leq n)$.

v) Both θ and the coefficients of the v_i's and p are integer polynomials in the coefficients of $\{f_1, \ldots, f_n\}$ and they can be evaluated from them by a non-scalar straight-line program within size of order $d^{O(n)}$, depth of order $O(n \log_2 d)$, and parameters in $\mathbb{Z}$ of height bounded by $d^{O(n)}$.

Proof. We apply proposition 27 to compute the element $\theta \in R$ and the polynomials $v_i \in R[Z]$: Since by hypothesis the ideal $\mathcal{J} := (f_1, \ldots, f_n)$ is radical, we have $\theta X_i - v_i(\gamma) \in \mathcal{J}$, and that $\deg v_i \leq \deg V \leq d^n$.

Moreover, defining

$$F_i(Z) := \theta^d f_i\big(\frac{1}{\theta}v_1(Z), \ldots, \frac{1}{\theta}v_n(Z)\big)$$

we observe that $F_i(\gamma) \equiv f_i(X_1, \ldots, X_n) \pmod{\mathcal{J}} \qquad (1 \leq i \leq n)$.
Now consider a greatest common divisor $p \in R[Z]$ of $F_1, \ldots, F_n$ and let us see that p verifies the required conditions:
The polynomial p is (a non-zero scalar multiple of) the minimal polynomial of γ modulo $\mathcal{J}$, and coincides with the characteristic polynomial of the homothesy η_γ since both polynomials have the same degree (for $\mathcal{J}$ is a zero-dimensional radical ideal and γ separates its points). Therefore Bezout's inequality (cf. [26]) implies that $\deg p \leq d^n$. This shows items 1 and 3.
We are going now to compute such a polynomial p.
First we will compute the coefficients of the F_i's from the coefficients of $f_1, \ldots, f_n$. (In fact we will compute slightly different polynomials G_i in order to keep all the operations in R.)
Fix i, and suppose

$$f_i := \sum_{|\mu| \leq d} a_\mu X_1^{\mu_1} \cdots X_n^{\mu_n} \qquad (1 \leq i \leq n)$$

then:

$$F_i(Z) := \sum_{0 \leq j \leq \delta} a_j X^j = \sum_{|\mu| \leq d} \theta^{d-|\mu|} a_\mu v_1(Z)^{\mu_1} \cdots v_n(Z)^{\mu_n}.$$

As $\deg v_j = d^{O(n)}$ $(1 \leq j \leq n)$ we observe that $\delta := \deg F_i = d^{O(n)}$ also (in fact we can take d^{3n}).
Now we interpolate in $\{0, \ldots, \delta\}$ by means of a Vandermonde matrix A to evaluate the coefficients of F_i:

$$\begin{pmatrix} 1 & 0 & 0 & \ldots & 0 \\ 1 & 1 & 1 & \ldots & 1 \\ 1 & 2 & 2^2 & \ldots & 2^\delta \\ \vdots & \vdots & \vdots & \ldots & \vdots \\ 1 & \delta & \delta^2 & \ldots & \delta^\delta \end{pmatrix} \begin{pmatrix} a_0 \\ a_1 \\ a_2 \\ \vdots \\ a_\delta \end{pmatrix} = \begin{pmatrix} F_i(0) \\ F_i(1) \\ F_i(2) \\ \vdots \\ F_i(\delta) \end{pmatrix}.$$

We define $G_i := (\det A) F_i$ and we compute the coefficients of G_i by means of the adjoint matrix of A. This can be done for all i, $1 \leq i \leq n$, in size $d^{O(n)}$, depth $O(n \log_d)$ and parameters of height of order $d^{O(n)}$ (compute first the entries of A, $\det A$ and $Adj(A)$, simultaneously evaluate θ^j and $v_j(0)^k, \ldots, v_j(\delta)^k$ to finally compute F_i^*).

We conclude computing, as in lemma 11, a greatest common divisor p of $G_1, \ldots, G_s$. □

4.7 Height of isolated points of affine varieties

We are now in conditions to show the bounds for the height of the isolated points of an affine algebraic variety $V := \{f_1 = 0, \ldots, f_s = 0\} \subseteq \mathbb{C}^n$, where the polynomials $f_i \in \mathbb{Z}[X_1, \ldots, X_n]$ have degree and height bounded by d and η respectively.

Corollary 6 *Let $f_1, \ldots, f_r$ $(r \leq n)$ and g be polynomials in $\mathbb{Z}[X_1, \ldots, X_n]$ of degree and height bounded by $d \geq n$ and η respectively, and let V denote the algebraic affine variety of $\mathbb{C}^n$ defined by:*

$$V := \{f_1 = 0, \ldots, f_s = 0\}.$$

Then V has at most d^n isolated points whose heights verify:

$$\log_2 ht(P) \leq d^{O(n)}(\log_2 s + \log_2 \eta)$$

Proof. First of all following the same analysis of [26], it is possible to find nonnegative integers λ_{ij} of height at most d^n, such that the isolated points of V are some of the isolated points of the algebraic set W determined by the equations:

$$\begin{cases} F_1(X_1, \ldots, X_n) := \lambda_{11} f_1(X_1, \ldots, X_n) + \cdots + \lambda_{1s} f_s(X_1, \ldots, X_n) \\ \quad\vdots \\ F_n(X_1, \ldots, X_n) := \lambda_{n1} f_1(X_1, \ldots, X_n) + \cdots + \lambda_{ns} f_s(X_1, \ldots, X_n) \end{cases}$$

The isolated points of W are a locally complete intersection so there exist polynomials $q_i(X_i)$ of degree bounded by d^n such that for every such an isolated point $P = (x_1, \ldots, x_n) \in W$ $q_i(x_i) = 0$ holds. These polynomials q_i have integer coefficients, which can be obtained by means of a non-scalar straight-line program of size $sd^{O(n)}$, depth $O(n \log_2 d)$ and parameters of height d^n in the coefficients of the f_i's. Thus, we conclude that for each i:

$$ht(q_i) = (s\,\eta)^{d^{O(n)}}.$$

Now let K be an algebraic number field and M_K a proper set of absolute values over K. As in [38] the height of an affine point $P := (x_1, \ldots, x_n) \in K^n$ is defined in terms of the projective point $(y_0 : \cdots : y_n) = (1 : x_1 : \cdots : x_n) \in P_n(K)$:

$$ht(P) := \prod_{\nu \in M_K} \sup_i \|y_i\|_\nu.$$

We have the following inequalities:

$$\begin{aligned} ht(P) &\leq \prod_{\nu \in M_K} \sup_i \max\{1, \|x_i\|_\nu\} \\ &\leq \prod_{i=1}^{n} \prod_{\nu \in M_K} \max\{1, \|x_i\|_\nu\}. \end{aligned}$$

As observed in [38], the last term is exactly the measure of the minimal polynomial of x_i over the integers. Since the measure of univariate polynomials is a multiplicative function, combining with Landau's inequality (cf. [37]) we have:

$$ht(P) \leq \prod_{i=1}^{n} M(q_i) \leq \prod_{i=1}^{n} \|q_i\|_2 \leq \prod_{i=1}^{n} d^n ht(q_i).$$

Therefore

$$\log_2 ht(P) = d^{O(n)}(\log_2 s + \log_2 \eta)$$

as desired (and in our conditions there are at most d^n isolated points P). □

4.8 Triviality test based on isolated points

As in [23, Théorème 3.5] one can design a probabilistic procedure that tests whether an ideal $(f_1, \ldots, f_s)$ is the trivial ideal in $\mathbb{Q}[X_1, \ldots, X_n]$. In fact the algorithm determines the dimension of the variety $V = \{f_1 = 0, \ldots, f_s = 0\}$. The method is based on the following fact: if the algebraic set V has dimension r, there are generic linear forms $\gamma_1, \ldots, \gamma_r$ such that the variety V_r given by

$$V_r = \{f_1 = 0, \ldots, f_r = 0, \gamma_1 = 0, \ldots, \gamma_r = 0\}$$

is of dimension 0 and that r is maximal with this property.

The procedure of [23] is based on the following scheme:

- Guess non-deterministically a list $\mathcal{Q}$ of $t = d^{O(n)} s^{O(1)}$ nonnegative integers of $\mathbb{Z}^{n(n+1)+n}$ of height at most $d^{O(n)} \log_2 s^{O(1)}$ (as in corollary 19). This guess $\mathcal{Q}$ will be a candidate for a questor set in all steps of the procedure.

- For all $1 \leq r \leq n$, introduce an $r \times (n+1)$ matrix of variables $M = (T_{i\,j})$ and the linear mappings:

$$\gamma_i := T_{i\,1} X_i + \cdots + T_{i\,n} X_n + T_{i\,n+1}$$

- Next introduce a new list $S_1, \ldots, S_n$ of variables the polynomials F_i of [22], defined as:

$$F_i := f_1 + S_i f_2 + S_i^2 f_3 + \cdots + S_i^{s-1} f_s + S_i^s \gamma_1 + \cdots + S_i^{s+r-1} \gamma_r$$

- Choose concrete values in $\mathcal{Q}$ and determine concrete polynomials $F_1, \ldots,$ F_n. For each selection, let W_r be the algebraic subset of $\mathbb{C}^n$ given by $F_1 = 0, \ldots, F_n = 0$. The isolated points of V_r are isolated points of some of the W_r because this property holds generically (cf. [23]).

- For the same selection of constants, applying proposition 27, we can obtain a linear form γ that separates the isolated points of W_r (if there is any). Moreover, we can obtain polynomials $p, v_1, \ldots, v_n$ and a constant θ, evaluable by non-scalar straight-line programs as in proposition 27, such that $p(\gamma)$ and $\theta X_i - v_i(\gamma)$ vanishes on the isolated points of W_r.

- Finally, for the polynomials $f_1, \ldots, f_s, \gamma_1, \ldots, \gamma_r$ we can substitute the variables X_i by $\frac{1}{\theta}v_i(\gamma)$ and, multiplying by the accurate power of θ, we obtain some univariate polynomials $p(Z), G_1(Z), \ldots, G_s(Z), L_1(Z), \ldots,$ $L_r(Z)$. The solutions of the equations they define determine the isolated points of V_r.

- If for every choice in $\mathcal{Q}$, the gcd of the polynomials $p(Z)$, G_i's and L_i's is 1 and if $\mathcal{Q}$ is a correct test sequence, we conclude that V_r contains no isolated point.

- Computing the maximal r such that V_r contains isolated points, we obtain the dimension of V and, in particular, we know whether V is empty or not.

This procedure tests non-zero equality (and hence requires evaluation) of non-scalar straight-line programs of size $sd^{O(n)}$, depth $O(n \log_2 d + \log_2 s)$ and uses parameters in $\mathcal{Q}$ of controlled height. The evaluation can be simulated by a Turing machine as in corollary 20. The correct acceptation depends on the quality of $\mathcal{Q}$ as questor set. Again by corollary 19, we can conclude:

Corollary 28 *There is a bounded-error probablistic Turing machine which decides whether an ideal described by generators $f_1, \ldots, f_s \in \mathbb{Z}[X_1, \ldots, X_n]$ of degree d and height η is trivial or not in time $s^{O(1)}d^{O(n)} \log_2 \eta^{O(1)}$. The error probability is bounded by $\frac{1}{u^{t/6}}$, where $u = d^{O(n)}$ and $t = d^{O(n)}$. Hence the problem of deciding the triviality of such an ideal belongs to the bounded probability polynomial time class of complexity BPP* (cf. [1, chapter 6] for notation).

5 Division in reduced complete intersections

In this section we state a crucial step for the description and computation of the wanted output polynomials in the effective Nullstellensatz and the membership problem: a division theorem which considers the problem of computing the quotient of a division in a ring which is a reduced complete intersection. This quotient will be technically the result of the action of a matrix whose entries can be computed by a non-scalar straight-line program of appropriate size, depth and parameters. The computation of this matrix instead of the computation of the concrete quotient will be essential for us in order to obtain the right heights for the output polynomials in the two mentionned results.

The main ingredient for the division theorem is a duality technique based on the existence of traces for Gorenstein algebras.

5.1 Notation and assumptions

Throughout this section we shall use the following notation and assumptions:

- n and r are nonnegative integers with $1 \leq r \leq n$, $R := \mathbb{Z}[X_1, \dots, X_{n-r}]$, $A := \mathbb{Q}[X_1, \dots, X_{n-r}]$ and $K := \mathbb{Q}(X_1, \dots, X_{n-r})$.

- $f_1, \dots, f_r$ are polynomials in $\mathbb{Z}[X_1, \dots, X_n]$ of degree and height bounded by $d \geq n$ and η, respectively, which define a reduced complete intersection ideal $\mathcal{J}$ in $\mathbb{Q}[X_1, \dots, X_n]$, and such that the following extension is integral

$$\mathbb{Q}[X_1, \dots, X_{n-r}] \hookrightarrow B := \mathbb{Q}[X_1, \dots, X_n]/\mathcal{J}.$$

- B denotes the factor ring $\mathbb{Q}[X_1, \dots, X_n]/\mathcal{J}$. For any polynomial $h \in \mathbb{Q}[X_1, \dots, X_n]$ we denote by $\overline{h}$ its class in B; also $\deg^o h$ will be its degree in the *free variables* $X_1, \dots, X_{n-r}$ and $\deg^* h$ its degree in the variables $X_{n-r+1}, \dots, X_n$.

Let us observe that under these assumptions $K \otimes_A B$ is a reduced zero-dimensional K-algebra.

5.2 The division theorem

Theorem 29 *Under the assumptions 5.1, let $f, g \in \mathbb{Z}[X_1, \dots, X_n]$, $\deg f \leq d$, $ht(f) \leq \eta$, be polynomials such that $\overline{f}$ is not a zero divisor and that $\overline{f} \mid \overline{g}$ in B. Then there exist a polynomial $q \in R[X_{n-r+1}, \dots, X_n]$ and an element $\xi \in R$ such that:*

i) $\xi g \equiv qf \pmod{\mathcal{J}}$ *and* $\xi | q$ *in* $\mathbb{Q}[X_1, \dots, X_n]$.

ii) $\deg^* q \leq r\,(d-1)$.

iii) $\max\{\deg^o q, \deg^o \xi\} = \deg^o g + (1 + \deg^* g) d^{O(r)}$.

iv) $\max\{\log_2 ht(q), \log_2 ht(\xi)\} = d^{O(r)} \deg g^{O}(1) \max\{\log_2 \eta, \log_2(ht(g))\}$.

This result will be an inmediate consequence of the following, more technical, result:

Lemma 30 *Under the assumptions 5.1, let $f, g \in \mathbb{Z}[X_1, \dots, X_n]$, $\deg f \leq d$, be polynomials such that $\overline{f}$ is not a zero divisor and that $\overline{f} \mid \overline{g}$ in B. Then there exist a matrix Q with coefficients in R and an element $\xi \in R$ such that if $q \in R[X_{n-r+1}, \dots, X_n]$ denotes the polynomial whose coefficients are obtained applying Q to the coefficients of g (in adequate bases of monomials):*

i) $\xi g \equiv qf \pmod{\mathcal{J}}$ *and* $\xi | q$ *in* $\mathbb{Q}[X_1, \dots, X_n]$.

ii) $\deg^* q \leq r\,(d-1)$.

iii) *The coefficients of Q and ξ are integer polynomials in the coefficients in R of $f_1, \ldots, f_r$ and f and can be computed from them by a non-scalar straight-line program of size $d^{O(r)} + O(\deg^* g)$, depth $O(r \log_2 d + \log_2(\deg^* g))$, and parameters in $\mathbb{Z}$ of height bounded by $d^{O(r)}$.*

The proof of this result is mainly inspired by the developments of [18]. Technically, we represent the polynomials by the matrix of the homothesies (or multiplication) they define on the K-vector space $K \otimes_A B$. Then, instead of computing the matrix of a trace through the computation of a dual basis as in [18], we perform the division procedure by means of adjoint matrices and polynomials as in [48, lemmas 9 and 15].

The encoding of polynomials as matrices is a consequence of the computation of a basis of $K \otimes_A B$, in terms of a primitive element chosen as in section 4. Moreover, the equations defining the primitive element are essential to compute this matrix of multiplication.

The computation of the trace in this Gorenstein algebra will be done by computing a pseudo-Jacobian determinant (the element of the duality) as in [35, example F.19 and corollary F.10].

We begin by recalling the basic facts of trace theory we need. (We thank [48] from where we textually borrowed the main part of paragraph 5.3 below.)

5.3 Basic general trace theory

The definition of the trace. We consider the ring B as an A-algebra and we denote by B^* the dual space $\mathrm{Hom}_A(B, A)$. The A-module B^* admits a natural structure of B-module in the following way: For any pair (b, β) in $B \times B^*$ the product $b \cdot \beta$ is the A-linear application of B^* defined by $(b.\beta)(x) := \beta(bx)$, for each x in B.

Our assumptions about A and B allow to show that the B-modules B and B^* are isomorphic (see [35, example F.19 and corollary F.10]) and therefore B^* can be generated by a single element. A generator σ of B^* is called a *trace* of B over A.

Under our hypothesis we have the additional property that B is a finite free A-module whose rank will be denoted by D (cf. [24]). Fix for the moment a basis of this module; each element $b \in B$ defines, by multiplication, a square matrix $M_b \in A^{D \times D}$. If we denote by $\mathrm{trace}(M_b)$ the trace of the matrix M_b, the application $b \mapsto \mathrm{trace}(M_b)$ defines (independently of the basis of B) an element of B^* called the *usual trace* and denoted by Tr.

Unfortunately the usual trace is not always a generator of B^* (in other words the usual trace is not necessarily a trace).

The trace associated to a regular sequence. Let us consider now the ring $B \otimes_A B$ which is both an A-algebra and a B-bialgebra (with right and left multiplications).

Let $\mu : B \otimes_A B \to B$ be the morphism of A-algebras (or B-bialgebras) defined by $\mu(b \otimes b') := bb'$. Denote by $\mathcal{K}$ the kernel of μ. It is easy to show that $\mathcal{K}$ is the ideal generated by all the elements $b \otimes 1 - 1 \otimes b$, where b ranges over B (see for example [32, proposition 1.3]).

From the fact that $\mathrm{Ann}_{B\otimes_A B}(\mathcal{K})(b \otimes 1 - 1 \otimes b) = 0$ for all $b \in B$, one infers that the induced structures of right and left B-modules over $\mathrm{Ann}_{B\otimes_A B}(\mathcal{K})$ coincide. In other words, if $\sum_i b_i \otimes b_i'$ belongs to $\mathrm{Ann}_{B\otimes_A B}(\mathcal{K})$ and b is an element of the ring B we have: $\sum_i bb_i \otimes b_i' = \sum_i b_i \otimes bb_i'$. Moreover, it is possible to show that $\mathrm{Ann}_{B\otimes_A B}(\mathcal{K})$ is a cyclic B-module ([35, corollary F.10]).

Let us consider the application $\Phi : B \otimes_A B \to \mathrm{Hom}_A(B^*, B)$ defined by

$$\Phi(\sum_i b_i \otimes b_i')(\beta) := \sum_i b_i \; \beta(b_i'),$$

where $b_i, b_i' \in B$ and $\beta \in B^*$. Since B is free as A-module, Φ is an isomorphism that identifies $\mathrm{Ann}_{B\otimes_A B}(\mathcal{K})$ and $\mathrm{Hom}_B(B^*, B)$.

For each generator $\Delta := \sum_m a_m \otimes b_m$ of the B-module $\mathrm{Ann}_{B\otimes_A B}(\mathcal{K})$ the element $\Phi(\Delta)$ is a generator of $\mathrm{Hom}_B(B^*, B)$ and then there exists a uniquely determinated $\sigma_\Delta \in B^*$ such that $\Phi(\Delta)(\sigma_\Delta) = 1$. One deduces immediately that σ_Δ is a trace for B (which is called the *trace associated to* Δ).

From the definitions of Φ, Δ and σ_Δ we have the following "*trace formula*" for all $b \in B$:

$$b = \sum_{1 \le m \le M} \sigma_\Delta(b\, b_m)\, a_m. \tag{7}$$

In particular, we observe that $a_1, \dots, a_M$ is a system of generators of the A-module B. By means of the element Δ it is possible to obtain a relation between the trace σ_Δ and the "usual trace" Tr: more precisely (see [35, corollary F.12]):

$$\mu(\Delta) \cdot \sigma_\Delta = \mathrm{Tr} \tag{8}$$

In terms of elements of B this formula says that for all $b \in B$ the equality $\sigma_\Delta(\mu(\Delta)b) = \mathrm{Tr}(b)$ holds.

In the case of reduced complete intersections rings B, there is a classical procedure to determine an element Δ generating $\mathrm{Ann}_{B\otimes_A B}(\mathcal{K})$ as B-module. This method, known by Jacobi, can also be seen in [35, corollary E.19 and example F.19]:

Now let us introduce new variables $Y_{n-r+1}, \dots, Y_n$. For every polynomial $f \in \mathbb{Q}[X_1, \dots, X_n]$ we consider the polynomial $f^{(Y)}$ obtained by replacing the variables X_j by Y_j for $n-r+1 \le j \le n$. Hence, we have the canonical isomorphism of A-algebras:

$$B \otimes_A B \cong \mathbf{A}[X_{n-r+1}, \dots, X_n, Y_{n-r+1}, \dots, Y_n]/(f_1, \dots, f_r, f_1^{(Y)}, \dots, f_r^{(Y)}). \tag{9}$$

The polynomials $f^{(Y)} - f$ belong to the ideal $(Y_j - X_j,\ n-r+1 \leq j \leq n)$. Then there exist polynomials l_{ij} such that the following equalities hold for $1 \leq i \leq r$:

$$f_i^{(Y)} - f_i := \sum_{1 \leq j \leq r} l_{ij} \cdot (Y_{n-r+j} - X_{n-r+j}). \tag{10}$$

Since both sequences are regular of the same length (and maximal) the class of the Jacobian determinant $\Delta = \det(l_{ij})$ in $B \otimes_A B$ generates $\mathrm{Ann}_{B \otimes_A B}(\mathcal{K})$ by means of the identification (9).
This Jacobian (or pseudo-Jacobian) determinant can be easily computed in terms of non-scalar straight-line programs. First of all, the polynomials l_{ij} can be computed in the following way:

$$f_i^{(Y)} - f_i :=$$
$$\sum_{j=n-r+1}^{n} \left(f_i(X_1, \dots, X_{j-1}, Y_j, \dots, Y_n) - f_i(X_1, \dots, X_j, Y_{j+1}, \dots, Y_n)\right),$$

therefore,

$$l_{ij} = \left(f_i(X_1, \dots, X_{j-1}, Y_j, \dots, Y_n) - f_i(X_1, \dots, X_j, Y_{j+1}, \dots, Y_n)\right) / (Y_j - X_j).$$

Thus, setting

$$f_i := \sum_{|\mu| \leq d} a_{i\mu} X_1^{\mu_1} \cdots X_n^{\mu_n}$$

we conclude that for all $1 \leq i \leq r$ and $n - r + 1 \leq j \leq n$:

$$\begin{aligned} l_{ij} &= \sum_{|\mu| \leq d} a_{i\mu} X_1^{\mu_1} \cdots X_{j-1}^{\mu_{j-1}} Y_{j+1}^{\mu_{j+1}} \cdots Y_n^{\mu_n} \left(\frac{Y_j^{\mu_j} - X_j^{\mu_j}}{Y_j - X_j} \right) \\ &= \sum_{|\mu| \leq d} \left(a_{i\mu} X_1^{\mu_1} \cdots X_{j-1}^{\mu_{j-1}} Y_{j+1}^{\mu_{j+1}} \cdots Y_n^{\mu_n} \cdot \sum_{k=0}^{\mu_j - 1} Y_j^k X_j^{(\mu_j - 1) - k} \right) \end{aligned}$$

(which defines a non-scalar straight-line program of depth $O(r \log_2 d)$, size $d^{O(r)}$, and parameters in $\{-1, 0, 1\}$).
We conclude computing the polynomial $\Delta = \det(l_{ij}) \in R[X_{n-r+1}, \dots, X_n, Y_{n-r+1}, \dots, Y_n]$ by means of the Berkowitz algorithm (proposition 9). The polynomial Δ, of degree $r(d-1)$, is then computable by means of a non-scalar straight-line program of size $d^{O(r)}$, depth $O(\log_2 d)$ using only the coefficients of the f_i's as parameters.

Finally interpolation allows us to compute the coefficients α_μ in R of Δ:

Lemma 31 *There is a non-scalar straight-line program Γ which computes from the coefficients of $f_1, \dots, f_r$ an element $\alpha \in \mathbb{Z} \setminus \{0\}$ and $\alpha'_\mu \in R$ such that*

$\alpha_\mu = \frac{\alpha'_\mu}{\alpha}$ *(and* $\alpha | \alpha'_\mu$ *in* R*). The size of* Γ *is of order* $d^{O(r)}$*, its depth of order* $\ell(\Gamma) = O(r \log_2 d)$ *and the parameters introduced in* Γ *are of height bounded by* $d^{O(1)}$.

In other words, we have (see also [18, §3.4]):

Proposition 32 *There exist polynomials* a_m, b_m $(1 \leq M \leq (rd)^{2r})$ *in* $\mathbb{Z}[X_1, \ldots, X_n]$ *satisfying* $\deg(a_m) + \deg(b_m) \leq r(d-1)$ $(1 \leq m \leq M)$ *such that* $\sum_m \bar{a}_m \otimes \bar{b}_m$ *is a generator of* $\mathrm{Ann}_{B \otimes_A B}(\mathcal{K})$ *and* $\overline{\Delta} = \sum_m \bar{a}_m \bar{b}_m$*. The coefficients of* a_m *are integer polynomials in the coefficients of the* f_i*'s and can be computed from them by a straight-line program of size* $d^{O(r)}$*, depth* $O(r \log_2 d)$*, and parameters of height* $d^{O(1)}$*. The polynomials* b_m *are simply monomials* $X_{n-r+1}^{\mu_{n-r+1}} \cdots X_n^{\mu_n}$.

Proof. If $\det(l_{ij}) = \sum \alpha_\mu X_{n-r+1}^{\mu_{n-r+1}} \cdots X_n^{\mu_n} Y_{n-r+1}^{\mu'_{n-r+1}} \ldots Y_n^{\mu'_n}$ we write

$$a_\mu := \alpha \alpha_\mu X_{n-r+1}^{\mu_{n-r+1}} \cdots X_n^{\mu_n} \qquad \text{and} \qquad b_\mu := X_{n-r+1}^{\mu'_{n-r+1}} \cdots X_n^{\mu'_n}$$

(where α comes from the previous lemma). □

Definition 33 *The trace associated to the generator of* $\mathrm{Ann}_{B \otimes_A B}(\mathcal{K})$ *introduced in proposition 32 will be called the trace associated to the regular sequence* $f_1, \ldots, f_r$ *and we will denote it by* σ_Δ *or by* σ *if there is no ambiguity.*

Let us observe that in this case relation (8) gives

$$\overline{\Delta} \cdot \sigma \;=\; \mathrm{Tr}. \tag{11}$$

5.4 Computing matrices of homothesies

Under assumptions 5.1 and from theorem 22 we can assume that are given:

- A linear form $\gamma := \gamma_{n-r+1} X_{n-r+1} + \cdots + \gamma_n X_n \in \mathbb{Z}[X_{n-r+1}, \ldots, X_n]$ of height $d^{O(r)}$ such that $\bar{\gamma}$ is a primitive element of the zero-dimensional K-algebra $K \otimes_A B$ of rank $D \leq d^r$ (thus $\mathcal{B} := \{1, \ldots, \bar{\gamma}^{D-1}\}$ is a basis of $K \otimes_A B$).

- A polynomial $p \in R[Z]$ of degree $D - 1 \leq d^r$ such that $p(\gamma) \in \mathcal{J}$.

- Elements $\theta \in R$ and $v_j \in R[Z]$ $(n-r+1 \leq j \leq n)$ of degree bounded by d^r such that $\theta X_j - v_j(\gamma) \in \mathcal{J}$ $(n-r+1 \leq j \leq n)$.

- The degrees of p, θ, and v_j $(n-r+1 \leq j \leq n)$ in the variables $X_1, \ldots, X_{n-r}$ is bounded by $d^{O(r)}$.

Under these conditions, we have

Lemma 34 *Set $\tau := \alpha^D\theta$, where α is the leading coefficient of p. There is a non-scalar straight-line program which computes from the coefficients of p and v_j $(n-r+1 \leq j \leq n)$ and from θ the coordinates (in R) of the matrices $M_{\tau\gamma}$ and $M_{\tau X_j}$ $(n-r+1 \leq j \leq n)$. The straight-line program has size of order $d^{O(r)}$, depth $O(r\log_2 d)$, and parameters $\{-1,0,1\}$.*

Proof. Observe that the matrix $M_{\alpha\gamma}$ has coordinates in R given by the coefficients of p.
From $M_{\alpha\gamma}$ we can compute:

$$M_{\tau X_j} = \alpha^D M_{\theta X_j} = \alpha^D v_j(\frac{1}{\alpha}M_{\alpha\gamma}).$$

To do this, set $v_j := \sum_{k=0}^{D} b_{jk}Z^k$ and define $V_j := \sum_{k=0}^{D}(\alpha^{D-k}b_{jk})Z^k$. Then, we have:

$$M_{\tau X_j} = V_j(M_{\alpha\gamma}).$$

To compute the powers $\{1,\alpha,\dots,\alpha^D\}$ requires size $D+1$ and depth $\log_2 D + 1$. Simultaneously we can compute the matrix $M_{\alpha\gamma}$ adding size $O(D^2)$. In additional depth 1 and size $r(D+1)+1$ we compute the products $\alpha^{D-k}b_{jk}$ and τ. Next in size $O(D^4)$ and depth $\log_2 D+2$, we compute all powers $\{M_{\alpha\gamma}^k : k=0,\dots,D\}$.
For every j, $n-r+1 \leq j \leq n$, we compute in additional size $O(D^3)$ and non-scalar depth 2 the matrix $v_j(M_{\alpha\gamma})$, which implies a total size $O(r\,D^3) = d^{O(r)}$ and the same depth. □

This lemma allows us to compute the matrix M_h for any polynomial $h \in R[X_{n-r+1},\dots,X_n]$. It is crucial for us to observe here that if $\deg^* h$ is fixed, then the computation of M_h is linear in the coefficients of h:

Lemma 35 *Under the same assumptions of lemma 34, let $h \in R[X_{n-r+1},\dots, X_n]$ and let $N := \deg^* h$ and $\tau := \alpha^D\theta$ (where α is the leading coefficient of p). Then, there is a non-scalar straight-line program which computes from the coefficients of h, p, v_j $(n-r+1 \leq j \leq n)$ and from θ the coefficients of the matrix $M_{\tau^N h}$. The straight-line program has size $d^{O(r)} + O(\deg^* h)$, depth $O(r\log_2 d + \log_2(\deg^* h))$ and parameters $\{-1,0,1\}$ in $\mathbb{Z}$.*

Proof. Set $h := \sum_{|\mu|\leq N} a_\mu X_{n-r+1}^{\mu_{n-r+1}} \cdots X_n^{\mu_n}$. Then

$$M_{\tau^N h} := \sum_{|\mu|\leq N} a_\mu\, \tau^{N-|\mu|}\, M_{\tau X_{n-r+1}}^{\mu_{n-r+1}} \cdots M_{\tau X_n}^{\mu_n}$$

The bounds for the size and depth of the straight-line program which design the coordinates of the matrix $M_{\tau^N h}$ are then straightforward. □

Let us recall now that we have a trace σ associated to the regular sequence $f_1, \ldots, f_r$ (which is a generator of the B-module B^*) such that for all $h \in B$, $\sigma(\Delta h) = \mathrm{Tr}(h)$ holds.

Proof of lemma 30.
The condition that f is not a zero-divisor in B implies that if $\mathcal{X}_f := T^D + \alpha_{D-1}T^{D-1} + \cdots + \alpha_0 \in A[T]$ is the characteristic polynomial of the homothesy η_f (which belongs effectively to $A[T]$, see [48, corollary 2]) then $\alpha_0 = (-1)^D \det \eta_f$ is a non-zero element of A. Therefore, if we define

$$f^* := f^{D-1} + \alpha_{D-1} f^{D-2} + \cdots + \alpha_2 f + \alpha_1.$$

We observe that $\overline{f}\,\overline{f^*} = (-1)^{D+1}\alpha_0$ in B. Now, the condition $\overline{f}|\overline{g}$ in B implies there exists $q_1 \in \mathbb{Q}[X_1, \ldots, X_n]$ such that $g \equiv q_1 f \pmod{\mathcal{J}}$, and therefore $g f^* \equiv (-1)^{D+1}\alpha_0 q_1 \pmod{\mathcal{J}}$.
We would like to define simply $q := f^* g$ and $\xi = (-1)^{D+1}\alpha_0$ but unfortunately this definition doesn't agree with our requirements on degrees and with the fact that everything must belong to $\mathbb{Z}[X_1, \ldots, X_n]$. However, we are going to keep essentially this idea in mind and to use the trace formula to reduce the degree of q:
From formula 7, $f^* g = \sum_{1 \le m \le M} \sigma(f^* g\, b_m)\, a_m$ and from relation 11, $\overline{\Delta}\,\sigma = \mathrm{Tr}$; therefore, if we define Δ^* analogously as f^*, we have:

$$(\Delta^* \Delta) f^* g = \sum_{1 \le m \le M} \mathrm{Tr}(\Delta^* f^* g\, b_m)\, a_m$$

Now we need to multiply all polynomials by factors in R in order to keep all the computations in $R[X_{n-r+1}, \ldots, X_n]$.
Let N be a bound for $\deg^* f$, $\deg^* \Delta$, $\deg^* b_m$, and $\deg^* g$, and set $f' := \tau^N f$, $\Delta' := \tau^N \Delta$, $g' := \tau^N g$, and $b'_m := \tau^N b_m$ (recall that τ is defined as in lemma 34).
Observe that:

- $M_{f'}$, $M_{\Delta'}$, $M_{g'}$, and $M_{b'_m}$ have their coefficients in R
- $f'^* = \tau^{N(D-1)} f^*$ and $M_{f'^*} = (-1)^{D+1} Adj(M_{f'})$ (analogously for Δ').

We define:

$$q := \sum_{1 \le m \le M} \mathrm{Tr}(\Delta'^* f'^* g' b'_m)\, a_m = \sum_{1 \le m \le M} \mathrm{Tr}\big(Adj(M_{\Delta'}) Adj(M_{f'})\, M_{g'}\, M_{b'_m}\big)\, a_m$$

and

$$\xi := (\Delta'^* \Delta')(f'^* f') = \det M_{\Delta'} \det M_{f'}.$$

The bound on $\deg^* q$ is then clear from the definition.
The matrix Q appears from the fact that if $\deg^* g$ is fixed the coefficients of $M_{g'}$ are linear (and uniform) in the coefficients of g (see the proposition above) and

so do the coefficients of $Adj(M_{\Delta'})Adj(M_{f'})\,M_{g'}\,M_{b'_m}$; we conclude observing that the trace is also linear and combining with the coefficients of the a_m's. Finally it is clear that $\xi \mid q$ in $\mathbb{Q}[X_1,\dots,X_n]$ since $g' \equiv q_1 f' \pmod{\mathcal{J}}$ implies that

$$\begin{aligned} \mathrm{Tr}(\Delta'^* f'^*\, g'\, b'_m) &= \mathrm{Tr}(\Delta'^* f'^*\, f'\, q_1\, b'_m) \\ &= f'^*\, f'\sigma(\Delta'\Delta'^*\, q_1 b'_m) \\ &= (\Delta'^*\Delta')(f'^*\, f')\sigma(q_1 b'_m) \end{aligned}$$

(where all equalities are true polynomial equalities since every term belongs to A and $A \cap \mathcal{I} = \mathcal{J} = (0)$). □

Proof of theorem 29.
From the straight-line program which computes ξ and the matrix Q in R from the coefficients of $f_1,\dots,f_r$and f we construct a straight-line program which evaluates ξ and q from the variables $X_1,\dots,X_n$ simply using as parameters the coefficients of $f_1,\dots,f_r,f,g$ as follows:
We precompute the coefficients in R of $f_1,\dots,f_r,f$ (which are of degree bounded by d) and of g (of degree $\deg^o g$) in size $O(d^{n-r}+\deg^o g^{n-r})$, non-scalar depth $\max\{\log_2 d, \log_2(\deg^o g)\}$ and parameters $\max\{\eta, ht(g)\}$. We also precompute all powers $X_{n-r+1}^{\alpha_{n-r+1}}\cdots X_n^{\alpha_n}$ with $|\alpha| \le \deg^* q$. Then we multiply the coefficients of the matrix Q by the coefficients in R of g and at the end combine the result with the corresponding powers of $X_{n-r+1},\dots,X_n$ computed above.
We conclude applying proposition 15. □

6 Preparation of the input data

This section is devoted to two preprocessing techniques: Bertini's theorem and Noether normalization.

6.1 Bertini's theorem.

An essential technique used to prepare our polynomials is given by Bertini's theorem.
Let $f_1,\dots,f_r \in \mathbb{Z}[X_1,\dots,X_n]$. We shall analyze two cases:

i) The ideal $(f_1,\dots,f_r)$ is the trivial ideal in $\mathbb{Q}[X_1,\dots,X_n]$ (r arbitrary).

ii) $r \le n$ and $(f_1,\dots,f_r)$ is a proper ideal of dimension $n-r$.

For the first case we can make, as in [26] or [12], a reduction to at most $n+1$ equations but having some additional properties.

Lemma 36 *(see also [48, proposition 18] and [15, proposition 1.2]) Let $f_1, \dots, f_r$ be polynomials with integer coefficients, of degree bounded by d, such that $1 \in (f_1, \dots, f_r)$ in $\mathbb{Q}[X_1, \dots, X_n]$. Then there exist $t \leq n$ and an integer matrix*

$$Q = \begin{pmatrix} \lambda_{1\,1} & \cdots & \lambda_{1\,r} \\ \vdots & & \vdots \\ \lambda_{t+1\,1} & \cdots & \lambda_{t+1\,r} \end{pmatrix}$$

such that if $f'_i := \lambda_{i\,1} f_1 + \cdots + \lambda_{i\,r} f_r$ $(1 \leq i \leq t+1)$, the following properties hold:

- $1 \in (f'_1, \dots, f'_{t+1})$.
- $f'_1, \dots, f'_{j-1}, f'_{j+1}, \dots, f'_{t+1}$ *is a regular sequence for all index j,* $1 \leq j \leq t+1$.
- $ht(Q) \leq (d+1)^n$.

Proof. We proceed inductively constructing polynomials $\{f'_1, \dots, f'_k\}$ such that:

- Either $(f'_1, \dots, f'_k) = (1)$ or $\dim(f'_1, \dots, f'_k) = n - k$.
- For all $S := \{j_1, \dots, j_s\} \subset \{1, \dots, k\}$ of cardinality strictly less than k, the ideal $\mathcal{J}_S := (f'_{j_1}, \cdots, f'_{j_s})$ has dimension $n - s$ (we suppose here $j_1 < j_2 < \ldots < j_s$).
- The linear combinations $f'_1, \dots, f'_k$ are obtained using parameters of height bounded by $(d+1)^n$.

For $k = 1$, either some of the f_i's is a unit in $\mathbb{Q}$ or we take $f'_1 := f_1$.
Suppose now that $f'_1, \dots, f'_k$ are constructed.
If they verify that $1 \in (f'_1, \dots, f'_k)$ we conclude $t := k - 1$.
Otherwise for every $S \subseteq \{1, \dots, k\}$, the variety $V_S \in \mathbb{C}^n$ defined by the ideal $\mathcal{J}_S := (f'_j, j \in S)$ is equidimensional. Denote by $\mathcal{C}_S$ the set of all irreducible components of V_S, and for every $C \in \mathcal{C}_S$ choose a point $x_C \in C$.
Now let us define the polynomial:

$$F(T_1, \dots, T_r) := \prod_{S \subseteq \{1, \dots, k\}} \prod_{C \in \mathcal{C}_S} (T_1 f_1(x_C) + \cdots + T_r f_r(x_C))$$

Since $1 \in (f_1, \dots, f_r)$, the Bezout inequality implies that F is a non-zero polynomial of degree at most $(d+1)^k$. Thus there exist nonnegative integers $\lambda_1, \dots, \lambda_r$ of absolute height at most $(d+1)^k$ such that $F(\lambda_1, \dots, \lambda_r) \neq 0$.
We define $f'_{k+1} := \lambda_1 f_1 + \cdots + \lambda_r f_r$. By construction, f'_{k+1} is not a zero divisor modulo $\mathcal{J}_S$, for all $S \subseteq \{1, \dots, k\}$.
We have the two following possibilities:

- $1 \notin (f'_1, \ldots, f'_{k+1})$: In this case the ideals $\mathcal{J}_S + (f'_{k+1})$ are all proper ideals of dimension $n - \sharp(S)$ (and $(f'_1, \ldots, f'_k)$ is of dimension $n-k$ by hypothesis) and we continue with our inductive procedure.
- $1 \in (f'_1, \ldots, f'_{k+1})$. Then there exists S of minimal cardinality t such that $1 \in \mathcal{J}_S + (f'_{k+1})$.
  The ideal $\mathcal{J}_S + (f'_{k+1})$ verifies our requirements. □

Therefore both cases (i) and (ii) can be reduced to the following proposition:

Proposition 37 *Let $f_1, \ldots, f_r \in \mathbb{Z}[X_1, \ldots, X_n]$ $(r \leq n+1)$ be polynomials of degree bounded by $d \geq n$ such that one of the two following conditions is satisfied:*

- *$1 \in (f_1, \ldots, f_r)$ and $f_1, \ldots, f_{j-1}, f_{j+1}, \ldots, f_r$ is a regular sequence for all j, $1 \leq j \leq r$.*
- *$r \leq n$ and $(f_1, \ldots, f_r)$ is an ideal of dimension $n - r$.*

Then there exists a matrix $Q \in \mathcal{M}_{(r-1)\times r}(\mathbb{Z})$ of height $d^{O(n)}$ such that the polynomials:

$$\begin{pmatrix} f'_1 \\ \vdots \\ f'_{r-1} \end{pmatrix} = Q \begin{pmatrix} f_1 \\ \vdots \\ f_r \end{pmatrix}$$

verify:

i) $(f'_1, \ldots, f'_{r-1}, f_r) = (f_1, \ldots, f_r)$.

ii) $f'_1, \ldots, f'_{r-1}$ is a regular sequence.

iii) For all k, $1 \leq k \leq r-1$, $(f'_1, \ldots, f'_k)$ defines a radical ideal in $\mathbb{Q}[X_1, \ldots, X_n]$.

Proof. For all j, $1 \leq j \leq r$, we define

$$\mathcal{U}_j := \{x \in \mathbb{C}^r : f_j(x) \neq 0\}$$

and the morphism:

$$\begin{array}{rccl} \Phi_j : & \mathcal{U}_j & \longrightarrow & \mathbb{C}^{r-1} \\ & x & \longmapsto & \left(\frac{f_1(x)}{f_j(x)}, \ldots, \frac{f_{j-1}(x)}{f_j(x)}, \frac{f_{j+1}(x)}{f_j(x)}, \ldots, \frac{f_r(x)}{f_j(x)}\right) \end{array}$$

If $(f_1, \ldots, f_r)$ is a proper ideal of dimension $n-r$, it is shown in [36] that $\{f_1, \ldots, f_r\}$ is an independent set in $\mathbb{C}[X_1, \ldots, X_n]$ and there is no non-zero homogeneous polynomial $F \in \mathbb{C}[Y_1, \ldots, Y_r]$ such that $F(f_1, \ldots, f_r) = 0$. This simply implies that $\mathrm{Im}(\Phi_j)$ is a constructible subset of $\mathbb{C}^{r-1}$ of dimension $r-1$.

On the other hand, if $f_1, \ldots, f_r$ verifies the first condition, for every j ($1 \leq j \leq r$) the ideal $(f_1, \ldots, f_{j-1}, f_{j+1}, \ldots, f_r)$ has dimension $n - (r-1)$ and f_j is not a zero divisor over it. This implies that the sequence $f_1, \ldots, f_{j-1}, f_{j+1}, \ldots, f_r$ defines an ideal of dimension $n - (r-1)$ in $\mathbb{C}[\mathcal{U}_j] := \mathbb{C}[X_1, \ldots, X_n]_{f_j}$. Now, applying the theorem on the dimension of the Fibers to Φ_j we can conclude that the image $Im(\Phi_j)$ is a constructible subset of $\mathbb{C}^{r-1}$ of dimension $r-1$.

Therefore in both cases the morphisms Φ_j are dominant and we can apply Bertini's theorem in the characteristic zero case (cf. [33, corollary 6.7]).
Thus there is a Zariski open subset in the space of matrices $Q = (\lambda_{i\,j}) \in \mathcal{M}_{(r-1)\times r}(\mathbb{C})$ such that

$$\begin{cases} f'_1 & := & \lambda_{1\,1} f_1 & + & \cdots & + & \lambda_{1\,r} f_r \\ \vdots & & & & & & \\ f'_{r-1} & := & \lambda_{r-1\,1} f_1 & + & \cdots & + & \lambda_{r-1\,r} f_r \end{cases}$$

verify that for all j, $1 \leq j \leq r$, $(f'_1, \ldots, f'_{r-1})$ is a regular sequence which defines a radical ideal in $\mathbb{C}[\mathcal{U}_j]$ (moreover, the same assertion is true for $(f'_1, \ldots, f'_k)$ for all k, $1 \leq k \leq r-1$).
We now intersect this Zariski open set of matrices with the $(r-1)$ open sets defined by the following polynomial conditions:

$$\det(\lambda_{i\,j})_{1 \leq i,j \leq k} \neq 0 \qquad (1 \leq k \leq r-1).$$

We claim that $f'_1, \ldots, f'_{r-1}$ verify the desired conditions:

(i) The condition on the determinant of order $r-1$ implies that $(f_1, \ldots, f_{r-1}, f_r) - (f'_1, \ldots, f'_{r-1}, f_r)$.

(ii) We will show that $f'_1, \ldots, f'_{r-1}$ is a regular sequence by induction:

- $f'_1 \neq 0$.
- $1 < k \leq r-2$. $(f'_1, \ldots, f'_{k-1})$ is a regular sequence by hypothesis. Let $(f'_1, \ldots, f'_{k-1}) = \bigcap Q_t$ be the primary decomposition of the ideal (where for all t, $\dim Q_t = n - (k-1)$ by the hypothesis of regularity). Now suppose that $f'_1, \ldots, f'_k$ is not a regular sequence: thus f'_k is either a zero-divisor modulo $(f'_1, \ldots, f'_{k-1})$ or the ideal $(f'_1, \ldots, f'_k)$ is the trivial one. The second case cannot hold since the extension of this ideal to every localization $\mathbb{C}[\mathcal{U}_j]$ is a proper ideal. This means that there exists a prime ideal $\mathcal{P}$ associated to $(f'_1, \ldots, f'_{k-1})$ such that $f'_k \in \mathcal{P}$. Therefore there should exist an index $j \geq k+1$ such that $f_j \notin \mathcal{P}$ since otherwise $\mathcal{P}$ would contain the ideal $(f'_1, \ldots, f'_k, f_{k+1}, \ldots, f_r)$; and by the hypothesis on the determinant of order k, this would imply that $(f_1, \ldots, f_r) \subset \mathcal{P}$. This leads to the contradiction $n - (k-1) = \dim \mathcal{P} \leq \dim (f_1, \ldots, f_r) = n - r$ or -1. Finally let $j \geq k+1$ be such that $f_j \notin \mathcal{P}$: the extension of the ideal $\mathcal{P}$ to the localization $\mathbb{C}[\mathcal{U}_j]$ is therefore also an associated prime ideal of $(f'_1, \ldots, f'_{k-1}) \subset \mathbb{C}[\mathcal{U}_j]$, and this implies that $f'_k \in \mathcal{P}$ is a zero-divisor in $\mathbb{C}[\mathcal{U}_j]/(f'_1, \ldots, f'_{k-1})$ which contradicts the fact that $f'_1, \ldots, f'_k$

is a regular sequence in $\mathbb{C}[\mathcal{U}_j]$. We conclude that $f'_1, \dots, f'_k$ must be a regular sequence.

(iii) Now let us verify that $(f'_1, \dots, f'_k)$ is a radical ideal, $1 \leq k \leq r-1$: Suppose that $g \in \mathrm{rad}(f'_1, \dots, f'_k)$, i.e., there exists N such that $g^N \in (f'_1, \dots, f'_k)$. We can view this property in $\mathbb{C}[\mathcal{U}_{k+1}], \dots, \mathbb{C}[\mathcal{U}_r]$, where it is a radical ideal, i.e., for M big enough, $g\, f_{k+1}^M, \dots, g\, f_r^M \in (f'_1, \dots, f'_k)$. Now if $g \notin (f'_1, \dots, f'_k)$, it means that $f_{k+1}, \dots, f_r$ belong all to the same prime ideal $\mathcal{P}$ associated to $(f'_1, \dots, f'_k)$, and we conclude as before.

Now, we are going to see how to determine the integers $\lambda_{i\,j} \in \mathbb{Z}$ of the matrix Q. Let us introduce some new variables $T_{i\,j}$, $1 \leq i \leq r-1$, $1 \leq j \leq r$, and consider the polynomial ring $R := \mathbb{Z}[T_{i\,j}]$, its quotient field K, and an algebraic closure $\mathbb{K}$.

Since Bertini holds generically, the polynomials in $R[X_1, \dots, X_n]$ defined as

$$F_i := T_{i\,1} f_1 + \cdots + T_{i\,r} f_r \qquad (1 \leq i \leq r-1)$$

verify that for every $1 \leq k \leq r-1$ the ideal $(F_1, \dots, F_k)$ is smooth of dimension $n-k$ in $K[X_1, \dots, X_n]_{f_j}$ for all $1 \leq j \leq r$, and the same result is true for almost every specialization $T_{i\,j} \longrightarrow \lambda_{i\,j}$. Our aim is to determine deterministically integer values $\lambda_{i\,j}$ $(1 \leq i \leq r-1, 1 \leq j \leq r)$ such that the polynomials F_i obtained specializing $T_{i\,j}$ in $\lambda_{i\,j}$ verify this smoothness property, simultaneously with the conditions $\det(\lambda_{i\,j})_{1 \leq i,j \leq k} \neq 0 \quad (1 \leq k \leq r-1)$.

Let $J_k = (\frac{\partial F_i}{\partial X_j})$ denote the Jacobian matrix of the $(F_1, \dots, F_k)$'s and let $\Delta_1^{(k)}, \dots, \Delta_{s_k}^{(k)}$ denote its $k \times k$ minors.

We recall the following Jacobian criterion (see [41, theorem 30.3 and proof]):

Let $\mathcal{O}$ be a regular local ring and $f_1, \dots, f_r \in \mathcal{O}$ be such that both conditions $(f_1, \dots, f_r) \neq (1)$ and $(f_1, \dots, f_r, \Delta_1, \dots, \Delta_s) = (1)$ hold (where Δ_j denote the maximal minors of the Jcobian matrix of the f_i's). Then $\mathcal{O}/(f_1, \dots, f_r)$ is a regular ring.

This criterion shows that the smoothness property of $(F_1, \dots, F_k)$ $(1 \leq k \leq r-1)$ is guaranteed when for every $1 \leq k \leq r-1$, $1 \leq j \leq r$ the following two conditions (in terms of triviality of ideals) hold:

- The ideal $(F_1, \dots, F_k)$ is not the trivial ideal in $K[X_1, \dots, X_n]_{f_j}$.
- The ideal $((F_1, \dots, F_k, \Delta_1^{(k)}, \dots, \Delta_{s_k}^{(k)})$ is the trivial ideal in $K[X_1, \dots, X_n]_{f_j}$.

(Note that the regularity condition of $F_1, \dots, F_{r-1}$ in $K[X_1, \dots, X_n]_{f_j}$ cannot be simply recovered from the regularity in the localizations at maximal ideals containing $(F_1, \dots, F_{r-1})$: we need the conditions for $1 \leq k \leq r-1$.)

Let Y denote a new variable. We can restate these conditions as:

i) The ideal $(F_1, \ldots, F_r, 1 - Yf_j)$ is not the trivial ideal in $K[X_1, \ldots, X_n, Y]$.

ii) For every k, $1 \leq k \leq r-1$, the ideal $(F_1, \ldots, F_k, \Delta_1^{(k)}, \ldots, \Delta_{s_k}^{(k)}, 1 - Yf_j)$ is the trivial ideal in $K[X_1, \ldots, X_n, Y]$.

Now we analyze conditions on the λ_{ij}'s which suffice to guarantee properties i) and ii) above. This can be obtained from the triviality test of [23].
For the first condition, there is an arithmetic network $\mathcal{N}$ of size $d^{O(n)}$, including polynomials in $Q[T_{ij}]$ of degree $d^{O(n)}$ which evaluates a formula equivalent to the property "the ideal $(F_1, \ldots, F_r, 1 - Yf_j)$ is not the trivial ideal in $K[X_1, \ldots, X_n, Y]$". This property holds generically on the variables T_{ij}. Therefore the formula evaluated by the network must contain an open conjunction, i.e., a conjunction of the kind:

$$G_1(T_{i,j}) \neq 0 \wedge \cdots \wedge G_R(T_{i,j}) \neq 0$$

where the degree of the G_j's is $d^{O(n)}$ and R is of order $d^{O(n)}$.
On the other hand, the same method allows to get an arithmetic network $\mathcal{N}'_k$ of size $d^{O(n)}$ which contains polynomials of degree $d^{O(n)}$ in the variables T_{ij} and evaluates a formula equivalent to the property "the ideal $(F_1, \ldots, F_k, \Delta_1^{(k)}, \ldots, \Delta_{s_k}^{(k)}, 1 - Yf_j)$ is the trivial ideal in $K[X_1, \ldots, X_n, Y]$". Again, since this holds generically on the T_{ij}'s, for every k, $1 \leq k \leq r-1$, the network $\mathcal{N}'_k$ contains an open conjunction:

$$H_1^{(k)}(T_{i,j}) \neq 0 \wedge \cdots \wedge H_{R_k}^{(k)}(T_{i,j}) \neq 0$$

Again, R_k is of order $d^{O(n)}$ and these polynomials have degree at most $d^{O(n)}$. Combining all these conjunctions, we obtain a single polynomial $G(T_{ij})$ (the product of all of them) whose degree is $d^{O(n)}$ and the condition:

$$G(T_{i,j}) \neq 0$$

implies the desired properties. We conclude choosing $\lambda_{ij} \in \mathbb{Z}$ such that $G(\lambda_{ij}) \neq 0$. Clearly the height of these integers λ_{ij} can be bounded by $d^{O(n)}$. Finally we apply a strategy as in section 4.8 (where we study the triviality of the ideals introducing generic linear forms) in order to find a matrix Q verifying the announced bounds. □

6.2 Simultaneous Noether normalization

We will use our analysis on questor sets and the sequential method described in [14] to obtain a simultaneous Noether normalization for a regular sequence $f_1, \ldots, f_r$ of polynomials in $\mathbb{Z}[X_1, \ldots, X_n]$. This simultaneous normalization is performed by a linear change of variables given by an upper triangular matrix $Q \in \mathcal{M}_n(\mathbb{Z})$ such that the new variables $\{X_1', \ldots, X_n'\}$ are in Noether position with respect to the regular sequence, i.e., for every $1 \le i \le r$,

$$\mathbb{Q}[X_1', \ldots, X_{n-i}'] \hookrightarrow \mathbb{Q}[X_1', \ldots, X_n']/(f_1, \ldots, f_i)$$

is an integral ring extension.

Proposition 38 *Let $f_1, \ldots, f_r \in \mathbb{Z}[X_1, \ldots, X_n]$ be polynomials of degrees bounded by d defining a regular sequence in $\mathbb{Q}[X_1, \ldots, X_n]$. Then there exists an upper triangular matrix $Q \in \mathcal{M}_n(\mathbb{Z})$ such that (up to a reorder of the variables) the variables*

$$\begin{pmatrix} X_1 \\ \vdots \\ X_n \end{pmatrix} = Q \begin{pmatrix} X_1' \\ \vdots \\ X_n' \end{pmatrix}$$

are in Noether position with respect to the given regular sequence. The height of the matrix Q is of order $d^{O(n)}$. In the particular case where all the ideals $(f_1, \ldots, f_i)$, $1 \le i \le r$, are radical ideals, the bound for $ht(Q)$ can be restricted to $d^{O(r)}$.

Proof. Here we apply essentially the analysis of [14], where the Noether normalization of an ideal is constructed inductively. Let us denote by V_i the zero set of the ideal $(f_1, \ldots, f_i)$ $(1 \le i \le r)$ in $\mathbb{C}^n$. Without loss of generality, since $f_1, \ldots, f_r$ is a regular sequence, we can reorder the variables $X_1, \ldots, X_n$ in such a way that they are independent with respect to the ideals $(f_1, \ldots, f_r), (f_1, \ldots, f_{r-1}), \ldots, (f_1)$ (i.e., $\mathbb{Q}[X_1, \ldots, X_{n-r}] \cap (f_1, \ldots, f_r) = (0), \ldots, \mathbb{Q}[X_1, \ldots, X_{n-1}] \cap (f_1) = (0)$). For every i and every $j \in \{1, \ldots, i-1\}$ we consider the mapping $\pi_{ij} : \mathbb{C}^n \to \mathbb{C}^{n-i+1}$ given by the projection onto the coordinates $(x_1, \ldots, x_{n-i}, x_{n-j})$. The image of V_i by π_{ij} is an hypersurface given as the zero set of a non-null polynomial $p_{ij} \in \mathbb{C}[X_1, \ldots, X_{n-i}, X_{n-j}]$ of degree bounded by $\deg(V_i)$, i.e., bounded by d^i (cf. [26, Remark 4]). This means that there exists $N \in \mathbb{N}$ big enough such that $\pi_{ij}^N \in (f_1, \ldots, f_i)$. The effective Nullstellensatz ([10],[12], [34] and [17]) says now that we can choose $N \le \max\{3, d\}^n$ (in the particular case where $(f_1, \ldots, f_i)$ is radical we take $N = 1$). In this way we show that there exists a non-null polynomial $g_{ij} \in (f_1, \ldots, f_i) \cap \mathbb{Q}[X_1, \ldots, X_{n-i}, X_{n-j}]$ of degree bounded by d^{n+i} (d^i in the radical case).
We are looking for new variables $X_1', \ldots, X_n'$ such that:

- $X_{n-r+1}', \ldots, X_n'$ are integer with respect to $X_1', \ldots, X_{n-r}'$ modulo $(f_1', \ldots, f_r')$ (where f_j' denote the new polynomial obtained performing the change of variables),
- $X_{n-r+2}', \ldots, X_n'$ are integer with respect to $X_1', \ldots, X_{n-r}', X_{n-r+1}'$ modulo $(f_1', \ldots, f_{r-1}')$,

⋮

- and finally X'_n is integer with respect to $X'_1, \ldots, X'_{n-1}$ modulo (f'_1).

We proceed inductively, and construct the matrix Q column by column, by introducing at every step a change of coordinates.

– *First step:* we consider the dependent variable X_{n-r+1} with respect to X_1, ..., X_{n-r}: There is a non-null polynomial $g_1 \in \mathbb{Q}[X_1, \ldots, X_{n-r}, X_{n-r+1}]$ which belongs to $(f_1, \ldots, f_r)$. Let G_1 be the (non-null) homogeneous component of maximal degree of g_1 (observe that $G_1(X_1, \ldots, X_{n-r}, 1) \neq 0$). The polynomial G_1 has degree bounded by d^{n+r} and can be trivially evaluated by a non-scalar straight-line program of depth $O((n + r) \log_2 d)$. and so does $G_1(X_1, \ldots, X_{n-r}, 1)$.
Therefore we can apply corollary 19 (about the height of questor sets) to affirm that there exist integers $\lambda_1^{(1)}, \ldots, \lambda_{n-r}^{(1)} \in \mathbb{Z}$ of height of order $O(d^{3(n+r)})$ such that:

$$G_1(\lambda_1^{(1)}, \ldots, \lambda_{n-r}^{(1)}, 1) \neq 0.$$

We make the following change of variables:

$$X_i := \begin{cases} X_i^{(1)} + \lambda_i^{(1)} X_{n-r+1}^{(1)} & \text{for } i = 1, \ldots, n-r \\ X_i^{(1)} & \text{for } i = n-r+1, \ldots, n \end{cases}.$$

This change verifies that $g_1(X_1, \ldots, X_{n-r+1}) =: g_1^{(1)}(X_1^{(1)}, \ldots, X_{n-r+1}^{(1)})$ is monic in $X_{n-r+1}^{(1)}$, since its principal coefficient in $X_{n-r+1}^{(1)}$ is exactly $G_1(\lambda_1^{(1)}$, $\ldots, \lambda_{n-r}^{(1)}, 1)$. Thus $X_{n-r+1}^{(1)}$ is integral with respect to $X_1^{(1)}, \ldots, X_{n-r}^{(1)}$ modulo $(f_1^{(1)}, \ldots, f_r^{(1)})$.
(This first change of variables corresponds to an identity matrix unless for the first $n - r$ elements of the column $n - r + 1$ which correspond to $\lambda_j^{(1)}$.)

Observe that in this procedure the degrees of the new polynomials $f_1^{(1)}, \ldots,$ $f_r^{(1)}$ which define the ideal didn't change.

– *Second step:* Because of the assumptions on the order of the variables, and the independence of $X_1^{(1)}, \ldots, X_{n-r}^{(1)}$, there exist (as in the first step) non-null polynomials $g_2 \in \mathbb{Q}[X_1^{(1)}, \ldots, X_{n-r+2}^{(1)}] \cap (f_1^{(1)}, \ldots, f_{r-1}^{(1)})$ and $h_2 \in \mathbb{Q}[X_1^{(1)}, \ldots, X_{n-r}^{(1)}, X_{n-r+2}^{(1)}] \cap (f_1^{(1)}, \ldots, f_r^{(1)})$. Let G_2 and H_2 be the homogeneous components of maximal degree of g_2 and h_2 respectively. and consider the polynomial

$$G := G_2(X_1^{(1)}, \ldots, X_{n-r+1}^{(1)}, 1).H_2(X_1^{(1)}, \ldots, X_{n-r}^{(1)}, 1).$$

This polynomial is evaluable by a straight-line program of length $\log_2(2\,d^{n+r})$.

As in the first step, there exist integers $\lambda_1^{(2)}, \ldots, \lambda_{n-r+1}^{(2)} \in \mathbb{Z}$ of height of order $O(d^{3(n+r)})$ such that:

$$G_2(\lambda_1^{(2)}, \ldots, \lambda_{n-r+1}^{(2)}, 1) \neq 0 \quad \text{and} \quad H_2(\lambda_1^{(2)}, \ldots, X_{n-r}^{(2)}, 1) \neq 0.$$

And we make the change of variables:

$$X_i^{(1)} := \begin{cases} X_i^{(2)} + \lambda_i^{(2)} X_{n-r+2}^{(2)} & \text{for } i = 1, \ldots, n-r+1 \\ X_i^{(2)} & \text{for } i = n-r+2, \ldots, n \end{cases}.$$

Thus the polynomial $g_2^{(2)}(X_1^{(2)}, \ldots, X_{n-r+2}^{(2)}) := g_2(X_1^{(1)}, \ldots, X_{n-r+2}^{(1)})$ is monic in $X_{n-r+2}^{(2)}$, i.e., $X_{n-r+2}^{(2)}$ is integral with respect to $X_1^{(2)}, \ldots, X_{n-r+1}^{(2)}$ modulo $(f_1^{(2)}, \ldots, f_{r-1}^{(2)})$. Similarly $h_2^{(2)}(X_1^{(2)}, \ldots, X_{n-r}^{(2)}, X_{n-r+2}^{(2)})$ is monic in $X_{n-r+2}^{(2)}$, i.e., $X_{n-r+2}^{(2)}$ is integral with respect to $X_1^{(2)}, \ldots, X_{n-r}^{(2)}$ modulo $(f_1^{(2)}, \ldots, f_r^{(2)})$. Finally, by a transitive argument, we conclude that $X_{n-r+1}^{(2)}$ is also integral with respect to $X_1^{(2)}, \ldots, X_{n-r}^{(2)}$ modulo $(f_1^{(2)}, \ldots, f_r^{(2)})$. The proof continues by induction, repeating the same arguments. (In the worst case we have to deal with r polynomials of degree d^{n+r}.) We obtain finally the upper triangular matrix obtained by filling its last r columns with the respective $\lambda_i^{(j)}$. □

7 Nullstellensatz and membership problem

7.1 Describing the output polynomials

Finally we are able to show our two main results, theorem 1 and 2, already stated in the introduction. The first one concerns the effective Nullstellensatz:

Theorem 1 (Effective Nullstellensatz) *Let $f_1, \ldots, f_s$ be polynomials in $\mathbb{Z}[X_1, \ldots, X_n]$ of degree and height bounded by $d \geq n$ and η, respectively, which don't share any zero in $\mathbb{C}^n$. Then there exists a straight-line program of size $s\, d^{O(n)}$, depth $O(n \log_2 d)$, and parameters of order $\max\{d^{O(n)}, \eta\}$ which computes an integer $a \in \mathbb{Z} - \{0\}$ and polynomials $g_1, \ldots, g_s \in \mathbb{Z}[X_1, \ldots, X_n]$ such that:*

$$a = g_1 f_1 + \cdots + g_s f_s.$$

This theorem admits straightforward the following corollary, simply applying proposition 15 which relates degrees of polynomials and heights of coefficients with the size, depth, and parameters of the corresponding straight-line program:

Corollary 4 *Let $f_1, \ldots, f_s$ be polynomials in $\mathbb{Z}[X_1, \ldots, X_n]$ of degree and height bounded by $d \geq n$ and η, respectively, which don't share any zero in $\mathbb{C}^n$. Then there exist an integer $a \in \mathbb{Z} - \{0\}$ and polynomials $g_1, \ldots, g_s \in \mathbb{Z}[X_1, \ldots, X_n]$ such that:*

- $a = g_1 f_1 + \cdots + g_s f_s$
- $\deg g_i = d^{O(n)} \quad (1 \leq i \leq s)$
- $\max\{\log_2 ht(a), \log_2 ht(g_1), \ldots, \log_2 ht(g_s)\} = d^{O(n)} \max\{\log_2(\eta), \log_2 s\}$.

The next result deals with membership and representation problem for an ideal $(f_1, \ldots, f_r)$, $r \leq n$ of dimension exactly $n - r$:

Theorem 2 (Membership problem) *Let $f_1, \ldots, f_r$ $(r \leq n)$ and g be polynomials in $\mathbb{Z}[X_1, \ldots, X_n]$ of degree and height bounded by $d \geq n$ and η, respectively. Assume that $(f_1, \ldots, f_r)$ is a proper ideal in $\mathbb{Q}[X_1, \ldots, X_n]$ of dimension $n - r$ and that $g \in (f_1, \ldots, f_r)$ holds. Then there exists a straight-line program of size $d^{O(n)}$, depth $O(r \log_2 d)$ and parameters of order $\max\{d^{O(n)}, \eta\}$ which computes an integer $a \in \mathbb{Z} - \{0\}$ and polynomials $g_1, \ldots, g_r \in \mathbb{Z}[X_1, \ldots, X_n]$ such that*

$$a\, g = g_1 f_1 + \cdots + g_r f_r.$$

Corollary 5 *Let $f_1, \ldots, f_r$ $(r \leq n)$ and g be polynomials in $\mathbb{Z}[X_1, \ldots, X_n]$ of degree and height bounded by $d \geq n$ and η, respectively. Assume that $(f_1, \ldots, f_r)$ is a proper ideal in $\mathbb{Q}[X_1, \ldots, X_n]$ of dimension $n - r$ and that $g \in (f_1, \ldots, f_r)$ holds. Then there exist an integer $a \in \mathbb{Z} - \{0\}$ and polynomials $g_1, \ldots, g_r \in \mathbb{Z}[X_1, \ldots, X_n]$ such that*

- $a\, g = g_1 f_1 + \cdots + g_r f_r$
- $\deg g_i = d^{O(r)} \quad (1 \leq i \leq r)$
- $\max\{\log_2 ht(a), \log_2 ht(g_1), \ldots, \log_2 ht(g_r)\} = d^{O(r)} n \log_2(\eta)$.

The proof of the two results is similar: It starts with the preparation of the input polynomials (section 6) in order to apply recursively the division lemma (lemma 30). Here we are going to treat first the proof of theorem 2, and at the end of the proof we will sketch briefly the different steps of the proof of the Nullstellensatz.

Proof. *(Membership problem)* First of all we prepare our input polynomials $f_1, \ldots, f_r$ applying proposition 37 (Bertini theorem) to $f_1, \ldots, f_r$ and proposition 38 (simultaneous Noether normalization) to the obtained $f_1, \ldots, f_{r-1}$ which are all radical ideals. We also call $f_1, \ldots, f_r$ the output of this procedure. For all i, $0 \leq i \leq r - 1$, set

$$\begin{aligned} \mathcal{J}_i &:= (f_1, \ldots, f_i) \\ R_i &:= \mathbb{Z}[X_1, \ldots, X_{n-i}] \\ A_i &:= \mathbb{Q}[X_1, \ldots, X_{n-i}] \\ B_i &:= A_i[X_{n-i+1}, \ldots, X_n]/\mathcal{J}_i \end{aligned}$$

Our preliminar input preparation insures that for all i, $0 \leq i \leq r-1$, $A_i \hookrightarrow B_i$ is in Noether position, and that $\mathcal{J}_i$ is a reduced complete intersection ideal. The mathematical idea to perform the division is straightforward and will suggest the computational treatment; we apply the division theorem (theorem 29) as follows:

- $f_r \mid g \pmod{\mathcal{J}_{r-1}}$ implies that we can compute $\xi_r \in R_{r-1}$ and $q_r \in R_{r-1}[X_{n-r+2}, \ldots, X_n]$ such that $\xi_r \mid q_r$ in $\mathbb{Q}[X_1, \ldots, X_n]$ and $\xi_r g \equiv q_r f_r \pmod{\mathcal{J}_{r-1}}$. This means that $\xi_r \left(g - \frac{q_r}{\xi_r} f_r\right) \in \mathcal{J}_{r-1}$ which is an equidimensional ideal of dimension $n-(r-1)$ which doesn't meet R_{r-1}. Therefore $g - \frac{q_r}{\xi_r} f_r \in \mathcal{J}_{r-1}$, i.e., $f_{r-1} \mid g - \frac{q_r}{\xi_r} f_r \pmod{\mathcal{J}_{r-2}}$. In other words,

$$\xi_r f_{r-1} \mid \xi_r g - q_r f_r \pmod{\mathcal{J}_{r-2}}.$$

- Now set $g_{r-1} := \xi_r g - q_r f_r$ and divide it by $\xi_r f_{r-1}$. We can compute $\xi_{r-1} \in R_{r-2}$ and $q_{r-1} \in R_{r-2}[X_n - r + 3, \ldots, X_n]$ such that $\xi_{r-1} g_{r-1} \equiv q_{r-1} \xi_r f_{r-1} \pmod{\mathcal{J}_{r-2}}$.

- Recursively we obtain

$$\xi_1 \cdots \xi_r g = q_1 \xi_2 \cdots \xi_r f_1 + \xi_1 q_2 \xi_3 \cdots \xi_r f_2 + \cdots + \xi_1 \ldots \xi_{r-1} q_r f_r$$

  (where for all i, $\xi_i \mid q_i$ in $\mathbb{Q}[X_1, \ldots, X_n]$)

and we conclude by one division, keeping denominators in $\mathbb{Z}$ at the left side of the equality.

Unfortunately this simple procedure is not allowed for our purpose since it leads to a straight-line program of depth essentially $O(n^2 \log_2 d)$, due to the fact that each q_i (which can be computed in depth $O(n \log_2 d)$) depends recursively on q_{i+1}. This procedure would yield to bounds of order $d^{O(n^2)}$ for the logarithmic height of the g_i's.

We will use here our more technical lemma 30: The crucial point for us is that it constructs a matrix Q which applied linearly on the coefficients of the dividend g gives the coefficients of the quotient q, and this depending only on $\deg^* g$ and on the coefficients of the f_i's.

Set:

$$\begin{array}{lcl} g_r & := & g \\ g_{r-1} & := & \xi_r g_r - q_r f_r \\ g_{r-2} & := & \xi_{r-1} g_{r-1} - \xi_r q_{r-1} f_{r-1} \\ & \vdots & \\ g_{r-i} & := & \xi_{r-i+1} g_{r-i+1} - \xi_{r-i+2} \cdots \xi_r q_{r-i+1} f_{r-i+1} \\ & \vdots & \\ g_1 & := & \xi_2 g_2 - \xi_3 \cdots \xi_r q_2 f_2 \end{array}$$

Observe that the fact that $R_{r-i} \subset R_{r-i-1}$ implies that each q_{r-i}, g_{r-i} viewed as a polynomial with coefficients in R_{r-i-1} verifies that $\deg^* q_{r-i} \leq (r-1)d$ and $\deg^* g_{r-i} \leq rd$.

In the sequel, we denote by $(h)_k$ the vector of coefficients of a polynomial h viewed as a polynomial in $R_k[X_{n-k+1}, \ldots, X_n]$.

Now we compute independently from $g_1, \ldots, g_r$ the following items:

- All the coefficients in $R_0, \ldots, R_{r-1}$ of the polynomials $f_1, \ldots, f_r$, and the coefficients in R_{r-1} of the polynomial g (in size $O(d^n)$, depth $O(\log_2 d)$, and parameters of height bounded by η).

- All the values $\xi_j \in R_{j-1}$ and all the matrices $Q_j \in R_{j-1}[X_{n-j+2}, \ldots, X_n]$ which verify that $Q_j((g_j)_{j-1}) = ((q_j)_{j-1})$ (in depth $O(r \log_2 d)$, size $d^{O(r)}$, and parameters of height $d^{O(r)}$).

- Matrices T_j $(1 \leq j \leq r)$ which transform the coefficients of a polynomial $h \in R_{j-1}[X_{n-j+2}, \ldots, X_n]$ of degree bounded by $(r-1)d$ on the coefficients of the polynomial $f_j h$ of degree $\leq rd$ (Size $d^{O(r)}$ and depth $O(\log_2 d)$).

- Matrices B_j $(1 \leq j \leq r)$ which transform the coefficients in R_{j-1} of a polynomial h of degree rd on its coefficients when viewed as a polynomial with coefficients in R_{j-2}. (Here B_1 will denote the identity matrix.) (Same size and depth.)

All these computations can then be performed by means of a straight-line program of size $d^{O(n)}$, depth $O(r \log_2 d)$, and parameters of height $\max\{d^{O(r)}, \eta\}$, and observe that the obtained value $d^{O(n)}$ for the size comes only from a precomputation of the coefficients of the input polynomials in the rings R_i.
We have

- $$\begin{aligned} (g_{r-1})_{r-2} &= B_r(\xi_r g_r - f_r q_r)_{r-1} \\ &= (B_r(\xi_r Id - T_r Q_r))\,(g_r)_{r-1} \end{aligned}$$
- $$\begin{aligned} (g_{r-2})_{r-3} &= B_{r-1}(\xi_{r-1} g_{r-1} - \xi_r f_{r-1} q_{r-1})_{r-2} \\ &= (B_{r-1}(\xi_{r-1} Id - \xi_r T_{r-1} Q_{r-1}))\,(g_{r-1})_{r-2} \\ &= (B_{r-1}(\xi_{r-1} Id - \xi_r T_{r-1} Q_{r-1}))\,(B_r(\xi_r Id - T_r Q_r))\,(g_r)_{r-1} \end{aligned}$$

etc.

Thus if we define for $1 \leq j \leq r$:

$$N_j := B_j(\xi_j Id - \xi_{j+1} \cdots \xi_r T_j Q_j)$$

we obtain:

$$(g_j)_{j-1} = (N_{j+1} \cdots N_r)(g_r)_{r-1}.$$

This fact allows to compute independently all the coefficients of the polynomials $g_j \in R_{j-1}[X_{n-j+2}, \ldots, X_n]$, as a product of matrices applied to $(g)_{r-1}$. Finally, for each g_j we compute $(q_j)_{j-1} = Q_j((g_j)_{j-1})$.
Without changing essentially size, depth and height of the parameters, as in the proof of the division lemma 30, we construct a straight-line program that evaluates from the variables $X_1, \ldots, X_n$ the polynomials $\xi_1, \ldots, \xi_r, q_1, \ldots, q_r$.
We obtain

$$\xi_1 \cdots \xi_r g = q_1 \xi_2 \cdots \xi_r f_1 + \xi_1 q_2 \xi_3 \cdots \xi_r f_2 + \cdots + \xi_1 \cdots \xi_{r-1} q_r f_r$$

(where for all i, $\xi_i \mid q_i$ in $\mathbb{Q}[X_1, \ldots, X_n]$), and we conclude applying proposition 21 (Vermeidung von Divisionen) in order to obtain $a \in \mathbb{Z}$ and $g'_1, \ldots, g'_r \in \mathbb{Z}[X_1, \ldots, X_n]$ such that

$$ag = g'_1 f_1 + \cdots + g'_r f_r.$$

The last step of the proof consists in recovering the expression of the original input polynomials from $f_1, \ldots, f_r$, by means of the inverses of the matrices of lemmas 37 and 38. This procedure adds size n, one step depth, and parameters of height $d^{O(r)}$. (It also slightly modifies the constant a multiplying it by the determinant of the concerned matrices.)
Therefore the total size of our straight-line program is of order $d^{O(n)}$, its depth is $O(r \log_2 d)$, and the parameters are of height $\max\{d^{O(r)}, \eta\}$.

(Nullstellensatz) The only difference is that our Bertini preparation allows to suppose that our input polynomials are $f_1, \ldots, f_{r+1}$ with $r \leq n$, where for all $i \leq r$, $(f_1, \ldots, f_i)$ are reduced complete intersection ideals and f_{r+1} is not a zero-divisor modulo $(f_1, \ldots, f_r)$. We begin the algorithm setting $g := 1$. □

7.2 A probabilistic Turing machine

We end this work describing briefly a probabilistic Turing machine that solves the effective Nullstellensatz problem (the one for the membership problem being analogous). We follow the model of [1, chapter 6].

Corollary 3 *The effective Nullstellensatz and the membership problem to complete intersection ideals are in the BPP-class of bounded error-probabilistic polynomial time machines, i.e., there is a probablistic Turing machine that solves both problems within time polynomial in the dense representation of the input polynomials. In particular, both problems can be solved in polynomial space counting also "external memory" requirements.*

Proof. The probabilistic Turing machine makes a guessing of a list $\mathcal{Q}$ of $t = d^{O(n)}$ nonnegative integer vectors of height $u = d^{O(n)}$ (which represents a candidate for a correct test sequence as in corollary 19). The probability of error for the procedure is bounded by $\frac{1}{u^{t/6}}$ as a consequence of the bounds in corollary 19.

Then the Turing machine tests the triviality of our ideal as in 4.8. If the ideal is trivial, it proceeds as in subsection 6.1 to determine the parameters for Bertini's theorem. These parameters are chosen from $\mathcal{Q}$. As observed there, the parameters are chosen by means of some triviality tests of simple ideals.

Now it proceeds as in [23] with generic linear changes of coordinates to get the simultaneous Noether normalization. The parameters for the linear change of coordinates are again obained from $\mathcal{Q}$. They are less precise than those obtained in subsection 6.2, but they also can be obtained in polynomial time.

The method described in section 4 allows to obtain from $\mathcal{Q}$ a primitive element for every Noether extension and to determine the bases of the extensions.

Following the steps in section 5 it describes the matrices for the division in the reduced complete intersections.

Finally, it performs as in subsection 7.1 above the iterated matrix product and applies the avoiding division method to obtain the encoding of the output.

Observe that there are only some steps where the Turing machine makes use of the guessing. In these steps, if no concrete candidate is acceptable, the machine rejects the input. Hence, the probability of error is simply the probability that $\mathcal{Q}$ is not a correct test sequence.
Since the binary size of all intermediate results never increases over $d^{O(n)}$ (cf. proposition 15), the whole procedure runs in polynomial time and corollary 3 is shown. □

Acknowledgments. We are indebted to Joos Heintz for having strongly inspirated this work, and to Pablo Solernó for many helpful suggestions. The first author also thanks the members of the Mathematics Department of the University of Cantabria (Santander) for their hospitality during April, 1994, when last details of this paper were discussed.

References

[1] J. L. Balcazar, J. L. Díaz, J. Gabarró. "*Structural Complexity I*". EATCS Mon. on Theor. Comp. Sci. **11**. Springer, 1988.

[2] C. Berenstein, A. Yger. "Effective Bézout identities in $\mathbb{Q}[X_1, \ldots, X_n]$". *Acta Math.*, **166** pp. 69–120, 1991.

[3] C. Berenstein, A. Yger. "Une Formule de Jacobi et ses conséquences". *Ann. Sci. E.N.S., 4^{ieme} série*, **24** pp. 363–377, 1991.

[4] S. J. Berkowicz."On computing the determinant in small parallel time using a small number of processors", *Inf. Proc. Letters* **18** pp. 147–150, 1984.

[5] A. Borodin, J. von zur Gathen, J. Hopcroft. "Fast parallel matrix and gcd computations", *Proc. 23th Annual Symp. FOCS* pp. 65–71, 1982.

[6] A. Borodin."On Relating Size and Depth to Time and Space". *SIAM J. on Comp.* **6** pp. 733–744, 1977.

[7] J.-B. Bost, H. Gillet, C. Soulé. "Un analogue arithmétique du théorème de Bézout", *C.R. Acad. Sci. Paris, t.* **312**, *Série I* pp. 845–848, 1991.

[8] J.-B. Bost, H. Gillet, C. Soulé. "Heights of projective varieties and positive Green forms", *Manuscrit I.H.E.S.*, 1993.

[9] J. Briançon."On Euclid's algorihtm and the computation of polynomial greatest common divisors", *J. ACM* **18** pp. 478–504, 1971.

[10] D. W. Brownawell. "Bounds for the degree in the Nullstellensatz", *Annals of Math.* **126** pp. 577–591, 1987.

[11] D. W. Brownawell. "Local Diophantine Nullstellen inequalities", *J. Amer. Math. Soc.* **1** pp. 311–322, 1988.

[12] L. Caniglia, A. Galligo, J. Heintz. "Borne simplement exponentielle pour les degrés dans le théorème des zéros sur un corps de caractéristique quelconque", *C.R. Acad. Sci. Paris, t.* **307** *Série I* pp. 255–258, 1988.

[13] J. Canny. "Some algebraic and geometric problems in PSPACE", *Proc. 20-th Ann. ACM Symp. Theory of Computing* pp. 460–467, 1988.

[14] A. Dickenstein, N. Fitchas, M. Giusti, C. Sessa. "The membership problem for unmixed polynomial ideals is solvable in single exponential time", *Discrete Appl. Math.* **33** pp. 73–94, 1991.

[15] A. Dickenstein, C. Sessa. "Résidus de formes méromorphes et cohomologie modérée", *Géométrie Complexe, Prépublication Université Paris 7*, 1994.

[16] M. Elkadi. "Bornes pour le degré et les hauteurs dans le problème de division", to appear in *Michigan Math. J.*, 1993.

[17] N. Fitchas, A. Galligo."Nullstellensatz effectif et conjecture de Serre (théorème de Quillen-Suslin) pour le Calcul Formel", *Math. Nach.* **149**, pp. 231–253, 1990.

[18] N. Fitchas, M. Giusti, F. Smietanski. "Sur la complexité du théorème des zéros", *Manuscript Ecole Polytechnique*, 1993.

[19] C. B. García and W. I. Zangwill. "Pathways to solutions, fixed points and equilibria", *Prentice-Hall, N. J.*, 1981.

[20] J. von zur Gathen. "Parallel algorithms for algebraic problems", *Proc. 13-th Conf. MFCS.* Springer LN Comp. Sci. **233**, pp. 93–112, 1986.

[21] A. O. Gel'fond. *Transcendental and Algebraic Numbers*, New York: Dover, 1960.

[22] M. Giusti, J. Heintz. "Algorithmes – disons rapides – pour la décomposition d'une variété algébrique en composantes irréductibles et équidimensionelles", *Progress in Math.* **94**, *T.Mora and C.Traverso Eds., Birkhäuser* pp. 169–193, 1991.

[23] M. Giusti, J. Heintz. "La détermination des points isolés et de la dimension d'une variété algébrique peut se faire en temps polynomial", to appear in *Proc. Int. Meeting on Commutative Algebra, Cortona*, 1991.

[24] M. Giusti, J. Heintz, J. Sabia. "On the efficiency of effective Nullstellensätze", *Computational Complexity* **3** pp. 56–95. 1993.

[25] D. Iu. Grigor'ev, N. N. Vorobjov. "Solving systems of polynomial inequalities in subexponential time", *J. Symb. Comp.* **5** pp. 37–64, 1988.

[26] J. Heintz. "Fast quantifier elimination over algebraically closed fields", *Theoret. Comp. Sci.* **24**, pp. 239–277, 1983.

[27] J. Heintz. "On the computational complexity of polynomials and bilinear mappings", *Proc. AAECC-5, Springer LNCS* **356**, pp. 269–300, 1989.

[28] J. Heintz, J. Morgenstern. "On the intrinsic complexity of elimination theory". *J. of Complexity* **9**, pp. 471–498, 1993.

[29] J. Heintz, C. P. Schnorr. "Testing Polynomials wich are easy to compute", *Proc. 12th Ann. ACM Symp. on Computing*, pp. 262–268, 1980.

[30] J. Heintz and M. Sieveking. "Lower Bounds for polynomials with algebraic coefficients", *Theoret. Comp. Sci.* **11**, pp. 321–330, 1980.

[31] G. Hermann. "Die Frage der endlich vielen Schritte in der Theorie der Polynomideale", *Math. Ann.* **95**, pp. 736–788, 1926.

[32] B. Iversen. *Generic Local Structures in Commutative Algebra*, Lect. Notes in Math. **310**, Springer-Verlag, 1973.

[33] J. P. Jouanolou. *Théorèmes de Bertini et applications*, Progress in Math. **42**, Birkhäuser, 1983.

[34] J. Kollár. "Sharp Effective Nullstllensatz", *J. of A.M.S.* **1**, pp. 963–975, 1988.

[35] E. Kunz. *Kähler Differentials*, Advanced Lectures in Math., Vieweg-Verlag, 1986.

[36] E. Kunz. *Introduction to Commutative Algebra and Algebraic Geometry*, Birkhäuser, 1985.

[37] E. Landau. "Sur quelques théorèmes de M. Petrovic relatifs aux zéros des fonctions analytiques", *Bull. S.M.F.*, t. **33**, pp. 251–261, 1905.

[38] S. Lang. *Fundamentals of Diophantine Geometry*, Springer-Verlag, 1983.

[39] S. Lang. *Transcendental Number Theory*, Addison-Wesley, 1966.

[40] D. Lazard. "Résolution des systèmes d'équations algébriques", Theor. Comp. Sci. **15**, pp. 77–110, 1981.

[41] H. Matsumura. *Commutative Ring Theory*, Cambridge University Press, 1986.

[42] E. Mayr. "Membership problem to polynomial ideals is exponential space complete", preprint.

[43] E. Mayr, A. Meyer. "The complexity of the word problem for commutative semigroups ", Advances in Math. **46**, pp. 305–329, 1982.

[44] M. Mignotte. *Mathématiques pour le Cacul Formel*, Presses Univ. de France, 1989.

[45] J. L. Montaña, L. M. Pardo. "Lower Bounds for Arithmetic Networks", *AAECC* vol. **4**, pp. 1–24, 1993.

[46] J. L. Montaña, L. M. Pardo, T. Recio. "The non-scalar model of complexity in computational semialgebraic geometry", *Progress in Mathematics* **94**, *T.Mora and C.Traverso eds., Birkhäuser*, pp. 346–362, 1991.

[47] D. W. Masser, G. Wüstholz. "Fields of large transcendence degree generated by values of elliptic functions", *Invent. Math.* **72**, pp. 407–463, 1971.

[48] J. Sabia, P. Solernó. "Bounds for traces in complete intersections and degrees in the Nullstellensatz", to appear in *AAECC Journal*, 1993.

[49] T. Schneider. *Einführung in die transzendenten Zahlen*, Springer-Verlag, Berlin, 1957.

[50] C. P. Schnorr. "Improved lower bounds on the number of multiplications/divisions which are necessary to evaluate polynomials". *Theor. Comp. Sci.* **7**, pp. 251–261, 1978.

[51] I. R. Shafarevich. "Basic Algebraic Geometry". Springer-Verlag, 1977.

[52] M. Shub, S. Smale. "On the intractability of Hilbert's Nullstellensatz and an algebraic version of P=NP?". *IBM Research Report 19624 (86196)*, 1994.

[53] V. Strassen. "Polynomials with rational coefficients which are hard to compute". *SIAM J. Comput.* **3**, pp. 128–149, 1974.

[54] V. Strassen. "Vermeidung von Divisionen", *Crelle J. Reine Angew. Math.* **264**, pp. 184–202, 1973.

[55] V. Strassen. "Algebraic Complexity Theory". *Hand Book of Theoretical Computer Science*, ch. **11**, pp. 634–671, 1990.

T. Krick (`krick@dm.uba.ar`)

Departamento de Matemática, Universidad de Buenos Aires. 1428 Buenos Aires (Argentina).

L. M. Pardo (`pardo@matsun1.unican.es`)

Departamento de Matemáticas, Estadística y Computación. Universidad de Cantabria, Santander 39071 (Spain).

Progress in Mathematics, Vol. 143,

An effective method to classify nilpotent orbits

P. Littelmann*

Introduction

Let $\rho : G \to GL(V)$ be a representation of a reductive algebraic group G defined over $\mathbb{C}$. A simple example for such a situation is the natural action of $SL_2(\mathbb{C})$ by coordinate substitution on the vector space $R_n := \mathbb{C}[x,y]_n$ of binary forms of degree n. The G-invariant polynomial functions on V are an important tool to study the orbit structure of the group in the representation space. It was one of the highlights in invariant theory before Hilbert when Gordan proved in 1868 that the ring of invariants is finitely generated in the example considered above. But even for this "simple-looking" example, a complete description of the invariants and the orbits exists only for the cases $n \leq 6$ and $n = 8$.

In general, a detailed investigation of the ring of invariants or the orbit structure of a representation seems to be hopeless. We call a representation "nice" if such a detailed investigation is possible. Of course, a precise definition of what "nice" means depends on the kind of information one is interested in. Here we take as a model for a "nice" representation the action of $GL_n(\mathbb{C})$ on $M_n(\mathbb{C})$ by conjugation, so a nice representation should have the following properties:

$(*)$ The ring of invariants $\mathbb{C}[V]^G$ is free and every fiber of the map $\pi : V \to \operatorname{Spec}\mathbb{C}[V]^G$ induced by the inclusion $\mathbb{C}[V]^G \hookrightarrow \mathbb{C}[V]$ is the union of a finite number of G-orbits. Further, there exists a finite group W and a subspace $\mathfrak{c} \subset V$ (the *Cartan subspace*) on which W acts, such that a G-orbit in V meets $\mathfrak{c}$ if and only if the orbit is closed, and the intersection of $G.x$ with $\mathfrak{c}$ is one W-orbit.

For the $GL_n(\mathbb{C})$-action on $M_n(\mathbb{C})$, the ring of invariants is generated by the elementary symmetric functions in the eigenvalues of a matrix: the second point corresponds to the fact that for a given set of eigenvalues there exist only a finite number of conjugacy classes having these eigenvalues (Jordan normal form); and the last point corresponds to the fact that a conjugacy class is closed if and only if the matrix is diagonalisable. The group W is here the symmetric group permuting the order of the eigenvalues.

A somewhat surprising fact is that almost all "nice" representations ([7], [10]) can be constructed as follows: There exists a reductive algebraic group $\tilde{G}$ and an automorphism θ of $\tilde{G}$ such that G is the connected component of the fixed point group of θ, and $d\theta$ induces a grading $\mathfrak{G} = \cdots \mathfrak{G}_{-1} \oplus \operatorname{Lie} G \oplus \mathfrak{G}_1 \cdots$ of the Lie algebra $\mathfrak{G}$ of $\tilde{G}$ such that V is isomorphic as G-module to $\mathfrak{G}_1$ (see

(*) Supported by Schweizerischer Nationalfonds.

section 1). A representation which can be obtained in this a way is called a *θ-representation.*

The representations one can construct as a θ-representation are classified ([6], [7]), the classification is closely related to the construction of affine Kac-Moody algebras. Because of the Cartan subspace $\mathfrak{c}$, the problem of classifying the closed orbits in a θ-representation is "reduced" to a finite group problem. Using the Jordan decomposition in $\mathfrak{G}$, the problem of finding the non-closed orbits can be transformed to the problem of finding nilpotent orbits in a "smaller" θ-representation. So to get an orbit classification for θ-representations, we are "reduced" to classify the nilpotent orbits, i.e., those orbits such that $0 \in \overline{Gx}$. For this problem a very nice method has been developed by Kac and Vinberg ([5], [11]), which transforms this problem to a combinatorial problem involving root systems and Weyl groups. A very good environment to deal with such problems is the software package LiE developed by a group at the CWI in Amsterdam.

The idea of the method is to associate to every nilpotent orbit a Lie algebra (*carrier algebra*), which characterizes this orbit uniquely. Further, Kac and Vinberg give precise conditions on a Lie algebra to be the carrier algebra of a nilpotent orbit. So the first step in our program (written in LiE) is to find a "good" list of candidates for the carrier algebras, i.e., the list should include a priori all carrier algebras, but the program should not spend to much time to clear out those algebras that are not carrier algebras.

The algorithm we have developed to get such a list of good candidates works for all θ-representations, but due to limitations of LiE, the program we have written does not yet work for all θ-representations. We have extended the method such that the program yields now the following additional information: For every nilpotent orbit an explicit representative $x \in V$ is given, the dimension of the orbit Gx is calculated, the structure of the stabilizer G_x is determined (i.e., the reductive part of the stabilizer is described in terms of its root system and the dimension of the center, for the unipotent radical the dimension is determined and the structure of its Lie algebra as module with respect to action of the semisimple part $(G_x)_{ss}$ of G_x is given), and the embedding of the stabilizer in $GL(V)$ is described (i.e., the structure of V as $(G_x)_{ss}$-module is determined).

In the first part of the article we give a short recollection of the main properties of θ-representations. In the second part we present a rather detailed exposition of the theoretical background of the program, i.e., we describe the carrier algebra method of Kac and Vinberg and how to extend this method to obtain the additional information on the stabilizer. In section 5, we describe how to "transform the method into an algorithm", i.e., we give a sketch of how the program performs the different steps of the method described in the sections before. The main point is, of course, how to get a reasonable list of candidates for the carrier algebras. At the end we present for a small example (the orbits of $GL_7(\mathbb{C})$ in $\bigwedge^3 \mathbb{C}^7$) the output of the program.

1 θ-representations and graded Lie algebras

1.0 The aim of this section is to recall the definition of θ-representations and its main properties. For a more detailed discussion we refer to the article of Vinberg [12], for the classification of the θ-representations we refer to [6], or the book [7] of Kac.

1.1 Suppose $\tilde{G} \subset GL_n(\mathbb{C})$ is a connected reductive algebraic group. For $h \in GL_n(\mathbb{C})$ let $\operatorname{int} h$ be the inner automorphism $g \mapsto hgh^{-1}$. We consider automorphisms θ of $\tilde{G}$ of the following special form: Either θ is of finite order and $\theta(g) = \operatorname{int} h(g)$ for some $h \in GL_n(\mathbb{C})$, or there exists a one-parameter subgroup $\lambda : \mathbb{C}^* \to \tilde{G}$, and $\theta = \theta_t$ denotes the family of inner automorphisms $\theta_t(g) = \operatorname{int} \lambda(t)(g)$ for $t \in \mathbb{C}^*$. We write G for the connected component of the fixed point group $\tilde{G}^\theta$ of θ.

If θ is of order m, then let $\mathfrak{G}_i$ be the eigenspace of $d\theta$ for the eigenvalue ξ^i, where $\xi = \exp(2\pi i/m)$. If θ is associated to a one parameter subgroup, then set

$$\mathfrak{G}_i := \{g \in \mathfrak{G} \mid \operatorname{Ad} \lambda(t)(g) = t^i g, t \in \mathbb{C}^*\}.$$

In any case, $\mathfrak{G} = \bigoplus_{i\in\mathbb{Z}_m} \mathfrak{G}_i$ (where $m = \infty$ in the last case) is a graded Lie algebra.

Note that if $\mathfrak{G}$ is semisimple and $\tilde{G}$ is simply connected, then any $\mathbb{Z}_m$-grading can be obtained in this way. In fact, any finite-order automorphism of $\tilde{G}$ can be realized in the way above. If $\mathfrak{G}$ is just reductive, then this is only true if the grading of the center is induced by an automorphism which is given by integral matrices for some basis of the center. We call a grading that is obtained in the way above an *algebraic* grading.

The subgroup $G \subset \tilde{G}$ acts via the restriction of the adjoint representation on $\mathfrak{G}$, and obviously each graded component $\mathfrak{G}_i$ is a G-stable submodule. The corresponding representation $\rho : G \to GL(\mathfrak{G}_i)$ is called a *θ-representation.*

1.2 For $X, Y \in M_n(\mathbb{C})$ denote by $B(X, Y) := Tr(XY)$ the trace form. The restriction of this form to $\mathfrak{G}$ is non degenerate since $\mathfrak{G}$ is reductive [3]. Further, since θ is induced by an inner automorphism of $GL_n(\mathbb{C})$, the form is θ-invariant. It is now easy to check that $B(\mathfrak{G}_i, \mathfrak{G}_j) = 0$ unless $i+j \equiv 0 \bmod m$. In particular, the restriction of the form to $\mathfrak{g} := \mathfrak{G}_0$ is non degenerate, so $\mathfrak{g}$ (and G) is reductive [3].

Let $x = x_s + x_n$ be the Jordan decomposition of $x \in \mathfrak{G}_i$. The uniqueness of the decomposition implies that $x_s, x_n \in \mathfrak{G}_i$. For the corresponding G-orbits one knows:

(•) $\overline{Gx} \supset Gx_s$, Gx is closed if and only if $x = x_s$, and $0 \in \overline{Gx}$ if and only if $x = x_n$.

Further, if Gx is the unique closed orbit in $\overline{Gy}$, then there exists an element $y' \in Gy$ with Jordan decomposition $y' = x + y'_n$. Note that if $m = \infty$, then $x = x_n$ for all $x \in \mathfrak{G}_i$, $i \neq 0$: The image $\lambda(\mathbb{C}^*)$ of the one-parameter subgroup is contained in G, so obviously $0 \in \overline{\lambda(C^*)x}$.

There is a strong connection between the $\tilde{G}$-orbits in $\mathfrak{G}$ and the G-orbits in $\mathfrak{G}_i$: For $x \in \mathfrak{G}_i$, the intersection $\tilde{G}x \cap \mathfrak{G}_i$ is a smooth quasiaffine variety, and each irreducible component is one G-orbit [12]. Since $0 \in \overline{Gx}$ if and only if $0 \in \overline{\tilde{G}x}$ by (•), note that this implies in particular that G has only a finite number of nilpotent orbits since $\tilde{G}$ has only a finite number of nilpotent conjugacy classes in $\mathfrak{G}$.

1.3 Suppose now $x = x_s$, so Gx and $\tilde{G}x$ are closed. The centralizer $Z = \mathrm{Cen}_{\tilde{G}}\, x$ is a θ-stable Levi subgroup of $\tilde{G}$, and for $\mathfrak{z} = \mathrm{Cen}_{\mathfrak{G}}\, x$ we get a grading $\mathfrak{z} = \bigoplus_{i \in \mathbb{Z}_m} \mathfrak{z}_i$ such that $\mathfrak{z}_i \subset \mathfrak{G}_i$. Denote by Z_0 the connected component of Z^θ (which is the stabilizer of x in G). The natural action of Z_0 on $\mathfrak{z}_i$ is a θ-representation of Z_0.

If $y = x + z$ is a Jordan decomposition for $y \in \mathfrak{G}_i$, then $[x, z] = 0$ and hence $z \in \mathfrak{z}_i$. Further, $0 \in \overline{Zz}$ and hence $0 \in \overline{Z_0 z}$, and for two such elements z, z' in $\mathfrak{z}_i$ one has $G(x + z) = G(x + z')$ if and only if $Z^\theta z = Z^\theta z'$. So the classification of the orbits containing Gx in its closure reduces to the classification of the nilpotent Z^θ-orbits in $\mathfrak{z}_i$.

1.4 The Cartan subspace $\mathfrak{c} \subset \mathfrak{G}_i$ mentioned in the introduction is obtained as follows: There exists a Cartan subalgebra $\mathfrak{t}$ in $\mathfrak{G}$ that is θ-stable and such that $\mathfrak{c} = \mathfrak{t} \cap \mathfrak{G}_i$ is a maximal abelian subspace in $\mathfrak{G}_i$ consisting of semisimple elements. The Weyl group W of $\mathfrak{c}$ is the quotient group N/Z, where $N := \mathrm{Nor}_G\, \mathfrak{c}$ is the normalizer of $\mathfrak{c}$ and $Z := \mathrm{Cen}_G\, \mathfrak{c}$ is the centralizer of $\mathfrak{c}$. The group W is a finite group and generated by pseudo reflection. Further, the canonical restriction map $\mathbb{C}[\mathfrak{G}_1]^G \to \mathbb{C}[\mathfrak{c}]^W$ on the ring of invariant functions is an isomorphism.

2 $\mathbb{Z}$-graded reductive Lie algebras

2.0 Let $\mathfrak{G} = \bigoplus_{i \in \mathbb{Z}} \mathfrak{G}_i$ be an algebraic $\mathbb{Z}$-grading of a reductive Lie algebra induced by a one-parameter subgroup $\lambda : \mathbb{C}^* \to \tilde{G}$. We need to fix some notation and to recall a few properties of these gradings [6]. By 1.2, $\mathfrak{g} := \mathfrak{G}_0$ is a reductive subalgebra. Fix a Cartan subalgebra $\mathfrak{t}$ of $\mathfrak{g}$. Note that $\mathfrak{t}$ is also a Cartan subalgebra of $\mathfrak{G}$: Any Cartan subalgebra $\mathfrak{t}' \supset \mathfrak{t}$ of $\mathfrak{G}$ commutes with $\mathrm{Lie}\,\lambda(\mathbb{C}^*) \subset \mathfrak{t}$ and is hence contained in $\mathfrak{g}$.

2.1 Let Φ be the root system of $\mathfrak{G}$. The $\mathbb{Z}$-grading induces a function $\deg : \Phi \to \mathbb{Z}$, where $\deg \beta := i$ if the corresponding root space $\mathfrak{G}_\beta$ is contained in $\mathfrak{G}_i$. We say that a basis Δ of Φ is compatible with the grading if $\deg \alpha \geq 0$ for all $\alpha \in \Delta$. Such a basis can be found as follows: Fix a set of positive roots Φ_0^+ of the root system Φ_0 of $\mathfrak{g}$. Then the set $\Phi^+ := \Phi_0^+ \cup \{\beta \mid \mathfrak{G}_\beta \subset \mathfrak{G}_i, i > 0\}$ is a set of positive roots for Φ. The corresponding basis Δ of the root system is then compatible with the grading, and Δ has a natural decomposition $\Delta = \bigcup \Delta_i$, where Δ_i is the set of simple roots such that $\deg \alpha = i$. This decomposition is called the characteristic of the grading.

On the other hand, if $\deg : \Delta \to \mathbb{N}$ is any function, then extend deg additively to a function $\deg : \Phi \to \mathbb{Z}$. This function induces naturally a $\mathbb{Z}$-grading $\mathfrak{G} = \bigoplus \mathfrak{G}_i$, where $\mathfrak{G}_i$ is the sum of all root spaces $\mathfrak{G}_\beta$ such that $\deg \beta = i$. The grading is obviously algebraic.

2.2 Let G be the connected subgroup of $\tilde{G}$ such that $\operatorname{Lie} G = \mathfrak{g}$. Since all elements in $\mathfrak{G}_i$ are nilpotent for $i \neq 0$, the group G has only a finite number of orbits in $\mathfrak{G}_i$ for $i \neq 0$, in particular G has an open and dense orbit.

An important class of $\mathbb{Z}$-gradings are associated to the Jacobson-Morozov triple corresponding to a nilpotent element x of $\mathfrak{G}$: This triple $\langle x, y, h\rangle$ spans a Lie algebra isomorphic to $\mathfrak{sl}_2$, where $h \in \mathfrak{G}$ is a semisimple element and $y \in \mathfrak{G}$ is nilpotent such that $[x, y] = h$. Further, $[h, x] = 2x$ and $[h, y] = -2y$. The elements h and y are not uniquely determined, but two such triples $\langle x, y, h\rangle$ and $\langle x', y', h'\rangle$ are conjugate under $\tilde{G}$ if and only if $\tilde{G}x = \tilde{G}x'$ [4].

The eigenvalues of $\operatorname{ad} h$ are integers, so $\mathfrak{G} = \bigoplus_{i\in\mathbb{Z}} \mathfrak{G}_i$ is a $\mathbb{Z}$-grading of $\mathfrak{G}$, where $\mathfrak{G}_i$ denotes the eigenspace corresponding to the eigenvalue $i \in \mathbb{Z}$. The characteristic of this grading is an invariant of the conjugacy class $\tilde{G}x$, and it determines the orbit uniquely. Further, the intersection of $\tilde{G}x$ with $\mathfrak{G}_2$ is the dense orbit $Gx \subset \mathfrak{G}_2$, and the centralizer $\mathfrak{G}_x$ of x is a subalgebra of $\bigoplus_{i\geq 0} \mathfrak{G}_i$, $\mathfrak{g}_x$ being the reductive part of it [4].

Suppose $\mathfrak{G} = \bigoplus_{i\in\mathbb{N}} \mathfrak{G}_i$ is a $\mathbb{Z}$-grading with characteristic $\Delta = \bigcup \Delta_i$, and let $c \in \mathfrak{t} \cap [\mathfrak{G}, \mathfrak{G}]$ be such that $\alpha(c) = \deg \alpha$ for all $\alpha \in \Delta$. Then the characteristic is associated to a nilpotent orbit if and only if (see [6]):

($\bullet$) $\Delta = \Delta_0 \cup \Delta_1 \cup \Delta_2$, and if $z \in \mathfrak{G}_2$ is an element of the dense G-orbit, then $\mathfrak{g}_z$ is reductive and $B(\mathfrak{g}_z, c) = 0$.

In general there exists no purely combinatorial criteria to decide whether a given characteristic corresponds to a nilpotent orbit or not. A special class of gradings are:

Definition A grading is called *distinguished* if $\Delta_i = \emptyset$ for $i \geq 3$ and $\dim \mathfrak{G}_2 = \dim([\mathfrak{G}, \mathfrak{G}] \cap \mathfrak{G}_0)$. The last condition can also be described as follows: For an element $z \in \mathfrak{G}_2$ in the dense G-orbit, $\mathfrak{g}_z$ is just the center of $\mathfrak{G}$.

Note that by ($\bullet$) a distinguished grading is always associated to a nilpotent conjugacy class, the corresponding orbit is also called distinguished. These gradings have been classified (independently) by Bala and Carter [1], [2] and Vinberg [12]. The distinguished gradings are even, i.e., $\Delta = \Delta_0 \cup \Delta_2$.

3 The carrier algebra and the mini-carrier algebra

3.0 Let $\mathfrak{G} = \bigoplus_{m\in\mathbb{Z}_m} \mathfrak{G}_m$ be a $\mathbb{Z}_m$-grading of a reductive Lie algebra and let $G \to GL(\mathfrak{G}_i)$ be a corresponding θ-representation. Without loss of generality we suppose in the following that $i = 1$. We associate to a nilpotent orbit $Gx \subset \mathfrak{G}_1$ the so-called *carrier algebra* $\mathfrak{l}$ [5]. The algebra $\mathfrak{l}$ is a $\mathbb{Z}$-graded reductive Lie algebra such that the grading is distinguished and $x \in \mathfrak{l}_2$ is an element of the corresponding distinguished nilpotent conjugacy class of $\mathfrak{l}$. To determine $x \in \mathfrak{l}_2$ explicitly, the mini-carrier algebra method [11] turns out to be very useful.

3.1 We fix a Cartan subalgebra $\mathfrak{t} \subset \mathfrak{g}$. Let $x \in \mathfrak{G}_1$ be a nilpotent element. Replacing x by a G-conjugate if necessary, we may assume that $\mathfrak{t}_{\langle x\rangle}$ is a Cartan subalgebra in the normalizer $\mathfrak{g}_{\langle x\rangle}$ of the line $\langle x\rangle$ passing through x. Note that the centralizer $\mathfrak{l}_0 := \mathrm{Cen}_{\mathfrak{g}}\, \mathfrak{t}_{\langle x\rangle}$ of $\mathfrak{t}_{\langle x\rangle}$ is a Levi subalgebra of $\mathfrak{g}$. Choose $t \in \mathfrak{t}_{\langle x\rangle}$ generic such that $\mathfrak{l}_0 = \mathrm{Cen}_{\mathfrak{g}}\, t$ and $[t, x] = 2x$. We set $\mathfrak{k}_{2i} := \{g \in \mathfrak{G}_{i \bmod m} \mid [t, g] = 2ig\}$ for $i \in \mathbb{Z}$ and $\mathfrak{k} := \bigoplus_{i\in\mathbb{Z}} \mathfrak{k}_{2i}$.

Obviously we have $\mathfrak{k}_0 = \mathfrak{l}_0$ and $\mathfrak{k}$ is a $\mathbb{Z}$-graded subalgebra of $\mathfrak{G}$. Further, $\mathfrak{k}$ is reductive: Let λ, μ be eigenvalues of $\mathrm{ad}\, t$ and let $\mathfrak{G}_{k,\lambda}$ and $\mathfrak{G}_{l,\mu}$ be the corresponding eigenspaces. Then $B(\mathfrak{G}_{k,\lambda}, \mathfrak{G}_{l,\mu}) = 0$ unless $k + l \equiv 0 \bmod m$ and $\lambda = -\mu$. Since $\mathfrak{G}$ is reductive this means that the restriction of B to $\mathfrak{k}_{-i} \oplus \mathfrak{k}_i$ is non degenerate, which implies $\mathfrak{k}$ is reductive.

Choose a basis of the root system of $\mathfrak{k}$ compatible with the grading. Let δ_0 be the set of simple roots of degree 0 and denote by δ_2 the simple roots of degree 2 such that $\mathfrak{k}_\alpha$ is contained in the $\mathfrak{k}_0$-module generated by x. The subalgebra $\mathfrak{l}$ of $\mathfrak{k}$ generated by $\mathfrak{l}_0$ and the root spaces $\mathfrak{k}_\alpha$ and $\mathfrak{k}_{-\alpha}$ for all $\alpha \in \delta_2$ is the carrier algebra:

Definition The Lie algebra $\mathfrak{l}$ is called a *carrier algebra* of the orbit Gx.

Lemma 3.2 *The carrier algebra $\mathfrak{l}$ is reductive, the induced $\mathbb{Z}$-grading is distinguished and $x \in \mathfrak{l}_2$ is an element of the corresponding nilpotent conjugacy class of $\mathfrak{l}$. In particular, if $H = \mathrm{Cen}_G\, \mathfrak{t}_{\langle x\rangle}$, then $\overline{Hx} = \mathfrak{l}_2$. Further, $\mathfrak{l}_0$ is a Levi subalgebra of $\mathfrak{g}$ and $\mathfrak{l}_{2n} \subset \mathfrak{G}_{n \bmod m}$.*

Proof It remains only to prove that the $\mathbb{Z}$-grading is distinguished and $x \in \mathfrak{l}_2$ is an element of the corresponding nilpotent conjugacy class of $\mathfrak{l}$. Set $\tilde{\mathfrak{l}} := [\mathfrak{l}, \mathfrak{l}]$ and denote by $c \in \mathfrak{t} \cap \tilde{\mathfrak{l}}$ the unique element such that $\alpha(c) = 0$ for $\alpha \in \delta_0$ and $\alpha(c) = 2$ for $\alpha \in \delta_2$. By construction we know that $\mathfrak{t}_x$ is a Cartan subalgebra of $\mathfrak{g}_x$ and hence a Cartan subalgebra of $(\mathfrak{l}_0)_x$. Now $\mathfrak{t}_x$ commutes with $\mathfrak{l}_0$ and hence also with $\mathfrak{l}_2$. So $\mathfrak{t}_x$ is the center of $\mathfrak{l}$ and $\tilde{\mathfrak{l}}_x$ consists only of nilpotent elements.

Since $x \in \mathfrak{l}_2$ is a nilpotent element of $\tilde{\mathfrak{l}}$, there exists a Jacobson-Morozov triple $\langle y, h, x\rangle$ such that $h \in \tilde{\mathfrak{l}}_0$ [3]. Since h normalizes x, h is an element of the center of $\tilde{\mathfrak{l}}_0$. In particular, h and c commute, so $[h - c, x] = 0$ implies $h = c$. Hence the $\mathbb{Z}$-grading is induced by $\mathrm{ad}\, h$ and corresponds to the nilpotent conjugacy class of x in $\mathfrak{l}$. By 2.2 (•), this means $(\tilde{\mathfrak{l}}_0)_x$ is reductive and hence $(\tilde{\mathfrak{l}}_0)_x = 0$, which implies that the grading is distinguished.

Lemma 3.3 *Let $Gx, Gy \subset \mathfrak{G}_1$ be two nilpotent orbits and let $\mathfrak{l}, \mathfrak{k} \supset \mathfrak{t}$ be carrier algebras associated to the orbits. Then $Gx = Gy$ if and only if there exists an element $\tau \in W$ such that $\tau(\mathfrak{l}) = \mathfrak{k}$.*

Proof Let $H_1, H_2 \subset G$ be such that $\operatorname{Lie} H_1 = \mathfrak{l}_0$ and $\operatorname{Lie} H_2 = \mathfrak{k}_0$. If $\tau(\mathfrak{l}) = \mathfrak{k}$, then $\tau(H_1) = H_2$, and since $\overline{H_1 x} = \mathfrak{l}_2$ we get $\tau(\overline{H_1 x}) = \mathfrak{k}_2 = \overline{H_2 y}$. But this implies $Gx = Gy$. Now if $gxg^{-1} = y$, then we may further assume that $g\mathfrak{t}_{\langle x\rangle}g^{-1} = \mathfrak{t}_{\langle y\rangle}$. But this implies by the construction of $\mathfrak{l}$ and $\mathfrak{k}$ that $gH_1g^{-1} = H_2$ and hence $g\mathfrak{l}_2 g^{-1} = \overline{gH_1xg^{-1}} = \overline{H_2 y} = \mathfrak{k}_2$. So $g\mathfrak{l}g^{-1} = \mathfrak{k}$, and replacing g by gg' for some $g' \in H_1$ we may further assume that $g\mathfrak{t}g^{-1} = \mathfrak{t}$. The class of g in W is the desired element τ in W.

3.4 Denote by Φ the root system of $\mathfrak{g}$ and by $\Phi_0 \subset \Phi$ the roots of $\mathfrak{l}_0$. Consider the set δ_2' of all differences $\gamma - \beta$, where $\gamma, \beta \in \delta_2$. Since $\mathfrak{t}_x$ is the center of $\mathfrak{l}$ note that $\mathfrak{t}_{\langle x\rangle}$ is the intersection of the kernels $\ker \eta$, where $\eta \in \delta_0 \cup \delta_2'$. Now $\mathfrak{l}_0$ is the centralizer of $\mathfrak{t}_{\langle x\rangle}$, so $\delta := \delta_0 \cup \delta_2$ satisfies the following equality because every root in the intersection is a root in the centralizer of $\mathfrak{t}_{\langle x\rangle}$: (Z) $\langle \delta_0 \cup \delta_2' \rangle_{\mathbb{Q}} \cap \Phi = \Phi_0$.

Theorem *The map which associates to a nilpotent orbit of G in $\mathfrak{G}_1$ the W-conjugacy class of its carrier algebra induces a bijection between the set of nilpotent G-orbits in $\mathfrak{G}_1$ and the W-conjugacy class of reductive, $\mathbb{Z}$-graded subalgebras $\mathfrak{l} = \bigoplus_{i\in\mathbb{Z}} \mathfrak{l}_i$ of $\mathfrak{G}$ such that the grading is distinguished, $\mathfrak{l}_{2n} \subset \mathfrak{G}_{n \bmod m}$, $\mathfrak{l}_0 \subset \mathfrak{g}_0$ is a Levi subalgebra, and the basis $\delta = \delta_0 \cup \delta_2$ of the root system of $\mathfrak{l}$ satisfies the condition (Z).*

Proof By Lemma 3.3 and 3.4 the map is well-defined and injective. Suppose now $\mathfrak{l} \subset \mathfrak{G}$ has the properties above. Choose an element $x \in \mathfrak{l}_2$ such that $[\mathfrak{l}_0, x_2] = \mathfrak{l}_2$. Since the grading is distinguished, we know that $\mathfrak{t}_{\langle x\rangle}$ is the intersection of the kernels $\operatorname{Ker}\alpha$, $\alpha \in \delta_0$ and $\operatorname{Ker}(\gamma - \beta)$, $\gamma, \beta \in \delta_2$. The condition (Z) implies now that $\mathfrak{l}_0 = \operatorname{Cen}_G \mathfrak{t}_{\langle x\rangle}$, and hence $\mathfrak{t}_{\langle x\rangle}$ is a Cartan subalgebra of $\mathfrak{g}_{\langle x\rangle}$. It follows now that $\mathfrak{l}$ is precisely the carrier algebra associated to the orbit Gx in 3.1.

3.5 To get a representative of the orbit one has to find an element $x \in \mathfrak{l}_2$ such that $[\mathfrak{l}_0, x] = \mathfrak{l}_2$. The mini-carrier algebra method developed by Vinberg [11] reduces this also nearly completely to a combinatorial problem of root systems and Weyl groups. Suppose $x \in \mathfrak{l}_2$ is such that $[\mathfrak{l}_0, x] = \mathfrak{l}_2$. Let $\mathfrak{k} \subset \mathfrak{l}$ be a subalgebra such that $\mathfrak{t} \subset \mathfrak{k}$ and for the induced grading of $\mathfrak{k}$ we have $x \in \mathfrak{k}_2$. Denote by K the adjoint group with Lie algebra $[\mathfrak{k}, \mathfrak{k}]$, and let K_0 be the connected subgroup of K with Lie algebra $[\mathfrak{k}, \mathfrak{k}] \cap \mathfrak{k}_0$.

Definition The Lie algebra $\mathfrak{k}$ is called a *mini-carrier algebra* for $\mathfrak{l}$ if $(K_0)_x = 1$.

To construct a mini-carrier algebra for $\mathfrak{l}$ set $\mathfrak{n} := [\mathfrak{l}, \mathfrak{l}]$ and let N be the adjoint group with Lie algebra $\mathfrak{n}$. Denote by N_0 the connected subgroup of N with Lie algebra $\mathfrak{n}_0$. If $(N_0)_x = 1$, then $\mathfrak{l}$ itself is a mini-carrier algebra. Else we know that $(N_0)_x$ is a finite group since the grading of $\mathfrak{l}$ is distinguished. Let $\mathfrak{k}$ be the subalgebra of $\mathfrak{n}$ consisting of the fixed points of $(N_0)_x$. Then the induced grading $\mathfrak{k} = \bigoplus_{i\in\mathbb{Z}} \mathfrak{k}_i$ is distinguished and $x \in \mathfrak{k}_2$ is an element of the corresponding nilpotent conjugacy class of $\mathfrak{k}$. Further, let K be the adjoint group with Lie algebra $\mathfrak{k}$ and denote by K_0 the connected subgroup of K with

Lie algebra $\mathfrak{k}_0$. By construction, $(K_0)_x = 1$, i.e., $\mathfrak{k}+\mathfrak{t}$ is a mini-carrier algebra for $\mathfrak{l}$. Summarizing we have:

Lemma *For every $\mathbb{Z}$-graded semisimple Lie algebra $\mathfrak{l}$ such that the grading is distinguished, there exists a mini-carrier algebra.*

3.6 The types and the gradings of the possible mini-carrier algebras $\mathfrak{k}$ for a $\mathbb{Z}$-graded Lie algebra $\mathfrak{l}$ can be found in [11]. In almost all cases $\mathfrak{k}_0 = \mathfrak{t}$ is just the Cartan subalgebra. In general $\mathfrak{k}_0$ is at most a direct sum of its center and simple algebras of type $\mathfrak{sl}_2$, so it is in most cases much easier to find an element $x \in \mathfrak{k}_2$ such that $[\mathfrak{k}_0, x] = \mathfrak{k}_2$ than to find and element $x \in \mathfrak{l}_2$ such that $[\mathfrak{l}_0, x] = \mathfrak{l}_2$.

The mini-carrier algebras for $\mathfrak{l}$ are not necessarily conjugate by W. But the type of a subalgebra $\mathfrak{k}$ of $\mathfrak{l}$ is sufficient to decide whether $\mathfrak{k}$ is a mini-carrier algebra or not:

Lemma *Let $\mathfrak{k} \supset \mathfrak{t}$ be a reductive subalgebra of $\mathfrak{l}$ such that the type and the induced grading corresponds in Table II in [11] to a mini-carrier algebra of $\mathfrak{l}$. Then $\mathfrak{k}$ is a mini-carrier algebra for $\mathfrak{l}$.*

Proof Denote by c the unique element in $\mathfrak{t} \cap [\mathfrak{l}, \mathfrak{l}]$ such that $\alpha(c) = \deg \alpha$ for $\alpha \in \delta$, and let $x \in \mathfrak{k}_2$ be such that $[\mathfrak{k}_0, x] = \mathfrak{k}_2$. Further, let $h \in \mathfrak{t} \cap [\mathfrak{k}, \mathfrak{k}]$ be the unique element inducing the grading of $\mathfrak{k}$. Note that $\mathrm{rk}[\mathfrak{k}, \mathfrak{k}] = \mathrm{rk}[\mathfrak{l}, \mathfrak{l}]$ ([11], Table II), so $c \in [\mathfrak{k}, \mathfrak{k}]$ and $[c - h, x] = 0$. Since the grading of $\mathfrak{k}$ is distinguished, this implies $c = h$. But since $h = c$ is also the semisimple element in an Jacobson-Morozov triple for x, this means that $x \in \mathfrak{k}_2 \subset \mathfrak{l}_2$ is an element of the nilpotent conjugacy class of $\mathfrak{l}$ corresponding to the grading.

4 The stabilizer of a nilpotent element

4.0 Let $\mathfrak{G} = \bigoplus_{m \in \mathbb{Z}_m} \mathfrak{G}_m$ be a $\mathbb{Z}_m$-grading of a reductive Lie algebra and let $G \to GL(\mathfrak{G}_1)$ be the corresponding θ-representation. For a nilpotent orbit $Gx \subset \mathfrak{G}_1$ let $\mathfrak{l}$ be a carrier algebra such that $x \in \mathfrak{l}_2$. We show now how to use the carrier algebra to determine the structure of stabilizer $\mathfrak{g}_x$ of x.

4.1 Set $\mathfrak{n} := [\mathfrak{l}, \mathfrak{l}]$ and let $c \in \mathfrak{n}_0$ be the element inducing the grading, i.e., c is the unique element in $\mathfrak{n}_0 \cap \mathfrak{t}$ such that $\alpha(c) = \deg a$ for all simple roots of $\mathfrak{l}$. We have already seen in 3.2 that if $\langle y, h, x\rangle$ is a Jacobson-Morozov triple for x such that $h \in \mathfrak{n}_0$, then $h = c$.

Recall that the eigenvalues of c are integers. Denote by $\mathfrak{G}^j$ and $\mathfrak{g}^j$ the eigenspaces of $\operatorname{ad} c$ in $\mathfrak{G}$ and $\mathfrak{g}$ corresponding to the eigenvalue j, and set $\mathfrak{G}^j_i := \mathfrak{G}^j \cap \mathfrak{G}_i$. Since the map

$$\operatorname{ad} x : \bigoplus_{j \geq 0} \mathfrak{G}^j \longrightarrow \bigoplus_{j \geq 2} \mathfrak{G}^j$$

is surjective [9], we know that $\operatorname{ad} x : \bigoplus_{j\geq 0} \mathfrak{g}^j \longrightarrow \bigoplus_{j \geq 2} \mathfrak{G}^j_1$ is surjective. Further, $\mathfrak{G}_x$ and $\mathfrak{g}_x$ are $\operatorname{ad} c$ stable, $\mathfrak{G}_x \subset \bigoplus_{j\geq 0} \mathfrak{G}^j$, and $\mathfrak{G}_x \cap \mathfrak{G}^0$ is reductive (2.2).

So

$$\mathfrak{g}_x \subset \bigoplus_{j\geq 0} \mathfrak{g}^j, \quad \text{and} \quad \mathfrak{g}_x \cap \mathfrak{g}^0 \text{ is reductive.}$$

4.2 This enables us to determine the dimension and the character of $\mathfrak{g}_x$ with respect to the Cartan subalgebra $\mathfrak{t}_x \subset \mathfrak{g}_x$: Set $\mathfrak{g}^{\geq n} = \bigoplus_{j\geq n} \mathfrak{g}^j$ and $\mathfrak{G}_1^n := \bigoplus_{j\geq n} \mathfrak{G}_1^j$. By the surjectivity of $\operatorname{ad} x$ we get $\dim \mathfrak{g}_x = \dim \mathfrak{g}^{\geq 0} - \dim \mathfrak{G}_1^{\geq 2}$. But note that the map is equivariant with respect to the action of $\mathfrak{t}_x$ on $\mathfrak{g}$ and $\mathfrak{G}_1$: $[t,[g,x]] = [[t,g],x]$ for $t \in \mathfrak{t}_x$ and $g \in \mathfrak{g}$. So we get for the corresponding $\mathfrak{t}_x$-characters: $\operatorname{Char} \mathfrak{g}_x = \operatorname{Char} \mathfrak{g}^{\geq 0} - \operatorname{Char} \mathfrak{G}_1^{\geq 2}$. To determine the type of the reductive $(\mathfrak{g}_x)_{red}$ part of $\mathfrak{g}_x$ we have just to find a root system whose character is $\operatorname{Char}(\mathfrak{g}_x)_{red} = \operatorname{Char} \mathfrak{g}^0 - \operatorname{Char} \mathfrak{G}_1^2$.

5 The program

5.0 The first problem which arises if one wants to apply the method (by hand or in form of a program) is to find candidates for the carrier algebras. Since such an algebra is completely determined by its roots, this amounts to finding pairs (Δ_0, Λ_0), where Δ_0 is a subset of the simple roots of $\mathfrak{g}$ and Λ_0 is a subset of the set of weights Λ in $\mathfrak{G}_1$. This pair should have the property that $\Delta_0 \cup \Lambda_0$ is a basis for a root system satisfying the condition (Z), and the grading defined by $\deg \alpha = 0$ for $\alpha \in \Delta_0$ and $\deg \alpha = 2$ for $\alpha \in \Lambda_0$ should be distinguished.

The method we propose in the following to get a "good" list of candidates for pairs (Δ_0, Λ_0) works for all cases, but for gradings induced by finite-order automorphisms we would need to work in the setting of Kac-Moody algebras. Since the language LiE (in which we implemented the method) provides the routines for calculations in the Weyl group and root system only for finite-dimensional Lie algebras, we consider in the following only the case where $\mathfrak{G} = \bigoplus_{i\in\mathbb{Z}} \mathfrak{G}_i$ is a $\mathbb{Z}$-grading. Further, without loss of generality the θ-representation is $\rho : G \to GL(\mathfrak{G}_1)$.

5.1 How to get the list of candidates Let $\tilde{W}$ be the Weyl group of $\mathfrak{G}$ and W (as before) the Weyl group of $\mathfrak{g} = \mathfrak{G}_0$. Let Δ be a basis for the root system of $\mathfrak{G}$ compatible with the grading. We associate to a coset $\tau \in W\backslash\tilde{W}$ the following pair:

- Choose $w \in \tilde{W}$ minimal (with respect to the Bruhat order) such that $w \equiv \tau \bmod W$. If there exists a root $\alpha \in w(\Delta)$ such that $\deg \alpha > 1$, then we set $\Delta_0 = \Lambda_0 = \emptyset$. Else we set $\Delta_0' := \{\alpha \in w(\Delta) \mid \alpha \in \Delta, \deg \alpha = 0\}$ and $\Lambda_0 := \{\alpha \in w(\Delta) \mid \deg \alpha = 1\}$. Clearly, if $\Delta_0' \cup \Lambda_0 \neq \emptyset$, then it is a basis for a root system Φ'. Denote by Δ_0 the subset of Δ_0' obtained by omitting all roots such that the irreducible component of Φ' containing α is orthogonal to Λ_0.

Denote by $\mathfrak{L}$ the list of pairs (Δ_0, Λ_0) obtained by running through the set $W\backslash\tilde{W}$. Since $\Delta_0 \cup \Lambda_0$ is a subset of a basis of the root system of $\mathfrak{G}$, note that

$\Delta_0 \cup \Lambda_0$ is obviously a basis for a root system and it satisfies the condition (Z). So to be a basis of the root system of a carrier algebra, it remains to check whether the grading $\deg\alpha = 0$ for $\alpha \in \Delta_0$ and $\deg\alpha = 2$ for $\alpha \in \Lambda_0$, is distinguished.

The LiE-package provides a very fast and efficient algorithm to traverse the set $W\backslash\tilde{W}$ without storing problems even in cases like $\mathfrak{G} = \mathrm{E}_8$ (where $|\tilde{W}| = 696729600$). Having in mind the important fact that for candidates of the form $\Delta_0 \cup \Lambda_0$ above, the otherwise time-consuming check of the property (Z) and to find out whether $\Delta_0 \cup \Lambda_0$ is a basis of a root system are not needed, it makes sense to put the remaining part of the program inside the big loop which traverses the set $W\backslash\tilde{W}$. Another advantage of this approach is the check to find out when two carrier algebras are in the same W-conjugacy class. Due to the choice of the minimal representative $w \in \tilde{W}$ of a class in $W\backslash\tilde{W}$, this check is very simple (see 5.3). We should point out that LiE chooses automatically a minimal representative for a coset. The most important point is of course:

Theorem *In every W-conjugacy class of carrier algebras there exists a pair $(\Delta_0, \Lambda_0) \in \mathfrak{L}$ such that $\Delta_0 \cup \Lambda_0$ is the basis for the root system of a carrier algebra in that class.*

Proof Let $\mathfrak{l}$ be a carrier algebra and let $\pi = \Delta_0 \cup \Lambda_0$ be a basis of its root system compatible with the grading. Choose $x \in \mathfrak{l}_2$ such that $[l_0, x] = \mathfrak{l}_2$, and let $\langle y, h, x\rangle$ be a Jacobson-Morozov triple such that $y \in \mathfrak{l}_{-2}$ and $h \in \mathfrak{t} \cap [\mathfrak{l}, \mathfrak{l}]$ is the unique element such that $\alpha(h) = \deg\alpha$ for $\alpha \in \pi$. Further, let $c \in \mathfrak{t} \cap [\mathfrak{G}, \mathfrak{G}]$ be the unique element inducing the grading of $\mathfrak{G}$, i.e., $\gamma(c) = \deg\gamma$ for $\gamma \in \Delta$.

Claim $Z := \mathrm{Cen}_{\mathfrak{G}}\, 2c - h = [l, l] \oplus k \oplus \mathfrak{s}$, where $\mathfrak{s} \subset \mathfrak{t}$ and $\mathfrak{k}$ is a semsimple Lie algebra such that $[k, [l, l]] = 0$.

Proof of the claim. The natural grading of Z induced by the (integral) eigenvalues of $\mathrm{ad}\, h$ is even: If α is a root of Z such that $\alpha(h)$ is odd, then necessarily $\alpha(2c)$ would be odd too which is not possible. Extend the basis $\pi = \Delta_0 \cup \Lambda_0$ to a basis $\tilde{\pi}$ of the root system of Z. Choose $\gamma \in \tilde{\pi} - \pi$ such that $\deg\gamma = 0$ (i.e., $\gamma(h) = 0$). Since then $\gamma(c) = 0$ too, we know that γ is a root of $\mathfrak{g}$.

Now γ is a simple root and hence $[\mathfrak{G}_\gamma, \mathfrak{G}_{-\alpha}] = 0$ for all $\alpha \in \pi$. But this implies $[\mathfrak{G}_\gamma, \mathfrak{l}_{-2}] = 0$, and by $\mathfrak{s}l_2$-representation theory ($y \in \mathfrak{l}_2$) we have $[\mathfrak{G}_\gamma, x] = 0$ too. And since the grading is distinguished we get

$$[\mathfrak{G}_\gamma, [\mathfrak{l}, \mathfrak{l}] \cap \mathfrak{l}_0] = [\mathfrak{G}_\gamma, [\mathfrak{l}_{-2}, x]] = 0, \quad [\mathfrak{G}_\gamma, l_2] = [\mathfrak{G}_\gamma, [[\mathfrak{l}, \mathfrak{l}] \cap \mathfrak{l}_0, x]] = 0.$$

But $[l, l]$ is generated by $[\mathfrak{l}, \mathfrak{l}] \cap \mathfrak{l}_0$, $\mathfrak{l}_{-2}$ and $\mathfrak{l}_2$, so $[\mathfrak{G}_\gamma, [l, l]] = 0$. This proves the decomposition for Z_0. But the grading is induced by a Jacobson-Morozov triple for x, so $Z_2 = [Z_0, x] = [\mathfrak{l}_0, x] = \mathfrak{l}_2$. Since the grading is even, this implies $\tilde{\pi} = \pi \cup \pi'$, where the roots in π' are of degree 0 and orthogonal to the roots in π, which proves the claim.

We come now back to the proof of the theorem. Let $\tilde{\pi} = \pi \cup \pi'$ be a basis of the root system Φ_Z of Z as above, and denote by Φ_Z^+ the corresponding set of positive roots. For the root system Φ of $\mathfrak{G}$ let $\Phi^{>0}$ be the set of roots such that $\gamma(h-2c) > 0$, then $\Phi_Z^+ \cup \Phi^{>0}$ is a set of positive roots for Φ. Choose $w \in \tilde{W}$ such that $w(\Delta)$ is the corresponding basis of the root system, note that $\tilde{\pi} \subset w(\Delta)$.

To prove the theorem we have to show that $\gamma(c) < 0$ for $\gamma \in w(\Delta) - \tilde{\pi}$. So suppose $\gamma \in w(\Delta) - \tilde{\pi}$ is such that $\gamma(c) \geq 0$. Since $\gamma(h-2c) > 0$, this implies that $\gamma(h) > 0$, and hence by $\mathfrak{sl}_2$-representation theory we get $[\mathfrak{G}_\gamma, y] \neq 0$. It follows that there exists a root $\beta \in \Phi_Z^+$ such that $[\mathfrak{G}_\gamma, \mathfrak{G}_{-\beta}] \neq 0$, which cannot be since γ is a simple root but $\gamma \notin \tilde{\pi}$. This proves that $\tilde{\pi}$ is the set of roots of degree zero and degree one in $w(\Delta)$.

To finish the proof of the theorem, let Φ^+ be the set of positive roots of Φ corresponding to the choice of the basis Δ. Denote for $w \in \tilde{W}$ by $\Phi^+(w)$ the set of positive roots in $w(\Phi^+)$ and by $\Phi^+(0,w)$ the set of positive roots in $w(\Phi^+)$ of degree 0. The length of w is the number $\sharp\Phi^+ - \sharp\Phi^+(w)$. Note that if we replace w by $w'w$ for some $w' \in W$, then the number $\sharp(\Phi^+(w) - \Phi^+(0,w))$ does note change, only $\sharp\Phi^+(0,w)$ changes. So obviously w is a minimal representative of its class in $W\backslash\tilde{W}$ if and only if $\Phi^+(0,w)$ is set of all positive roots of degree 0. But in this case we have clearly that the set of roots of degree zero in $w(\Delta)$ is a subset of Δ, which finishes the proof.

5.2 Distinguished pairs For a pair (Δ_0, Λ_0) let $\mathfrak{l}$ be the corresponding subalgebra of $\mathfrak{G}$. To check whether the grading $\deg\alpha = 0$ for $\alpha \in \Delta_0$, $\deg\alpha = 2$ for $\alpha \in \Lambda_0$, is distinguished, there are two possibilities. Since the distinguished gradings are classified, there would be the possibility to check by a routine which consists of a mixture of a program for the checking in the classical case and a table for the exceptional groups. But it turned out that in general the dimension function provided by LiE is sufficiently fast. A routine just checks whether for the induced grading on $[\mathfrak{l}, \mathfrak{l}]$ the dimension of the degree 0 part is equal to the dimension of the degree 2 part.

5.3 Representative of the W-conjugacy class of a carrier algebra Though of course $w(\Delta)$ and $w'(\Delta)$ are not conjugate under W if w, w' do not belong to the same class in $W\backslash\tilde{W}$, the associated pairs $\{\Delta_0, \Lambda_0\}$ and $\{\Delta_0', \Lambda_0'\}$ still might be. So we have to check when do they belong to the same W-conjugacy class of carrier algebras.

Let $\mathfrak{l}$ and $\mathfrak{l}'$ be the carrier algebras having $\Delta_0 \cup \Lambda_0$, respectively $\Delta_0' \cup \Lambda_0'$, as basis of its root system. Let $h \in \mathfrak{t} \cap [\mathfrak{l}, \mathfrak{l}]$ be (as in 5.1) the unique element such that $\alpha(h) = \deg\alpha$ for $\alpha \in \Delta_0 \cup \Lambda_0$. We consider in fact h as a linear function on the root lattice of $\mathfrak{l}$. Now LiE provides a simple routine to determine the extension of h to a linear function on the root lattice of $\mathfrak{G}$ induced by the embedding $\mathfrak{l} \hookrightarrow \mathfrak{G}$. Similarly, let h' be the linear function on the root lattice of $\mathfrak{G}$ associated to the pair $\{\Delta_0', \Lambda_0'\}$.

Lemma $\mathfrak{l}$ *and* $\mathfrak{l}'$ *are in the same* W*-conjugacy class if and only if* $h = h'$.

Proof The proof of Theorem 5.1 shows that h determines $\mathfrak{l}$ uniquely: Let c be the element inducing the grading of $\mathfrak{G}$, then we obtain $\mathfrak{l}$ from the centralizer Z of $h - 2c$ by omitting all root spaces such that the roots are orthogonal to the roots in $Z \cap \mathfrak{G}_1$. So $\mathfrak{l}$ and $\mathfrak{l}'$ are in the same W-conjugacy class if and only if h and h' are conjugate under W. But we have seen in 5.1 that if $\Phi^+(0)$ is the set of positive roots of $\mathfrak{g}$ (i.e., the set of positive roots of $\mathfrak{G}$ of degree 0), then $\beta(h), \beta(h') \geq 0$ for all $\beta \in \Phi^+(0)$. But then h and h' can only be conjugate under W if $h = h'$.

5.4 Representative of the orbit To determine an explicit representative of the orbit, we use the mini-carrier method described in section 3. We included a routine that uses Table II in [11] to determine explicitly for every carrier algebra a corresponding mini-carrier algebra. This is done by a small routine for the Lie algebras of classical type and by a "table" in the exceptional case.

To determine a representative of the corresponding orbit, it suffices to do this for every simple component of the mini-carrier algebra. Suppose now $\mathfrak{m}$ is such a simple component. If the grading is such that $\deg \alpha = 2$ for all simple roots of the component, then a representative is just any vector in $\mathfrak{m}_2$ which has non-zero components in all root spaces in $\mathfrak{m}_2$. In the other cases (which can only be if $\mathfrak{m}$ is of exceptional type or type $\mathtt{D}$) one has to choose "by hand" a representative in each case. The program contains a small routine which does this choice for the type $\mathtt{D}$ case, and for the exceptional types the program uses a table.

5.5 Stabilizer Let $\mathfrak{l}$ be a carrier algebra. We have already described in section 4 how to determine the structure of the stabilizer. Let h be the semisimple element in the Jacobson-Morozov triple $\langle y, h, x\rangle$. Note that we have already determined h in 5.3. Let $\mathfrak{g}^0$ be the centralizer of h in $\mathfrak{g}$ and denote by $\mathfrak{G}_1^2$ the set of elements $z \in \mathfrak{G}_1$ such that $[h, z] = 2z$. Consider the virtual character $C := \mathrm{Char}\, \mathfrak{g}^0 - \mathrm{Char}\, \mathfrak{G}_1^2$.

Now LiE provides a routine that calculates the restriction of a (virtual) character to the center of a Levi subalgebra. By construction we know that the center Z of $\mathfrak{l}$ is a Cartan subalgebra in the stabilizer $\mathfrak{g}_x$. In section 4 we have already seen that hence $\mathrm{res}_Z\, C$ is the character of the reductive part of the $\mathfrak{g}_x$. To determine the structure of the root system, a routine fixes a set of positive roots and chooses a set of indecomposable roots and hence a basis of the root system. Again using the knowledge of $\mathrm{res}_Z\, C$, it is possible to calculate the Cartan matrix of the root system: For two simple roots α, γ a routine just determines the maximal numbers such that $\alpha + k\gamma$ respectively $l\alpha + \gamma$ are roots. A third routine uses then the Cartan matrix to recognize the type of the root system. The dimension of the central torus is now just the dimension of the center of the carrier algebra $\mathfrak{l}$ minus the rank of the root system. These data are obtained by standard routines included in LiE.

To determine the embedding of the semisimple part $S\mathfrak{g}_x$ of $\mathfrak{g}_x$ in $GL(\mathfrak{G}_1)$, we need to describe $\mathfrak{G}_1$ as $S\mathfrak{g}_x$-module. To get the character of $\mathfrak{G}_1$ as $S\mathfrak{g}_x$-module we compute first the character of $\mathfrak{G}_1$ as $\mathfrak{G}$-module: It is just the sum over the roots of degree one. As pointed out before, LiE provides a routine to calculate the restriction of that character to the center of $\mathfrak{l}$, which is a Cartan subalgebra in $\mathfrak{g}_x$. To translate this information into a character for $S\mathfrak{g}_x$, a routine computes for every weight λ in the character and any simple root α of $S\mathfrak{g}_x$ the maximal numbers such that $\lambda + k\alpha$ respectively $\lambda - l\alpha$ are weights of the representation. Then $k - l$ is the coefficient of ω_α for the description of λ as sum of the fundamental weights of $S\mathfrak{g}_x$. Once we have the character of $\mathfrak{G}_1$ as $S\mathfrak{g}_x$-module, LiE provides the routines to determine the highest weights of the irreducible representations of $S\mathfrak{g}_x$ in $\mathfrak{G}_1$.

Similarly we find the structure of the nilpotent radical of $\mathfrak{g}_x$. By section 4, we know the character is just the restriction of the virtual character

$$\mathrm{Char} \bigoplus_{j\geq 1} \mathfrak{g}^j - \mathrm{Char} \bigoplus_{j\geq 3} \mathfrak{G}_1^j$$

to the center of $\mathfrak{l}$. As above one transforms this character into a character of an $S\mathfrak{g}_x$-module, and LiE provides then the routines to determine the highest weights of the irreducible representations of $S\mathfrak{g}_x$ in the nilpotent radical of $\mathfrak{g}_x$.

5.6 Example The representation of $GL_7(\mathbb{C})$ on $\bigwedge^3 \mathbb{C}^7$ can be constructed as follows: Consider the Lie algebra of type $\mathtt{E}_7$ with $\mathbb{Z}$-grading $\deg \alpha_2 = 1$ and $\deg \alpha = 0$ else (in the Bourbaki enumeration [3]). The corresponding θ-representation is then just the representation of $GL_7(\mathbb{C})$ considered above. On a Sun 4 the program needs about 20 seconds ($|W\backslash\tilde{W}| = 576$) to get the following orbit classification:

1) $x = 0112111 + 0112210 + 1111111 + 1112110 + 1122100$,
$\mathfrak{g}_x = \mathtt{G}_2$, $\dim Gx = 35$,
Embedding: $\mathfrak{G}_1 = [0,0] + [1,0] + [2,0]$.

2) $x = 0112211 + 1112111 + 1112210 + 1122110$,
$\mathfrak{g}_x = 2\mathtt{A}_1\mathtt{T}_1 + 8\mathbb{C}$, $\dim Gx = 34$, $R_n\mathfrak{g}_x = [1 \mid 3]$,
Embedding: $\mathfrak{G}_1 = 2[0 \mid 0] + [0 \mid 2] + [0 \mid 4] + 2[1 \mid 1] + [1 \mid 3] + [2 \mid 2]$.

3) $x = 0112221 + 1111111 + 1123210$,
$\mathfrak{g}_x = 2\mathtt{A}_1\mathtt{T}_2 + 10\mathbb{C}$, $\dim Gx = 31$, $R_n\mathfrak{g}_x = 2[0 \mid 0] + 2[0 \mid 1] + 2[1 \mid 0]$,
Embedding: $\mathfrak{G}_1 = 7[0 \mid 0] + 4[0 \mid 1] + 4[1 \mid 0] + 3[1 \mid 1]$.

4) $x = 0112221 + 1112211 + 1122111 + 1123210$,
$\mathfrak{g}_x = \mathtt{A}_2\mathtt{T}_1 + 12\mathbb{C}$, $\dim Gx = 28$, $R_n\mathfrak{g}_x = [0,1] + [1,0] + [2,0]$,
Embedding: $\mathfrak{G}_1 = 3[0,0] + X[0,1] + [0,2] + 2[1,0] + [1,1] + [2,0]$.

5) $x = 1111111 + 1123210$,
$\mathfrak{g}_x = 2\mathtt{A}_2\mathtt{T}_2 + 6\mathbb{C}$, $\dim Gx = 26$, $R_n\mathfrak{g}_x = [0,0 \mid 1,0] + [0,1 \mid 0,0]$,
Embedding: $\mathfrak{G}_1 = [0,0 \mid 0,0] + [0,0 \mid 0,1] + [0,1 \mid 0,1] + [0,1 \mid 1,0] + [1,0 \mid 0,0] + [1,0 \mid 1,0]$.

6) $x = 1112221 + 1122211 + 1123210$,
$\mathfrak{g}_x = \mathtt{A}_2\mathtt{T}_2 + 14\mathbb{C}$, $\dim Gx = 25$, $R_n\mathfrak{g}_x = 2[0,1] + [1,1]$.
Embedding: $\mathfrak{G}_1 = 4[0,0] + [0,2] + 3[1,0] + 2[1,1]$.

7) $x = 0112221 + 1112211 + 1122111$,
$\mathfrak{g}_x = \mathtt{C}_3\mathtt{T}_1 + 6\mathbb{C}$, $\dim Gx = 21$, $R_n\mathfrak{g}_x = [1,0,0]$,
Embedding: $\mathfrak{G}_1 = [0,0,0] + [0,0,1] + [0,1,0] + [1,0,0]$.

8) $x = 1122221 + 1123211$, $\mathfrak{g}_x = \mathtt{B}_2\mathtt{A}_1\mathtt{T}_2 + 14\mathbb{C}$, $\dim Gx = 20$,
$R_n\mathfrak{g}_x = [0,0 \mid 1] + [0,1 \mid 0] + [0,1 \mid 1]$,
Embedding: $\mathfrak{G}_1 = 2[0,0 \mid 0] + [0,0 \mid 1] + 2[0,1 \mid 0] + [0,1 \mid 1] + [1,0 \mid 0] + [1,0 \mid 1]$.

9) $x = 1123321$,
$\mathfrak{g}_x = \mathtt{A}_3\mathtt{A}_2\mathtt{T}_1 + 12\mathbb{C}$, $\dim Gx = 13$, $R_n\mathfrak{g}_x = [0,0,1 \mid 1,0]$,
Embedding: $\mathfrak{G}_1 = [0,0,0 \mid 0,0] + [0,0,1 \mid 0,0] + [0,1,0 \mid 1,0] + [1,0,0 \mid 0,1]$.

10) $x = 0$.

We use here the following conventions: The sequences in the first row correspond to positive roots in $\mathtt{E}_7$, we use here again the Bourbaki notation. A representative of the orbit is then any element in $\mathfrak{G}_1$ which has non-zero components precisely in those weight spaces corresponding to the roots listed in the first row.

The reductive part of the stabilizer $\mathfrak{g}_x$ is described in terms of its root system and the dimension of the center ($\mathtt{T}_n$ means an n-dimensional center). In addition we give the dimension of the nilpotent radical $R_n\mathfrak{g}_x$ of $\mathfrak{g}_x$ ($n\mathbb{C}$ means an n-dimensional nilpotent radical), the dimension of the G-orbit Gx, and we describe the structure of $R_n\mathfrak{g}_x$ as $S\mathfrak{g}_x$-module. For a simple root system we use the Bourbaki enumeration for the fundamental weights.

In the row "Embedding" we describe how $\mathfrak{G}_1$ decomposes as module with respect to the action of the semisimple part of the stabilizer.

If one takes instead the grading $\deg\alpha_4 = 1$ and $\deg\alpha = 0$ else, then the same computer needs about 4 minutes to obtain the orbit classification, this time for the action of $GL_4(\mathbb{C}) \times GL_3(\mathbb{C}) \times GL_2$ on $\mathbb{C}^4 \otimes \mathbb{C}^3 \otimes \mathbb{C}^2$ (there are 24 orbits). In this case the size of the set $W\backslash\tilde{W}$ is 10.080. Another example is the 64-dimensional $Spin_{14} \times \mathbb{C}^*$-representation, which is obtained by the grading $\deg\alpha_1 = 1$ and $\deg\alpha = 0$ else of the simple Lie algebra $\mathtt{E}_8$. In this case the size of the set $W\backslash\tilde{W}$ is 2160, and the same computer needs again about 4 minutes to find the 10 orbits.

References

[1] Bala, P., Carter, R. W., Classes of unipotent elements in simple algebraic groups I, *Math. Proc. Camb. Phil. Soc.* **79** (1976), 401–425.

[2] Bala, P., Carter, R. W., Classes of unipotent elements in simple algebraic groups II, *Math. Proc. Camb. Phil. Soc.* **80** (1976), 1–18.

[3] N. Bourbaki, "Groupes et algèbres de Lie," Hermann, Paris, (1968).

[4] Dynkin, E. B., Semisimple subalgebras of semisimple Lie algebras, *Amer. Math. Soc. Transl.* **6** (1957), 111–244.

[5] Gatti, V., Viniberghi, E., Spinors of a 13-domensional space, *Adv. Math.* **30** (1978), 137–155.

[6] Kac, V., Some remarks on nilpotent orbits, *J. Algebra* **64** (1980), 190–213.

[7] Kac, V., " Infinite dimensional Lie algebras," Birkhäuser Verlag, Basel–Boston, (1983).

[8] Kostant, B., The principle three-dimensional subgroup and the Betti numbers of a complex simple Lie algebra, *Amer. J. Math.* **81** (1959), 973–1032.

[9] Springer, T. A., Steinberg, R., "Conjugacy classes: Seminar on algebraic groups and related finite groups," Springer Lecture Notes, Springer Verlag, Berlin-Heidelberg-New York, **131** (1970).

[10] Littelmann, P., Koreguläre und äquidimensionale Darstellungen, *J. Algebra* **123** (1989), 193–222.

[11] Vinberg, E. B., On the classification of the nilpotent elements of a graded Lie algebra, *Math. USSR-Izv.* **10** (1976), 463–495.

[12] Vinberg, E. B., The Weyl group of a graded Lie algebra, *Sov. Math. Dokl.* **16** (1975), 1517–1520.

P. Littelmann (`littelma@math.u-strasbourg.fr`)
IRMA, Université Louis Pasteur, 7 rue René Descartes, F-67084 Strasbourg Cedex, France

References

Progress in Mathematics, Vol. 143, © 1996 Birkhäuser Verlag Basel/Switzerland

Some algebraic geometry problems arising in the field of mechanism theory

J.-P. Merlet

1 Introduction

Mechanism theory deals with *kinematic chains*, i.e., rigid bodies (called *links*) connected by *joints*. These joints enable motion between the links and are characterized by the type of motion they allow. The main joints allow only one motion, either a rotation around a given axis (*rotary joint*) or a translation along one given axis (*prismatic joint*). More complex joints can be constructed with these basic joints, e.g., the ball-and-socket joint enabling every rotation around a point. Note that a finite set of parameters defines the status of the joint. For example, for a rotary joint the rotation angle fully defines the joint. The independent parameters of the joints will be called the *articular coordinates* of the mechanism.

Among all the joints in a mechanism, some may be actuated (i.e., a motor enables to change the value of the parameter of the joint) and the other joints are passive. The parameters of the actuated joints are the *input* of the mechanism.

It is usual to distinguish two special links in a mechanism: the base link, which is fixed to the ground, and the end-effector, whose position/orientation is the *output* of the mechanism. The number of parameters required to define the position/orientation of the end-effector is called the number of *degrees of freedom (DOF)* of the mechanism. For planar mechanisms this number is at most 3 (two translations and one rotation) and 6 (three translations and three rotations) for spatial mechanisms. The parameters describing the the position/orientation of the end-effector are called the *generalized cartesian coordinates* of the mechanism.

In order to control all the n degrees of freedom of the end-effector, it is necessary that the mechanism has at least n independent actuated elementary joints. A mechanism with $m > n$ independent actuated joints is called a redundant mechanism, but we wil restrict ourselves to the case where $m = n$.

We introduce now the *degree of connection* of a link as the number of rigid bodies connected to the link. At this point we may define two kinds of kinematics chains:

- open-loop kinematic chains: all the links have a degree of connection of two except the base and the end-effector, which each has a degree of connection equal to one.

- closed-loop kinematic chains: at least one link has a degree of connection greater or equal to three, or the degree of connection of the end-effector is greater than one.

Figure 1 shows examples of open- and closed-loop planar kinematics chains with only rotary joints.

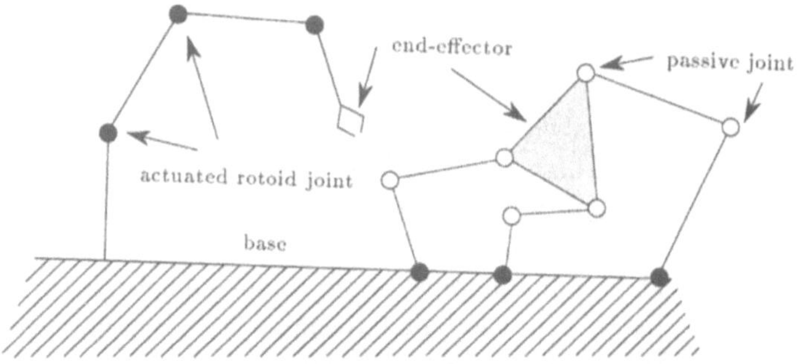

Figure 1: An open-loop planar kinematic chain left, and a closed-loop kinematic chain right

2 Classical problems in the field of mechanism theory

When studying a mechanism, classical problems arise:

- determining the realtion between the articular coordinates of the actuated joints and the generalized cartesian coordinates. The *direct kinematics* consists in determining the generalized cartesian coordinates for a given set of articular coordinates. The *inverse kinematic* problem is finding the opposite relation.

- determining the *workspace* of the end-effector, i.e., the limited region which can be reached by the end-effector owing to constraints of the joints (for example, a limited range for a prismatic actuator or no intersection between the links).

- determining the *singular configurations* of the mechanism. In the geheral case there is a linear relationship between the actuator velocities and the velocities of the end-effector, but this realtion is no more valid for some special configurations of the mechanism.

- the *design* or *synthesis* problem: for a given type of mechanism finding the geometry of the mechanism such that the trajectory of the end-effector contains a given set of points, called the *precision points*.

The above problems involve solving algebraic equations or inequalities which are deduced from the geometry of the mechanism. All of them are of great practical interest: the kinematics problems are to be solved to control the mechanism, the workspace and sythesis problems are useful for the design of robots, and the singularities are essential since a mechanism can suffer a breakdown when crossing a particular configuration.

3 Some results

Very elegant results have been established in the past, especially for planar mechanisms. I will mention a result established by Freudenstein which has shown that it is always possible to design a planar mechanism such that the trajectory of the end-effector (or at least a part of it) corresponds to a given algebraic curve [1].

3.1 The four-bar mechanism

Another well-studied mechanism is the so-called *four-bar mechanism* (figure 2). Basically the end-effector is a triangle with two vertices connected by two segments to the base. At the extremities of the segments, rotary joints enable a rotation.

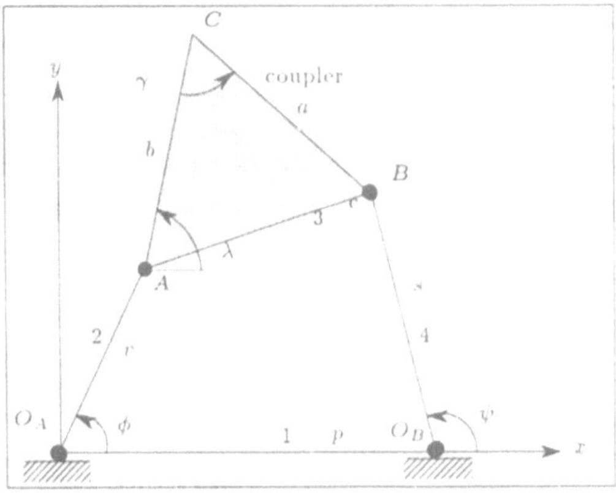

Figure 2: The four-bar mechanism

Note that both the base and the end-effector have a degree of connection of two, and, consequently, the four bar mechanism is a closed-loop mechanism. If the joint at O_A is actuated, then point C of the end-effector will describe a curve (the *coupler curve*) which happens to be a sextic [5] which has at most four double points. This sextic has a full circularity, i.e., it intersects the imaginary circle three times at the circular imaginary points. Although this mechanism is one of the most simple planar mechanisms, the equation of the sextic is rather complex. For example, the synthesis problem with three or four precision points has been solved [4], but only recently has a solution to the nine-precision-point synthesis been proposed [16]. Solvin this synthesis is equivalent to solving a system of four fourth-degree polynomials in four unknowns. The solution proposed by Wampler uses the numerical method called *continuation*, which has been succesfully used for solving kinematics problems [15].

3.2 The 6R manipulator

As for spatial kinematics chains, a classical problem is the resolution of the inverse kinematic of the 6R robot. Most industrial robots are open-loop kinematic chains with only rotary joints. A 6-DOF robot will have 6 such joints, all actuated. The articular coordinates are the 6 rotation angles θ_i of the joints, and the generalized cartesina coordinates are the 3 coordinates of the center of the end-effector together with the three angles describing the orientation of the end-effector. If the θ_i angles are known it is easy to calculate the generalized cartesian coordinates, but the opposite problem is far more complex; only recently has the final result been established. First it has been shown that by combining the algebraic equations from the direct kinematic it is possible to obtain a 16th-order polynomial in one unknown [12], which can be solved numerically and in some cases leads to 16 real roots. Reducing a system of algebraic equations to a polynomial in one unknown is a common practice in kinematics. This method drawbacks, however, as a complex polynomial is obtained from an initial set of usually simple equations. Consequently, some numerical problems may arise when computing the coefficients of the final polynomial. Another drawback is that no closed-form solution will be obtained. An alternative has been proposed by Wampler that uses the continuation method on the original set of equations to get all the solutions [14].

3.3 Parallel robots

3.3.1 Planar mechanisms

The most difficult problems in the field of mechanism theory appear when dealing with closed-loop mechanisms. However, new results have been obtained recently. These advances will be illustrated by examples dealing with a special kind of robot: the *parallel manipulator*. Figure 3 shows an example of planar parallel manipulator, called the 3-*RPR* manipulator, as the end-effector is linked to the ground by three identical chains consisting of a rotary joint (R) connected to the ground, followed by an active prismatic joint (P) which is connected to the platform by a rotary joint (R).

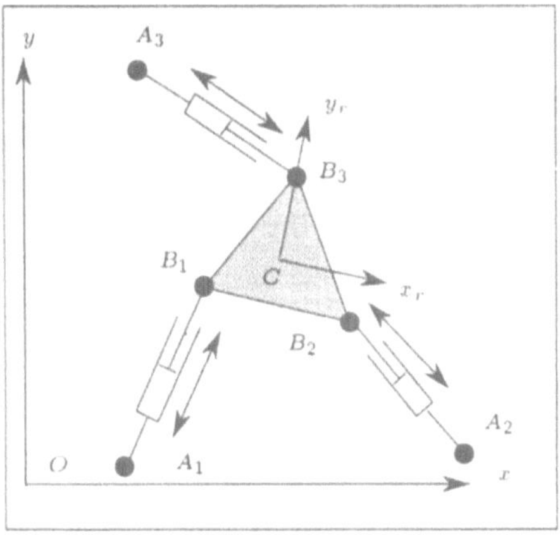

Figure 3: The 3-RPR parallel manipulator:
The end-effector is the gray triangle

By changing the length of the three prismatic actuators, it is possible to control the position and orientation of the end-effector. Consequently, this closed-loop mechanism is a three-DOF robot. Consider the inverse kinematics problem for this mechanism, i.e., find the link lengths (the distance between the points A_i, B_i) for a given position/orientation of the end-effector. As the end-effector location is known, it is easy to determine the coordinates of the B_i points in the reference frame. By construction the location of the A_i are known. Consequently, computing the norm of the vector A_iB_i is straightforward. Solving the direct kinematic is far more complex, and basically the problem can be reduced to solving a set of three algebraic equations in three unknowns. First of all, the solution may be not unique. The maximum number of solutions can be established in the following manner. Suppose we disconnect the mechanism at one joint (say B_3). We get therefore two mechanisms: a four-bar mechanism $(A_1B_1B_3^1B_2A_2)$ and a circular mechanism $(A_3B_3^2)$. A solution of the direct kinematic is obtained when the points B_3^1, B_3^2 are at the same location. Consequently, the solutions are the *real* intersection points of the coupler curve of the four-bar mechanism (i.e., a sextic) and a circle. Using Bezout's theorem we deduce that there are no more than 12 intersection points. But remember that the circularity of the coupler curve is 3 and the circularity of the circle is 1. Consequently, among the 12 intersection points, six are the circular imaginary points, and therefore the number of real intersection points is no more than 6.

The system of equations of the inverse kinematic can be combined to obtain 6th-order polynomial in one of the unknowns [3], and examples with 6 real solutions have been found. Figure 4 shows an example of such a configuration: the coupler curve intersects a circle in 6 different points. A particular example illustrates the influence of algebraic geometry in kinematics. Assume that the (A_1, A_2, A_3), (B_1, B_2, B_3) points lie on two lines. For this special case the degree of the direct kinematics polynomial is reduced to 3, but each root leads to two positions for the end-effector; thus there is also 6 potential solutions. But Sturm's method has enabled us to show that the polynomial has at most two real solutions, and consequently the direct kinematic has at most 4 solutions [2].

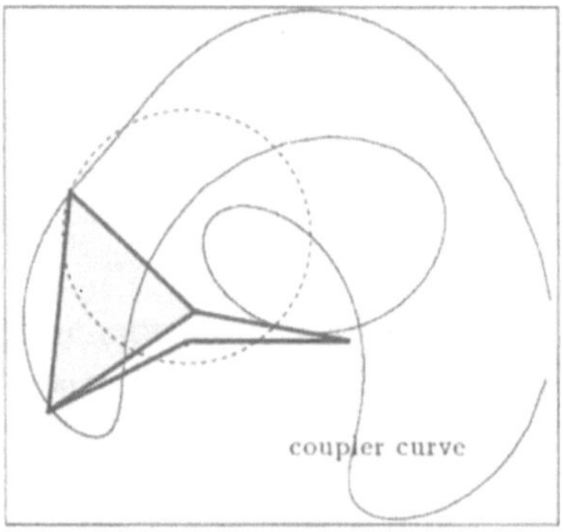

Figure 4: The coupler curve (in thin line) intersects a circle (the dash line) in 6 different locations.

3.3.2 Spatial mechanisms

The concept of parallel robots can be extended to build 6-DOF robots. Figure 5 shows an example of such a manipulator. The end -effector is a triangular platform which is connected to the ground by 6 legs. At each extremity of one leg, there is a ball-and-socket joint, and in each leg a prismatic actuator enablesone to change the leg length. Similarly to the planar robot by changing the leg lengths we can control the position/orientationof the platform. The inverse kinematic is straightforward and leads to a set of 6 algebraic equations, but the direct kinematic is much more complex. It is possible, however to show that there will be at most 16 solutions [10]. The basic idea of the demonstration is first to notice that each if the B_i is connected to two legs and consequently is able to move only on a circle. The mechanism can therefore be replaced by a new mechanism with only three legs, attached to the platform by a ball-and-socket joint and to the base by a rotary joint (figure 5, on right). Then we disconnect one of the legs at one of the B_i points (say B_3) and get two mechanisms: a circular mechanism and a mechanism called *RSSR* mechanism. Using one of Cayley's theorems [5] it is possible to show that the coupler surface of the *RSSR* (i.e., the surface described by B_3) is of order 16 with a circularity of 8. AS the solutions of the direct kinematics are the intersection points of the coupler surface and the circle, there cannot be more than 16 solutions. This claim has been verified, as it is possible to deduce from the inverse kinematic equation a 16th-order polynomial in one variable. This polynomial may have 16 real solutions [10].

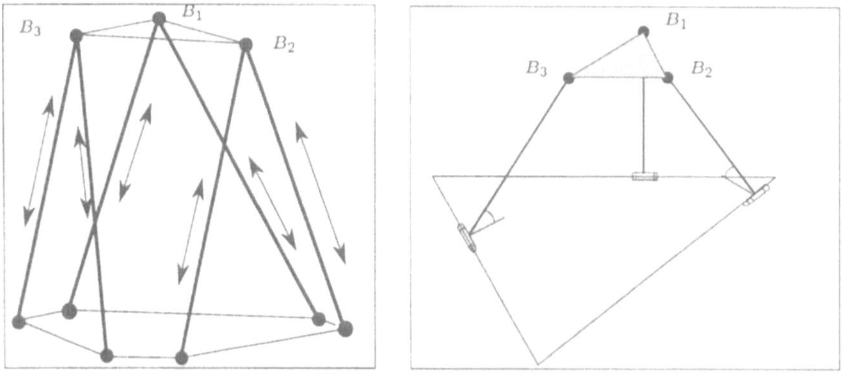

Figure 5: A 6-DOF parallel robot.

3.3.3 Singular configurations

We have seen that for parallel manipulators determining the leg lengths ρ when the cartesian coordinates $\mathbf{X}$ are known is straightforward. Let

$$\rho = F(\mathbf{X}).$$

From this relation it is easy to deduce the linear relation between the articular velocities $\dot{\rho}$ and the cartesian velocities $\dot{\mathbf{X}}$

$$\dot{\rho} = J^{-1}(\mathbf{X})\dot{\mathbf{X}}$$

where $J^{-1} = ((\partial F/\partial X))$ is a 6x6 matrix called the *inverse jacobian matrix* of the robot. For a given $\dot{\rho}$ there will be in general an unique $\dot{\mathbf{X}}$ except if J^{-1} is singular. The configurations where J^{-1} is singular are called the *singular configurations* of the robot: the velocity of the platform may be non-zero although the articular coordinates do not change, and consequently the robot can not be controlled. The determinant of the inverse jacobian matrix is also used for determining the forces in the legs τ for given forces and torques $\mathbf{F}$ acting on the platform as

$$\tau = J^{-T}\mathbf{F}.$$

A consequence is that the articular forces are obtained by dividing some quantity by the determinant of the inverse jacobian matrix. Consequently the forces increase as the robot comes close to a singular configuration which may cause a breakdown. For finding these configurations we may try to compute the determinant of J^{-1}. But the expression of this determinant is huge, and finding its roots is a difficult task. Another approach is to notice that the rows of J^{-1} are the Plücker vectors of the lines associated to the legs. If J^{-1} is singular then we have a linear dependency between the lines, which can occur for some special geometric configurations of the lines as stated by Grassmann geometry. The problem is thus reduced to find the cartesian coordinates such that these special configurations occur, and this leads to rather simple conditions [9].

3.4 Workspace of parallel manipulators

Consider a 6-DOF parallel manipulator. The position/orientation of its end-effector ia therefore described by a set of 6 parameters: the coordinates of its center and the three angles describing its orientation. In practice there is always a limitation on the leg lenghts and mechanical limitson the passive joints. Furthermore some positions of the end-effector cannot be reached as they will lead to an intersection between some legs. Consequently the position which can be reached by the center of the end-effector, the *workspace* of the robot, is a limited region of the physical space. But as we have to deal with a region in a 6-parameter space, there is no human-readable representation of the full workspace. However, it is usually admitted to represent a restriction of the full workspace by assigning some fixed values to some of the parameters. For example, it is usual to fix the orientation of the end-effector and one of the coordinates of the center of the end-effector (for example, the x, y coordinates of the center remain free). The reduced problem is now to compute the border of planar cross-sections of the workspace. This problem can be stated as an algebraic problem as we have to deal with algebraic inequalities in the unknowns. For example, the leg lengths can be expressed as polynomials in x, y which have to verify the limitation of the lengths. But the geometrical meaning of these polynomilas is used to simplify the problem and leads to compute the intersection of simple geometric objects: circles for the constraints on the leg length, polygons for the mechanical limits on the joints, and conics when dealing with the intersection of the legs [10]. Figure 6 shows an example of such cross-sections.

4 Open problems

4.1 Intersection of two coupler curves

The *maximal workspace* of a robot is the region which can be reached by the center of the end-effector with at least one orientation. Consider a planar 3-RPR parallel robot with some limitations on the leg lengths and assume that the center of the end-effector is one of its vertices (say B_3). It is easy to show that a location of B_3 will belong to the border of the maximal workspace if at least one of the leg length has an extremal value. Thus the border of the maximal workspace is the intersection of the trajectory obtained when one or two actuators have a maximal value. For only one actuator the extremal location of B_3 are circles, and when two have an extremal value the trajectory of B_3 is the coupler curve of the four-bar mechanism $A_1B_1B_3B_2A_2$.Therefore to compute the border we have to calculate the intersection of the coupler curves of the four-bar mechanisms whose link lengths are the combination of the extremal leg lengths.

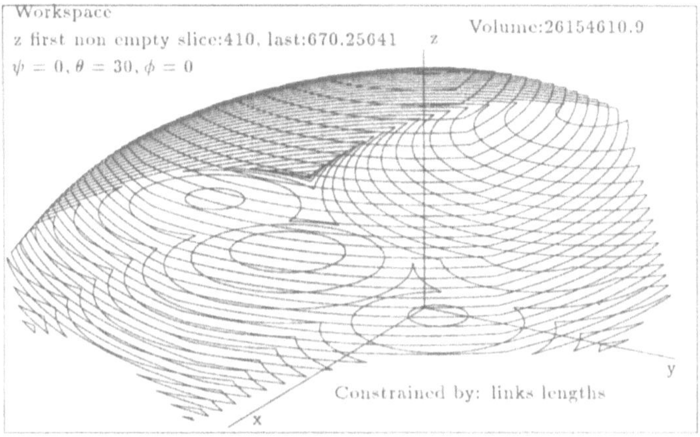

Figure 6: A 3D view of the workspace of a parallel manipulator for a fixed orientation of the end-effector.

Innocenti has shown that in general two coupler curves have at most 18 intersection points. He has then exhibited a 18x18 matrix whose determinant is a polynomial of 18th order and enables one to compute the solution of the intersection points [6]. Unfortunately Innocenti was not able to obtain this polynomial although the matrix is sparse. Instead he takes 19 particular values for the unknowns, numerically computes the determinant, and gets a linear system of 19 equations in the coefficients of the polynomial. But this method is not robust and is computer-intensive. For the maximal workspace problem the four-bar mechanisms have the same connection points to the ground and the same coupler but differ by the lengths of their links (figure 7). In that particular case it can be shown that there cannot be more than 12 intersection points, but finding an algorithm for computing the intersection is still an open problem.

4.2 Maximal workspace of a spatial parallel robot

The concept of maximal workspace can be extended to spatial parallel manipulators. It is assumed that the leg lengths are constrained to lie in some given intervals, and we want to determine all the possible position of the center of the end-effector which can be reached with at least one orientation. At this time no method has been proposed to solve this problem. The even simpler problem, *Is a given location of the center of the end-effector inside the maximal workspace?*, has no known solution. Basically this problem is to show that there exist three unknowns (the three orientation angles) which satisfy a set of 12 inequalities (the inverse kinematic equations).

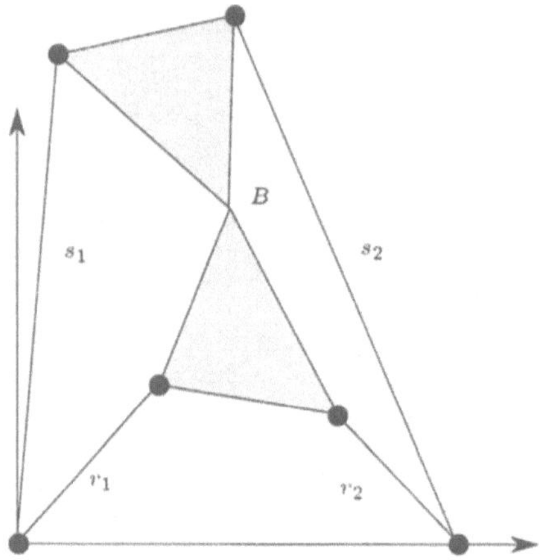

Figure 7: For determining the maximal workspace of a planar parallel manipulator, it is necessary to compute the intersection of the coupler curves of two four-bar mechanisms.

4.3 Direct kinematics of a general parallel manipulator

A general parallel manipulator is a mechanism with two bodies connected by 6 legs, whose attachment points are in general position on the bodies. It has been recently shown that the direct kinematic problem of a general parallel manipulator will have at most 40 solutions (complex and real) [7], [11], [13]. But no practical algorithm is known to determine these solutions. What is the maximum number of real solutions? In the case where the base points and platform points lie in two planes, an algorithm is known [8], but no configuration with more than 16 real solutions has been found.

Another problem can be stated as follows. Consider a parallel manipulator which has been mounted in some initial configuration. Assume now that an algorithm exists for determining all the solutions of the direct kinematic problem. Among the set of solutions, is there one unique solution which can be reached from the initial configuration without dismantling the manipulator or crossing a singular configuration?

4.4 Intersection of a cycloid and a circle

Not all problems in mechanism theory are algebraic. Here is an example of open problem which involves a non-algebraic curve. Assume that a planar 3-*RPR* parallel manipulator has to move from an initial configuration M_1 to a final configuration M_2, the intermediate configurations being obtained by linear interpolation. As the center of the platform translate along the segment M_1M_2, the platform will rotate around it, and consequently the vertices B_i of the platform will lie on an arc of cycloid. Assume now that we have some restriction on the leg lengths of the robot, i.e., the lengths ρ has to belong to the interval $[\rho_{min}, \rho_{max}]$. Consequently each B_i has to lie in an annular region with center A_i and radii ρ_{min}, ρ_{max}. Assume now that we want to verify that the trajectory of the robot is feasible, i.e., that there is no violation of the leg lengths constraints on the whole trajectory. We have thus to check that the arcs of cycloid lie fully inside the annular regions (figure 8). As we may assume that the initial and final points are valid positions, we therefore have to check that there is no intersection between the arcs of cycloid and the circles.

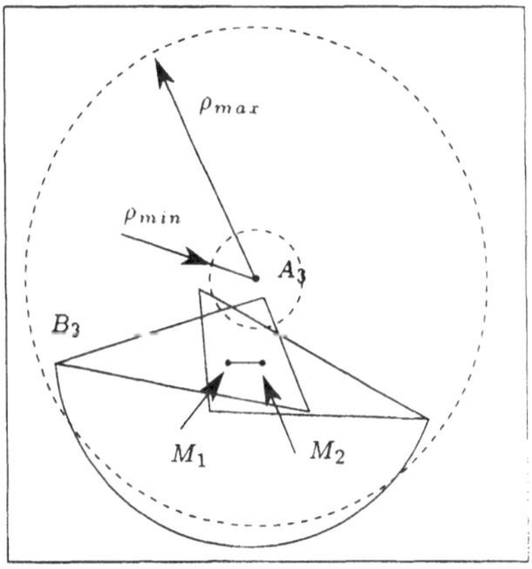

Figure 8: When the platform moves on the segment M_1M_2, point B_3 describes an arc of cycloid (in thin line). If the trajectory is feasible, this arc should lie inside an annular region centered in A_3 (in dashed line). Here the trajectory is not feasible.

5 Conclusions

Algebraic geometry plays an essential role in mechanism theory as most of the equations used to describe the geometric model of a mechanism are algebraic. The main problems that have to be solved are:

- find the number of real solutions and the solutions of a system of algebraic equations.
- find if there are some unknowns which satisfy a system of algebraic inequalities (although the values of the unknowns need not to be necessarily determined)
- being given a set of inequalities with n unknowns, find the border of the region of the n-parameter space for which the system of inequalities is satisfied.

An advantage of mechanism theory is that the algebraic equations have in general a geometric meaning which may often help to solve the problem. But it remains that the equations which are used in this field are often huge, even for the simplest mechanism, and consequently constitute an important challenge for algebraic geometry with a potentially huge field of applications.

References

[1] Freudenstein F. On the variety of motion generated by mechanisms. *Transaction of the ASME*, pp. 156–160, February 1962.

[2] Gosselin C. and Merlet J.-P.. On the direct kinematics of planar parallel manipulators: special architectures abd number of solutions. *Mechanism and Machine Theory*, March, 1994.

[3] Gosselin C., Sefrioui and Richard M. J. Solution polynomiale au problème de la cinématique directe des manipulateurs parallèles palns à 3 degrés de liberté. *Mechanism and Machine Theory*, 27(2): 107–119, March, 1992.

[4] Hartenberg R. S., Denavit J. Kinematic synthesis of linkages. MacGraw-Hill, 1964.

[5] Hunt K. H. Kinematic geometry of mechanisms. Clarendon Press, 1978.

[6] Innocenti C. Analytical determination of the intersection of two coupler-point curves generated by two four-bar linkages. In *Computational Kinematics* (eds J. Angeles, P. Kovacs and G. Hommel), pp. 251–262. Kluwer, 1993.

[7] Lazard D. Generalized Stewart Platform: How to compute with rigid motions?. In *IMACS*, pp. 85–88, Lille, May, 1993.

[8] Merlet J.-P. An algorithm for the forward kinematics of general parallel manipulators. In *ICAR*, pp. 1131–1135, Pise, June, 1991.

[9] Merlet J.-P. Singular configurations of parallel manipulators and Grassmann geometry. *International Journal of Robotics Research*, 8(5): 45–56, Ocotber, 1989.

[10] Merlet J.-P. Les robots parallèles. Hermès, Paris, 1990.

[11] Mourrain B. Enumeration problems in geometry, robotics and vision. This volume.

[12] Raghavan M. and Roth B. Kinematic analysis of the 6R manipulator of general geometry. In *5th Int. Symp. of Robotics Research*, pp. 263–270, Tokyo, 1990.

[13] Ronga F. and Vust T. Stewart platforms without computer. Preprint, 1992.

[14] Wampler C. and Morgan A. Solving the 6R inverse position problem using a generic-case solution methodology. *Mechanism and Machine Theory*, 26(1): 91–106, 1991.

[15] Wampler C., Morgan A. and Sommese A. J. Numerical continuation methods for solving polynomial systems arising in kinematics. *ASME J. of Mechanical Design*, 112: 59–68, March, 1990.

[16] Wampler C., Morgan A. and Sommese A. J. Complete solution of the nine-point path synthesis problem for four-bar linkages. *ASME J. of Mechanical Design*, 114: 153–159, March, 1992.

J.-P. Merlet (`jean-pierre.merlet@sophia.inria.fr`)

INRIA, 2004 Routes de Lucioles, BP 9306902 Sophia-Antipolis Cedex (France)

Progress in Mathematics, Vol. 143, © 1996 Birkhäuser Verlag Basel/Switzerland

Enumeration problems in geometry, robotics and vision

B. Mourrain*

1 Introduction

The objective of this work is to give intersection formulas in algebraic problems appearing in mechanics and vision. We begin with problems in $\mathbb{P}^2$ where distances are involved. The second section describes the case of surfaces in $\mathbb{P}^3$ which have a conic at infinity (called the umbilic) in common. In other words, we are interested here in intersection problems on the space of spheres. A formula is given for the number of common points outside this umbilic. The third section deals with the degree of varieties, corresponding to segments whose extremities are on two curves. Following this progression, the next section is devoted to intersection in the variety of displacements. In this section, we analyze precisely the ring of functions on the variety of displacements and give its multiplicity, which allows us to bound the number of solutions in the direct kinematic problem of a parallel robot and in the problem of reconstruction from points in vision.

2 Plane curves with circularity

In this part, we consider curves in $\mathbb{P}^2(\mathbb{C})$ ($\mathbb{P}^2(\mathbb{C})$ is the projective space on $\mathbb{C}$ of dimension 2), defined by equations in the variables x, y, t. The line at infinity is defined by $\mathbf{V}(t = 0)$, and on this line we consider two special points $I = (1 : \mathbf{i} : 0)$ and $J = (1 : -\mathbf{i} : 0)$. These points are on all circles of equations $x^2 + y^2 + 2a\,t\,x + 2\,b\,t\,y + c\,t^2 = 0$ (where $a, b, c \in \mathbb{C}$) and are called the *cyclic points*.

Definition 2.1 *Let C be a curve of $\mathbb{P}^2$. The circularity of C is the minimum of the multiplicities of I and J on C.*

If the curve is defined by a real equation, the two multiplicities are equal.

Proposition 2.2 *Let C be a curve of $\mathbb{P}^2$ defined by an equation of degree d:*

$$f(x, y, t) = f_0(x, y) + f_1(x, y)\,t + \cdots + f_d(x, y)\,t^d$$

with $f \in \mathbb{R}[x, y, t]$ (f_d is a constant). Let c_j be the greatest power of $x^2 + y^2$ dividing $f_j(x, y)$ so that $f_j(x, y) = (x^2+y^2)^{c_j} g_j(x, y)$ (with $gcd(g_j, x^2+y^2) = 1$). Let $f^*(x, y, u) = \sum_j u^{c_j} g_j(x, y)$. The circularity of C is

---

(*) Work partially supported by PoSSo, EEC ESPRIT BRA contract 6846.

1. *the minimum of* $\{c_j + j\}$,
2. *or equivalently* $\deg(f) - deg(f^*)$.

Proof. For the first point, the multiplicity of I is by definition equal to the valuation of f around this point (see [3] or [2]), considering this point as the origin. Taking $Y = y/x$, $T = t/x$, and $U = Y - \mathbf{i}$, we obtain a polynomial of the form

$$f(1, Y, T) = f(1, \mathbf{i} + U, T) = f_0(1, \mathbf{i} + U) + T f_1(1, \mathbf{i} + U) + \cdots + f_d T^d.$$

The multiplicity of I will be the valuation of this polynomial in U, T. Each real polynomial $f_j(1, Y)$ is of the form $(1 + Y^2)^{c_j} g_j(1, Y) = (Y - \mathbf{i})^{c_j} (Y + \mathbf{i})^{c_j} g_j(1, Y)$ with $g_j(1, \mathbf{i}) \neq 0$ by assumption. Thus

$$f(1, \mathbf{i} + U, T) = U^{c_0} h_0(U) + T\, U^{c_1}\, h_1(U) + \cdots + h_d\, T^d$$

where each polynomial $h_j(U)$ has a non-zero constant term. This implies that the least degree in U, T is the minimum of the degree of $T^j U^{c_j}$.

For the second point of the proposition, remark that the degree of each term $u^{c_j} g_j(x, y)$ is equal to $c_j + d - (2\, c_j + j) = d - (c_j + j)$ so that the maximal degree is equal to the $d - c$ where c, the minimum of $\{c_j + j\}$, is equal to the circularity. □

Proposition 2.3 *Let C_1 (resp. C_2) be a curve of $\mathbb{P}^2$ of degree d_1 (resp. d_2) and circularity c_1 (resp. c_2). Assume that these curves have only a finite number of points N (counted with multiplicity) in common in $\mathbb{A}^2$ (the affine space of dimension 2). Then*

$$N \leq d_1 d_2 - 2\, c_1\, c_2. \tag{1}$$

If the curves have no other points than I, J at infinity, if these two points have the same multiplicity on each curve, and if the curves are not tangent at I or J, equality holds in (1).

Proof. As these two curves meet in a finite set of points, according to the Bézout theorem (see [3], [2], or [11]), we have

$$d_1\, d_2 = \sum_{P \in C_1 \cap C_2} m(P, C_1 \cap C_2),$$

where $m(P, C_1 \cap C_2)$ is the multiplicity of intersection of C_1 and C_2 at P. We know that I and J are of multiplicity c_1 (resp. c_2) on C_1 (resp. C_2) and thus appear at least with multiplicity $c_1\, c_2$ (see [3]) in the intersection (and more if the curves are tangent at these points). Therefore the number of points in $\mathbb{A}^2$ is

$$N \leq \sum_{P \in C_1 \cap C_2, P \neq I, P \neq J} m(P, C_1 \cap C_2) \leq d_1\, d_2 - 2\, c_1\, c_2.$$

Equality holds if the curves are not tangent at these points, if there are no other common points at infinity, and if the multiplicity of I and J are the same. □

Examples

- Consider the intersection of two distinct circles C_1, C_2 in the affine plane of equation $x^2 + y^2 - 2\,a_i\,x - 2b_i\,x + c_i = 0$ $(i \in \{1, 2\})$. This system is equivalent to

$$\begin{cases} u - 2\,a_1\,x - 2b_1\,x + c_1 = 0 \\ u - 2\,a_2\,x - 2b_2\,x + c_2 = 0 \\ u - (x^2 + y^2) = 0 \end{cases}$$

  and the number of common points is $2 \times 2 - 2 \times 1 \times 1 = 2$.

- Let C_1, C_2 be sextics of circularity 3. Then the number of common points in the affine part is at most $6 \times 6 - 2 \times 3 \times 3 = 18$.

This ends the easy case of curves of $\mathbb{P}^2$ which contains the cyclic points with multiplicities. Let us now consider the case of surfaces in $\mathbb{P}^3$ defined by constrains of distances.

3 Surfaces with circularity

In this section we are considering surfaces of $\mathbb{P}^3$, defined by homogeneous equations in the variables x, y, z, t. The plane at infinity will be $\mathbf{V}(t = 0)$, and on this plane we will consider a special curve $\Omega = \mathbf{V}(x^2 + y^2 + z^2 = 0, t = 0)$, which is called the *umbilic* or the *imaginary circle*.

Definition 3.1 *Let S be a surface of $\mathbb{P}^3$. The circularity of S is the multiplicity of Ω on S. We denote it $m_S(\Omega)$. In the same way, we can define the circularity of a curve C of $\mathbb{P}^3$ as $\sum_{P \in C \cap \Omega} m_C(P)$ where $m_C(P)$ is the multiplicity of P on C.*

Property 3.2 *Let S be a surface of $\mathbb{P}^3$ defined by an equation of degree d:*

$$f(x, y, z, t) = f_0(x, y, z) + f_1(x, y, z)\,t + \cdots + f_d(x, y. z)\,t^d$$

($f_d(x, y, z) = f_d$ is a constant). Let c_j be the greatest power of $x^2 + y^2 + z^2$ that divides $f_j(x, y, z)$ so that $f_j(x, y, z) = (x^2 + y^2 + z^2)^{c_j} g_j(x, y, z)$. Let $f^*(x, y, z, u) = \sum_j u^{c_j} g_j(x, y, z)$. The circularity of S is

1. *the minimum of $\{c_j + j\}$,*
2. *or equivalently $deg(f) - deg(f^*)$.*

Proof. Let us prove the first point. The multiplicity of Ω on S is the valuation of f around a generic point $(1, Y_0, Z_0, 0)$ of Ω (see [3], [10], [11, p. 388–394]). Taking again $Y = y/x$, $Z = z/x$, $T = t/x$ and $Y = Y_0 + U$. $Z = Z_0 + U$, $T = W$ (with $1 + Y_0^2 + Z_0^2 = 0$), we obtain a polynomial of the form

$$\begin{aligned} f(1, Y, Z, T) &= f(1, Y_0 + U, Z_0 + V, T) = \\ &= f_0(1, Y_0 + U, Z_0 + V) + T f_1(1, Y_0 + U. Z_0 + V) + \cdots + f_d T^d. \end{aligned}$$

As $1+Y_0^2+Z_0^2=0$ and $f_j(x,y,z)=(x^2+y^2+z^2)^{c_j}g_j(x,y,z)$, the valuation of $f_j(1,Y_0+U,Z_0+V)$ in U,V is at least c_j. Moreover, we can choose $(1,Y_0,Z_0,0)$ in a non-empty open subset of Ω such that $g_j(1,Y_0,Z_0)\neq 0$ (otherwise g_j would be divisible by $x^2+y^2+z^2$), so that the valuation of $f_j(1,Y_0+U,Z_0+V)$ in U,V is exactly c_j.

Finally, the valuation of $f(1,Y_0+U,Z_0+V,W)$ is the minimum of the valuation c_j+j of each term $T^j f_j(1,Y_0+U,Z_0+V)$ (because no cancelation between such terms can appear), which proves that the circularity of S is the minimum of $\{c_j+j\}$.

To prove the second point, remark that the degree of a term $u^{c_j}g_j(x,y,z)$ is $c_j+(d-j-2\,c_j)=d-j-c_j$ and the maximal degree is $d-c$ where $c=\min_j(c_j+j)$. □

Theorem 3.3 *Let S_1 (resp. S_2, S_3) be a surface of $\mathbb{P}^3$ of degree d_1 (resp. d_2, d_3) and circularity c_1 (resp. c_2,c_3). Assume that these surfaces have only a finite number N of points in common in $\mathbb{A}^3$. Then*

$$N\leq d_1d_2d_3-2\,(c_1\,c_2\,d_3+c_1\,c_3\,d_2+c_2\,c_3\,d_1)+6\,c_1\,c_2\,c_3.$$

In the case where $d_i=2\,c_i$ for some i (one of the surfaces is of maximal circularity), we have $N\leq 2\,(d_1-c_1)\,(d_2-c_2)\,(d_3-c_3)$.

We need the following lemma:

Lemma 3.4 *Let $f(x,y,z,t)$ be a homogeneous polynomial of degree d and circularity c, and $f^*(x,y,z,u)$ be the polynomial defined above. By $\bar{f}(x,y,z,u,v)$ we denote the homogenization of the polynomial f^* in v. The valuation of $\bar{f}(X,Y,Z,1,X^2+Y^2+Z^2)$ at the origin is $d-2\,c$.*

Proof. The polynomial $\bar{f}(x,y,z,u,v)$ of degree $d-c$ is of the form:

$$\bar{f}(x,y,z,u,v)=\sum_j u^{c_j}g_j(x,y,z)v^{d-c-(d-\,c_j-j)}=\sum_j u^{c_j}g_j(x,y,z)v^{c_j+j-c},$$

where $g(x,y,z)$ is a homogeneous polynomial of degree $d-2\,c_j-j$. Consequently, the valuation of $\bar{f}(X,Y,Z,1,X^2+Y^2+Z^2)$ is the minimum of the degree of $g_j(X,Y,Z)(X^2+Y^2+Z^2)^{c_j+j-c}$ which is $d-2\,c_j-j+2\,(c_j+j-c)=d-2\,c+j$. The minimal degree is obtained for $j=0$ so that the valuation is $d-2\,c$. □

Proof of the theorem. Let $f_i(x,y,z,t)$ be the equations of S_i in $\mathbb{P}^3$. The intersection of these surfaces in $\mathbb{A}^3$ is defined by the system

$$\left\{\begin{array}{l} f_1(x,y,z,1)=0\\ f_2(x,y,z,1)=0\\ f_3(x,y,z,1)=0\end{array}\right.$$

which is equivalent in $\mathbb{A}^4$ to

$$\left\{\begin{array}{l} f_1^*(x,y,z,u)=0 \\ f_2^*(x,y,z,u)=0 \\ f_3^*(x,y,z,u)=0 \\ u-(x^2+y^2+z^2)=0. \end{array}\right. \qquad (*)$$

After homogenization of this system (in (x,y,z,u,v)), we obtain 3 equations of the form

$$\bar{f}_i(x,y,z,u,v)=\sum_{j=0}^{d_i} g_{i,j}(x,y,z)\,u^{c_{i,j}}\,v^{c_{i,j}+j-c_i}=0 \quad \text{for } i\in\{1,2,3\}$$

(of degree d_i-c_i) and the equation $u\,v-(x^2+y^2+z^2)=0$ (of degree 2), defining in $\mathbb{P}^4$ a variety V. The affine part of V is what we are looking for. By assumption, it is a finite set of points. The part at infinity is defined by $v=0, x^2+y^2+z^2=0$, and $\sum_{0\leq j\leq d_i, c_{i,j}+j=c_i} g_{i,j}(x,y,z)u^{c_{i,j}}=0$ for $i\in\{1,2,3\}$. We remark that this part at infinity might contain the point $\omega=(0{:}0{:}0{:}1{:}0)$.

According to Bézout's theorem, the degree of the affine part of V is bounded by the product of the degree of the equation minus the intersection multiplicity of the 4 hypersurfaces in $\mathbb{P}^4$ at the point ω. This multiplicity is obtained by considering the above system around the point ω, taking $X=x/u$, $Y=y/u$, $Z=z/u$, and $V=v/u$:

$$\left\{\begin{array}{l} \bar{f}_1(X,Y,Z,1,V)=0 \\ \bar{f}_2(X,Y,Z,1,V)=0 \\ \bar{f}_3(X,Y,Z,1,V)=0 \\ V-(X^2+Y^2+Z^2)=0 \end{array}\right. \quad \text{or} \quad \left\{\begin{array}{l} \bar{f}_1(X,Y,Z,1,X^2+Y^2+Z^2)=0. \\ \bar{f}_2(X,Y,Z,1,X^2+Y^2+Z^2)=0. \\ \bar{f}_3(X,Y,Z,1,X^2+Y^2+Z^2)=0. \\ V-(X^2+Y^2+Z^2)=0 \end{array}\right.$$

As the valuation of $\bar{f}_i(X,Y,Z,1,X^2+Y^2+Z^2)$ is $d_i-2\,c_i$ (according to the previous lemma), the multiplicity of intersection at ω is at least $(d_1-2\,c_1)\,(d_2-2\,c_2)\,(d_3-2\,c_3)$; thus the degree of the affine part of V is at most

$$\begin{aligned} &2\,(d_1-c_1)\,(d_2-c_2)\,(d_3-c_3)-(d_1-2\,c_1)\,(d_2-2\,c_2)\,(d_3-2\,c_3) \\ =\;& d_1d_2d_3-2\,(c_1\,c_2\,d_3+c_1\,c_3\,d_2+c_2\,c_3\,d_1)+6\,c_1\,c_2\,c_3. \end{aligned}$$

If one of the surfaces is of maximal circularity, then ω is not in the part at infinity of V and the degree of its affine part is at most $2\,(d_1-c_1)\,(d_2-c_2)\,(d_3-c_3)$. □

Examples

- If S_1, S_2, S_3 are spheres, then $d_i=2$, $c_i=1$ and the number of common points in the affine space is at most 2.

- Let S_1, S_2, S_3 be tori defined as the sets of points $X = (x, y, z)$ such that

$$\cos^2(A_iX \angle B_iX) = c_i$$

($\angle$ means the angle between the vectors), for given points A_i, B_i in $\mathbb{A}^3$ ($i \in \{1, 2, 3\}$). The equation of each S_i is of the form

$$\langle A_iX, B_iX\rangle^2 - c_i \langle A_iX, A_iX\rangle \langle B_iX, B_iX\rangle$$

$$= (1 - c_i)\langle X, X\rangle^2 - 2(1 - c_i)\langle X, X\rangle(\langle X, A_i\rangle + \langle X, B_i\rangle) + \overbrace{\cdots\cdots}^{\text{degree in } X \leq 2}$$

where $\langle\, ,\, \rangle$ is the standard inner-product in $\mathbb{A}^3$. The degree of these tori is 4 and the circularity is 2 because the term of degree 3 is divisible by $\langle X, X\rangle = x^2 + y^2 + z^2$. Therefore the number of common points in the affine space is at most $2 \times 8 = 16$.

3.1 A view on a blow-up

The previous section deals with equations in x, y, z and $x^2+y^2+z^2$ or, in other words, with equations in the space of spheres of $\mathbb{P}^3$. The enumeration of common points, which involves in this case the degree of each surface and the multiplicity of the umbilic, corresponds to what is called a problem of residual intersection (the surfaces have a curve in common at infinity) in modern intersection theory. Let us have a glimpse at this other world, described in another language in [10, p. 156], in order to see how it is related to the previous computations.

Consider, for instance, a point O in the plane $\mathbb{A}^2$ (O will be the origin of the plane) with a certain multiplicity on two given curves C_1, C_2. The multiplicity of intersection is not necessarily the product of the multiplicities of the point on each curve. It can be more if the curves have common tangents at O. To take care of this possibility and to analyze correctly the intersection multiplicity, one is led to add new information, which in this case is the slope of a line from O to a point $M = (x, y)$ of the plane. As a slope of a line is actually defined in $\mathbb{P}^1$, we are working in a space of dimension 3, which is $\mathbb{A}^2 \times \mathbb{P}^1$. The image of $\mathbb{A}^2$ is a surface in $\mathbb{A}^2 \times \mathbb{P}^1$ defined by $u\,y - v\,x = 0$ and the blowing-up algebra associated to it is $\mathbb{C}[x, y][u, v]/(u\,y - v\,x)$. If $x \neq 0$ and $y \neq 0$, the slope of the line is $s = \frac{v}{u} = \frac{y}{x}$. This construction yields a spurious line $O \times \mathbb{P}^1$, which is the "image" of the point O ("blowing-up" of O) where the slope is not defined. This line is called the *exceptional line*. By this construction, a curve of $\mathbb{A}^2$ through O is transformed into the union of the exceptional lines and another curve, called the *strict transform*. The way this strict transform meets the exceptional line gives us the limit of the slopes of cords OM when M tends to O. In other words, it gives the slope of the tangents at O. The intersection of the transforms of the two curves C_1, C_2 will then give us the exceptional line (counted with the product of the two multiplicities of O) and the intersection of the strict transforms will tell us "how much" the curves are tangent.

This can be extended naturally to surfaces (to any variety in fact) containing a given curve (here the umbilic) in the following way. The blowing-up algebra is $\mathbb{C}[x,y,z,t][u,v]/(up_2 - vp_1)$ where $p_1 = t$, $p_2 = x^2 + y^2 + z^2$ are two equations defining the umbilic. Remark that if we work in the open subset $v \neq 0$, $t = 1$, we have $U = \frac{u}{v} = x^2 + y^2 + z^2$ which is just the new variable that we added in the last section. The computation done there is in fact precisely the computation done in the blowing-up algebra when we assume that some variables do not vanish, that is, when we are in an open subset of this "blowing-up". For further information, see [11], [10][1].

4 Another application

We present here a theorem of Cayley and a simple proof using Bézout's theorem in a product of projective spaces. A reference for this theorem is [13] but no proof is available there.[2]

For this purpose, we need to use classical results of intersection theory. The classical Bézout theorem says that the intersection of two varieties V_1, V_2 of $\mathbb{P}^l$ and of degree d_1 (resp. d_2) and codimension δ_1 (resp. δ_2) is in general empty if $\delta_1 + \delta_2 > l$, and otherwise it is a variety of degree $d_1 d_2$ and codimension $\delta_1 + \delta_2$. This can be formalized in the following way: We associate to a variety V of $\mathbb{P}^l$ the element $d\,h^\delta$ of $\mathbb{Z}[h]$ where d is its degree and δ its codimension. Then the intersection corresponds to the product modulo the relation $h^{l+1} = 0$. The space $A[X] = \mathbb{Z}[h]/(h^{l+1})$ is called the Chow ring of $X = \mathbb{P}^l$. More generally, a Chow ring describing intersections of subvarieties on a non singular projective variety X can be constructed with the properties given in [11, p. 426]. To any subvariety V of X is associated a class $[V]$ in this Chow ring, and the intersection corresponds to the product of classes. Moreover, if the subvariety V is the union of two subvarieties $V = V' \cup V''$, then we have $[V] = [V'] + [V'']$.

Using the fact that if $Y = X \times X'$ then $A[Y] = A[X] \otimes_{\mathbb{Z}} A[X']$ (see [11]) we immediately see that the Chow ring of $\mathbb{P}^3 \times \mathbb{P}^3$ is $\mathbb{Z}[h_1, h_2]/(h_1^4, h_2^4)$ (see also [17, p. 196]). The class of a hypersurface of $\mathbb{P}^3 \times \mathbb{P}^3$ defined by a bi-homogeneous equation of degree (d_1, d_2) is $d_1 h_1 + d_2 h_2$.

Theorem 4.1 *Let C_1 and C_2 be curves of degree d_1, d_2 and circularity c_1, c_2 which have no common points at infinity. Let $\mathcal{L}^\circ$ be the set of lines (P_1, P_2) such that $||P_1 P_2||$ ($||\ ||$ is the Euclidean norm of vectors) is fixed and $P_1 \in C_1 \cap \mathbb{A}^3$, $P_2 \in C_2 \cap \mathbb{A}^3$, and let $\mathcal{L}$ be the closure of $\mathcal{L}^\circ$. Then $\mathcal{L}$ is of codimension 3 in the Grassmann Variety $\mathcal{G}_2$ of all lines and degree*

$$2 \times d_1 \times (d_2 - c_2) + 2 \times d_2 \times (d_1 - c_1).$$

[1] I would like to thank M. Merle for interesting and fruitful discussions on this topic.

[2] I would like to thank J.P Merlet, who pointed this result out to me.

Proof. Consider the couples $(P_1, P_2) \in \mathbb{P}^3 \times \mathbb{P}^3$ such that $P_1 \in C_1, P_2 \in C_2$ and $\|P_1P_2\|^2 = \rho^2$ ($\rho \in \mathbb{C}$). We are going to bound the degree of this subvariety, denoted by $\mathcal{V}$ of $\mathbb{P}^3 \times \mathbb{P}^3$.

The Chow ring of $\mathbb{P}^3 \times \mathbb{P}^3$ is generated by h_1, h_2 the class of $\mathbb{P}^2 \times \mathbb{P}^3$ and $\mathbb{P}^3 \times \mathbb{P}^2$, and we have $h_1^4 = 0$, $h_2^4 = 0$. We have $A[\mathbb{P}^3 \times \mathbb{P}^3] = \mathbb{Z}[h_1, h_2]/(h_1^4, h_2^4)$.

The class of the variety $C_1 \times \mathbb{P}^3$ is $d_1\, h_1^2$, the class of $\mathbb{P}^3 \times C_2$ is $d_2\, h_2^2$, and the variety defined by the equation $\|P_1P_2\|^2 - \rho^2 = [P_1, P_1] - 2 \times [P_1, P_2] + [P_2, P_2] - \rho^2 = 0$ corresponds to $2 \times h_1 + 2 \times h_2$.

The intersection of these three varieties corresponds to the couples of points defining the lines we are looking for. Its class in the Chow ring is

$$d_1h_1^2.d_2h_2^2.(2 \times h_1 + 2 \times h_2) \equiv 2d_1d_2(h_1^3h_2^2 + h_1^2h_2^3).$$

It is a variety of codimension 5 in $\mathbb{P}^3 \times \mathbb{P}^3$ that is a curve but it is not irreducible. In fact, if P_1 is any point of C_1 which is also on the umbilic, then the homogenization in t of the equation $\|P_1P_2\|^2 - \rho^2 = 0$ yields $[P_1, P_1]\, t_2^2 - 2\,[P_1, P_2]\, t_1t_2 + [P_2, P_2]\, t_1^2 - \rho^2t_1^2t_2^2 = 0$. This condition is satisfied when $t_1 = 0$ and $[P_1, P_1] = 0$ for any value of P_2. In other words, the variety $\mathcal{V}$ contains the curve $P_1 \times C_2$ for any points P_1 of C_1, which is also on the umbilic. This can also be done with P_2. As there are c_1 (resp. c_2) such curves $P_1 \times C_2$ (resp. $C_1 \times P_2$) (counted with multiplicities), the class of the affine part of $\mathcal{V}$ which is $\mathcal{L}$ is

$$2d_1d_2(h_1^3h_2^2+h_1^2h_2^3)-c_1d_2h_1^3h_2^2-c_2d_2h_1^2h_2^3 = 2d_1(d_2-c_2)h_1^3h_2^2+2d_2(d_1-c_1)h_1^2h_2^3.$$

Now consider the corresponding variety of lines in the Grassmann Variety $\mathcal{G}_2$ of $\mathbb{P}^3$. It is a curve in this space and its degree is given by the number of lines of this family which meet a given "generic" line (see [12, Vol. II, p. 359]). The intersection of a line (P_1, P_2) with another one corresponds to an equation of degree $h_1 + h_2$ in the Chow ring associated to $\mathbb{P}^3 \times \mathbb{P}^3$. This means that we have $(2d_1(d_2 - c_2)h_1^3h_2^2 + 2d_2(d_1 - c_1)h_1^2h_2^3)(h_1 + h_2) = 2d_1(d_2 - c_2) + 2d_2(d_1 - c_1)$ solutions which is the degree of the curve of lines. □

5 The variety of displacements

In this section, we are going to focus on the set $\mathcal{D}$ of displacements of the affine space $\mathbb{A} = \mathbb{C}^3$ (also denoted by $\mathbb{A}^3$). We want to describe as "effectively" as possible the algebra of functions on this variety. The objective is to give a complete system of rules which allows us to reduce any element of this ring to a normal form.

We note $(\,|\,)$ the standard inner product of $\mathbb{A}$ and $(U_1, \dots, U_k|V_1, \dots, V_k) = \det((U_i|V_j)_{1\leq i,j\leq k})$ for all vectors[3] $U_i, V_j \in \mathbb{A}$.

[3]The vector associated to an element A of $\mathbb{A}^n$ is the vector from the origin to the point A.

A displacement D of $\mathcal{D}$ is described by a couple (R, T) where R is a rotation satisfying the equations $R^T R = R R^T = Id_3$ (where Id_3 is the identity matrix of $\mathbb{A}$), $\det(R) = 1$ and $T = (t_1, t_2, t_3)$ is a translation. By $R_j = (r_{1,j}, r_{2,j}, r_{3,j})$ we denote the jth column of R, and by $R_i^* = (r_{i,1}, r_{i,2}, r_{i,3})$ we denote the ith row of R. These vectors satisfy

$$\det([R_1, R_2, R_3]) = 1 \text{ and } (R_i|R_j) = (R_i^*|R_j^*) = \delta_{i,j}. \tag{2}$$

In the applications coming from robotics and vision, the equations constraining the displacements often involve distances between points so that the coordinates of the vectors

$$R^T T, R_i, R_i^* \text{ and } (T|T)$$

naturally appear. Let us therefore introduce new variables $U = (u_1, u_2, u_3)$, ρ and the equations

$$\rho = (T|T), U = R^T T. \tag{3}$$

We easily see that $T = RU$ is a consequence of the previous relations. Up to now, we are working in a space $\mathbb{A}^{16}$, corresponding to the variables $\{r_{i,j}, t_i, u_i, \rho\}$. Let us introduce a new variable z of homogenization in order to work in $\mathbb{P}^{16}$. Setting $z = 0$ in the ideal generated by the previous equations we obtain the part at infinity of the variety of $\mathbb{P}^{16}$ that defines the ideal generated by the equations (2) and (3).

Lemma 5.1 *The part at infinity in $\mathbb{P}^{16}$ of the previous variety is the set of matrices $[R_1, R_2, R_3, T]$ of rank one and image in the "umbilic" $(X|X) = 0$.*

Proof. The part at infinity satisfies

$$\det(R) = (R_i|R_j) = (R_i^*|R_j^*) = 0, \ (R_i|T) = 0, \ (R_i^*|U) = 0, (T|T) = 0$$

(by putting $z = 0$ in the previous equations). This implies that at infinity the matrix $[R, T]$ is of rank 1. Then the matrix R is of rank 1. If not (R of rank two), its transpose matrix R^T is also of rank 2, and its kernel of dimension 1 contains the image of R (because $R^T R = 0$). This contradiction implies that R is at most of rank 1. Moreover, T is in the orthogonal of the image of R. If (R_1, R_2, R_3, T) is of rank 2, its orthogonal is of rank 1 but it contains R_1, R_2, R_3, T. Consequently, (R_1, R_2, R_3, T) has to be of rank 1, and for each vector V in the image of $[R_1, R_2, R_3, T]$, we have $(V|V) = 0$. □

Following the previous section, to blow up the part at infinity we introduce new variables for the vectors $R_i \wedge T$. Let us write $(k, l|i, 4)$ the variable corresponding to $(e_k, e_l|R_i, T) = r_{k,i} t_l - r_{l,i} t_k$ which is, up to a sign, a coordinate of $R_i \wedge T$. The role of the new variables is to "throw away" the component at infinity of the first variety. Note $r_{i,j} = (i|j), t_i = (i|4)$ so that we have the relations

$$(i|k)(j|4) - (i|4)(j|k) = (i, j|k, 4) \tag{4}$$

for $1 \leq i, j, k \leq 3$.

The set of variables that we are going to consider is

$$X = \{(i|j), (i|4),\ (i,i'|j,4),\ u_1,\ u_2,\ u_3,\ \rho\}_{1\leq i,j,i'\leq 3}. \tag{5}$$

Let us call $\mathcal{I}$ the ideal of $\mathbb{C}[X]$ generated by the equations (2), (3), and (4). These equations define in $\mathbb{A}^{25}$ a variety that we still note $\mathcal{D}$. We also note $\mathbb{C}[\mathcal{D}] = \mathbb{C}[X]/\mathcal{I}$. The set of matrices (R,T) corresponding to the "usual" description of displacements is in fact the projection of this variety $\mathcal{D}$ on the space of the first coordinates.

5.1 Straightening laws

In this part, we are going to exhibit a structure for $\mathbb{C}[\mathcal{D}]$ similar to what is called *an algebra with straightening laws* (see [4]).

For that, let recall some definitions:

Definition 5.2 *Let X be a set (of variables) and $\ll$ a partial order on X. A monomial $m \in X^*$ is* standard *if it can be written as $m = x_{i_1}\cdots x_{i_k}$ with $x_{i_1} \ll \cdots \ll x_{i_k}$.*

Definition 5.3 *Let I be an ideal of $\mathbb{C}[X]$. The algebra $A = \mathbb{C}[X]/I$ is called*

- an algebra with straightening laws *(ASL) if*
  - (H_1) *the standard monomials are a basis of $A = \mathbb{C}[X]/I$,*
  - (H_2) *for each couple of variables $x_1, x_2 \in X$, such that $m = x_1x_2$ is not standard, the decomposition of m in standard monomials involves monomials where one variable is strictly less than (for $\ll$) than x_1 and x_2 (see [4]).*
- a pseudo-algebra with straightening laws *if* (H_2) *is satisfied and* (H_1') *the standard monomials are a generating set of $A = \mathbb{C}[X]/I$ (as a $\mathbb{C}$-vector space).*

The first point specifies that the reduction is canonic, and the second that it is compatible with the partial order. In our case, we will take for X the set of variables defined in (5) with the partial order $\ll$ given by the Hasse diagram shown in figure 1.

Remark that the restriction of $\ll$ to the subset $X' = \{(i|j), (i|4),\ (i,i'|j,4)\}_{i,j,i'}$ is the partial order:

$$\begin{cases} (i|j) \ll (i'|j') & \text{if } i \leq i', j \leq j' \\ (i|j) \ll (i',i''|j',j'') & \text{if } i \leq i', j \leq j' \\ (i,j|k,l) \ll (i',j'|k',l') & \text{if } i \leq i', j \leq j', k \leq k', l \leq l' \end{cases}$$

which compares the "tableaux" index by index (this is the usual partial order on Young tableaux, defined for instance in [6]).

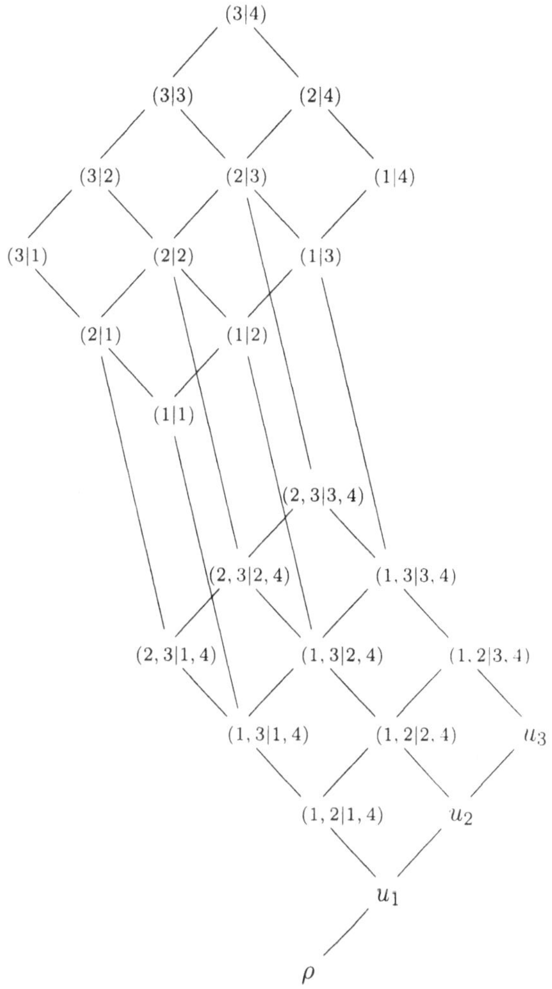

Figure 1

5.2 Relations in $\mathcal{I}$

The determinants $(i, j|k, 4)$ are up to a sign the coordinates of the vectors $R_k \wedge T$. To simplify the notation, denote by $w_{i,k}$ the ith coordinate of $R_k \wedge T$ and $w^*_{i,k}$ the ith coordinate of $R^*_k \wedge U$ so that $w_{i,k} = (-1)^{i+1}(i', i''|k, 4)$ with $\{i, i', i''\} = \{1, 2, 3\}$. We will also note sometimes $u_i = (4|i)$.

Proposition 5.4 *The following relations are true modulo $\mathcal{I}$:*

(1) $(R_i|R_j) = \delta_{i,j}$, $(R^*_i|R^*_j) = \delta_{i,j}$, $(R_i|T) = u_i$, $(R^*_i|U) = t_i$. $(T|T) - \rho = 0$, $(U|U) - \rho = 0$,

(2) $(i, j|i', j') = (-1)^{k+k'}(k|k')$ *with* $\{i, j, k\} = \{i', j', k'\} = \{1, 2, 3\}$.

(3) $w_{i,j} = -w^*_{j,i}$,
(4) $w_{i,l}\,(j|m) - w_{j,l}\,(i|m) = u_m(k|l) - \delta_{l,m}(k|4)$ *where* $\{i,j,k\} = \{1,2,3\}$ *and* $1 \leq m \leq 4$,
(5) $w_{i,l}\,u_j - w_{j,l}\,u_i = -t_m(k|l) + \delta_{l,m}u_k$ *where* $\{i,j,k\} = \{1,2,3\}$,
(6) $w_{i,k}w_{j,l} = \begin{vmatrix} \delta_{k,l} & u_k & r_{j,k} \\ u_l & \rho & t_j \\ r_{i,l} & t_i & \delta_{i,j} \end{vmatrix}$, *with* $1 \leq i,j,k,i',j',k' \leq 3$.

Proof.

∘ The polynomials (1) are by definition in $\mathcal{I}$.

∘ As the determinant of R is one, and as $R^T \cdot R = Id$, a 2×2 minor of R is equal to its cofactor. Thus modulo I we have $(i,j|i',j') = \pm(k|k')$ with $1 \leq i,j,k,i',j',k' \leq 3$.

∘ As R is of determinant 1, we have

$$w_{i,j} = |R_j, T, e_i| = |R.e_j, R.U, R.R^*.e_i| = |e_j, U, R^*_i| = -w^*_{j,i},$$

or, in other words, $(i,i'|j,4) = (-1)^{i+j+1}(i'',4|j',j'')$ with $\{i,i',i''\} = \{j,j',j''\} = \{1,2,3\}$.

∘ Consider now the polynomials in (4). Using the formula of double vector product

$$(a \times b) \times c = (b|c)\,a - (a|c)\,b,$$

we have

$$(R_i \times T) \times R_j = (T|R_j)\,R_i - (R_i|R_j)T = u_j R_i - \delta_{i,j} T$$

and we obtain the relations (4) by taking one coordinate of this vector.

∘ In the same way, we obtain the relations (5) by considering $(R^*_i \times U) \times R^*_j$.

∘ For the relations (6), we have

$$\begin{aligned}(R_k \times T)_i (R_l \times T)_j &= |R_k, T, e_i||R_l, T, e_j| \\ &= \begin{vmatrix} (R_k|R_l) & (R_k|T) & (R_k|e_j) \\ (T|R_l) & (T|T) & (T|e_j) \\ (e_i|R_l) & (e_i|T) & (e_i|e_j) \end{vmatrix} = \begin{vmatrix} \delta_{k,l} & u_k & r_{j,k} \\ u_l & \rho & t_j \\ r_{i,l} & t_i & \delta_{i,j} \end{vmatrix}.\end{aligned}$$

The right-hand side of this relation is a combination of monomials, less than $(R_k \times T)_i (R_l \times T)_j$ for $\prec$. □

See [18, p. 144] for a similar result on the orthogonal group[4].

Proposition 5.5 *The algebra* $\mathbb{C}[\mathcal{D}]$ *is a pseudo-algebra with straightening laws with respect to* $\ll$.

[4] I would like to thank T. Recio, for some discussions on this hidden part of H. Weyl work, closely related to the following section.

Proof. Let M be the 3×4 matrix $[R, T]$ and denote by $(i_1, \ldots, i_k | j_1, \ldots, j_k)$ the minor of M of rows $1 \leq i_1 < \cdots < i_k \leq 3$ and columns $1 \leq j_1 < \cdots < j_k \leq 4$. The algebra $\mathbb{C}[(i_1, \ldots, i_k | j_1, \ldots, j_k)]_{i_l, j_m}$ has a natural structure of ASL where the partial order consists in comparing the first indices of rows and columns. These straightening laws allow us to write any polynomial of $\mathbb{C}[X']$ (X' is defined page 294) as a sum of standard monomials in $Y = \{(i_1, \ldots, i_k | j_1, \ldots, j_k)\}$. Remark now that any 3×3 minor of M is equivalent to a polynomial of lower degree in X. We have, for instance, $|R_i, R_j, T| = |e_i, e_j, U| = \pm u_k$ with $\{i, j, k\} = \{1, 2, 3\}$. Using these relations, any standard monomial in Y can be rewritten as a sum of monomials in X and standard in the variables X'. We now have to check that for any variable $x \in X'$ and $u \in \{u_1, u_2, u_3\}$ such that $m = x\,u$ is not standard, m can be rewritten in a standard way. This is done by checking by hand all non standard monomials of this type. We have, for instance,

$$(i|2)\,u_3 = w^*_{1,i} + (i|3)u_2 = -w_{i,1} + (i|3)u_2.$$

Non-standard monomials of the form $(i|j)\,u_k$ can be normalized in the same way. The other kind of non standard monomials is

$$(i, j|1, 4)u_3 = w_{k,1}u_3 = w^*_{1,k}u_3 + w^*_{3,k}u_1 - w^*_{3,k}u_1 = -w_{k,3}\,u_1 + t_m(k|l) - \delta_{l,m}\,u_k.$$

Remark that the right-hand side of these relations is a sum of monomials which always involves a variable strictly lower that the variables of the left-hand side. Using the relations (1) ...(6), we check in this way that all non standard monomials of the form $r_{i,j}\,r_{i',j'}, r_{i,j}\,u_k, r_{i,j}\,w_{i',j'}, w_{i,j}\,u_k$ can be reduced to a linear combination of standard monomials and that this reduction is compatible with $\ll$. In other words, $\mathbb{C}[X]/\mathcal{I}$ is a pseudo-algebra with straightening laws.

□

5.3 The multiplicity of $\mathbb{C}[\mathcal{D}]$

Let us define a graduation on $\mathbb{C}[X]$ by setting

$$deg(r_{i,j}) = 1, deg(t_i) = 1,\ deg(w_{i,j}) = 2,\ deg(u_i) = 2,\ deg(\rho) = 2.$$

Let $\prec$ be a total order, compatible with the partial order $\ll$. Take, for example, the lexicographic order on X^* refining the degree such that the variables are ordered from top to bottom and from left to right on the Hasse diagram. Note that if $x_1 x_2$ is not standard, it is also the leading term (for $\prec$) of an element of $\mathcal{I}$.

Let $\mathcal{I}^*$ be the monomial ideal generated by the initials of the polynomials of $\mathcal{I}$. The ideal $\mathcal{I}^*$ contains all the non standard monomials. It also contains the leading terms of the equations (2) and (3): $r_{3,j}r_{3,k}$, $r_{j,3}r_{k,3}$ (for $1 \leq j, k \leq 4$).

Define the Hilbert function of the algebra $\mathbb{C}[X]/\mathcal{I}$ to be the dimension of the vector space $(\mathbb{C}[X]/\mathcal{I})_{\leq n}$ of polynomials of $\mathbb{C}[X]/\mathcal{I}$ of degree lower than n: $H(n) := dim_k(\mathbb{C}[X]/\mathcal{I})_{\leq n}$.

Proposition 5.6 *The Hilbert function of* $\mathbb{C}[X]/\mathcal{I}$ *is also the Hilbert function of* $(\mathbb{C}[X]/\mathcal{I}^*)_{\leq n}$ *and the two* $\mathbb{C}$*-vector spaces* $(\mathbb{C}[X]/\mathcal{I})_{\leq n}$ *and* $(\mathbb{C}[X]/\mathcal{I}^*)_{\leq n}$ *are isomorphic.*

This classical lemma is true for any graduation on the algebra $\mathbb{C}[X]/\mathcal{I}$ (see [7]).
Proof. If a monomial m is the initial of an element of $\mathcal{I}$, it can be reduced (modulo $\mathcal{I}$) to a linear combination of monomials which are less than m (for $\prec$), so that a generating set of $\mathbb{C}[X]/\mathcal{I}$ consists of all monomials which are not in $\mathcal{I}$. Moreover, the monomials which are not leading terms are linearly independent in $\mathbb{C}[X]/\mathcal{I}$, otherwise a linear combination of them would be in $\mathcal{I}$ and one of them would be the leading term of an element of $\mathcal{I}$. Consequently, a basis of $\mathbb{C}[X]/\mathcal{I}$ is formed by monomials which are not in $\mathcal{I}^*$. These monomials also form a basis of $\mathbb{C}[X]/\mathcal{I}^*$. As any polynomial is rewritten as a combination of monomials of lower degree and not in $\mathcal{I}^*$, the Hilbert functions of the two algebras are the same. □

When n is large enough the function of $\mathbb{C}[X]/\mathcal{I}$ becomes a polynomial function (see [1]) and we can take its leading coefficient (as a polynomial in n), which is of the form $\frac{\delta}{d!}n^d$. Let us define the *multiplicity of* $\mathbb{C}[X]/\mathcal{I}$ as δ. This is the classical algebraic definition of a degree of a variety V when $\mathcal{I}$ is the ideal associated to the variety. This is also the multiplicity of $\mathbb{C}[X]/\mathcal{I}^*$.

Theorem 5.7 *The multiplicity of* $\mathbb{C}[\mathcal{D}]$ *is at most* 40.

The proof we are giving here is based on simple but "magic" computations on monomials. It is **yet another proof** for this problem. The first one by F. Ronga and T. Vust (see [16]) used intersection theory on vector bundles, Chern classes, etc. The second approach (see [14]) used Gröbner bases to compute very quickly the polynomials we have shown in the previous section and to deduce the Hilbert function. The third one (see [15]) uses an explicit computation of a resultant (by a computer algebra system) to conclude that the dimension is 40. The proof we are giving here requires neither software nor Chern class, but some patience.

Proof. We are going to decompose $\mathbb{C}[X]/\mathcal{I}^*$ (and also $\mathbb{C}[X]/\mathcal{I}$) into a sum of a components which are polynomial rings times a monomial so that no element of $\mathcal{I}^*$ appears in such component. The result of this computation look like the computation of Hilbert series, but here the variables are not replaced by a single variable z (giving the graduation).

In fact, we are concentrating only on the components of dimension 6 (which polynomial rings are generated by 6 variables). We call a *component* such a product of a monomial by a polynomial ring.

Let us first consider the subring of $\mathbb{C}[X]$ generated by the "free" variables, that is,

$$S = \mathbb{C}[\rho, u_1, u_2, r_{1,1}, r_{2,1}, r_{1,2}, r_{2,2}, t_1, t_2].$$

We are going to simplify it modulo $\mathcal{I}$. We use the following simple relation (called *Elliott relation*, see [8]):

$$\mathbb{C}[a, b] = \mathbb{C}[a\,b, a] \oplus b\,\mathbb{C}[a\,b, b] \tag{6}$$

which is equivalent to $\mathbb{C}[a] \oplus b\,\mathbb{C}[b]$ if $ab \equiv 0$. This relation yields a decomposition of $\mathbb{C}[\rho, u_1, r_{1,1}, r_{1,2}, r_{2,1}, r_{2,2}]$ of the form

$$\mathbb{C}[\rho, u_1, r_{1,1}, r_{1,2}, r_{2,2}] + r_{2,1}\,\mathbb{C}[\rho, u_1, r_{1,1}, r_{2,1}, r_{2,2}]$$

because $r_{2,1}r_{1,2}$ is not standard. Adding the variables t_1, t_2 the same process implies that $\mathbb{C}[\rho, u_1, r_{1,1}, r_{1,2}, r_{2,1}, r_{2,2}, t_1, t_2]$ is equivalent (modulo $\mathcal{I}^*$) to

$$\begin{aligned}\mathbb{C}[\rho, u_1, r_{1,1}, r_{1,2}, t_1, t_2] + r_{2,2}\,\mathbb{C}[\rho, u_1, r_{1,1}, r_{1,2}, r_{2,2}, t_2]\\ + r_{2,1}\,\mathbb{C}[\rho, u_1, r_{1,1}, r_{2,1}, r_{2,2}, t_2]\end{aligned}$$

because $r_{2,2}t_1$ and $r_{2,1}t_1$ are in $\mathcal{I}^*$.

Finally, adding the last variable u_2, we obtain the decomposition

$$\begin{aligned} S &= \mathbb{C}[\rho, u_1, u_2, r_{1,1}, r_{2,1}, r_{1,2}, r_{2,2}, t_1, t_2] \\ &\equiv \mathbb{C}[\rho, u_1, r_{1,1}, r_{1,2}, t_1, t_2] + u_2\,\mathbb{C}[\rho, u_1, u_2, r_{1,2}, t_1, t_2] \\ &\quad + r_{2,2}\,\mathbb{C}[\rho, u_1, r_{1,1}, r_{1,2}, r_{2,2}, t_2] + r_{2,2}u_2\,\mathbb{C}[\rho, u_1, u_2, r_{1,2}, r_{2,2}, t_2] \\ &\quad + r_{2,1}\,\mathbb{C}[\rho, u_1, r_{1,1}, r_{2,1}, r_{2,2}, t_2]. \end{aligned} \tag{7}$$

Remark that the number of components in this decomposition is also the number of increasing paths from u_1 to t_2 (on the subset $\{\rho, u_1, \dots, t_2\}$) (that is, 5) and that the dimension of each of these components is 6 (that is, the dimension of $\mathcal{D}$). Remark also that such a decomposition is not unique, for we can choose at each step a decomposition $\mathbb{C}[a] + b\,\mathbb{C}[b]$ or $a\,\mathbb{C}[a] + \mathbb{C}[b]$.

We consider now the subring $B_1 = \mathbb{C}[u_3, t_3, r_{3,3}, r_{3,2}, r_{3,1}, r_{2,3}, r_{1,3}]$ modulo $\mathcal{I}^*$. Since the monomials $r_{3,j}r_{3,k}$, $r_{j,3}r_{k,3}$ (for $1 \le j, k \le 4$) are in $\mathcal{I}^*$, this ring is equivalent to

$$\begin{aligned} B_1 \;\equiv\; & \mathbb{C} + \mathbb{C}\,u_3 + \mathbb{C}\,t_3 + \mathbb{C}\,r_{3,1} + \mathbb{C}\,r_{1,3} + \mathbb{C}\,r_{3,2} + \mathbb{C}\,r_{2,3} + \mathbb{C}\,r_{3,3} \\ & + \mathbb{C}\,r_{1,3}\,t_3 + \mathbb{C}\,r_{2,3}\,t_3 + \mathbb{C}\,r_{3,1}\,u_3 + \mathbb{C}\,r_{3,2}\,u_3 + \mathbb{C}\,t_3\,u_3. \end{aligned}$$

The last set of variables (which are not free modulo $\mathcal{I}^*$) is $\{w_{i,j}\}_{i,j}$. Because of relations (6) (prop. 5.4), the ring $B_2 = \mathbb{C}[w_{i,j}]_{i,j}$ is equivalent to $\sum_{i,j} \mathbb{C}\,w_{i,j}$.

The final step of this proof consists in taking the tensor product of S, B_1, B_2 (modulo $\mathcal{I}^*$) to obtain a decomposition of $\mathbb{C}[X]/\mathcal{I}^*$. Here we are interested only in the part of dimension 6 of such a decomposition. This implies that for each component of the decomposition (7), we only need to consider the variables of B_1 and B_2 which extend the corresponding increasing path. For instance the ring $\mathbb{C}[\rho, u_1, r_{1,1}, r_{1,2}, t_1, t_2] \otimes B_1 \otimes B_2$ is equivalent to

$$\begin{aligned}
&\mathbb{C}[\rho, u_1, r_{1,1}, r_{1,2}, t_1, t_2] \otimes (1 + t_3 + r_{2,3} + t_3 r_{2,3} \\
&+ \ (1,3|1,4) + (1,2|1,4) + (1,3|1,4)t_3 + (1,2|1,4)t_3 + (1,3|1,4)r_{2,3} \\
&+ \ (1,2|1,4)r_{2,3}) + \text{components of lower dimension.}
\end{aligned}$$

However $(1,3|1,4)t_3$ and $(1,2|1,4)r_{2,3}$ are the leading terms of relations (4) and (5) (prop. 5.4), so we can forget the corresponding parts, and the previous ring is a sum of 8 components. Finally, taking into account all these leading terms, we obtain the decomposition

$$\begin{aligned}
& S \otimes B_1 \otimes B_2 \\
\equiv \ & \mathbb{C}[\rho, u_1, r_{1,1}, r_{1,2}, t_1, t_2] \\
& \otimes(1 + t_3 + r_{2,3} + t_3 r_{2,3} + (1,3|1,4) + (1,2|1,4) + (1,2|1,4)t_3 \\
+ \ & (1,3|1,4)r_{2,3}) + u_2\, \mathbb{C}[\rho, u_1, u_2, r_{1,2}, t_1, t_2] \\
& \otimes(1 + t_3 + r_{1,3} + t_3 r_{2,3} + (1,3|2,4) + (1,2|2,4) + (1,2|2,4)t_3 \\
+ \ & (1,3|2,4)r_{1,3}) + r_{2,2}\, \mathbb{C}[\rho, u_1, r_{1,1}, r_{1,2}, r_{2,2}, t_2] \\
& \otimes(1 + t_3 + r_{1,3} + t_3 r_{1,3} + (1,3|1,4) + (1,2|1,4) + (1,2|1,4)t_3 \\
+ \ & (1,3|1,4)r_{1,3}) + r_{2,2}u_2\, \mathbb{C}[\rho, u_1, u_2, r_{1,2}, r_{2,2}, t_2] \\
& \otimes(1 + t_3 + r_{1,3} + t_3 r_{2,3} + (1,3|2,4) + (1,2|2,4) + (1,2|2,4)t_3 \\
+ \ & (1,3|2,4)r_{1,3}) + r_{2,1}\, \mathbb{C}[\rho, u_1, r_{1,1}, r_{2,1}, r_{2,2}, t_2] \\
& \otimes(1 + t_3 + r_{2,3} + t_3 r_{2,3} + (1,3|1,4) + (1,2|1,4) + (1,2|1,4)t_3 \\
+ \ & (1,3|1,4)r_{2,3}) + \text{components of lower dimension.}
\end{aligned}$$

In this decomposition we have $5 \times 8 = 40$ parts of dimension 6 and the other components of lower dimension. From this, it is immediate to deduce that the Hilbert polynomial of the quotient $\mathbb{C}[X]/\mathcal{I}^*$ is of the form $\frac{N}{6!}n^6 + h_5 n^5 + \cdots$ with $N \leq 40$. According to proposition 5.6 the multiplicity of $\mathbb{C}[\mathcal{D}]$ is at most 40. □

5.4 Application to a problem of robotics

We present here an application of the previous section to the direct kinematic problem of a parallel robot. The mechanism of this robot (also called Stewart platform or left hand) is the following. Consider six fixed points $(X_i)_{1 \leq i \leq 6}$ of a fixed solid $\mathcal{S}_X$ and six other points Z_i, attached to a moving solid $\mathcal{S}_Z$. The articulations between the two solids $\mathcal{S}_X$ and $\mathcal{S}_Z$ are extensible bars $(X_i, Z_i)_{1 \leq i \leq 6}$ with spherical joints. This mechanism is well suited for doing precise movements of adjustment. It is used also in flying simulators and appears in suspension links of cars (but with only one degree of freedom).

From a practical point of view, it is easy to measure the length of these articulations but much less obvious to determine the position of the solid $\mathcal{S}_Z$

knowing these lengths. A first step in solving this problem is: "knowing the length $\|X_i, Z_i\|^2 = \delta_i$, how many positions there are for a 'generic' platform?" Let us note $(Y_i)_i$ the position of the upper platform at a given time and $D = (R, T)$ the displacement from (Y_i) to (Z_i) (so that we have $Z_i = R\,Y_i + T$). The problem is equivalent to finding the solutions (over $\mathbb{C}$) of the system:

$$\begin{aligned}[R.Y_i + T - X_i, R.Y_i + T - X_i] &= \delta_i \\ &= [Y_i, Y_i] + [T.T] + [X_i, X_i] + 2\,[Y_i, R^T T] - 2\,[X_i.T] - 2\,[R.Y_i, X_i]\end{aligned}$$

for $1 \leq i \leq 6$. This system is linear in the variables of X. According to the previous section, we know that it has at most 40 solutions (over $\mathbb{C}$).

To complete the approach, we show that there are exactly 40 solutions in the generic case by exhibiting an example where we can find at least 40 different solutions.

Proposition 5.8 *The degree of* $\mathbb{C}[\mathcal{D}]$ *is at least* 40.

Proof. Consider the following configurations. Let $X_1 = [0, 0, 0]$, $X_6 = [2, 0, 0]$ and

$$X_2 = [\frac{1}{2}, -\frac{1}{2}, -1], X_3 = [\frac{3}{2}, -\frac{1}{2}, -1], X_4 = [\frac{3}{2}, \frac{1}{2}, -1], X_5 = [\frac{1}{3}, \frac{1}{2}, -1]$$

be fixed points in $\mathbb{A}$ and

$$Y_1 = [0, 0, 0], Y_2 = [1, 0, 1], Y_3 = [0, 1, 1], Y_4 = [-1, 0, 1], Y_5 = [0, -1, 0], Y_6 = Y_1$$

attached to the moving platform. The displacement (R, T) of this platform gives us the points $Z_i = R\,Y_i + T$. *We want to compute all possible displacement (solutions in* $\mathbb{C}$*) such that* $\|Z_i - Y_i\|^2 = \delta_i$. We take for instance

$$\delta_1 = 2,\ \delta_2 = 23/2,\ \delta_3 = 23/2,\ \delta_4 = 23/2,\ \delta_5 = 241/36,\ \delta_6 = 2.$$

This yields the system:

$$\begin{aligned}[R.Y_i + T - X_i, R.Y_i + T - X_i] &= \delta_i \\ &= [Y_i, Y_i] + [T, T] + [X_i, X_i] + 2\,[R.Y_i, T] - 2\,[X_i.T] - 2\,[R.Y_i, X_i] \\ &= [Y_i, Y_i] + [T, T] + [X_i, X_i] + 2\,[Y_i, U] - 2\,[X_i.T] - 2\,[R.Y_i, X_i].\end{aligned}$$

This system is linear in the set of variables X; therefore, if it has a finite number of solutions, this number of solution is at least (in fact exactly) the degree of $\mathcal{D}$.

Remark that because $Z_1 = Z_6$ and $\|X_1, Z_1\|^2 = \|X_6, Z_6\|^2 = 2$, we know that Z_1 is on the circle defined by $x = 1, y^2 + z^2 = 1$. We use the classical representation of rotations with quaternions:[5]

$$
\begin{array}{l}
\omega \; : \mathbb{P}^3 \backslash \mathbf{V}(a^2+b^2+c^2+d^2=0) \to SO_3 \\
(a,b,c,d) \mapsto R = \\
\frac{1}{a^2+b^2+c^2+d^2} \times \begin{pmatrix} a^2+b^2-c^2-d^2 & 2\,b\,c-2\,d\,a & 2\,b\,d+2\,c\,a \\ 2\,b\,c+2\,d\,a & a^2-b^2+c^2-d^2 & 2\,c\,d-2\,a\,b \\ 2\,b\,d-2\,c\,a & 2\,a\,b+2\,c\,d & a^2-b^2-c^2+d^2 \end{pmatrix}
\end{array} \tag{8}
$$

so that reducing each equation to the same denominator, the previous system is transformed into a homogeneous system of degree 2 in $\mathbf{q} = (a, b, c, d)$ with coefficients which are polynomials of degree 1 in T: If we look at the $2^{th}, \ldots, 5^{th}$ equations in (a, b, c, d), we have 4 polynomials homogeneous in these 4 variables which must have at least one common root. Using the method described in [15], we can compute their resultant in a, b, c, d. It is a polynomial in T which vanishes if and only if there exists a rotation such that the 4 equations are satisfied. Setting $x = 1, y = \frac{1-t^2}{1+t^2}, z = \frac{2\,t}{1+t^2}$ (Z_1 is on a circle) and reducing to the same denominator, we obtain the following polynomial in t of degree 40:

$$
\begin{array}{l}
1999838017108489289\,t^{40} + 4825092325063207512\,t^{39} + 25591605368925301144\,t^{38} \\
+ 49412508531502879288\,t^{37} + 94338129683192039026\,t^{36} + 100240961798861174680\,t^{35} \\
+ 16597248736182035880\,t^{34} - 118197965991748728264\,t^{33} - 412172958797937298563\,t^{32} \\
- 713877700408239650784\,t^{31} - 823165922491391120672\,t^{30} \\
- 910939538430197488224\,t^{29} - 821520400353846151560\,t^{28} \\
- 458613961567107873632\,t^{27} + 57554126909561412704\,t^{26} \\
+ 1210375736546818623264\,t^{25} + 3283908982716218990018\,t^{24} \\
+ 5723500447433301901584\,t^{23} + 7583544515108833961680\,t^{22} \\
+ 7510028139203118254928\,t^{21} + 4321406032357634284108\,t^{20} \\
- 1651271769827903795696\,t^{19} - 8336732191251700599568\,t^{18} \\
- 13102461964513308633904\,t^{17} - 13975383485284377843310\,t^{16} \\
- 10564032960330005219808\,t^{15} - 4512673491739135774368\,t^{14} \\
+ 1514652059205820223904\,t^{13} + 5505673827065320024632\,t^{12} \\
+ 6757049249522328815008\,t^{11} + 5788010131725221987424\,t^{10} \\
+ 3832583473196284161568\,t^{9} + 2009647163759307870261\,t^{8} \\
+ 803719762218156787288\,t^{7} + 202341398520619750232\,t^{6} + 13298239527064533816\,t^{5} \\
- 6354878613375129966\,t^{4} - 2775897814417637352\,t^{3} - 1476611419815432 56\,t^{2} \\
+ 12536074683882424\,t + 11543005543897265.
\end{array}
$$

[5]This map is surjective on $\mathbb{C}$: Given a matrix R satisfying the equations $R\,R^T = R^T R = Id, \det(R) = 1$, we want to search (a, b, c, d) such that $a^2+b^2+c^2+d^2 = 1$ and $\omega(a, b, c, d) = R$. We see that $\frac{1}{4}(Tr(R) + 1)$ defines a up to a sign and that $\frac{1}{4}(R - R^T)$ which is an antisymmetric matrix, defines (b, c, d) up to the same sign so that the possible solutions (q or $-q$) are determined by $Tr(R)$ and $R - R^t$. Let $c = \frac{1}{2}(Tr(R) - 1)$ and $t_R = \frac{1}{2}(R - R^T)$ and v the vector such that $t_R(x) = v \wedge x$. We check that $(v|v) + c^2 = 1$.

We first assume that $c \neq -1$. Using this identity, "Rodriguez representation" of R and the rule of double vector product, we have (on $\mathbb{C}$)

$$
R = (1 - cos(\theta))u \otimes u + cos(\theta)\, Id + t_v = \frac{1}{1+c} t_v \circ t_v + Id + t_v
$$

(where θ is the angle of rotation). Remark that this is the Cayley parametrization of rotations. We immediately deduce that R and $\omega(q)$ are identical, for they have the same "cosines" and the same axis of rotation.

Now if $c = -1$, we have $a = 0$. In this case, $R = R^t$. If u is a unitary vector of the axis of rotation, we check that $R = 2u \otimes u - Id$ so that taking $(b, c, d) = (u_1, u_2, u_3)$, we see that $\omega(0, b, c, d) = R$.

Approximations of the solutions of these equations are

$$\begin{aligned}\big[&1., 1.0691339, 1.274397037, 1.42735636, 0.1017980095 \pm 2.642074408\,\mathbf{i},\\ &-0.0958801865 \pm 0.1812183212\,\mathbf{i}, 0.1674502267 \pm 0.0235791365\,\mathbf{i},\\ &-0.150235581 \pm 0.2612522182\,\mathbf{i}, 0.1915807034 \pm 0.7774948275\,\mathbf{i},\\ &-0.4993601843 \pm 0.7378197715\,\mathbf{i}, -0.6288016023 \pm 0.610477651\,\mathbf{i},\\ &-0.01532341299 \pm 1.019869943\,\mathbf{i}, 0.3198204635 \pm 1.25748005\,\mathbf{i},\\ &-0.6858938694 \pm 1.743050505\,\mathbf{i}, 0.9887899948 \pm 0.8661843231\,\mathbf{i},\\ &-0.9104205059 + 0.1639209135\,\mathbf{i}, -0.9104205032 - 0.1639209151\,\mathbf{i},\\ &-0.966582795 - 0.1140359517\,\mathbf{i}, -0.9665827934 + 0.1140359527\,\mathbf{i},\\ &-0.8413731333 + 0.4674405879\,\mathbf{i}, -0.8413731338 - 0.4674405878\,\mathbf{i},\\ &-0.01229812098 + 0.995078441\,\mathbf{i}, -0.01229811926 - 0.9950784417\,\mathbf{i}\\ &-0.9457523385 + 0.342443233\,\mathbf{i}, -0.9457523389 - 0.34244432324\,\mathbf{i},\\ &0.1645190391 - 1.023140473\,\mathbf{i}, 0.1645190403 + 1.023140468\,\mathbf{i},\\ &0.2261488542 + 1.052751221\,\mathbf{i}, 0.2261488542 - 1.052751217\,\mathbf{i}\big].\end{aligned}$$

These computations have been done in MAPLE using some elementary functions on polynomials to compute the resultant and the function TRAUBJEN to find approximate solutions of the previous equation (with six digits at least). Each value of t gives us a position of T and at least one solution in R. □

5.5 Application to a problem of vision

In this section, we describe the problem of reconstruction from points with calibrated cameras and see how it is related to the previous section. We consider the following problem: *Given N points P_i in 3-dimensional space $\mathbb{A}_3$, we know their images on two cameras and we want to determine the displacement between the two cameras.* The image of a point is the projection of this point on the retinal plane from the center of the camera. We assume that we know the coordinates of the image in a referential attached to the first (resp. second) camera. Taking a referential such that the retinal planes are given by $z = 1$, $z' = 1$, the coordinates of the points in the first (resp. second) referential are $p_i = (x_i, y_i, 1)$ (resp. $p'_i = (x'_i, y'_i, 1)$). This is equivalent to saying that the cameras are calibrated (see figure 2).

Note $D = (R, T)$ the displacement of the second frame relative to the first one. The images p'_i in the second retinal plane correspond in the first referential to the vectors $Rp'_i + T$. We want to find the possible displacements, knowing the coordinates of pairs of images in the retinal planes. Scaling the scene will not change the image in the retinal planes and the two symmetric positions of a retinal plane with respect to the center of projection give the same images in this plane. According to this freedom, a *reconstruction* will be

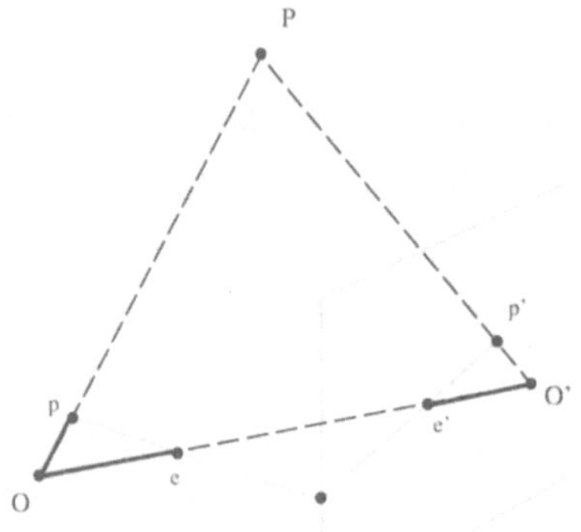

Figure 2

a set of displacements $R_{V,W} = \{(R,T)|T = \lambda\, V, R_3 = \mu\, W\}$ (R_3 is the normal to the second retinal plane). This corresponds to the set of all displacements which will give the same position for the second camera, up to the symmetries of our problem.

We recall here a theorem for this solved problem of reconstruction. See [5] and [9] for two different points of view on the question. The proof we present is a third one, obtained by application of the previous section.

Proposition 5.1 *Given* 5 *points "in generic position", the number of possible reconstructions is at most* 10.

Proof. As the vectors $OO' = T, OP_i = \lambda\, p_i, O'P_i = \lambda' p_i'$ are in a same plane, we must have

$$|p_i, T, R\, p_i' + T| = p_i^T\, T \wedge R\, p_i' = 0. \tag{9}$$

These equations are linear in the coordinates of $R_i \wedge T$, that is with our notations, in $(w_{i,j})$. The matrix $(w_{i,j})$ is also called an *essential matrix.* Scaling the vector of translation does not change the system of equations (9). To fix this degree of freedom, we add a new equation

$$(T|T) - 1 = 0 \text{ or } \rho - 1 = 0. \tag{10}$$

Here again, it is a linear equation in the set of variables X of the previous section. As the space of displacements is of dimension six, it is enough to consider six equations to obtain a finite number of solutions. This means that we can consider only five images of points ($N = 5$). According to prop. (5.7) and Bézout's theorem, this system has 40 solutions (R, T).

Let us note σ_T the symmetry around the axis of direction T. Remark that $\forall x \in \mathbb{A}, -T \wedge \sigma_T R x = T \wedge R x$, so that if (R, T) is a solution of (9) and (10), we also have $(R, -T)$ and $(\sigma_T R, T)$, $(\sigma_T R, -T)$. These four solutions correspond to the same reconstruction, for these four cases are obtained by taking the symmetric of the retinal planes with respect to the centers of projection. Consequently, there are at most 10 possible reconstructions when 5 points are projected on two retinal planes. □

6 Conclusion

In this paper, some results on enumeration of solutions in problems coming from mechanics, robotics, or vision are presented. The approach we followed does not require sophisticated tools from algebraic geometry but rather is based on simple calculus with polynomials and the novelty of the results is more in the proofs we give. Following a parallel between change of variables and blowing-up, we show how simple algebra (related to the part at infinity of the variety) helps us to simplify the enumeration of solutions. This approach applies to the varieties of displacements for which a structure similar to a so-called algebra with straightening laws is presented. This structure allows us to compute the multiplicity of this representation, a result which can be used immediately in the direct kinematic problem of a parallel robot and in the problem of reconstruction in vision.

It also raises some questions for possibles generalizations:

- How can we use, in a general context, such changes of variables in order to simplify the algebraic situation?
- Can this be helpful to compute more easily a graded part of an ideal (called Gröbner bases)?

We find this parallelism between simple algebraic transformations and blowing-up an interesting field for future investigation, which could help to avoid blind algorithmic methods that do not lead to satisfactory answers in effective algebraic geometry.

References

[1] M. F. Atiyah and I. G. MacDonald. *Introduction to Commutative Algebra*. Addison-Wesley, 1969.

[2] E. Brieskorn and H. Knörrer. *Plane Algebraic Curves*. Birkhäuser Verlag, 1986.

[3] A. Chenciner. *Courbes algébriques planes*. Publications mathématiques de l'Université Paris VII, 1978.

[4] C. DeConcini, D. Eisenbud, and C. Procesi. Hodge algebras. *Astérisque*, 91, 1982.

[5] M. Demazure. Sur deux problèmes de reconstruction. *Rapport de Recherche INRIA*, 882, 1988.

[6] J. Désarménien, Joseph P. S. Kung, and Gian-Carlo Rota. Invariant theory, Young bitableaux, and combinatorics. *Advances in Mathematics*, 27(1):63–92, January 1978.

[7] D. Eisenbud. *Algebra, a view towards geometry.* To appear, 1993.

[8] E. B. Elliott. On linear homogeneous diophantine equations. *Quart. J. Pure Appl. Math.*, 34:348–377, 1903.

[9] O. Faugeras and S. Maybank. Motion from point matches: Multiplicities of solutions. *Int. J. of Computer Vision*, 4:225–246, 1990.

[10] William Fulton. *Intersection theory.* Springer-Verlag, 1984.

[11] R. Hartshorne. *Algebraic Geometry.* Springer-Verlag, 1977.

[12] W. Hodge and D. Pedoe. *Methods of algebraic geometry.* Cambridge University Press, 1952.

[13] K. H. Hunt and E. J. F. Primrose. Assembly configurations of some parallel-actuated manipulators. *Mech. Mach. Theory*, 1–12, 1992.

[14] D. Lazard. Generalized Stewart platform: How to compute with rigid motions? In *IMACS-SC'93*, 1993.

[15] B. Mourrain. The 40 generic positions of a parallel robot. In M. Bronstein, editor, *ISSAC'93*, ACM press, pages 173–182, Kiev (Ukraine), July 1993.

[16] F. Ronga and T. Vust. Stewart platforms without computer. *Preprint*, 1992.

[17] I. R. Shafarevitch. *Basic Algebraic Geometry.* Springer-Verlag, 1974.

[18] H. Weyl. *The Classical Groups, their invariants and representations.* Princeton University Press, 1939.

B. Mourrain (`Mourrain@sophia.inria.fr`)

INRIA, 2004 Routes de Lucioles, BP 9306902 Sophia-Antipolis Cedex (France)

Progress in Mathematics, Vol. 143, © 1996 Birkhäuser Verlag Basel/Switzerland

Mixed monomial bases

P. Pedersen[1], B. Sturmfels[2]

1 Introduction

Given a system of n generic Laurent polynomials

$$f_i(x) \quad = \quad \sum_{q \in \mathcal{A}_i} c_{iq}\, x^q; \qquad q = (q_1, \ldots, q_n); \qquad x^q = x_1^{q_1} x_2^{q_2} \cdots x_n^{q_n} \qquad (1.1)$$

with support sets $\mathcal{A}_i \subset \mathbb{Z}^n$, we consider the ring

$$A := K[x_1, x_1^{-1}, \ldots, x_n, x_n^{-1}]/(f_1, \ldots, f_n),$$

where K is the field $\mathbb{Q}(\{c_{iq}\})$. The K-dimension of A equals the number of toric roots $\{\, x \in (\mathbb{C}^*)^n \;:\; f_i(x) = 0, \quad 1 \leq i \leq n \,\}$. By Bernstein's theorem [Ber], this number equals the mixed volume $\mathcal{MV}(P_1, \ldots, P_n)$ of the Newton polytopes $P_i := conv(\mathcal{A}_i)$. The objective of this note is to construct explicit K-bases for A, using the combinatorial technique of mixed subdivisions of the Minkowski sum $P := P_1 + \cdots + P_n$.

The mixed volume estimate for the number of toric roots of a system of polynomials is frequently much better than the Bezout bound, because it takes into account finer information about the combinatorics of the supporting set of monomials. It is easy to construct examples of families of polynomials of fixed support for which the ratio of the Bezout bound to the mixed volume bound is exponentially large (in n). In many applications, e.g., inverse kinematics, one would like sharp estimates for the number of solutions of a system of polynomial equations, and one knows a priori that the solutions are toric.

Following [GKZ1],[PS],[CE],[HS],[St1],[St2], we consider toric deformations of (1.1),

$$f_i(x,t) \quad := \quad \sum_{q \in \mathcal{A}_i} c_{iq}\, x^q\, t^{\omega_i(q)}, \qquad i = 1, \ldots, n. \qquad (1.2)$$

The weights $\omega_i(q)$ determine polytopes $\widehat{P}_i := \{\, (q, \omega_i(q)) \;:\; q \in P_i \}$ in $\mathbb{R}^{n+1}$. Let $\widehat{P} = \widehat{P}_1 + \cdots + \widehat{P}_n$ denote the Minkowski sum of the lifted polytopes. The *lower convex hull* $\widehat{\Delta}$ of $\widehat{P}$ is the collection of facets $\widehat{F}$ of $\widehat{P}$ whose inner normal

(1) Partially supported by NSF grant CCR-9258533 (NYI).

(2) Partially supported by a David and Lucile Packard Fellowship and NSF grants DMS-9201453 and DMS-9258547 (NYI).

has positive last component. Each such facet has the form $\widehat{F} = \widehat{F}_1 + \cdots + \widehat{F}_n$, where $\widehat{F}_i$ is a face of $\widehat{P}_i$. We say that $\widehat{F}$ is a *mixed facet* if $dim(\widehat{F}_i) = 1$ for $i = 1, \ldots, n$. Suppose $\pi : \mathbb{R}^{n+1} \to \mathbb{R}^n$ projects to the first n coordinates, then $\Delta := \{\pi(\widehat{F}) : \widehat{F} \in \widehat{\Delta}\}$ is a subdivision of P. Each cell C of Δ has the form $\pi(\widehat{F}) = F_1 + \cdots + F_n$ where $\pi(\widehat{F}_i) = F_i$ is a subpolytope of P_i. We assume that the weights ω are *sufficiently generic*, meaning that $\dim(C) = \dim(F_1) + \cdots + \dim(F_n)$, for every $C \in \Delta$. In this case Δ is called a *mixed subdivision* of P. The projection of a mixed facet of $\widehat{\Delta}$ is a *mixed cell* of Δ. The sum of the volumes of the mixed cells of any mixed subdivision Δ equals the *mixed volume* $\mathcal{MV}(P_1, \ldots, P_n)$. This is an integer which does not depend on the choice of subdivision. For details on mixed subdivisions, mixed volumes, and sparse polynomial systems, see [Bet],[HS],[St2] and the references given there.

Mixed cells are parallelotopes. If one considers them "half-open", then their volume equals their number of lattice points. In order to view the mixed cells half-open in a consistent manner, we follow the method of Canny and Emiris [CE] by displacing Δ to $\Delta + \delta$, with $\delta \in \mathbb{R}^n$ generic, and then counting all (now strictly interior) lattice points.

Theorem 1.1. *Let $f_1, \ldots, f_n$ be generic Laurent polynomials with Newton polytopes $P_1, \ldots, P_n$, respectively. The monomials corresponding to the lattice points lying in the mixed cells of any generically displaced mixed subdivision $\Delta + \delta$ of $P = P_1 + \cdots + P_n$ form a vector space basis for the quotient ring $A = K[x_1, x_1^{-1}, \ldots, x_n, x_n^{-1}] / (f_1, \ldots, f_n)$.*

We remark that an alternative proof of theorem 1.1 has been provided by Emiris and Rege [ER].

Consider a mixed cell $C = E_1 + \cdots + E_n$, where E_j is the one-dimensional subpolytope $[q_{j1}, q_{j2}]$ with $q_{jk} \in \mathcal{A}_j$. Then C supports the *binomial system*

$$c_{11}x^{q_{11}} + c_{12}x^{q_{12}} \quad = \quad \cdots \quad = \quad c_{n1}x^{q_{n1}} + c_{n2}x^{q_{n2}} \quad = \quad 0. \tag{1.3}$$

We use the following notations consistently from here on:

$$b_C \quad := \quad (-c_{12}/c_{11}, \ldots, -c_{n2}/c_{n1}), \tag{1.4}$$

$$U_C \quad := \quad \text{the } (n \times n)\text{-matrix with column vectors } \quad q_{j1} - q_{j2}, \tag{1.5}$$

$$a_C \quad := \quad \sum \{ q : q \in (C + \delta) \cap \mathbb{Z}^n \}. \tag{1.6}$$

We shall refer to the case when $b_C = \underline{1} := (1, 1, \ldots, 1)$ as the *unit* case.

There is a natural representation of the coordinate algebra A in $\mathrm{Hom}_K(A, A)$ which maps $f \mapsto$ (multiplication by f). This defines a trace $Tr(\cdot)$ on A, and a bilinear form $B : A \times A \to K$, $(g, h) \mapsto \mathrm{Tr}(g \cdot h)$. We call B the *trace form* and represent it by a symmetric matrix. The rank of B equals the number of distinct roots of (1.1). If the field K contains the reals $\mathbb{R}$, then the signature of B equals the number of distinct real roots (see [BW],[PRS]). Our second result concerns the asymptotic behavior of the trace form.

Theorem 1.2. *With respect to the basis arising from a displaced mixed subdivision as in theorem 1.1, the trace form of the deformed system (1.2) is a matrix polynomial in t,*

$$B(t) \quad = \quad B_0\, t^d\, (1 + o(1)),$$

where $d = \sum_C \gamma_C \cdot a_C$, $\det(B_0) \; = \; \prod_C \mathrm{Vol}(C)^{\mathrm{Vol}(C)} \cdot b_C^{2U_C^{-1}a_C}$. *and* $(\gamma_C, 1)$ *is the vector supporting the mixed facet above* C. *(The sum and product are over all mixed cells of* Δ.*)*

The expression $\mathrm{Vol}(C)^{\mathrm{Vol}(C)}$ plays a prominent role in the fundamental work of Gel'fand-Kapranov-Zelevinsky on $\mathcal{A}$-discriminants [GKZ1],[GKZ2]. (See Section 5 for further information regarding the relationship to their work.)

Throughout this paper we employ the following generalized multi-index notation: For a row vector $x = (x_j)$ and a matrix $U = (u_{ji})$, we let x^U denote the row vector whose i-th entry equals the monomial $x_1^{u_{1i}} \cdots x_n^{u_{ni}}$. This definition reduces to the usual multi-index notation when U is a column vector, and it reduces to a vector of powers of t when $x = t$ is a scalar and U is a row vector. One has the relation $(x^U)^V = x^{(UV)}$.

2 Binomial systems

Suppose $A = K[x_1^{\pm 1}, \ldots, x_n^{\pm 1}]\,/\,I$ is a finite-dimensional reduced K-algebra of dimension d, and X the variety in $(\mathbb{C}^*)^n$ defined by an ideal I of generic polynomials as above. (We are using the fact that X is reduced.) Then a set of d monomials $\{\, x^q \; : \; q \in M \,\}$ is a basis of A if and only if the $d \times d$-matrix $S := (\xi^q)_{\xi \in X,\, q \in M}$ is non-singular. To establish our results we may also use the matrix $S^T S$, whose entries lie in the ground field K.

In this section we prove theorem 1.1 in the "local case" where I is generated by the binomial system (1.3). This system can be rewritten in the form

$$-(c_{11}/c_{12}) \cdot x^{q_{11}-q_{12}} \quad = \quad \cdots \quad = \quad -(c_{n1}/c_{n2}) \cdot x^{q_{n1}-q_{n2}} \quad = \quad 1, \quad (2.1)$$

or more compactly $x^{U_C} = b_C$ according to the notations (1.4), (1.5). Here Δ consists of a single mixed cell C, and our basis-to-be is $M := (C + \delta) \cap \mathbb{Z}^n$.

Lemma 2.1. *Let* Λ *denote the subgroup of* $\mathbb{Z}^n$ *generated by the edges* $v_1, \ldots, v_n$ *of* C. *Then the points in* M *are coset representatives for the finite abelian group* $G := \mathbb{Z}^n/\Lambda$.

Proof: Let $C = \{v_0 + \sum \lambda_i v_i \; : \; 0 \le \lambda_i \le 1\}$, where $v_0 \in \mathbb{Z}^n$. By the genericity of δ $C + \delta$ has no lattice points on its boundary. Given $q, q' \in M$, we have $q - q' = \sum \mu_i v_i$, where $0 \le \mu_i < 1$ are unique rational numbers. Therefore $q \not\equiv q' \pmod{\Lambda}$. On the other hand, any $x \in \mathbb{Z}^n$ may be expressed as $v_0 + \delta + \sum \mu_i v_i$, and this expression is normalized by extracting the integer parts of the μ_i to put it in the form $v' + v$, where $v' \in \Lambda$, and $v \in M$. $\square$

Lemma 2.2. *With the notations of lemma 2.1, the group algebra* $\mathbb{C}[G]$ *is isomorphic to the Laurent polynomial ring* $\mathbb{C}[x_1^{\pm 1}, \ldots, x_n^{\pm 1}] / (x^{v_1}-1, \ldots, x^{v_n}-1)$.

Proof: Define a homomorphism

$$\begin{array}{rcl} \phi \ : \ \mathbb{Z}^n & \longrightarrow & \mathbb{C}[x_1^{\pm 1}, \ldots, x_n^{\pm 1}] / (x^{v_1} - 1, \ldots, x^{v_n} - 1) \\ (a_1, \ldots, a_n) & \mapsto & x_1^{a_1} \cdots x_n^{a_n}. \end{array}$$

Then the kernel of ϕ has a $\mathbb{Z}$-basis $v_1, \ldots, v_n$, and these generate Λ. □

Lemma 2.3. *In the unit case* $b_C = \underline{1}$, *the set* X *of roots of (2.1) may be identified with the group* $\mathrm{Hom}(G, \mathbb{C}^*)$ *of characters of* $G = \mathbb{Z}^n / \Lambda$.

Proof: We have the identifications

$$\begin{array}{rcl} \mathrm{Hom}_{\mathrm{grp.}}(G, \mathbb{C}^*) & \cong & \mathrm{Hom}_{\mathrm{alg.}}(\mathbb{C}[G], \mathbb{C}) \\ & \cong & \mathrm{spec}_m(\mathbb{C}[G]) \\ & \cong & \mathrm{spec}_m(\mathbb{C}[x_1^{\pm 1}, \ldots, x_n^{\pm 1}] / (x^{v_1} - 1, \ldots, x^{v_n} - 1)). \end{array}$$

The first line follows from general principles of character theory (see [CR], §10). The second line follows from the identification of maximal ideals with kernels of homomorphisms into fields. The last line follows by lemma 2.2. □

Proposition 2.4. *In the unit case, the matrix* $S = (\xi^q)_{\xi \in X, q \in M}$ *is non-singular, and* $\det(S^T S) = \pm \mathrm{Vol}(C)^{\mathrm{Vol}(C)}$.

Proof: By lemma 2.1 and lemma 2.3, the matrix S may be viewed as the matrix of characters $\xi \in X = \mathrm{Hom}(G, \mathbb{C}^*)$ evaluated at coset representatives $q \in M$ of $\mathbb{Z}^n$ modulo Λ. Applying the second orthogonality relation for group characters (which sums across characters; see [CR], §31) in the abelian case, the matrix $\overline{S}^T S$ is a diagonal matrix with entries $|G|$ on the diagonal. On the other hand, the matrix S differs from its complex conjugate $\overline{S}$ only by some permutation of the columns corresponding to the automorphism $q \mapsto -q$ of G. Therefore, $\det(S^T S) = \pm |G|^{|G|}$. On the other hand, $|G|$ equals the number of lattice points in $C + \delta$, which equals the volume of C. □

Corollary 2.5. *For all choices of non-zero coefficients in (2.1), the matrix* $S = (\xi^q)_{\xi \in X, q \in M}$ *is non-singular. With* b_C, U_C, *and* a_C *defined as in (1.4)–(1.6), we have*

$$\det(S^T S) \ = \ \pm \mathrm{Vol}(C)^{\mathrm{Vol}(C)} \cdot b_C^{2\, U_C^{-1} a_C}. \tag{2.2}$$

Proof: Suppress the subscript C. Let $\mathrm{diag}(y_1, \ldots, y_n)$ denote the diagonal matrix with y_i's on the diagonal, and make the change of variables $x = \mathrm{diag}(y_1, \ldots, y_n) \cdot b^{U^{-1}}$. Then

$$b \ = \ (\mathrm{diag}(y_1, \ldots, y_n) \cdot b^{U^{-1}})^U \ = \ y^U \cdot b^{U^{-1} U} \ = \ y^U \cdot b,$$

or, equivalently, $y^U = \underline{1}$. Thus after the change of variables we are in the unit case. The general matrix entry ξ^q of S differs from the same entry for the unit case as follows:

$$(\mathrm{diag}(\xi_1, \ldots, \xi_n) \cdot b^{U^{-1}})^q \; = \; \xi^q \cdot b^{(U^{-1} q)}. \tag{2.3}$$

We can factor the constant $b^{(U^{-1} q)}$ out of the column labelled q. The aggregate factor is

$$b^{(U^{-1} \sum_{q \in M} q)} \quad = \quad b^{U^{-1} a}. \qquad \square$$

3 General sparse systems

The roots of (1.2) are algebraic functions $x(t)$ of the parameter t. Their Puiseux series for small t may be determined from the ansatz

$$x(t) \quad = \quad y \cdot t^{\gamma} \, (1 + o(1)) \; := \; (y_1 t^{\gamma_1}, \ldots, y_n t^{\gamma_n}) \, (1 + o(1)), \tag{3.1}$$

where y is a vector of complex variables and $\gamma = (\gamma_1, \ldots, \gamma_n)$ is a vector of rational numbers. Substituting (3.1) into the i-th equation of (1.2), one obtains terms

$$c_{iq} \, y^q \, t^{\langle \gamma, \, q \rangle + \omega_i(q)} \, (1 + o(1)). \tag{3.2}$$

The exponent of t in (3.2) may be expressed as $\langle \, (\gamma, 1), \; (q, \omega_i(q)) \, \rangle$, that is to say, the value of the linear functional $(\gamma, 1)$ on the lifting of some point $q \in \mathcal{A}_i$. Lowest-order terms are determined by faces $\widehat{F}_{i,\gamma}$ of $\widehat{P}_i$ on which $(\gamma, 1)$ is minimized. Let $F_{i,\gamma} := \pi(\widehat{F}_{i,\gamma})$. We refer to each

$$f_{i,\gamma}(y) \quad := \quad \sum_{q \in \pi(\pi^{-1}(\mathcal{A}_i) \cap \widehat{F}_{i,\gamma})} c_{iq} \, y^q, \quad i = 1, \ldots, n \tag{3.3}$$

as the *degeneration* of f_i with respect to the linear functional γ under the lifting ω_i. (We include only monomials which lift to the face $\widehat{F}_{i,\gamma}$.) The following result is proved in lemmas 3.1 and 3.2 of [HS].

Lemma 3.1. *The system $f_{1,\gamma}(y) = \cdots = f_{n,\gamma}(y) = 0$ has a solution $y \in (\mathbb{C}^*)^n$ if and only if $(\gamma, 1)$ supports a mixed facet $\widehat{F}$ of the lower envelope $\widehat{\Delta}$ of $\widehat{P}$. For γ fixed, the number of solutions y equals the volume of the mixed cell $C = \pi(\widehat{F})$.*

This lemma shows that all relevant degenerations (3.3) are binomial systems of the form shown in (1.3) and (2.1). Every pair $\gamma \in \mathbb{Q}^n$ and $y \in (\mathbb{C}^*)^n$ as in lemma 3.1 contributes a branch (3.1) to the vector-valued algebraic function $x(t)$ defined by (1.2). Taking the sum over all mixed facets in $\widehat{\Delta}$, one finds that the total number of branches equals the mixed volume $\mathcal{MV}(P_1, \ldots, P_n)$. This technique was used in [HS] to give a new proof and algorithm for Bernstein's theorem. We are now prepared to prove our main result.

Proof of theorem 1.1: Suppose $\{ C_j : j = 1, \ldots, r \}$ are the mixed cells of the subdivision Δ, and let $M_j := (C_j + \delta) \cap \mathbb{Z}^n$ and $M := M_1 \cup \cdots \cup M_r$. We need to show that the matrix $S := (\xi^q)_{\xi \in X,\, q \in M}$ is non-singular, where $X := \{ \xi \in (\mathbb{C}^*)^n : f_1(\xi) = \cdots = f_n(\xi) = 0 \}$. If the determinant of S were zero, then this identity would continue to hold for the branches $x(t)$ in (3.1). Writing Ω for the set of all branches, the corresponding matrix is $S(t) = (x(t)^q)_{x(t) \in \Omega,\, q \in M}$. We will prove $\det(S(t)) \neq 0$ by showing that the lowest-order term in t is non-zero. We do this by showing that to lowest order the matrix $S(t)$ is block diagonal, and each block is of the type considered in the previous section.

Let $(\gamma_i, 1)$ be the inner normal vector of a mixed facet $\widehat{F}_i$ lying over the mixed cell C_i. Let Ω_i denote the set of branches $x_i(t) = y_i\, t^{\gamma_i}\,(1 + o(1))$ arising from C_i as in lemma 3.1. The vector y_i is a solution of the binomial system determined by the edges of C_i. We consider the matrix $\underline{S}(t)$ obtained from $S(t)$ by multiplying the *column* labelled $q \in M_j$ by $t^{\omega_j(q)}$. This has the effect of multiplying $\det(S(t))$ by a factor t^d, where $d = \sum_{j=1}^{r} \sum_{q \in M_j} \omega_j(q)$. We have $\det(\underline{S}(t)) =$

$$
\begin{array}{c|cccc|}
 & q_1 \in M_1 & q_2 \in M_2 & \ldots & q_r \in M_r \\
x_1 \in \Omega_1 & (t^{\omega_1(q_1)} x_1(t)^{q_1}) & (t^{\omega_2(q_2)} x_1(t)^{q_2}) & \ldots & (t^{\omega_r(q_r)} x_1(t)^{q_r}) \\
x_2 \in \Omega_2 & (t^{\omega_1(q_1)} x_2(t)^{q_1}) & (t^{\omega_2(q_2)} x_2(t)^{q_2}) & \ldots & (t^{\omega_r(q_r)} x_2(t)^{q_r}) \\
\vdots & \vdots & \vdots & \ddots & \vdots \\
x_r \in \Omega_r & (t^{\omega_1(q_1)} x_r(t)^{q_1}) & (t^{\omega_2(q_2)} x_r(t)^{q_2}) & \ldots & (t^{\omega_r(q_r)} x_r(t)^{q_r})
\end{array}
\qquad (3.4)
$$

Let S_{ij} denote the block corresponding to Ω_i and M_j. The typical entry of S_{ij} looks like

$$t^{\omega_j(q_j)} x_i(t)^{q_j} \;=\; t^{\omega_j(q_j)} (y_i\, t^{\gamma_i}\,(1 + o(1)))^{q_j} \;=\; y_i^{q_j}\, t^{\langle\, (\gamma_i, 1),\; (q_j, \omega_j(q_j))\,\rangle}\,(1 + o(1)).$$

The order in t of this expression is abbreviated

$$a_{ij}(q) \quad := \quad \langle\, (\gamma_i,\, 1),\;\; (q,\, \omega_j(q))\,\rangle; \quad q \in M_j.$$

We note that $a_{ii}(q)$ has a constant value a_{ii} for $q \in M_i$. Any mixed facet $\widehat{F}_j \neq \widehat{F}_i$ of the lower envelope $\widehat{\Delta}$ of $\widehat{P}$ lies above $\widehat{F}_i$ in the direction $(\gamma_i,\, 1)$. There can be no "ties" in the values of $(\gamma_i,\, 1)$ on vertices appearing on faces which $\widehat{F}_i$ shares with $\widehat{F}_j$, since all the vertices we consider are *strictly interior* to the mixed cells of the displaced subdivision. In more precise terms, we have $a_{ii} < a_{ij}(q)$, $\forall q \in M_j$, $j \neq i$. Therefore the components of the diagonal block S_{ii} have strictly smaller order in t than any other entries in their rows. We may extract from the *rows* corresponding to Ω_i a factor $t^{a_{ii}}$. This leaves terms in the S_{ii} block of order $(1 + o(1))$, and terms in the off-diagonal blocks S_{ij} with positive exponents $a_{ij}(q) - a_{ii} > 0$. The complete power of t extracted from (3.4) is

$$\sum_{i=1}^{r} |M_i| \cdot a_{ii} \;=\; \sum_{i=1}^{r} \sum_{q \in M_i} \gamma_i \cdot q + \omega_i(q) \;=\; \sum_{i=1}^{r} \sum_{q \in M_i} \gamma_i \cdot q + \sum_{i=1}^{r} \sum_{q \in M_i} \omega_i(q).$$

The last term cancels the extra factor t^d we had introduced when passing from $S(t)$ to $\underline{S}(t)$. We therefore have, to lowest order in t,

$$\begin{aligned} \det(S(t)) &= \left\{ \prod_{i=1}^{r} \det(S_{ii}) \right\} t^{\sum_{i=1}^{r} \sum_{q \in M_i} \gamma_i \cdot q} + \cdots = \\ &= \left\{ \prod_{i=1}^{r} \det(S_{ii}) \right\} t^{\sum_{C} \gamma_C \cdot a_C} + \cdots . \end{aligned}$$

Here γ_C denotes the vector which selects the mixed cell C. By Proposition 2.5, each of the factors $\det(S_{ii})$ is non-zero. This concludes the proof of theorem 1.1. □

4 Trace forms and real roots

Let $B(\cdot, \cdot)$ denote the trace form (cf. [PRS],[BW]) of the finite-dimensional reduced K-algebra A, and let $X = Spec_m(A)$ as before. It is known that the trace of a polynomial is the sum of its values at all roots. Therefore the trace form can be written as

$$B(f, g) \;=\; Tr(f \cdot g) \;=\; \sum_{\xi \in X} f(\xi) g(\xi), \qquad \text{for } f, g \in A. \tag{4.1}$$

Proof of theorem 1.2: Let M be the basis of A which was established in theorem 1.1, and let $S = (\xi^q)_{\xi \in X, q \in M}$ as before. Given any element $f = \sum_{q \in M} \lambda_q x^q$ in A, with column vector of coefficients λ, then $S \cdot \lambda$ equals the column vector $(f(\xi) : \xi \in X)$. Therefore,

$$\mu^T \cdot (S^T S) \cdot \lambda \;=\; (S\mu)^T \cdot (S\lambda) \;=\; B(f, g), \quad \text{where} \quad g = \sum_{q \in M} \mu_q x^q.$$

In other words, the matrix $S^T S$ represents the trace form B with respect to the basis M.

By the results of the previous sections, the matrix polynomial $B(t)$ is asymptotically

$$B(t) = S(t)^T S(t) = \begin{bmatrix} S_{11}^T S_{11} & 0 & \ldots & 0 \\ 0 & S_{22}^T S_{22} & \ldots & 0 \\ \vdots & \vdots & \ddots & \vdots \\ 0 & 0 & \ldots & S_{rr}^T S_{rr} \end{bmatrix} \cdot t^{2 \cdot \sum_i \gamma_i \cdot a_i} + \cdots \tag{4.2}$$

From the second orthogonality relation of character theory we saw that, up to a constant factor, each block $S_{ii}^T S_{ii}$ is a permutation of the diagonal matrix $\mathrm{diag}(\mathrm{Vol}(C_i), \ldots, \mathrm{Vol}(C_i))$. By Proposition 2.5, the leading term of the matrix polynomial $B(t)$ has determinant $\pm \prod_C \mathrm{Vol}(C)^{\mathrm{Vol}(C)} b_C^{2U_C^{-1} a_C}$. □

For the remainder of this section we suppose that K contains $\mathbb{R}$, and that the coefficients of the input system (1.1) are real. An important theorem of computational algebra (cf. [Pe],[PRS],[BW]) states that the number of real roots $\# X(\mathbb{R})$ counted without multiplicity equals the signature of the trace form B.

We now come to the problem of determining the *asymptotic number of real roots* of the system (1.2), meaning the number of roots $x \in (\mathbb{R}^*)^n$ for any sufficiently small fixed real value of $t > 0$. For each mixed cell C of Δ, we consider the finite abelian group $G_C := \mathbb{Z}^n / \Lambda$, where Λ is the lattice generated by the edges of C. Consider the decomposition into invariant factors,

$$G_C \;:=\; \mathbb{Z}^n / \Lambda \;\cong\; \mathbb{Z}/n_1\mathbb{Z} \oplus \cdots \oplus \mathbb{Z}/n_k\mathbb{Z} \tag{4.3}$$

with $n_1 \,|\, n_2 \,|\, \cdots \,|\, n_k$. Let $p(C)$ denote the number of even invariant factors n_i of G_C. The following result was proved in [St1] for the special case of unmixed systems.

Proposition 4.1. *The asymptotic number of real roots of (1.2) is at most* $\sum_C 2^{p(C)}$.

Proof: To lowest order in t, the signature of $S^T S$ is the sum of the signatures of the blocks $S_{ii}^T S_{ii}$. Writing C for the i-th mixed cell and $S_C = S_{ii}$, we need to show that the signature of $S_C^T S_C$ is bounded above by $2^{p(C)}$.

We fix C and consider the invariant decomposition (4.3). Let W_j denote the character table of the cyclic group $\mathbb{Z}/n_j\mathbb{Z}$. It is easy to see that

$$W_j^T W_j \;=\; \begin{bmatrix} n_j & 0 & \ldots & 0 \\ 0 & 0 & \ldots & n_j \\ \vdots & \vdots & \iddots & \vdots \\ 0 & n_j & \ldots & 0 \end{bmatrix}, \tag{4.4}$$

because the automorphism $x \mapsto -x$ of $\mathbb{Z}/n_j\mathbb{Z}$ leaves 0 invariant, and cycles the remaining elements as shown. In the matrix (4.4) there are either one or two diagonal entries depending on whether n_i is odd or even. The signature of such a matrix equals the number of non-zero diagonal entries (cf. [Sh], p. 12). Hence the signature of $W_j^T W_j$ is 2 if n_j is even, and 1 if n_j is odd. Let W be the character table of G_C. This is the tensor product of the character tables W_j, and therefore

$$W^T W = (W_1 \otimes \cdots \otimes W_r)^T (W_1 \otimes \cdots \otimes W_r) = (W_1^T W_1) \otimes (W_2^T W_2) \otimes \cdots \otimes (W_r^T W_r).$$

Since the signature of real symmetric matrices is multiplicative with respect to tensor products, the signature of W^TW equals $2^{p(C)}$.

Now, the matrix S_C is obtained from W by multiplying each column by a scalar as in (2.3), and it can be written as $S_C = W_1' \otimes \cdots \otimes W_r'$, where W_i' is obtained from W_i by multiplying each column by a non-zero scalar. Then $(W_i')^T W_i'$ is obtained from $W_i^T W_i$ by multiplying each entry by a non-zero scalar, preserving symmetry of the matrix and non-negativity of the signature. The new symmetric matrix has signature either 0 or 2 if n_j is even, while it remains 1 if n_j is odd. Therefore the signature of

$$S_C^T S_C = (W_1' \otimes \cdots \otimes W_r')^T (W_1' \otimes \cdots \otimes W_r') = ((W_1')^T W_1') \otimes \cdots \otimes ((W_r')^T W_r')$$

is bounded above by the signature of W^TW, which is $2^{p(C)}$. □

5 Two problems

We close with two remarks which suggest directions for future research.

(1) Our theorem 1.1 is non-trivial in the sense that, even for generic coefficients c_{iq}, not all collections of $\mathcal{MV}(P_1, \ldots, P_n)$ monomials form a basis for A. (For instance, the set $\{1, x^4y^2, x^3y^3, x^2y^4\}$ is a non-base modulo the system $\{a_0 + a_1x + a_2xy + a_3y,\ b_0 + b_1x^2y + b_2xy^2\}$.) It would be interesting to find a combinatorial characterization of the non-bases. This problem can be rephrased as follows: The data $\mathcal{A}_1, \ldots, \mathcal{A}_n \subset \mathbb{Z}^n$ define a matroid of rank $\mathcal{MV}(P_1, \ldots, P_n)$ on the (infinite) set $\mathbb{Z}^n$, by independence of monomials in A. The idea is to study this matroid. For instance, when is it uniform?

(2) Given any monomial basis for A, we can express the trace form B by a symmetric matrix, and its determinant $\det(B)$ is well-defined rational function of degree 0 in the coefficients c_{iq}. The most important divisor of the numerator of this rational function is the $\mathcal{A}$-*discriminant* $D_{\mathcal{A}}$ due to Gel'fand, Kapranov, and Zelevinsky [GKZ1],[GKZ2]. Here $\mathcal{A} = \cup_{i=1}^n e_i \times \mathcal{A}_i \subset \mathbb{Z}^{2n}$ is obtained by the *Cayley trick* from (1.1). It would be very interesting to understand all other factors, and to see how they depend on the choice of monomial basis. In light of our theorem 1.2, we conjecture that the leading coefficient of the $\mathcal{A}$-discriminant $D_{\mathcal{A}}$ with respect to ω is precisely $\pm \prod_C \mathrm{Vol}(C)^{\mathrm{Vol}(C)}$, where C runs over all mixed cells of the mixed subdivision Δ. This would provide a refinement of Thm 3D.2 in [GKZ1].

6 References

[BW] Becker, E., Wörmann, T.: "On the trace formula for quadratic forms", Recent Advances in Real Algebraic Geometry and Quadratic Forms, Proceedings of the RAGSQUAD Year, Berkeley 1990-1991, W.B. Jacob, T.-Y. Lam, R.O. Robson (editors), *Contemporary Mathematics*. **155** (1993) pp. 271–291.

[Ber] Bernstein, D. N.: "The number of roots of a system of equations", *Functional Analysis and its Applications* **9** (1975), pp. 1–4.

[Bet] Betke, U.: "Mixed volumes of polytopes": *Archiv d. Mathematik* **58** (1992), pp. 388–391.

[CE] Canny, J., Emiris, I.: "An Efficient Algorithm for the Sparse Mixed Resultant", in Proc. AAECC-10, edited by G. Cohen, T. Mora and O. Moreno", Springer Lecture Notes in Computer Science 263 (1993), pp. 89–104.

[CR] Curtis, C.W., I. Reiner, I.: "Representation Theory of Finite Groups and Associative Algebras", Wiley Interscience Publishers, 1962.

[ER] Emiris, I.Z., Rege, A.: "Monomial bases and polynomial system solving", submitted to ISSAC 1994.

[GKZ1] Gelfand, I. M., Kapranov, M. M., Zelevinsky, A. V.: "Discriminants of polynomials in several variables and triangulations of Newton polytopes", *Algebra i analiz (Leningrad Math. J.)* **2** (1990) pp. 1–62.

[GKZ2] Gelfand, I. M., Kapranov, M. M., Zelevinsky, A. V.: "Discriminants, Resultants and Multidimensional Determinants", Birkhäuser, Boston, 1994.

[HS] Huber, B., Sturmfels, B.: "A polyhedral method for solving sparse polynomial systems", *Mathematics of Computation*, to appear.

[Pe] Pedersen, P.: "Multivariate Sturm Theory", in Proc. AAECC-9, New Orleans 1991, Springer Lect. Notes in Comp. Sci. 537 (1991), pp. 318–332.

[PRS] Pedersen, P., Roy, M. F., Szpirglas, A.: "Counting real zeros in the multivariate case", in *"Computational Algebraic Geometry"* (eds. F. Eyssette, A. Galligo), Proceedings MEGA-92, Birkhäuser, 1992, pp. 203–223.

[PS] Pedersen, P., Sturmfels, B.: "Product formulas for resultants and Chow forms", *Mathematische Zeitschrift* **214** (1993) pp. 377–396.

[Sch] Scharlau, W.: "Quadratic and Hermitian Forms", Springer, Grundlehren der mathematischen Wissenschaften **270**, 1985.

[St1] Sturmfels, B.: "The asymptotic number of real zeros of a sparse polynomial system", to appear in Proceedings of the Workshop on "Hamiltonian and Gradient Flows: Algorithms and Control", (ed. A.Bloch), Fields Institute, Waterloo, Ontario, March 1992.

[St2] Sturmfels, B.: "On the Newton polytope of the resultant", *Journal of Algebraic Combinatorics* **3** (1994) pp. 207–236.

P. Pedersen (`paul@cs.cornell.edu`)
Department of Computer Science, Cornell University, Ithaca, NY 14853 (USA).

B. Sturmfels (`bernd@math.cornell.edu`)
Department of Mathematics, Cornell University, Ithaca NY 14853 (USA).

Progress in Mathematics, Vol. 143, © 1996 Birkhäuser Verlag Basel/Switzerland

The complexity and enumerative geometry of aspect graphs of smooth surfaces

S. Petitjean

1 Introduction

Informally, the aspect graph [KvD79] (also called view graph) is a qualitative, viewer-centered representation which enumerates all possible appearances of an object. More formally, choosing a camera model (orthographic-parallel or perspective-central-projection) and a viewpoint determines the aspect of an object (i.e., the structure of the observed line-drawing). The range of possible viewpoints can be partitioned into maximal connected sets (regions) that yield identical aspects. The change in aspect at the boundary between regions is called a visual event. The maximal regions and the associated aspects form the nodes of an aspect graph, whose arcs correspond to the visual event boundaries between adjacent regions.

Before going further on, let us give an example assuming orotographic projection [Pet92]. Figure 1.a shows a shaded view of a squash-shaped object described by a quartic surface whose implicit equation is

$$4y^4 + 3xy^2 - 5y^2 + 4z^2 + 6x^2 - 2xy + 2x + 3y - 1 = 0$$

Figure 1.b shows a line drawing of the same object. Parabolic and flecnodal curves have been drawn on the surface to reveal its strucutre by separating its convex, hyperbolic, and concave parts. As will be shown in the next sections, these curves play a crucial role in the construction of the aspect graph. Figures 1.c and 1-d show the aspect graph of the transparent squash. Under orthographic projection, the range of possible viewpoints is two-dimensional and can be modeled as a unit view sphere. This sphere has been partitioned into a number of maximal regions such that points within a region see the same aspect of the object. These regions are delineated by a number of curves called visual event curves. The characteristic aspects associated with each region are shown in Figure 1.d. Here, we used a longitude/latitude parameterization of the sphere to display all the aspect-graph regions in a single picture. Finally, the aspect graph itself is displayed in Figure 2.

Recently, algorithms have been proposed and implemented for constructing the aspect graph of smooth and piecewise-smooth objects. Some of these algorithms rely on numerical methods to solve systems of polynomial equations (like homotopy continuation for instance [Mor87]); the others use symbolic methods (cylindrical algebraic decomposition [Col75] and Gröbner bases

[Buc85]). Symbolic methods provide infinite accuracy at every stage (which is important since computations that deal with algebraic or semialgebraic sets are unstable), but at the expense of high computational costs.

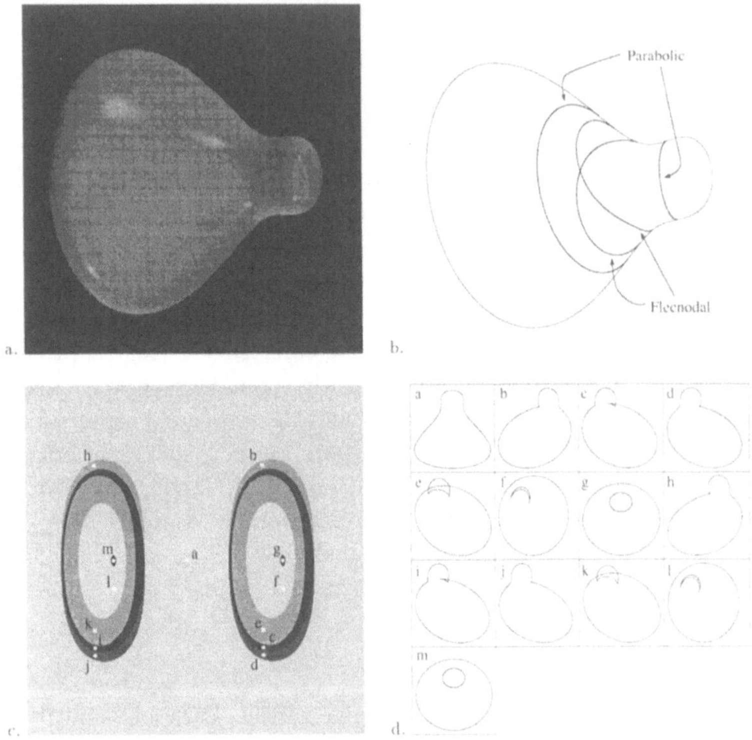

Fig. 1. a. A squash-shaped object. b. Its parabolic and flecnodal curves. c. The aspect graph of the transparent squash in parameter space. d. The corresponding aspects.

Algorithms for constructing the view graph of solids of revolution have been described and implemented: [EB89] and [KP90] use numerical methods, and more recently [RvE92] use a symbolic approach. For objects bounded by smooth surfaces, it was recognized early that singularity theory offered a complete catalog of visual events [Ker81], [Pla81], [Arn83]. which could be used to elucidate the structure and formation of aspect graphs [KvD79], [CW85]. Assuming orthographic projection, these ideas have been implemented for smooth objects bounded by parameterized algebraic surfaces in [PK90] (numerical) and [Rie92] (symbolic). For implicit algebraic surfaces, [Pet92] reprots the full implementation of an algorithm using numerical methods and assuming orthographic projection, while [Rie93a] presents a symbolic algorithm for both parallel and central projection.

The full classification of the visual events for the case of piecewise-smooth algebraic surfaces[1] observed under perspective projection has only been completed very recently [Rie93b], after the works of [Rie87] and [Tar91], and a symbolic algorithm is reported.

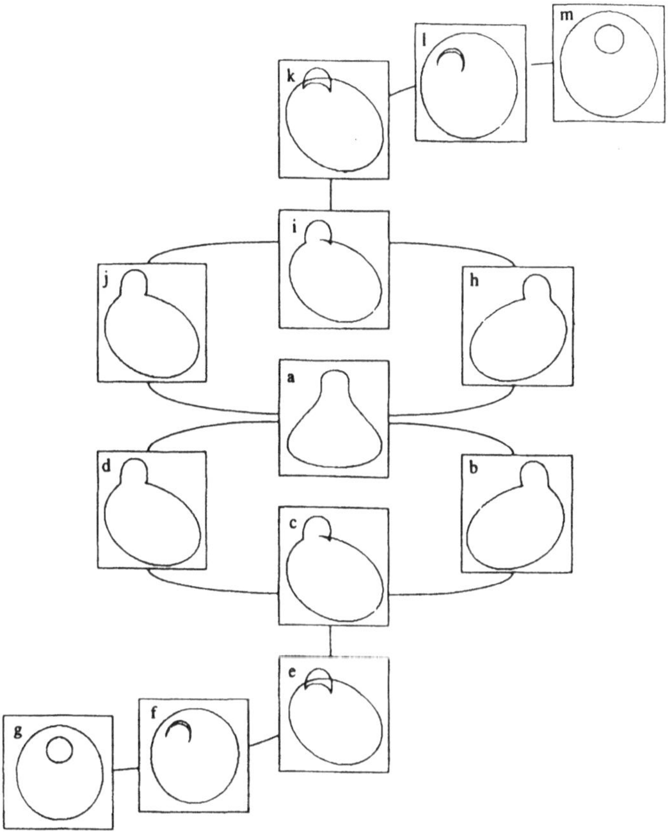

Fig. 2. The aspect graph of a squash-shaped object.

Most of the work achieved thus far on aspect graphs merely gives local, or at most semiglobal, information about the visual event surfaces, and brings up many theoretical questions of a global kind. Among these questions are: Is there an upper bound for the number of nodes of the aspect graph of an object bounded by a smooth algebraic surface in terms of the degree d of its defining equation? The answer is yes, and the best-known bound is $O(d^{12})$ for orthographic projection and $O(d^{18})$ for perspective projection [Rie93b]. Actually this result comes as a special case of a formula obtained for the number of nodes

[1]Surfaces consisting of smooth algebraic surfaces intersecting pairwise transversally along crease curves and having isolated triple points

of the viewgraph of a piecewise-smooth surface X consisting of n smooth algebraic surfaces of degree at most d. This bound is $O(n^{K \dim \nu} n^{6 \dim \nu})$, where ν is the viewspace (so $\dim \nu = 2$ if parallel projection is assumed, and $\dim \nu = 3$ for perspective projection), and $K = 3$ if X is topologically equivalent to a polyhedral surface having $O(n)$ edges, vertices, and faces, and $K = 6$ otherwise [Rie93b].

Many questions related to the geometry of the viewspace as cut out by the visual event surfaces have already been answered qualitatively for generic C^∞-surfaces [Ker81], [Arn83], [Rie90]. What we propose to do here is to work on the quantitative or enumerative geometry of the entities appearing in the construction of aspect graphs of smooth surfaces. Our main task will be to compute some numerical invariants of the principal organizers of the partitioning of viewspace, i.e., the visual event surfaces, and to give properties of their sets of singular points.

The rest of this presentation is organized as follows. Elements of singularity theory and visual events are introduced in Section 2. Then, preliminary results needed for our computations are proved in Section 3. Section 4 shows how the study of the visual events can be achieved through the understanding of contacts of linear spaces with the surface. Further projective characters of the visual event surfaces along with properties of their singular curves are given in Section 5. Section 6 compares our results with others that have appeared recently in the literature. Finally, Section 7 discusses some of the issues raised by our results and sketches future research directions.

2 Visual events

For a given viewpoint, the line drawing of a smooth surface (the *outline* or *image contour*) is the projection onto the image of the *contour generator* (or *occluding contour*), defined as the set of points where the line of sight grazes the surface. Like any other plane curve, the outline may have a certain number of singularities. *Singularity theory* tell us exactly which ones can be expected, and how the contour structure changes as a function of viewpoint. This is precisely the information needed to construct the aspect graph.

In this section, we will first informally review some results of singularity theory and their application to the classification of stable and accidental viewpoints (so-called *visual events*). We also introduce the theory of *stationary multiple-points* developed in [Col86], [Col87] that will allow us to later study the contacts of linear spaces with a smooth surface.

2.1 Singularity theory

In 1955, Whitney published a paper [Whi55] that laid to the foundations for the theory of singularities of smooth mappings. Since then, singularity theory has been further developed by many researchers, especially [Tho56], [Arn83], [Arn84].

Here we are interested in a particular smooth mapping, the projection of a surface onto some image plane. The contour generator consists of the singular points of this mapping. Combining results of [Whi55] and [Mat73], we infer that for "almost all" (*generic*) observation points, the contour generator of a smooth embedded surface is a regular curve formed by *fold* points and a discrete set of *cusp* points; anf the outline is a curve whose only singularities are a discrete set set of ordinary double points, formed by the transversal superposition of the projection of two fold points, and ordinary cusps, formed by the projection of cusp points. Mather's result [Mat73] asserts that the set of viewpoints for which the preceding result does not hold has measure 0.

Figure 3 illustrates this theorem. Note that the definition of the projection and its singularities does not take into account the transparency or opaqueness of the observed surface. For a transparent object, both branches of the contour are visible at either a cusp or a double point. For an opaque object, only one branch of the cusp is visible (yielding a *contour termination*), while the double point actually becomes a *t-junction*.

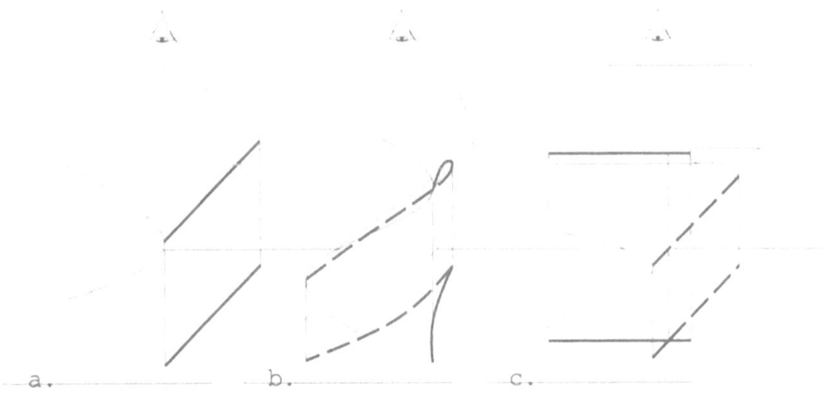

Fig. 3. Stable singularities. a. Fold. b. Cusp. c. t-junction. The occluding and image contours correspond to an observer looking at the surface from above. The contours seen by the observer are shown as thick lines, their hidden parts being dashed.

Generically (for all cases but some exceptional ones) only these singularities are encountered. All other singularities dissapear under small perturbations of the object or of the direction of projection, while the types listed above are *stable*, i.e., every nearby mapping has a similar singularity at an appropriate nearby point. Whitney also proved that every singularity of a smooth mapping of a surface onto a plane splits into t-junctions and cusps after an appropriate small perturbation of the point of view.

2.2 Nongeneric observation directions

If the point of view is generic, then the only singularities of the outline are ordinary cusps and double points. However, if one chooses it in a special way, one may also obtain some nongeneric projections of a smooth surface. These singularities are often classified according to their *codimension*. Consider a k-germ π and a singularity α of π. Then the codimension of α is the difference between the dimensions of the space of k-germs and of the orbit of π (the set of germs yielding "equivalent"-diffeomorphic-views). Stable types of the previous section are often referred to as being of codimension 0, since they need no control to be observed. Singularities of codimension greater than 0 partition the view space in a finite number of cells, separated by the *view bifurcation set*.

For "almost all" embedded C^∞-surfaces, it is sufficient to detect the degenerate singularities of codimension less than the dimension of the viewspace considered. (More degenerate singularitiescan only be observed for very special surfaces.) We shall refer to such surfaces as being *projection-generic*, which means that their family of projections is versal (for the type of projection considered). However, the case of algebraic surfaces is slightly different: There are open subsets of the space of algebraic surfaces of low degree for which projection-genericity fails [Rie93b].

We shall consider only projection-generic C^∞-surfaces in what follows, since not much can be said about the geometry of bifurcation sets of general smooth algebraic surfaces.

In the case of perspective projection, the stable views accupy volumes in a three-dimensional space of viewpoints. The boundaries of these volumes are formed by transitional views of codimension 1, 2, and 3, i.e., surfaces, lines, and points. They have been studied in [Arn83], [Rie90].

If we consider orthographic projection, then we represent the range of viewpoints by a unit view sphere of viewing directions. In this case, the stable views occupy areas on the view sphere. The boundaries of these patches are formed by transitional views of codimension 1 and 2, i.e., lines and points (the visual events) [Ker81]. Thus, orthographic projection fails to capture singularities of codimension 3. Subsequently, the partitioning of viewpoint spaces assuming perspective projection is more complex, and the number of views of the object is larger. (The partitioning of the view sphere can be thought of as obtained by the intersection of the partitioning of a three-dimensional viewspace with a sphere at infinity. Only those cells at infinity are taken into account.)

In the next two sections, we show how some of these singularities relate to the local shape of the observed surface, using a classification due to [Pla81].

2.3 A classification of surface points

We saw previously that the line of sight is tangent to the surface at an occluding contour point. Generally speaking, we say that a tangent vector in some point *has contact of order n* with the surface (or *is of order n*) when the directional derivatives of order i of the surface equation in the direction of the tangent are zero for all $i < n$, and nonzero for $i = n$. All surface points have an infinity of tangents of order 2 in their tangent plane. By definition, a tangent of order 3 or greater is called an *asymptotic tangent*, and its direction at the point of tangency an *asymptotic direction*. *Elliptic* points do not have asymptotic tangents, *parabolic* points have one, and *hyperbolic* points have two. It can be shown that the line of sight is of order 2 (ordinary tangential) at a fold point and of order 3 at a cusp point [Pla81].

We have just characterized the surface points where the stable singularities may occur in terms of the local geometry of the surface [doC'76]. In the next section, we will show that higher-order singularities also correspond to high-order contacts between the line of sight and the surface, so, first, we are going to give a complete classification of all surface points according to the order of contacts of their tangents.

A smooth curve in the plane may have any order of contact with its tangent, but for projection-generic surfaces, tangents of order greater than 5 do not exist. All points of a projection-generic surface can be divided into the following eight classes according to the orders of their tangents (following [Pla81]), as illustrated in Figure 4:

(1) the *elliptic domain* where all tangents are of order 2.

(2) the *hyperbolic domain* where each point has two asymptotic tangents. The envelope of these tangents are called the *asymptotic curves*.

These two regions are separated by a common boundary:

(3) the *curve of parabolic points*, where each point has a single asymptotic tangent. (We shall denote the set of parabolic points by P in what follows.)

Within the domain of hyperbolicity there is a special curve:

(4) the *flecnodal curve*, formed by the inflections of the projection of the asymptotic curves onto the tangent plane, where each point has a tangent of order 4 or greater. (The flecnodal curve is denoted by S in the following.)

On this curve are singled out some special special isolated points of three types:

(5) the *biflecnodes*, which are points of double inflection of the asymptotic lines, with a tangent of order 5 (S_5 in what follows.)

(6) the points of inflection of both asymptotic curves, with two tangents of order 4 ($S_{4,4}$ in what follows). These points are the points of self-intersection of the flecnodal curve.

(7) the *godrons*, which are points of tangency of the parabolic and flecnodal curves, with a tangent of order 4 (P_4 in the next sections).

Finally, there is one more type of special isolated points on the parabolic curve:

(8) the *glutterpoints*, which are stationary points of the asymptotic tangents.

For projection-generic surfaces, the parabolic curve is nonsingular, and the only singularities of the flecnodal curve are the double points (6).

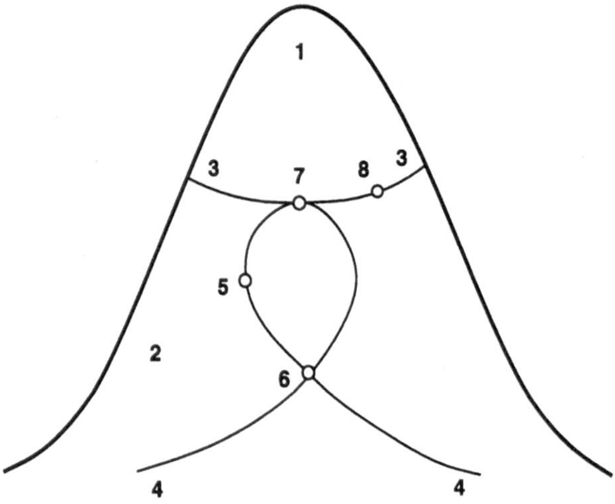

Fig. 4. Classification of points on a surface according to the orders of their tangent. (Adapted from Arnold's book [Arn84].)

Note that the definitions of these entities do not depend on the projection-genericicty of the surface, but their dimension and regularity do.

2.4 The visual events of projection-generic surfaces

We are now in a position to define the visual events of a projection-generic C^∞-surface. We first discuss the local visual events, which occur when the line of sight grazes the surface at a single point. Then we turn our attention to *multilocal* events, which occur when the line of sight has high-order contact with the surface in two or more distinct points. These events govern the changes in contour structure due to occlusion.

For now, we assume perspective projection, and we shall see at the end how to derive the visual events for orthographic projection from the perspective case.

Codimension 1 singularities. In [Ker81], it is proved that there is a fundamental relation between the points of the surface we have just defined and the line of sight that singularizes them, and it is showed that there are 3 one-codimensional local singularities (Figure 5). They are the *lip*, the *beak-to-beak*, and the *swallowtail* catastrophes (from the french *lèvre*, *bec-à-bec*, and *queue d'aronde*; these terms were coined by Thom).

In a lip transition (denoted L), a closed contour appears out of nowhere, with the formation of two cusps (Figure 5.a). In a beak-to-beak transition (denoted B), two distinct portions of the contour, each having a cusp, meet at a point in the image; after meeting, the contour splits into two new segments: the connectivity of the contour changes, and the two cusps disappear (Figure 5.b). Finally, during a swallowtail transition, a smooth image contour forms a singularity, and then breaks off into two cusps and a t-junction (Figure 5.c).

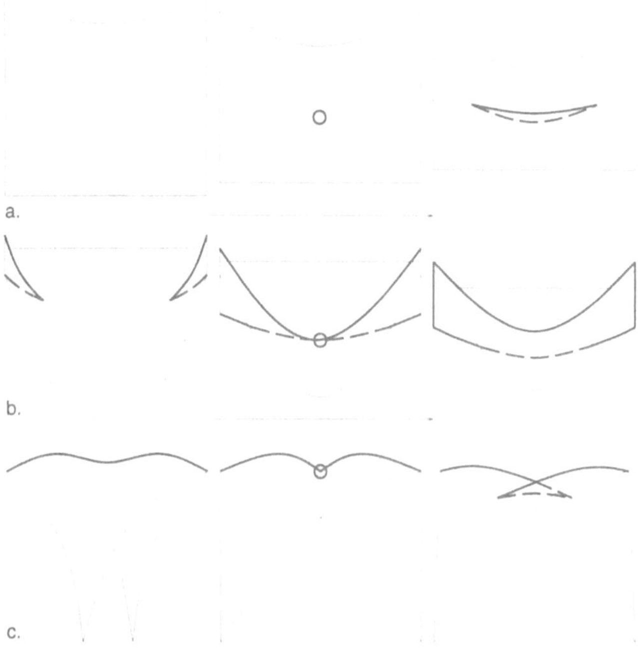

Fig. 5. Local events. a. Lip. b. Beak-to-beak. c. Swallowtail. The transitions take place from left to right, the singular points being indicated by small circles. AS before, the comtours actually seen by the observer are shown as thick lines, and their hidden parts are dashed.

A swallowtail occurs when the viewing direction is an asymptotic tangent at some point of a flecnodal curve of the surface; both beak-to-beak and lip events occur when the viewing direction is an asymptotic tangent at a parabolic point, the separation occurring at gutterpoints ([Arn84], [Koe86], [Koe90]). The asymptotic tangents at parabolic points form a ruled surface (the tangents are called the *generators* of *fibers*), which enjoys the further property that it is also *developable* (meaning that consecutive generators always intersect) [Wea27] page 44. The asymptotic tangents at flecnodal points form a *scroll*, i.e., a non-developable ruled surface.

We use the following notation to describe multilocal singularities. (It is consistent with the one used in [Ker81] and in [Rie90].) If F and C stand for fold and cusp, +, ++, +++, denote contact of order 1, 2, and 3 (or respectively *transverse superposition*, tangential contact, and inflectional contact), then the notation $F+F$ denotes the transverse superposition between two folds. In the case of a singular apparent contour (the tangent vector is not welldefined), the notation is an extension of the one for nonsingular contours. In this sense, the notation $F++C$ means a tangential contact between a fold and the *limiting tangent line* of a cusp.

There are three types of multilocal events of codimension 1 [Ker81]: *tangent crossing*, *cusp crossing*, and *triple point* (Figure 6). A tangent crossing (denoted $F++F$ or TC) occurs when two fold points project onto the same image point and share a common tangent (Figure 6.a). A cusp crossing ($F+C$ or CC for short) occurs when a cusp and a fold point project onto the same image point (Figure 6.b). Finally, a triple point ($F+F+F$ or TP) is formed when three fold points project transversally onto the same point (Figure 6.c).

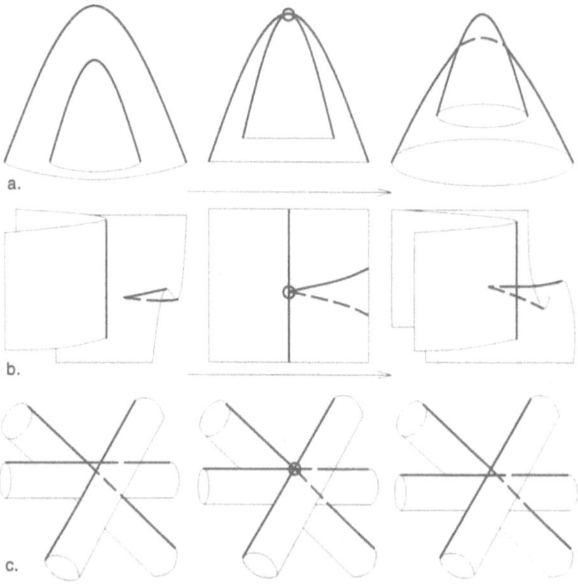

Fig. 6. Multilocal events. a. Tangent crossing. b. Cusp crossing. c. Triple point.

A 1-codimensional multilocal event occurs when the line of sight grazes the surface in n different points (n is equal to 2 or 3). It corresponds to a ruled surface in $\mathbb{P}_3$ obtained by sweeping the line of sight while maintaining n-point contact with X, this surface being tangent to X along a family of n curves, the corresponding viewing directions being the directions joining two of the contact points. The ruled surface generated by the rays singularizing tangent crossing points has the property that it is developable [Koe86].

Codimension 2 singularities. According to [Arn84], there are three kinds of local singularities of codimension 2. To see the first, one has to look at the surface from a generic point of the *cuspidal edge* (also called the *edge of regression*) of the parabolic developable (this surface is denoted by P_r). It is a *goose* singularity. The second of these is realized by projection from a generic point of an asymptotic line through a godron (called a *gull* or *ruffle*), and the third is realized by projection from a generic point of an asymptotic tangent through a biflecnode (a *butterfly*).

Kergosien gave a catalog of 2-codimensional multilocal singularities. They are: $F+++F$, $F+B$, $F+L$, $F+S$, $F++F+F$, $F+F+C$ (2 kinds), $F++C$, and $F+F+F+F$.

Codimension 3 singularities.. A catalog of local singularities of codimension 3 also exists. Each singularity corresponds to a discrete set of points in $\mathbb{P}^3$. According to [Rie90], there are five kinds of local 3-codimensional singularities, but only some of them can be realizedas central projections of a projection-generic rigid surface [Arn83]. The first is observed from special points of the asymptotic line through a godron. Another singularity is observed from a cusp of the cuspidal edge of P_r. And this edge goes to infinity when trying to observe a gutterpoint, since locally on the surface asymptotic directions are parallel. Finally, when looking at the surface from special points of an asymptotic line through a biflecnode another singularity is observed.

Not much is known about multilocal singularities of codimension 3. Only those of these singularities that are bilocal have been studied in [Rie88].

The orthographic case. The 1- and 2-codimensional visual events on the view sphere correspond to the visual event surfaces and curves defined above for perspective projection. These curves can be obtained by intersecting the partitioning of a three-dimensional viewpoint space by a sphere at infinity. Alternatively, the partitioning of the view sphere is the *spherical representation* of the partitioning obtained for perspective projection, defined as the arrangement of points in which radii through the center of the sphere and parallel to rays of the visual events meet the surface of the sphere.

2.5 Multiple-point theory and genericity considerations

Let X be a smooth m-dimensional variety in $\mathbb{P}^n$. Out interest is to determine the set of linear spaces having prescribed contact with X, and the loci of points where this contact takes place. To achieve this, we use modern *multiple-point theory* to give the derivations of several formulas for the classes of lines having specified contact with smooth projective surfaces.

The general approach was made precised by [Kle81], [Kle90] using the intersection theory of [Ful84]. He developed techniques to compute an enumerative formula for a class whose support is the locus of r-fold points of a suitably generic map. The technique was then enhanced by [Col86], [Col87] to derive formulas for the loci of *stationary multiple-points*.

The idea is to study the map $\bar{q} : F_X \to G$, where

$$F_X = \{(x, l) \in X \times G / x \in l \cap X\},$$

G being the Grassmannian of lines in $\mathbb{P}^n$. $\bar{q}$ is induced by projection onto the second factor. The dimension of F_X is $n + m - 1$ and that of G is $2n - 2$ so that the codimension of $\bar{q}$ is $n - m - 1$.

Now the construction is as follows. Let r be an integer, $\mathbf{a}$ a partition of r, meaning that $\mathbf{a} = (a_1, \ldots, a_k)$ is a k-uple of positive integers such that $a_1 \leq \ldots \leq a_k$ and $\sum a_j = r$. The singularity subscheme of stationary r-fold points, denoted $S_{\mathbf{a}}(q)$ consists of points (x_1, l_1) in F_X such that there are $r - 1$ other points $(x_2, l_2), \ldots, (x_r, l_r)$ of F_X all having the same image under $\bar{q}$ and also such that the first a_1 points lie infinitely near each other, the next a_2 points lie infinitely near each other, and so on. From this description of $S_{\mathbf{a}}$, Colley establishes the following result [Col86].

Proposition 2.1 *Let $\bar{q}$ be the projection onto G and $\bar{p}$ the projection onto X. Then*

- *The locus $\bar{q}(S_{\mathbf{a}})$ is the locus of lines in $\mathbb{P}^n$ incident with X in at least k points $x_1, \ldots, x_k$ with order of contact at x_i at least a_i.*
- *The locus $\bar{p}(S_{\mathbf{a}})$ is the locus of points $x_1, \ldots, x_k$ on X supporting lines l in $\mathbb{P}^n$ such that the order of contact of l at x_i is at least a_i.*

Colley defines an intersection-theoretic class $\mathbf{l}(\mathbf{a})$ for computing the cycle on $S_{\mathbf{a}}$, and develops a tecnique for computing $\mathbf{l}(\mathbf{a})$ in terms of the Chern classes of the normal sheaf of F_X in G. A shortcoming of the method is that this technique computes the class corresponding to ordered tuples rather than unordered ones, and thus $\mathbf{l}(\mathbf{a})$ is a computable constant times the cycle on $S_{\mathbf{a}}$. To get to the result for unordered tuples, one needs to evaluate this "overcount" factor, $p(\mathbf{a})$. Suppose we write $\mathbf{a}$ as follows:

$$\mathbf{a} = (a_1, \ldots, a_{q_1}, a_{q_1+1}, \ldots, a_{q_1+q_2}, \ldots, a_{q_1+\ldots+q_p})$$

where

$$a_1 = \cdots = a_{q_1} < a_{q_1+1} = \cdots = a_{q_1+q_2} < \cdots < a_{q_1+\ldots+q_{p-1}+1} = \cdots = a_{q_1+\ldots+q_p}$$

Suppose no we set $p(\mathbf{a}) = (q_1 - 1)!q_2! \cdots q_p!$. Then $\mathbf{l}(\mathbf{a})$ is $p(\mathbf{a})$ times the fundamental cycles on $S_{\mathbf{a}}$.

To be able to use Kleiman and Colley's theory on $\bar{q}$, one has to make an assumption about the variety X, namely that it is *generic*. The setup just defined allows to give a precise definition of what we mean by the genericity of a surface.

Definition 2.1 *A smooth m-dimensional variety X in $\mathbb{P}^n$ is generic if the closure of the locus of lines incident with X in k points $x_1, \ldots, x_k$ of X with order of contact a_i at x_i either is empty or has the expected codimension in G, namely $(r-1)(n-m-1) + \sum(a_i - 1)$, where $\mathbf{a} = (a_1, \ldots, a_k)$ is a partition of r.*

The reader might well wonder how genericity is related to projection-genericity. We will not however try to explore the interplays between the two notions. In deed, we will see in Section 6 how to get rid of the genericity assumption for all of our results. We will even see that projection-genericity can be raised for the formulas giving the degrees of the visual event surfaces.

3 The standard diagram

Having introduced the entities constituting an aspect graph of a smooth projection-generic surface, it is now time to turn our attention to the study of the visual events. Our main goal in this section will be to compute the Chern classes of the normal sheaf $\mathcal{N}_{F_X/G}$, from which we will be able to use Colley's enumerative formulas. Note that the path followed in this section is similar to that found in [leB78], [Kul83].

3.1 The setup

Let V be a vector space over $\mathbb{C}$, $\dim V = 4$, and let $\mathbb{P}^3 = \mathbb{P}(V)$ be the corresponding projective space. The investigation of the singularities of an embedding $X \subset \mathbb{P}^3$ is done in the domain of the *standard diagram*, consisting of the manifold F_3 of *flags* (a flag consists of a point A and a line l, such that $A \in l$), of the F_X manifold, and of the projections of F_3 and F_X onto the Grassmann manifold G of lines in $\mathbb{P}^3$. In what follows, we shall sometimes use the symbol $*$ for the class of a point. Note that both F_X and G are 4-dimensional, while F_3 has dimension 5.

Let p be the projection map of F_3 onto $\mathbb{P}^3$ and q the projection map of F_3 onto G. It is a classical fact [Ful84] that p canonically identifies F_3 with the projective space bundle $\mathbb{P}(\mathcal{T}_{\mathbb{P}^3})$, $\mathcal{T}_{\mathbb{P}^3}$ being the tangent sheaf to $\mathbb{P}^3$, and also that F_3 is isomorphic to $\mathbb{P}(T)$, where T is the tautological subbundle on G. Let us write $F_X = p^{-1}(X)$ and $\bar{q} = q|_{F_X}$. We have the following diagram:

$$\begin{array}{ccccc} & & F_3 & \xrightarrow{p} & \mathbb{P}^3 \\ & \stackrel{q}{\swarrow} & & & \\ G & & \cup i & & \cup j \\ & \stackrel{\nwarrow}{\bar{q}} & & & \\ & & F_X & \xrightarrow{\bar{p}} & X \end{array}$$

We thus have $j^* p_* = \bar{p}_* i^*$.

3.2 The Chow ring of a Grassmann manifold

Using simple arguments, the Chow ring $A(G)$ can be described completely with the help of Schubert cycles. Since we shall make use of these in the next sections, let us see how this turns out.

Let γ_1 be the cycle in G of lines meeting a given line, and γ_2 the cycle of lines contained in a given plane. We then have the following description of $A(G)$:

$$\begin{aligned} A^0(G) &= \mathbb{Z}G, \\ A^1(G) &= \mathbb{Z}\gamma_1, \\ A^2(G) &= \mathbb{Z}\gamma_1^2 \oplus \mathbb{Z}\gamma_2, \\ A^3(G) &= \mathbb{Z}\gamma_1\gamma_2, \\ A^4(G) &= \mathbb{Z} * . \end{aligned}$$

We have the relations $\gamma_1^3 = 2\gamma_1\gamma_2$, $\gamma_2^2 = \gamma_1^2\gamma_2 = *$, and $\gamma_1^4 = 2*$. Note that $\gamma_1^2 - \gamma_2$ is the cycle of lines passing through a given point. Now, let T denote again the tautological subbundle on G, and R the universal quotient bundle. Both have rank 2. Then we have the exact sequence:

$$0 \longrightarrow T \longrightarrow V_G \longrightarrow R \longrightarrow 0$$

where V_G is the trivial vector bundle of rank 4 on G. Then we have:

$$0 \longrightarrow \mathrm{Hom}(T,T) \longrightarrow \mathrm{Hom}(T,V_G) \longrightarrow \mathrm{Hom}(T,R) \longrightarrow 0$$

or equivalently, since $\mathcal{T}_G$ identifies with $\mathrm{Hom}(T,R)$,

$$0 \longrightarrow T \otimes \check{T} \longrightarrow 4\check{T} \longrightarrow \mathcal{T}_G \longrightarrow 0$$

$\check{T}$ being the dual of T. From this, and the fact that the Chern classes of $\check{T}$ are γ_1 and γ_2, we get the Chern classes of $\mathcal{T}_G$:

$$CP_t(\mathcal{T}_G) = 1 + 4\gamma_1 t + 7\gamma_1^2 t^2 + 12\gamma_1\gamma_2 t^3 + 6\gamma_2 t^4$$

where CP_t denotes the Chern polynomial.

3.3 F_3 as the projectivization of $\mathcal{T}_{\mathbb{P}^3}$

Let $\xi = c_1(\Theta_{F_3}(1))$, h a hyperplane class in $\mathbb{P}^3$, and H the pullback to F_3 of h. Then results found in [Gro58] allow to know the behaviour of the Chow ring $A(F_3)$ with respect to the direct image p_*. We obtain

$$p_*\xi^2 = 1_{\mathbb{P}^3}, \quad p_*\xi = 0, \quad \xi^3 + 4\xi^2 H + 6\xi H^2 + 4H^3 = 0.$$

This, in turn, means that

$$\begin{aligned} p_*(\xi^2 H) &= p_*(\xi^2 \cdot p^*h) = p_*\xi^2 \cdot h = h, \\ p_*(\xi H^2) &= p_*(\xi \cdot p^*h^2) = p_*\xi \cdot h^2 = 0, \\ p_*(H^3) &= p_*(p^*h^3) = p_*1 \cdot h^3 = 0, \end{aligned}$$

so finally $p_*(\xi^3) = -4h$.

Now writing down the two exact sequences for the projectivization [Kle76]

$$\begin{aligned} 0 \longrightarrow \mathcal{T}_{F_3/\mathbb{P}^3} \longrightarrow \mathcal{T}_{F_3} \longrightarrow p^*\mathcal{T}_{\mathbb{P}^3} \longrightarrow 0, \\ 0 \longrightarrow \Theta_{F_3} \longrightarrow (p^*\mathcal{T}_{\mathbb{P}^3})(1) \longrightarrow \mathcal{T}_{F_3/\mathbb{P}^3} \longrightarrow 0, \end{aligned}$$

we get that $CP_t(\mathcal{T}_{F_3}) = CP_t(p^*\mathcal{T}_{\mathbb{P}^3})CP_t((p^*\mathcal{T}_{\mathbb{P}^3})(1))$. From this we obtain the first Chern classes of $\mathcal{T}_{F_3}$:

$$\begin{aligned} c_1(\mathcal{T}_{F_3}) &= 8H + 3\xi, \\ c_2(\mathcal{T}_{F_3}) &= 28H^2 + 20H\xi + 3\xi^2, \\ c_3(\mathcal{T}_{F_3}) &= 52H^3 + 50H^2\xi + 12H\xi^2. \end{aligned}$$

3.4 F_3 as a projective bundle on G

The Chow ring $A(F_3)$ is generated by ξ and H. Thus, if γ_1' is the pullback to F_3 of the first Schubert cycle γ_1, we must have $\gamma_1' = \alpha\xi + \beta H$. From the Euler exact sequence and the fact that F_3 can be realized as the projectivization of a rank 2 vector bundle on G, we must have $c_2(\mathcal{T}_{F_3/G}) = 0$. Using the exact sequence

$$0 \longrightarrow \mathcal{T}_{F_3/G} \longrightarrow \mathcal{T}_{F_3} \longrightarrow \mathcal{T}_G \longrightarrow 0,$$

we obtain that

$$c_2(\mathcal{T}_{F_3/G}) = 3(3\alpha^2 - 4\alpha + 1)\xi^2 + 2(9\alpha\beta - 16\alpha - 6\beta + 10)\xi H + (9\beta^2 - 32\beta + 28)H^2,$$

which is identically zero when $\alpha = 1$ and $\beta = 2$. Thus $\xi = \gamma_1' - 2H$. Similarly, one gets that $\gamma_2' = \xi H$, where $\gamma_2'$is the pullback of γ_2.

We thus conclude that

$$CP_t(\mathcal{T}_{F_3/G}) = 1 + (3\xi + 8H - 4\gamma_1')t = 1 - \xi t,$$

which translates into

$$CP_t(\bar{q}^*\mathcal{T}_G - \mathcal{T}_{F_3}) = \frac{1}{CP_t(\mathcal{T}_{F_3/G})} = \sum_{i=0}^{5} \xi^i t^i = \sum_{i=0}^{5} (\gamma_1' - 2H)^i t^i.$$

3.5 The Chow ring of F_X

Let now E be the pullback to F_X of the class of a hyperplane section e of X. Let also Γ_1 and Γ_2 respectively be the pullbacks to F_X of γ_1' and γ_2'. We first have the relation that:

$$\Gamma_2 = k^*\gamma_2' = k^*(H\xi + H^2) = k^*(H\gamma_1' - H^2) = \Gamma_1 E - E^2,$$

from which we get that $\Gamma_1^3 = 2\Gamma_1\Gamma_2 = 2\Gamma_1 E(\Gamma_1 - E)$. From pulling back the analogous relation on $A(X)$, we get that $E^3 = 0$. And since $\bar{q}$ has degree d (X is cut in d points by a line in general position, which means that d points of F_X project onto a single point of G.) and codimension 0, we have that

$$\bar{q}_*(\Gamma_1^{4-2k}\Gamma_2^k) = \bar{q}_*\bar{q}^*(\gamma_1^{4-2k}\gamma_2^k) = \gamma_1^{4-2k}\gamma_2^k \cdot \bar{q}_*(1_{F_X}) = d\gamma_1^{4-2k}\gamma_2^k,$$

so $\Gamma_1^4 = 2d*$, $\Gamma_1^2\Gamma_2 = d*$, and $\Gamma_2^2 = d*$. Finally, $\Gamma_1^2 E^2 = \Gamma_2^2 = d*$ and $\Gamma_1^3 E = 2d*$. Now, for the projection onto X, we have that:

$$\begin{aligned}
&\bar{p}_* E^k = \bar{p}_*\Gamma_1 = 0,\\
&\bar{p}_*\Gamma_1^2 = \bar{p}_* i^*(H^2 + H\xi + \xi^2) = j^* p_*(H^2 + H\xi + \xi^2) = j^* 1_{\mathbb{P}_3} = 1_X,\\
&\bar{p}_*(\Gamma_1^2 E) = \bar{p}_*(\Gamma_1^2 \cdot \bar{p}_* e) = \bar{p}_*\Gamma_1^2 \cdot e = e,\\
&\bar{p}_*(\Gamma_1 E^2) = \bar{p}_*(\Gamma_1 \cdot \bar{p}_* e^2) = \bar{p}_*\Gamma_1 \cdot e^2 = 0,\\
&\bar{p}_*\Gamma_1^3 = \bar{p}_*(2\Gamma_1^2 E - 2\Gamma_1 E^2) = 2e.
\end{aligned}$$

F_X is the projectivization of the restriction to X of $\mathcal{T}_{\mathbb{P}^3}$, and thus since

$$\mathcal{N}_{F_X/G} = \bar{q}^*\mathcal{T}_G - \mathcal{T}_{F_X} = i^*(q^*\mathcal{T}_G - \mathcal{T}_{F_3}) + \mathcal{N}_{F_X/F_3}$$

and $CP_t(\mathcal{N}_{F_X/F_3}) = 1 + dEt$, we get that

$$c_k = c_k(\mathcal{N}_{F_X/G}) = (\Gamma_1 + (d-2)E)(\Gamma_1 - 2E)^{k-1},$$

for $k = 1, \ldots, 4$.

3.6 Defining three manifolds

Let $(l \cdot X)_x$ denote the intersection index of the line l with the surface X at point x. If l_1 and l_2 are asymptotic directions at point x, $n_i = (l_i \cdot X)_x$, the pairs of numbers (n_1, n_2) are as follows: $(3,4)$ for S, $(4,4)$ for $S_{4,4}$ and $(3,5)$ for S_5; $(l \cdot X)_x = 3$ for $x \in P$ and $(l \cdot X)_x = 4$ for $x \in P_4$.

Let A_2, A_3, and A_4 be the following subsets of F_X:

- $A_2 = \{(x,l) \in X \times G / l \text{ is tangent to } X \text{ at } x\}$,
- $A_3 = \{(x,l) \in X \times G / l \text{ is an asymptotic line to } X \text{ at } x\}$,
- $A_2 = \{(x,l) \in X \times G / x \in S, (l \cdot X)_x \geq 4\}$.

We let the inclusions into F_X respectively be α_2, α_3, and α_4. Then Kulikov [Kul83] proves the following result:

Proposition 3.1

$$\mathcal{N}_{A_2/F_X} = \alpha_2^* \mathcal{N}_{F_X/G}$$
$$\mathcal{N}_{A_2/F_X} = L_1|_{A_3} + \alpha_3^* \mathcal{N}_{F_X/G} \qquad \mathcal{N}_{A_2/F_X} = L_1^2|_{A_4} + \alpha_4^* \mathcal{N}_{F_X/G}$$

where $L_1 = \Theta_{\mathbb{P}(\mathcal{T}_{F_X})}(1)$.

In the next sections, we are also going to need the following result:

Proposition 3.2 *If* $\bar{c} \in A(A_k)$ *is the restriction of* $c \in A(\mathbb{P}(\mathcal{T}_{F_X}))$, *and* $\alpha_k \colon A_k \subset F_X$ *is an embedding, then*

$$\alpha_{k*}(\bar{c}) = (c \cdot (c_1 + \tau) \cdots (c_1 + (k-2)\tau) \cdot (c_1 \cdot \tau^3 + c_2 \cdot \tau^2 + \cdots + c_4))_3$$

$()_3$ *meaning the coefficient of* τ^3 *in the polynomial.*

4 Calculating the singularities

We are now in a position to compute the cycles corresponding to the special curves on the surface giving birth to the visual events whose knowledge is necessary to compute the aspect graph. Most of the results given here are classical [Sal15], [Bak33], through the framework in which all the derivations are done uses the modern language of cycles, Chern classes, and stationary multiple-point theory. We shall see how this framework can be used in correlation with the aspect graph problem.

4.1 Properties of the asymptotic tangents

Before going to the results announced, let us say a few words about classical facts concerning asymptotic tangents.

Proposition 4.1 *The locus of points on* X *at which the asymptotic tangents meet an arbitrary line in space is cut out by the cycle* $(3d-4)e$, *where* e *is the class of a hyperplane section. The number of asymptotic tangents through a given point in space is* $d(d-1)(d-2)$. *The number of asymptotic tangents contained in a plane section of the surface is* $3d(d-2)$.

Proof. In the Chow ring of F_X, Γ_1 is the cycle of couples (x, l) such that l meets a given line. And Γ_2 is the cycle of couples such that l is contained in an arbitrary plane. Then, the cycle in F_X corresponding to points having an asymptotic direction (contact of order 3) is given by $l(3)$ (lines tangent at three coinciding points). And the projection of $l(3)\cdot\Gamma_1$ onto X gives the locus of points on X at which the asymptotic tangents meet an arbitrary line. Colley's technique allows to compute $l(3)$ in terms of the Chern classes of the virtual normal sheaf $\mathcal{N}_{F_X/G}$, and we find $l(3) = c_1^2 + c_2$. Replacing in terms of Γ_1 and E, we get:

$$l(3) = 2\Gamma_1^2 + (3d-8)\Gamma_1 E + (d-2)(d-4)E^2,$$

then

$$l(3) \cdot \Gamma_1 = (3d-4)\Gamma_1^2 E + (d^2 - 6d + 4)\Gamma_1 E^2,$$

and consequently $\bar{p}_*(l(3) \cdot \Gamma_1) = (3d-4)e$.

The number of asymptotic tangents through a given point in space is $\int l(3) \cdot (\Gamma_1^2 - \Gamma_2)$, where $\int$ denotes the degree of the cycle, and using properties of Scubert cycles, we find that this number is $d(d-1)(d-2)$. The number of asymptotic directions contained in a given plane section is given by $\int l(3) \cdot \Gamma_2$, i.e., $3d(d-2)$ (it is also the number of inflections of a plane curve of degree d).

4.2 Parabolic points

Proposition 4.2 *The set of parabolic points P is cut out on X by the cycle $4(d-2)e$.*

Proof. The set of parabolic points is the ramification curve of the map between A_3 and X. Since A_3 is 2-dimensional, we have (see Kleiman's work [Kle76]) $[P] = \bar{p}_* \alpha_{3*} c_1(\mathcal{N}_{A_3/X})$. Let us first look at the corresponding cycle in F_X, and we call it pa. Let i_3 denote the inclusion of A_3 in A_2. From the exact sequences

$$0 \longrightarrow \mathcal{N}_{A_3/F_X} \longrightarrow \mathcal{N}_{A_3/X} \longrightarrow \alpha_3^* \mathcal{N}_{F_X/X} \longrightarrow 0,$$
$$0 \longrightarrow \mathcal{N}_{A_3/A_2} \longrightarrow \mathcal{N}_{A_3/F_X} \longrightarrow i_3^* \mathcal{N}_{A_2/F_X} \longrightarrow 0,$$

and from Proposition 3.1, we get

$$pa = \alpha_{3*}(c_1(L_1|_{A_3}) + \alpha_3^*(c_1(\mathcal{N}_{F_X/X}) + 2c_1(\mathcal{N}_{F_X/G}))).$$

Proposition 3.2 allows to calculate the first part:

$$\alpha_{3*} c_1(L_1|_{A_3}) = (\tau \cdot (c_1 + \tau) \cdot (c_1\tau^3 + c_2\tau^2 + c_3\tau + c_4))_3 = c_1c_2 + c_3.$$

From previous results, we know that the fundamental cycle of A_3 in F_X is $l(3)$, so

$$pa = c_1c_2 + c_3 + (c_1(\mathcal{N}_{F_X/X}) + 2c_1) \cdot l(3),$$

from which we finally get that

$$pa = (d-2)(4\Gamma_1^2 E + 2(3d-8)\Gamma_1 E^2).$$

Pushing forward to X, we get $[P] = \bar{p}_* pa = 4(d-2)e$. (Otherwise said, the degree of the parabolic curve is obtained as $\int pa \cdot E = 4d(d-2)$.)

Proposition 4.3 *The ruled surface P_r generated by the asymptotic tangents at parabolic points has degree $2d(d-2)(3d-4)$.*

Proof. The degree of a surface in $\mathbb{P}^3$ is the number of its intersections with a line in general position. Now, a ruled surface is formed by it sgenerators, which means that the degree of a ruled surface is the number of its generators meeting a general line in space.

Thus, this degree can be computed by finding the number of these asymptotic tangents at parabolic points meeting a line in general position in $\mathbb{P}^3$. But this is simply

$$\int \bar{q}_* pa \cdot \gamma_1 = \int \bar{q}_*(pa \cdot \bar{q}^* \gamma_1) = \int pa \cdot \Gamma_1 = 2d(d-2)(3d-4).$$

4.3 Flecnodal points

Proposition 4.4 *The flecnodal curve S on a projection-generic surface of degree d is cut out by the cycle $(11d-24)e$.*

Proof. The locus of flecnodal points is given by the projection onto X of the class $l(4)$ (lines tangent at four coinciding points). Again, Colley's technique gives us $l(4) = c_1^3 + 3c_1c_2 + 2c_3$, or equivalently

$$l(4) = (11d-24)E\Gamma_1^2 + 2(3d^2 - 22d + 30)E^2\Gamma_1.$$

and we have that $[S] = \bar{p}_* l(4) = (11d-24)e$.

Proposition 4.5 *The degree of the flecnodal scroll S_r is $2d(d-3)(3d-2)$.*

Proof. Indeed the number of tangents having fourth-order contact with the surface and meeting a general line in space is given by the degree of $l(4) \cdot \Gamma_1 = 2d(d-3)3d-2)*$, so the degree of the ruled surface is

$$\int l(4) \cdot \Gamma_1 = 2d(d-3)(3d-2).$$

Proposition 4.6 *The number of biflecnodes on the surface is $\#S_5 = 5d(d-4)(7d-12)$.*

Proof. The locus of biflecnodes is given by the projection onto X of the class $l(5)$, and we have $l(5) = c_1^4 + 6c_1^2c_2 + c_2^2 + 10c_1c_3 + 6c_4$, and carrying the calculations $l(5) = 5d(d-4)(7d-12)*$, which gives for the projection

$$[S_5] = \bar{p}_* l(5) = 5d(d-4)(7d-12)*.$$

Proposition 4.7 *The number of godrons on the surface is $\#P_4 = 2d(d-2)(11d-24)$.*

Proof. Indeed, the godrons are the points of tangency of parabolic and flecnodal curves, which means in terms of cycles that the cycle $2[P_4]$ is cut out on X by the product of the cycles $[P]$ and $[S]$, so $\#P_4 = 2d(d-2)(11d-24)$.

Proposition 4.8 *The set of points of inflection of both asymptotic directions on X is cut out by the cycle $5(7d^2 - 28d + 30)e^2$, and so the number of such points is $\#S_{4,4} = 5d(7d^2 - 28d + 30)$.*

Proof. Indeed, the arithmetic genus of S is given by the adjunction formula:

$$2p_a(S) - 2 = \int [S]([S] + [K]),$$

where $[K] = (d - 4)e$, K being the canonical divisor on X, so $p_a(S) = 1 + 2d(3d - 7)(11d - 24)$. But only the singularities of S are the ordinary double points $S_{4,4}$, and A_4 can be viewed as the normalization of S (see [Kul83]), so we have $\#S_{4,4} = p_a(S) - g(A_4)$. Now, to compute the genus of A_4, we need the degree of its first Chern class. Let v be the push forward of this class to F_X. Then $v = \alpha_{4*}c_1(\mathcal{T}_{A_4})$ or

$$v = \alpha_{4*}(\alpha_4^* c_1(\mathcal{T}_{F_X}) - c_1(\mathcal{N}_{A_4/A_3}) - i_4^* c_1(\mathcal{N}_{A_3/A_2}) - \alpha_4^* c_1(\mathcal{N}_{F_X/G})),$$

which turns out to be ($l(4)$ is the fundamental cycle of A_4 in F_X)

$$v = (c_1(\mathcal{T}_{F_X}) - 3c_1) \cdot l(4) - 3(\tau \cdot (c_1 + \tau) \cdot (c_1 + 2\tau) \cdot (c_1\tau^3 + c_2\tau^2 + c_3\tau + c_4))_3,$$

and finally $v = (-62d^2 + 316d - 372)\Gamma_1^2 E^2$. Then the genus of A_4 is given by:

$$g(A_4) = 1 - \frac{1}{2}\int \bar{p}_* v = 1 + 186d - 158d^2 + 31d^3,$$

and we obtain that

$$\#S_{4,4} = p_a(S) - g(A_4) = 5d(7d^2 - 28d + 30).$$

4.4 Triple points

Proposition 4.9 *The locus of the points of contact of lines being ordinary tangents at three different places on the surface is cut out by the cycle $[TP] = \frac{1}{2}(d - 2)(d - 4)(d - 5)(d^2 + 5d + 12)e$.*

Proof. To study triple points, we use the formula for the lines being ordinary tangential at three different points of the surface, i.e., $l(2, 2, 2)$. After some very lengthly computations, we get that:

$$l(2, 2, 2) = c_1\bar{q}^*\bar{q}_*(c_1\bar{q}^*\bar{q}_*c_1 - 4c_1^2 - 2c_2) - 2(4c_1^2 + c_2)\bar{q}^*\bar{q}_*c_1 + 42c_1^3 + 54c_1c_2 + 24c_3.$$

Replacing with Γ_1 and E, we obtain:

$$l(2, 2, 2) = (d - 4)(d - 5)((d - 2)(d^2 + 5d + 12)\Gamma_1^2 E + 2(d + 6)(5 - 3d)\Gamma_1 E^2),$$

and we obtain the desired result by noting that the overcount factor in this case is $p(2, 2, 2) = 2$.

Proposition 4.10 *The degree of the triple point scroll TP_r is $\frac{1}{3}d(d-3)(d-4)(d-5)(d^2+3d-2)$.*

Proof. The degree of this surface is the number of these lines meeting a general line in space. We have:

$$l(2,2,2)\cdot\Gamma_1 = 2d(d-3)(d-4)(d-5)(d^2+3d-2)* .$$

Now, this is the number of such lines for ordered triplets of points. For unordered ones, we have to divide by 6 to get the result.

4.5 Cusp crossings

Proposition 4.11 *The locus of points having an ordinary tangent that is asymptotic at some other place of the surface is cut out on X by the cycle $(d-2)(d-4)(d^2+2d+12)e$*

Proof. Now, to study cusp crossings, we use the formula for the lines being ordinary tangential at one point and asymptotic at another point. i.e.. $l(2,3)$. From the iterative scheme of Colley, we get that:

$$l(2,3) = c_1\bar{q}^*\bar{q}_*(c_1^2+c_2) - 6c_1^3 - 12c_1c_2 - 6c_3.$$

In terms of Γ_1 and E, we have:

$$l(2,3) = (d-4)((d-2)(d^2+2d+12)\Gamma_1^2E + (-d^3+3d^2-38d+60)\Gamma_1E^2).$$

and we get the result, because there is no overcount in this case.

Proposition 4.12 *The degree of the ruled surface CC_r generated by the singular directions at cusp crossings is $d(d-3)(d-4)(d^2+6d-4)$.*

Proof. Indeed, the number of these lines meeting a general line in space is $l(2,3)\cdot\Gamma_1 = d(d-3)(d-4)(3d^2+6d-4)*$, and thus the result.

Proposition 4.13 *The locus of points having an asymptotic tangent that it is ordinary tangent at some other place is cut out on X by the cycle $(d-4)(3d^2+5d-24)e$.*

Proof. Indeed, it seems reasonable to think that we should be able to find a kind of "$l(3,2)$" formula, though the scheme developed by Colley does not allow such a construction. However, using the symmetry observed in the stationary multiple-point formulas, we can state that:

$$l(3,2) = (c_1^2+c_2)\bar{q}^*\bar{q}_*c_1 + \alpha c_1^3 + \beta c_1c_2 + \gamma c_3.$$

And using the fact that $l(3,2)\cdot\Gamma_1 = l(2,3)\cdot\Gamma_1$, we get that:

$$l(3,2) = (c_1^2+c_2)\bar{q}^*\bar{q}_*c_1 - 6c_1^3 - 12c_1c_2 - 6c_3,$$

$$\text{or}\quad l(3,2) = (d-4)((3d^2+5d-24)\Gamma_1^2E + (d^3-3d^2-32d+60)\Gamma_1E^2).$$

and thus the result.

4.6 Tangent crossings

Proposition 4.14 *The locus of points corresponding to a tangent crossing is cut out on X by the cycle $[TC] = (d-2)(d^3 - d^2 + d - 12)e$.*

Proof. Let $\check{X}$ denote the *dual surface*, i.e., the closure of the set of tangent planes to X. X being generic, $\check{X}$ is two-dimensional and therefore is a surface in $\check{\mathbb{P}}^3$, the projective space of planes in $\mathbb{P}^3$. The morphism $\check{\pi}: X \to \check{\mathbb{P}}^3$ which sends a point of X to its tangent plane is called the *dual map*. Note that if X is of degree d, one can show that $\check{X}$ is of degree $d(d-1)^2$ [Pie77], [Ful84]. This number also refers to the class of X, i.e., the number of tangent planes through an arbitrary point.

Let us use the following notation: $h = c_1(\Theta_{\mathbb{P}^3}(1))$, $\check{h} = c_1(\Theta_{\check{\mathbb{P}}^3}(1))$,, e the pullback to X of h, and $\check{e}$ the pullback to X of $\check{h}$. Now, using the facts that $c_1(\mathcal{T}_X) = (4-d)e$ and $c_2(\mathcal{T}_X) = (d^2 - 4d + 6)e^2$, and in view of the exact sequence

$$0 \longrightarrow \mathcal{T}_X \longrightarrow \check{\pi}^* \mathcal{T}_{\check{\mathbb{P}}^3} \longrightarrow \mathcal{N}_{X/\check{\mathbb{P}}^3} \longrightarrow 0,$$

we get that $c_1(\mathcal{N}_{X/\check{\mathbb{P}}^3}) = 4\check{e} - (4-d)e$, and

$$c_2(\mathcal{N}_{X/\check{\mathbb{P}}^3}) = 6\check{e}^2 - 4(4-d)\check{e} \cdot e + 2(5-2d)e^2.$$

Now, $A^1(X)$ is generated by e, so we must have $\check{e} = \alpha e$. But on the one hand

$$\deg_{\check{\mathbb{P}}^3} \check{e} = \int \check{e}^2 = \alpha^2 \int e^2 = \alpha^2 d,$$

and on the other hand $\deg_{\check{\mathbb{P}}^3} \check{e} = d(d-1)^2$, so we get that $\check{e} = (d-1)e$. Thus the Chern classes of $\mathcal{N}$ are $c_1(\mathcal{N}_{X/\check{\mathbb{P}}^3}) = (5d-8)e$, and $c_2(\mathcal{N}_{X/\check{\mathbb{P}}^3}) = 2(d-2)(5d-8)e^2$.

Now, the 1-dimensional singular set M_2 of $\check{\pi}$ is the set of pairs of points on X projecting onto the same point on $\check{X}$. (As such, it is given by Kleiman's technique as $l(1,1)$.) We have $[M_2] = \check{\pi}^* \check{\pi}_* [X] - c_1(\mathcal{N}_{X/\check{\mathbb{P}}^3})$, or equivalently, since $\check{\pi}_*[X] = d(d-1)^2 \check{h}$,

$$[M_2] = (d-2)(d^3 - d^2 + d - 4)e$$

Let $[D]$ be the cycle on $\check{X}$ corresponding to its 1-dimensional singular set. We have to be careful here since the ramification locus of $\check{\pi}$ is not finite [Pie77], which means that the 1-dimensional part of the singular set of $\check{X}$ is made of a cuspidal curve in addition to a nodal curve. Let $[D] = [D_b] + [D_c]$, where $D_b]$ is the cycle corresponding to the nodal curve, and $[D_c]$ the cycle associated to the cuspidal curve. This means that we have $[M_2] = [\Gamma_b] + 2[\Gamma_c]$. The cuspidal curve is generated by the parabolic set on X (the tangent plane is stationary when crossing the parabolic curve), and thus $[\Gamma_c] = [P] = 4(d-2)e$, so we can compute $[\Gamma_b] = [TC]$, the set of points corresponding to a visual event of type tangent crossing.

Proposition 4.15 *The curve TC has a cusps and b double points, where a and b are given by:*

$$a = 4d(d-2)(d-3)(d^3+3d-16)$$
$$b = \tfrac{1}{2}d(d-2)(d^7-4d^6+7d^5-45d^4+114d^3-111d^2+548d-960).$$

Proof. According to the geometric descriptions of [Sal15], pages 300–306, the intersections of D_b and $D_c$are of two types: The first yields a cusp of D_b and comes either from a transversal intersection between Γ_b and Γ_c or from a cusp of Γ_b, the second yields a cusp on D_c and comes from a tangential contact between Γ_b and Γ_c.

From this, it follows that Γ_b has cusps (let us say a of them) and ordinary double points (let us say b of them). And D_b will have a cusps and $b/3$ triple points. Now, the number of transversal intersections between Γ_b and Γ_c is a, while the tangential contact occurs at the godrons that we have already encountered (see [Pie77]). Thus:

$$\int [\Gamma_b] \cdot [\Gamma_c] = a + 2\beta,$$

so that with $\beta = \int [P_4]$, we have $a = 4d(d-2)(d-3)(d^3+3d-16)$.

Now, the set of triple points of $\check{\pi}$ is given by Kleiman's technique as $l(1,1,1)$ for this map, and the formula is

$$[M_3] = \check{\pi}^*\check{\pi}_*[M_2] - 2c_1(\mathcal{N}_{X/\check{\mathbb{P}}^3}) \cdot [M_2] + 2c_2(\mathcal{N}_{X/\check{\mathbb{P}}^3}),$$

and if t denotes the number of triple points of the dual map, then we have $6t = \int [M_3]$, and finally

$$t = \frac{1}{6}d(d-2)(d^7-4d^6+7d^5-21d^4+52d^3-39d^2+80d-96),$$

which means that the total number of triple points of $\check{X}$ is t. Now, the triple points on $\check{X}$ come from triple points on D_b $(b/3)$, from cusps on D_b (a), and from cusps on D_c (β), because again, the ramification locus is not finite. From this we get the relation

$$t = \frac{b}{3} + a + \beta,$$

or $b = 3(t - a - \beta)$, which gives

$$b = \frac{1}{2}d(d-2)(d^7-4d^6+7d^5-45d^4+114d^3-111d^2+548d-960).$$

Proposition 4.16 *The degree of the developable surface TC_r is $d(d-2)(d-3)(d^2+2d-4)$.*

Proof. A point of TC on X projects onto a point of D_b on $\check{X}$. Thus the number of lines generating TC_r and meeting a general line in space (i.e., the degree of TC_r) is the number of tangents to D_b meeting a general line in space. This number has been known classically as the rank of D_b, and it is given by the following formula [SR49], page 84:

$$q = 2\deg(D_b) + 2g(D_b) - 2 - a,$$

where a is the number of cusps on this curve. The degree of the nodal curve D_b is given by (TC is a 2-to-1 covering of D_b):

$$\begin{aligned} 2\deg(D_b) &= \int [TC]\cdot\check{e} = \int \check{\pi}_*([TC]\cdot\check{\pi}^*\check{h}) = \int \check{\pi}_*[TC]\cdot\check{h} = \\ &= (d-2)(d^3-d^2+d-12)\int \check{\pi}_* e\cdot\check{h}, \end{aligned}$$

and since

$$\int \check{\pi}_*\check{e}\cdot\check{h} = \int \check{e}^2 = d(d-1)^2$$

on the one hand, and

$$\int \check{\pi}_*\check{e}\cdot\check{h} = (d-1)int\check{\pi}_* e\cdot\check{h}$$

on the other hand, we get that:

$$\deg(D_b) = \frac{1}{2}d(d-1)(d-2)(d^3-d^2+d-12).$$

(Note that $\deg(D_b)$ is also the number of planes bitangent to X through a general point in space.)

Now we need to compute the genus of D_b. We relate it to the genus of Γ_b by the Riemann-Hurwitz formula [Har77], page 301, applied to the map $\check{\pi}_{\Gamma_b'} : \Gamma_b' \to D_b'$ between the normalization of Γ_b and D_b:

$$2(2g(D_b)-2) + \lambda = 2g(\Gamma_b) - 2,$$

λ being the number of ramification points of $\check{\pi}_{\Gamma_b'}$, i.e., β. Now, by the adjunction formula, we have that:

$$2p_a(\Gamma_b) - 2 = [\Gamma_b]\cdot([K] + [\Gamma_b]),$$

and we know that Γ_b has a cusps and b double points, so $g(\Gamma_b) = p_a(\Gamma_b) - a - b$. Putting everything together, we obtain the genus of D_b, and consequently its rank. We get:

$$q = d(d-2)(d-3)(d^2+2d-4).$$

5 Projective characters of visual event surfaces

To this point, we have shown how the properties of the curves on the surface giving birth to the visual event surfaces can be investigated. We have also obtained the degree of these surfaces. We now turn our attention to the study of the elementary projective characters of the visual event surfaces, and their singularities, and we show how these characters can be computed using those of the curves on which they are based. We start by inferring properties that are true for all kinds of ruled surfaces. We then move to study independently developable surfaces and scrolls, and show how the results apply to the study of the parabolic developable.

5.1 Ruled surfaces

The term *ruled surfaces* refers to two different entities. First to the bundle $Z = \mathbb{P}(\mathcal{E})$ of projective lines over a projective curve C (which we assume nonsingular for now) of genus g associated to a locally free sheaf $\mathcal{E}$ of rank 2. Second, the classical meaning refers to the image R of Z in $\mathbb{P}^3$ under a mapping ϕ with the following two properties: the restriction $Z \to R = \phi(Z)$ is birational and ϕ carries the fibers of Z onto lines in $\mathbb{P}^3$. Such a ϕ can be obtained as follows: first embed Z in a high-dimensional projective space so that the fibers of Z become lines and then project Z from a center in general position ([Kle76]).

Preliminary results. Now, let σ be the class of a hyperplane section of Z. Let $m : Z \to C$ be the projective morphism, $F = m^{-1}(x)$ be a *fiber* with $x \in C$ and let f be its class in $A^1(Z)$. Then one can show that $A^1(Z)$ is generated by f and σ, and that $f^2 = 0$ and $\int f\sigma = 1$ [Bea78]. From this, we have:

$$m^* c_1(\mathcal{E}) = -\mu_0 f, \qquad m^* c_1(\mathcal{T}_C) = (2 - 2g)f.$$

We make again use of results in [Gro58] to infer properties about the behaviour of the ring $A(Z)$. This gives $\sigma^2 - \mu_0 f\sigma = 0$, and $m_*\sigma = 1_C$.

Now, we construct the Chern classes of Z from thos of C and $\mathcal{E}$ with the following two exact sequences:

$$0 \longrightarrow \mathcal{T}_{Z/C} \longrightarrow \mathcal{T}_Z \longrightarrow m^*\mathcal{T}_C \longrightarrow 0,$$
$$0 \longrightarrow \Theta_Z \longrightarrow (m^*\mathcal{E})(1) \longrightarrow \mathcal{T}_{Z/C} \longrightarrow 0,$$

and carrying the calculations yields

$$CP_t(\mathcal{T}_Z) = 1 + (2\sigma + (2 - 2g - \mu_0)f)t + (4 - 4g) * t^2.$$

Thus if K_Z denotes the canonical divisor on Z, we have $[K_Z] = -2\sigma + (2g - 2 - \mu_0)f$ and

$$\int K_Z^2 = 4\int \sigma^2 + 4(2 - 2g - \mu_0) = 8(1 - g).$$

If π denotes the sectional genus of R, then we have $2\pi - 2 = \sigma \cdot ([K_Z] + \sigma)$, so that we get $\pi = g$.

Multiple points. To study the set of double points of ϕ, we first need the Chern classes of the virtual normal sheaf. From the exact sequence

$$0 \longrightarrow \mathcal{T}_Z \longrightarrow \phi^*\mathcal{T}_{\mathbb{P}^3} \longrightarrow \mathcal{N}_{Z/\mathbb{P}^3} \longrightarrow 0,$$

we get that

$$CP_t(\mathcal{N}_{Z/\mathbb{P}^3}) = 1 + (2\sigma + (2g - 2 + \mu_0)f)t) + (4g - 4 + 2\mu_0) * t^2.$$

The 1-dimensional singular set of R is made of a nodal curve D_b and a possible cuspidal curve D_c. Let us call it W. Pulling back $[W]$ to Z (see [Pie77]), we get that $[M_2] = [\Gamma_b] + 2[\Gamma_c]$ (where Γ_b and Γ_c are the curves on Z corresponding to D_b and D_c). But Kleiman's work allows to relate the class of $[M_2]$, the 1-dimensional singular set of ϕ, to invariants of the mappings as follows:

$$[M_2] = \phi^*\phi_*[Z] - c_1(\mathcal{N}_{Z/\mathbb{P}^3}),$$

so that we get $[M_2] = (\mu_0 - 2)\sigma + (2 - 2g - \mu_0)f$. Since the degree b of the nodal curve D_b is given by $2b = \int[\Gamma_b] \cdot \sigma$ and the degree c of the (possible) cuspidal curve D_c is $c = \int[\Gamma_c] \cdot \sigma$, we get that:

$$b + c = \frac{1}{2}(\mu_0 - 1)(\mu_0 - 2) - g.$$

The set of triple points of the mapping is given by:

$$[M_3] = \phi^*\phi_*[M_2] - 2c_1(\mathcal{N}_{Z/\mathbb{P}^3}) \cdot [M_2] + 2c_2(\mathcal{N}_{Z/\mathbb{P}^3}).$$

Then, if $6k = \int[M_3]$, we have:

$$k = \frac{1}{6}(\mu_0 - 4)((\mu_0 - 3)(\mu_0 - 2) - 6g).$$

Curves on a ruled surface. Another interesting fact that we can obtain is how to relate the genus of a curve drawn on R to that of the base curve C. Let M be a curve of multiplicity s (for instance, D_b has multiplicity 2), of degree v on R which meets each generator in w points. Let M' be the corresponding curve on Z, of degree v' and meeting each generator in w' points (a (v', w') curve for short). Then we have $v' = \int[M'] \cdot \sigma = sv$ and $w' = \int[M'] \cdot f$, and the cycle $[M']$ in $A^1(Z)$ is given by $[M'] = w'\sigma + (v' - \mu_0 w')f$. We first can obtain the number of intersections of a (v'_1, w'_1) curve M'_1 with a (v'_2, w'_2) curve M'_2 on Z. We have:

$$[M'_1] \cdot [M'_2] = v'_1 w'_2 + v'_2 w'_1 - \mu_0 w'_1 w'_2.$$

One can also relate the genus of M to that of C. Using the adjunction formula gives the arithmetic genus of M':

$$2p_a(M') - 2 = \int [M'] \cdot ([M'] + [K_Z]) = w'(2g - 2) + (2v' - \mu_0 w')(w' - 1).$$

This is indeed a formulation of the Riemann-Hurwitz theorem, and the quantity $\delta = (2v' - \mu_0 w')(w' - 1)$ is the number of generators singular for the mapping $M' \to C$ (C meant has a curve on Z), i.e., the number of generators touching the curve M' (equivalently, the number of generators whose contact with M' is other than tranversal in at least one point). Now, we relate the arithmetic genus of M' to its genus by noting the kind of singularities this curve has, and we relate the genus of M to that of M' through another use of the Riemann-Hurwitz theorem. We shall see in a moment how this applies to the computation of the genus of the nodal curve.

Let us now turn our attention to the definition of the class of immersion of a curve in a surface. Let $k : \bar{Z} \to Z$ be the blowup of the ramification ideal of ϕ, and $B \subseteq \bar{Z}$ the exceptional divisor. Let $\breve{\sigma}$ be the pullback to Z of the dual hyperplane class, let $v = k^*\sigma$ and $\breve{v} = k^*\breve{\sigma}$. Then [Pie77] shows that $\breve{v} = (\mu_0 - 1)v - [k^{-1}\Gamma_b] - 3[k^{-1}\Gamma_c] - [B]$, and that

$$\nu_2 = \int \breve{v} \cdot [B] = \int c_2(\mathcal{N}_{Z/\mathbb{P}^3}) = 2(\mu_0 + 2g - 2).$$

We define the *class of immersion* $\rho(A)$ of a curve A in R to be the number of planes tangent to R at points of A and going through a general point in space. As such, if A' is the curve on Z corresponding to A, it is given by

$$\rho(A) = \int \breve{v} \cdot [k^{-1}A'].$$

If A' is not singular, then:

$$\rho(A) = \int \breve{\sigma} \cdot [A'].$$

Intersection of two ruled surfaces. Let R_1 and R_2 be two ruled surfaces intersecting transversally along I. Suppose $[I_1]$ (resp. $[I_2]$) is the cycle in $A(Z_1)$ (resp. $A(Z_2)$) corresponding to I. Then $[I_1] = w_1\sigma_1 + (v_1 - \mu_0(R_1)w_1)f_1$, $[I_2] = w_2\sigma_2 + (v_2 - \mu_0(R_2)w_2)f_2$, and $v_1 = v_2 = \mu_0(R_1)\mu_0(R_2)$. Now, it is easy to realize that $w_1 = \mu_0(R_2)$ and $w_2 = \mu_0(R_1)$, so that $[I_1] = \mu_0(R_2)\sigma_1$ and $[I_2] = \mu_0(R_1)\sigma_2$. This means that I has

$$\int [I_1] \cdot [\Gamma_b(Z_1)] + \int [I_2] \cdot [\Gamma_b(Z_2)] = b_1\mu_0(R_2) + b_2\mu_0(R_1)$$

and

$$\int [I_1] \cdot [\Gamma_c(Z_1)] + \int [I_2] \cdot [\Gamma_c(Z_2)] = c_1\mu_0(R_2) + c_2\mu_0(R_1)$$

cusps.

5.2 Developable surfaces

A simply infinite algebraic system Δ of planes (dual of a curve, where duality has the same meaning as in Section 4.6) is called a *developable of planes.* For such a surface, it is convenient to introduce certain terms to represent notions derived by duality form a curve B. We shall call the line of intersection of two consecutive planes of the system a *focal line* or *generator* (it is the dual of a tangent line of B). A *focal point* of Δ is a point of intersection of three consecutive planes of the system (it is the dual of an osculating plane of B).

Any space curve C defines an *osculating developable*, generated by its osculating planes. And dually, any developable Δ has a *cuspidal edge*, locus of its focal points. We have the result that the osculating developable of the cuspidal edge of any developable Δ is Δ itself; [SR49] page 86. Let Δ be the osculating developable of C. Then by duality we have that the number of focal lines of Δ meeting a general line in space is the number of tangent lines to C meeting a general line in space, i.e., the rank r of C.

Now, we shall call a *developable surface* the set of generators of a developable of planes. By the definition of an edge of regression, we see that a developable of planes and its corresponding developable surface have the same edge of regression. Let D be the developable visual event surface constructed on the curve C on X. D enjoys the property that its tangent plane is constant along each generator, which means that the number of tangent planes to D passing through a general point in space, i.e., its rank μ_1, is equal to the number of planes tangent to X at points of C through a general point in space, i.e., $\rho(C)$ the class of immersion of C in X.

Let us assume C has genus g. Let D_c be the cuspidal curve on C, c its degree, r its rank, κ the number of cusps on D_c. Every generator of D is tangent to D_c ([Koe86], page 272), and so the points of C and D_c can be put in $(1,1)$ algebraic correspondence, which in turn means that the genus of the curves are equal (set $w = 1$ in the formula seen in the previous section). We thus have the following classical relation for the rank of D_c:

$$r = 2c + 2g - 2 - \kappa.$$

But now, we can think of D as the set of focal lines of the osculating developable Δ of D_c, in which case r is the number of generators of Δ meeting a general line in space, i.e., the degree μ_0 of D.

The class of D_c can be expresed by $n' = 3c + 6g - 6 - 2\kappa$, and this number refers to the number of its osculating planes meeting a general point in space, i.e., the number of planes in Δ meeting a general point in space, i.e., μ_1. Summarizing, we have

$$\kappa - 2c = 2g - 2 - \mu_0, \qquad 2\kappa - 3c = 6g - 6 - \mu_1.$$

Solving this system yields $c = 2\mu_0 - \mu_1 + 2g - 2$ for the degree of the cuspidal curve and $\kappa = 3\mu_0 - 2\mu_1 + 6g - 6$ for its number of cusps. And again the rank

μ_1 is computed as the class of immersion of C in X. Note that the number of tangent planes to D containing an arbitrary line (the class μ_2 of D) is equal to the number of planes tangent to X at points of C through a line, and this number is generically zero, i.e., $\mu_2 = 0$.

Now, one can express $[\Gamma_c]$ in terms of σ and f. Let us write $[\Gamma_c] = \alpha\sigma + \beta f$. Taking into account that $c = \int[\Gamma_c]\cdot\sigma = \alpha\mu_0 + \beta$ and that $\int[\Gamma_c]\cdot f = 1$, which gives $\alpha = 1$, we obtain $[\Gamma_c] = \sigma + (\mu_0 - \mu_1 + 2g - 2)f$. And since $[\Gamma_b] = [M_2] - 2[\Gamma_c]$, we have $[\Gamma_b] = (\mu_0 - 4)\sigma + (-3\mu_0 + 2\mu_1 - 6g + 6)f$. And b is given by:

$$b = \frac{1}{2}\int[\Gamma_b]\cdot\sigma = \frac{1}{2}\mu_0(\mu_0 - 7) + \mu_1 + 3 - 3g.$$

Note here that μ_0, μ_1, b, and c are linked by the following formula:

$$\mu_1 = \mu_0(\mu_0 - 1) - 2b - 3c.$$

Since $\int[\Gamma_b]\cdot f = \mu_0 - 4$, we infer that the nodal curve D_b meets each fiber in $\mu_0 - 4$ points.

Γ_c is a $(c, 1)$ curve on Z, which gives $p_a(\Gamma_c) = g$. Since $p_a(\Gamma_c) \geq g(\Gamma_c)$ and by Riemann-Hurwitz, $g(\Gamma_c) \geq g(D_c) = g$, we get that Γ_c is nonsingular and the map between the normalizations of Γ_c and D_c is unramified.

The intersections of D_b and D_c are in number (expressed in terms of b, c, and μ_0)

$$i = [\Gamma_b]\cdot[\Gamma_c] = 2b + (c - \mu_0)(\mu_0 - 4).$$

and these are counted with proper multiplicity. Now, D_b has κ' cusps (at its transversal intersections with D_c) and these two curves further intersect tangentially at the cusps of D_c (i.e., the tangent to D_b is the limiting tangent line to the cusp). This means that ($\kappa = \mu_0(\mu_0 - 4) - 2b$)

$$\kappa' = i - 2\kappa = 6b + (c - 3\mu_0)(\mu_0 - 4).$$

Now, D_b has j triple points, and this number is given by:

$$j = k - \kappa - \kappa' = \frac{1}{3}(\mu_0 - 8)(3b - \mu_0(\mu_0 - 4)).$$

And Γ_b has κ' cusps and $3j$ ordinary double points. Γ_b and Γ_c intersect transversally at κ' points, and tangentially at κ points.

The arithmetic genus of Γ_b is computed noting that it is a $(2b, \mu_0 - 4)$ curve on Z:

$$2p_a(\Gamma_b) - 2 = 2b(\mu_0 - 6) + 2(\mu_0 - c)(\mu_0 - 4).$$

and since $g(\Gamma_b) = p_a(\Gamma_b) - \kappa' - 3j$, we get that

$$2g(\Gamma_b) - 2 = -4b(\mu_0 - 6) + 2(\mu_0 - 4)(\mu_0(\mu_0 - 4) - 2c).$$

Applying again Riemann-Hurwitz to the map between the normalizations Γ'_b and D'_b of Γ_b abd D_b, we get that $2(2g(D_b)-2)+\kappa = 2g(\Gamma_b)-2$, or

$$2g(D_b) - 2 = b(13 - 2\mu_0) + \frac{1}{2}(\mu_0 - 4)(\mu_0(2\mu_0 - 9) - 4c).$$

Now, let us see what the pullback $\breve{\sigma}$ to Z of the dual hyperplane class $c_1(\Theta_{\breve{\mathbb{P}}^3}(1))$ is. Since $A(Z)$ is generated by σ and f, it must be expressible as a symbolic function of these. From what we have seen, we can state that $\breve{\sigma} = (\mu_0 - 1)\sigma - [\Gamma_b] - 3[\Gamma_c]$. In our case, this amounts to saying that $\breve{\sigma} = \mu_1 f$. The class of immersion of D_c in D is (Γ_c is nonsingular)

$$\rho(D_c) = \int \breve{\sigma} \cdot [\Gamma_c] = (\mu_0 - 1)c - i - 3\Gamma_c^2 = \mu_1$$

as expected. The class of immersion of D_b in D is (using Piene's formula)

$$\rho(D_b) = \int \breve{v} \cdot ([k^{-1}\Gamma_b] - [B]) = \mu_1(\mu_0 - 1) - 3\rho(D_c) - 2\nu_2 - \mu_2 =$$
$$= \mu_1(\mu_0 - 4) - 4\mu_0 - 8g + 8,$$

because we have seen that developable surfaces have the property that $\mu_2 = 0$.

The parabolic developable. The class of immersion of the parabolic curve P in X is the degree of the cuspidal curve on $\breve{X}$, i.e., $\mu_1(P_r) = \rho(P) = 4d(d-1)(d-2)$. The genus of P is obtained through the adjunction formula:

$$2g(P) - 2 = \int [P] \cdot ([P] + [K]) = 4d(d-2)(5d-12).$$

We have seen previously that $\mu_0(P_r) = 2d(d-2)(3d-4)$. Thus, the degree of the cuspidal curve is given by $c(P_r) = 4d(d-2)(7d-15)$, and this curve has $\kappa(P_r) = 10d(d-2)(7d-16)$ cuspa on it. The nodal curve has degree $b(P_r) = d(d-2)(18d^4 - 84d^3 + 128d^2 - 111d + 96)$. The number of cusps on D_b is

$$\kappa'(D_b) = 2d(d-2)(84d^4 - 460d^3 + 824d^2 - 641d + 360).$$

The number of triple points of the mapping is

$$k = \frac{2}{3}d(d-2)(3d^3 - 10d^2 + 8d - 2)(18d^4 - 84d^3 + 128d^2 - 109d + 92)$$

and so that D_b has

$$j(D_b) = \frac{2}{3}d(d-2)(3d^3 - 10d^2 + 8d - 4)(18d^4 - 84d^3 + 128d^2 - 181d + 256)$$

triple points.

The arithmetic genus of Γ_b is given by

$$2p_a(\Gamma_b)-2 = 4d(d-2)(3d-7)(18d^6-102d^5+218d^4-293d^3+333d^2-199d+56).$$

and the genus of D_b is given by

$$2g(D_b) - 2 = d(d-2)(156d^4 - 856d^3 + 1536d^2 - 1175d + 624).$$

5.3 Scrolls

Scrolls are nondevelopable ruled surfaces, which using the vocabulary introduced before means that the set of its focal points is finite. Thus, the only singularities of scrolls are a nodal curve, a finite number of *pinchpoints* (the focal points, at which the two generators coincide) and a finite number of triple points on this nodal curve.

Setting $c = 0$ in what we have seen in section 5.1 gives $b = \frac{1}{2}(\mu_0 - 1)(\mu_0 - 2) - g$ and $[\Gamma_b] = (\mu_0 - 2)\sigma + (2 - 2g - \mu_0)f$. From the previous section, we see that the rank μ_1 of a scroll is given by

$$\mu_1 = \mu_0(\mu_0 - 1) - 2b = 2\mu_0 + 2g - 2.$$

Then $\int[\Gamma_b]\cdot f = \mu_0 - 2$ (the nodal curve meets each generator in $\mu_0 - 2$ points). The number of pinchpoints ν_2 on this curve is the number of ramification points of ϕ (finite in this case), so

$$\nu_2 = \int c_2(\mathcal{N}_{Z/\mathbb{P}^3}) = 4g - 4 + 2\mu_0 = 2(\mu_1 - \mu_0).$$

And the number of triple points is k, which we have obtained earlier. The last quantity we want to compute is the class μ_2 of this surface. A tangent plane through a general line in space must contain a generator that meets this line, so that the number of planes tangent to a scroll and containing a general line is the number of its generators meeting a general line in space, i.e., $\mu_2 = \mu_0$.

To compute the genus of D_b, we first note that Γ_b is a $(2b, \mu_0 - 2)$ curve which gives, using a formula seen before:

$$2p_a(\Gamma_b) - 2 = (\mu_0 - 4)((\mu_0 - 1)(\mu_0 - 2) - 2g).$$

Now, D_b has k triple points, which means that Γ_b has $3k$ ordinary double points. Thus

$$2g(\Gamma_b) - 2 = 2p_a(\Gamma_b) - 2 - 6k = 2(\mu_0 + 2g - 2)(\mu_0 - 4).$$

Writing Riemann-Hurwitz for the map between the normalizations of Γ_b and D_b, we get:

$$2(2g(D_b) - 2) + \nu_2 = 2g(\Gamma_b) - 2.$$

$$\text{or} \qquad 2g(D_b) - 2 = (\mu_0 - 5)(\mu_0 + 2g - 2).$$

Then, the rank of this curve is given by:

$$r(D_b) = 2b + 2g(D_b) - 2 = 2(\mu_0 - 2)(\mu_0 - 3) + 2g(\mu_0 - 6).$$

From what we have seen for developable surfaces, we get that $\breve{\sigma} = (\mu_0 - 1)\sigma - [\Gamma_b]$, which gives in our case $\breve{\sigma} = \sigma + (\mu_0 + 2g - 2)f$. The class of immersion of D_b is

$$\rho(D_b) = \int \breve{v} \cdot ([k^{-1}\Gamma_b] - [B]) = \mu_1(\mu_0 - 1) - 2\nu_2 - \mu_2 = (\mu_0 - 2)(2\mu_0 - 5) + 2g(\mu_0 - 5),$$

because $\mu_2 = \mu_0$.

6 A few comments on the results

In this section, we want to briefly compare our results with those found by [Rie93b], applied to the case of a single smooth algebraic surface.

Our degree formulas for the components of the view bifurcation set give the same complexity orders for the visual event surfaces corresponding to multilocal singularities ($O(d^5)$ for cusp crossings and tangent crossings, $O(d^6)$ for triple points). As far as local events are concerned, our bounds are sharper ($O(d^3)$ versus $O(d^4)$). The complexity order on the number of distinct views in the graph is governed by the $O(d^6)$ bound for triple points. Using the Thom-Milnor formula [Mil64], we infer that the number of views is bounded by $O(d^{12})$ for the case of orthographic projection, and $O(d^{18})$ for the case of perspective projection.

The nice thing here is that we obtained exact formulas for the degrees of the visual event surfaces, but we had to impose genericity conditions on the original surface to be able to use Colley's stationary multiple-point theory. On the other hand, Rieger obtained asymptotic bounds, but without assuming genericity.

These genericity assumptions can be avoided, however. Another very powerful modern technique for generating enumerative formulas is through the use of *Hilbert schemes* [leB87]. They can be used to establish that enumerative formulas must be polynomials in d that are valid for arbitrary projection-generic surfaces. The precise determination of these polynomials is not necessary. Since Colley's technique also yields valid formulas that are polynomials in d, the Hilbert scheme polynomials must coincide with these.

Projection-genericity is necessary to ensure that the singularities we deal with have the right codimension. However, getting rid of this assumption will not affect our results on the degrees of the visual event surfaces, since these correspond to 1-codimensional events (i.e., they can not have a lower codimension).

7 Future research directions

In this paper, we dealt with theoretical questions related to the construction of aspect graphs of objects bounded by smooth projection-generic C^∞-surfaces. We were interested mainly in the visual event surfaces, because they separate the regions in view space that correspond to stable views. To this purpose, we introduced a formalism using the modern language of algebraic cycles and Chern classes, along with techniques from multiple-point theory and enumerative geometry. We were able to compute the elementary projective characters of the visual event curves on the surface and of the visual event surfaces. We also showed how the singularities of the visual event surfaces could be described.

Let us now conclude by briefly discussing some areas of current interest. First, we would like to study the two congruences related to the aspect graph problem. Indeed, rays giving birth to local visual events belong to the *asymptotic ray congruence*, the doubly infinite system of asymptotic rays, and those giving rise to multilocal visual events belong to the *bitangent ray congruence*, the system of bitangent lines to the surface. Those congruences are surfaces in both F_X and G, on which the visual event surfaces are represented by curves. Second, and in connection with the preceding, we would like to derive the results for the orthographic case from the perspective case. This involves building a "ray" mapping, which maps a singular ray to the point of intersection of the radii through the center of the sphere and parallel to this ray with the surface of the sphere. Through this map, a congruence would transform to a portion of the sphere, and a ruled surface to a singular curve on the sphere. The self-intersections of these curves are currently unknown.

References

[Arn83] Arnold, V. Singularities of Smooth Mappings. *Russian Mathematical Surveys*, 38(2):87–176 (1983).

[Arn84] Arnold, V. *Catastrophe Theory*. Springer-Verlag, Heidelberg (1984).

[Bak33] Baker, H. Introduction to the Theory of Algebraic Surfaces and Higher Loci. In *Principles of Geometry*, volume VI. Frederick Ungar Publishing Co. (1933).

[Bea78] Beauville, A. Surfaces algébriques complexes. In *Astérisque*, volume 54. Société Mathématique de France (1978).

[Buc85] Buchberger, B. Gröbner Bases: An Algorithmic Method in Polynomial Ideal Theory. In Bose, N., editor, *Multidimensional Systems Theory*. 184–232. Reidel, Dordrecht-Boston-Lancaster (1985).

[CW85] Callahan, J. and Weiss, R. A Model for Describing Surface Shape. In *Proceedings of IEEE Conference on Computer Vision and Pattern Recognition, San Francisco, CA (USA)*, 240–245 (1985).

[Col86] Colley, S. Lines Having Specified Contact with Projective Varieties. In *Proceedings of the 1984 Vancouver Conference in Algebraic Geometry*, volume 6, 47–70, American Mathematical Society (1986).

[Col87] Colley, S. Enumerating Stationary Multiple-Points. *Advances in Mathematics*, 66(2):149–170 (1987).

[Col75] Collins, G. Quantifier Elimination for Real Closed Fields by Cylindrical Algebraic Decomposition. In *Lecture Notes in Computer Science*, volume 33, 134–183, Springer-Verlag (1975).

[doC76] do Carmo, M. *Differential Geometry of Curves and Surfaces*. Prentice-Hall, Englewood Cliffs, New Jersey (1976).

[EB89] Eggert, D. and Bowyer, K. Computing the Orthographic Projection Aspect Graph of Solids of Revolution. In *Proceedings of the IEEE Workshop on Interpretation of 3D Scenes*, 102–108, Austin, Texas (1989).

[Ful84] Fulton, W. *Intersection Theory*. Ergebnisse der Mathematik und ihrer Grenzgebiete. Springer-Verlag (1984).

[Gro58] Grothendieck, A. Sur quelques propriétés fondamentales en théorie des intersections. In *Proceedings of Seminaire C. Chevalley*, E.N.S., Paris (1958).

[Har77] Hartshorne, R. *Algebraic Geometry*. Springer-Verlag (1977).

[Ker81] Kergosien, Y. La famille des projections orthogonales d'une surface et ses singularités. *C.R. Académie des Sciences de Paris*, 292:929–932 (1981).

[Kle76] Kleiman, S. The Enumerative Theory of Singularities. In Holm, P., editor, *Real and Complex Singularities*, 297–397. Sijthoff and Noordhoff International Publishers. Nordic Summer School, Oslo (1976).

[Kle81] Kleiman, S. Multiple-Point Formulas I: Iteration. *Acta Mathematica*, 147:13–49 (1981).

[Kle90] Kleiman, S. Multiple-Point Formulas II: The Hilbert Scheme. In Xambó-Descamps, S., editor, *Enumerative Geometry*, volume 1436 of *Lecture Notes in Mathematics*, 101–138, Springer-Verlag (1990).

[Koe86] Koenderink, J. An Internal Representation for Solid Shape Based on the Topological Properties of the Apparent Contour. In Richards, W. and Ullman, S., editors, *Image Understanding: 1985-86*, chapter 9, 257–285. Ablex Publishing Corporation, Norwood, NJ (1986).

[Koe90] Koenderink, J. *Solid Shape*. MIT Press, Cambridge, MA (1990).

[KvD79] Koenderink, J. and van Doorn, A. The Internal Representation of Solid Shape with Respect to Vision. *Biological Cybernetics*, 32:211–216 (1979).

[KP90] Kriegman, D. and Ponce, J. Computing Exact Aspect Graphs of Curved Objects: Solids of Revolution. *International Journal of Computer Vision*, 5(2): 119–135 (1990).

[Kul83] Kulikov, V. The Calculation of the Singularities of the Embedding of a Generic Algebraic Surface in the Projective Space $\mathbb{P}^3$. *Functional Analysis and Applications*, 17:176–186 (1983).

[leB78] le Barz, P. Géométrie énumerative pour les multisécantes. In Dold, A. and Eckmann, B., editors, *Variétés Analytiques Compactes*, volume 683 of *Lecture Notes in Mathematics*, 116–167, Springer-Verlag (1978).

[leB87] le Barz, P. Formules pour les trisécantes des surfaces algébriques. *L'Enseignement Mathématique*, 33:1–66 (1987).

[Mat73] Mather, J. Generic Projections. *Annals of Mathematics*, 98:226–245 (1973).

[Mil64] Milnor, J. On the Betti Numbers of Real Varieties. In *Proceedings of the American Mathematical Society*, volume 15, 275–280 (1964).

[Mor87] Morgan, A. *Solving Polynomial Systems using Continuation for Engineering and Scientific Problems*. Prentice Hall, Englewood Cliffs, New Jersey (1987).

[Pet92] Petitjean, S., Ponce, J. and Kriegman, D. Computing Exact Aspect Graphs of Curved Objects: Algebraic Surfaces. *International Journal of Computer Vision*, 9(3):231–255 (1992).

[Pie77] Piene, R. Some Formulas for a Surface in $\mathbb{P}^3$. In Dold, A. and Eckmann, B., editors, *Algebraic Geometry*, volume 687 of *Lecture Notes in Mathematics*, 196–235, Springer-Verlag (1977).

[Pla81] Platonova, O. Singularities of the Mutual Disposition of a Surface and a Line. *Russian Mathematical Surveys*, 36:248–249 (1981).

[PK90] Ponce, J. and Kriegman, D. Computing Exact Aspect Graphs of Curved Objects: Parametric Surfaces. Technical Report UIUCDCS-R-90-1579, University of Illinois-CS Department (1990).

[Rie87] Rieger, J. On the Classification of Views of Piecewise-Smooth Objects. *Image and Vision Computing*, 5:91–97 (1987).

[Rie88] Rieger, J. *Apparent Contours and their Singularities*. PhD thesis, Queen Mary College (1988).

[Rie90] Rieger, J. The Geometry of View Space of Opaque Objects Bounded by Smooth Surfaces. *Artificial Intelligence*, 44(1-2):1–40 (1990).

[Rie92] Rieger, J. Global Bifurcation Sets and Stable Projections of Nonsingular Algebraic Surfaces. *International Journal of Computer Vision*, 7(3):171–194 (1992).

[Rie93a] Rieger, J. Computing View Graphs of Algebraic Surfaces. *Journal of Symbolic Computation*, 11:1–14 (1993).

[Rie93b] Rieger, J. On the Complexity and Computation of View Graphs of Piecewise-Smooth Algebraic Surfaces. Technical Report FBI-HH-M-228/93, Universität Hamburg (1993).

[RvE92] Roy, M.-F. and van Effelterre, T. Aspects Graphs of Bodies of Revolution with Algorithms of Real Algebraic Geometry. In *Proceedings of AAGR'92*, Linz, Austria (1992).

[Sal15] Salmon, G. *A Treatise on the Analytic Geometry of Three Dimensions*, volume II. Dublin, 5th edition (1915).

[SR49] Semple, J. and Roth, L. *Introduction to Algebraic Geometry*. Clarendon Press (1949).

[Tar91] Tari, F. Projections of Piecewise-Smooth Surfaces. *Journal of London Mathematical Society*, 44:155–172 (1991).

[Tho56] Thom, R. Les singularités des applications différentiales. *Annales Institute Fourier*, 6:43-87 (1956).

[Wea27] Weatherburn, C. *Differential Geometry*. Cambridge University Press (1927).

[Whi55] Whitney, H. On singularities of Mappings of Euclidean Spaces. I. Mappings of the Plan into the Plane. *Annals of Mathematics*, 62(3):374–410 (1955).

S. Petitjean (petitjea@loria.fr)

CRIN/INRIA Lorraine, BP 239. 54506 Vandoeuvre Cedex (France)

Progress in Mathematics, Vol. 143, © 1996 Birkhäuser Verlag Basel/Switzerland

Aspect graphs of bodies of revolution with algorithms of real algebraic geometry

M.-F. Roy, T. Van Effelterre

1 Introduction

When we look at the image of a 3-D object onto a "retinal" plane under perspective projection, the abrupt changes of intensity in the image plane may have many different physical causes: the geometry of the object, the ambient illumination, texture, markings,etc. If we restrict the investigation to "geometric" image contours, we have to take two types of curves that lie on the object's surface into account: the *occluding contours*, which are the points where the tangent plane contains a line of view (these curves change with the viewpoint), and the *edges*, where the normal vector's direction is discontinuous (which are fixed). The *image contour* is the union of the projections of the occluding contours and of the edges onto the image plane. Eventually, the *visible contour* is the subset of the image contour which is really visible that is obtained after the removal of the occluded segments. A 3-D object has in general an infinite number of different images. However, an object whose surface boundaries are piecewise algebraic (as it is the case of any "CAD" object) has only a *finite* number of different visible contours under topological equivalence. From there comes the idea of representing a 3-D object with an *aspect graph* whose nodes are its qualitatively different visible contours (its *aspects*) and whose arcs are the transitions between aspects (its *visual events*) ([K1], [K2]). This type of *viewer-centered* representation of 3-D objects could be relevant for *visual object recognition* ([V2]).

Aspect graphs have already been designed for polyhedra, bodies of revolution, objects of *constructive solid geometry*, and algebraic boundary representations (splines), using numerical methods (see [V1] for references, and more specifically [EB1], [EB2], [KP] for aspect graphs of bodies of revolution). More recently, symbolic methods have been used to compute *exact* aspect graphs for smooth algebraic surfaces ([R1], [R2]), piecewise-smooth algebraic surfaces ([R3]) and assemblies of quadrics ([V1]). The last section of the present paper will address the *exact* computation of aspect graphs of bodies of revolution under perspective projection.

Results from *singularity theory* tell us how the image contours are generically, and how they may change. In the sequel, the 3-D objects that we consider have piecewise-smooth algebraic surface boundaries, the smooth patches intersect transversally (along edges) and at most 3 surface patches may intersect at isolated points (the vertices) ([R3]). Generically, the *image contours* of such 3-D objects are piecewise-smooth curves that may have only 6 different types of singular points: *Whitney cusps* (figure 1), *transversal crossings originating from 2 occluding contours* (figure 2), *vertices* (figure 3), *semi-cusps* (figure 4), *transversal crossings originating from 1 occluding contour and 1 edge* (figure 5), and *transversal crossings originating from 2 edges* (figure 6) ([W], [R1], [R2], [R3]). There are 19 different codimension-1 *visual events* by which the topology of the image contours may change ([Ar], [R1], [R2], [R3]). Petitjean ([Pe] in the same issue) describes the 2 types of generic singularities and the 6 different types of visual events that may occur for smooth algebraic surfaces. Here we take into account the extra types of generic singularities and visual event types involving the edges and the vertices that may occur when the surface boundaries are piecewise smooth algebraic ([R3]).

Under perspective projection, the *viewspace* is the 3-D euclidean space, and the codimension-1 visual events locus is a set of surfaces that tesselates the viewspace in a finite number of *cells*. The aspect of the object (i.e., the topology of its visible contour) remains the same in each cell and may change only when the viewpoint crosses a visual event surface. In this article we study the aspect graph of "bodies of revolution" (or so-called objects of the "pottery world") whose surface boundaries are obtained by revolving a piecewise polynomial curve around a symmetry axis. In section 2, we study the structure of the visual events locus of bodies of revolution, and we show that this locus is in fact the union of a finite number of hyperboloids and cones of revolution. In section 3, we study the complexity of the aspect graph and we show that the total number of aspects of a surface of revolution whose generating curve is polynomial of degree d is upper-bounded by $\frac{1}{2}C^2(d)$, where $C(d) = \frac{8}{3}d^3 + \frac{35}{3}d^2 + \frac{43}{3}d + \frac{31}{3}$. Eventually, in section 4, we show how to use techniques from *real algebraic geometry* in order to compute the aspect graph *exactly*.

2 Structure of the visual events locus

In general, visual events happen along (view)lines that have specific types of contacts with the surface. Visual events loci are therefore ruled surfaces, because if a visual event occurs at a given point of viewspace, it will also occur all along a viewline. Bodies of revolution are rotationally symmetric; therefore, if a visual event happens along a line, it will also happen along any line obtained by the rotation of this line around the symmetry axis. Two cases may occur:

a) The line along which a visual event occurs and the symmetry axis belong to a common plane: In this case we obtain a cone (whose axis is the symmetry axis of the body of revolution).

b) The line along which a visual event occurs is skew with respect to the symmetry axis: In this case, by rotation, we obtain an hyperboloid of revolution. For each type of event, the visual event locus is obtained by the rotation of a finite number of lines around the symmetry axis. We see in this way that the global visual events locus of a body of revolution is the union of a finite number of cones and hyperboloids of revolution (whose symmetry axis is the axis of the body of revolution). This gives us also the means to decrease the dimensionality of the viewspace; indeed, we may work with any plane containing the symmetry axis, and the "new" visual events locus is then the union of a finite number of hyperbolas and degenerate conics.

The surface boundary of the bodies of revolution that we study here may be parametrized by:

$$p : \mathbb{R}^2 \longrightarrow \mathbb{R}^3 : (t, \theta) \longrightarrow (r(t)cos\theta, r(t)sin\theta, t)$$

with $0 \leq \theta \leq 2\Pi$, $low \leq t \leq up$ and $r(t)$ polynomial in t.

This can be very easily generalized to bodies of revolution whose generating curve is piecewise polynomial. We will take the $z - x$ plane as viewspace. Such a 3-D object has 2 edges, which are the 2 bounding circles, and no vertices: Thus, the image contours may only have 5 different types of singularities (vertices may not occur), and there are potentially 16 different types of visual events that may occur (all except the 3 types of events that involve vertices). Visual events may involve 1 point on the surface (*local events*), 2 distinct points on the surface (*bilocal events*), or 3 distinct points on the surface (*trilocal events*). We will characterize here 3 types of visual events: *swallowtails* (local events), *tangent crossings originating from 2 occluding contours* (bilocal events), and *triple points originating from 3 occluding contours* (trilocal events) (see [Pe] in the same issue), but a similar characterization may be obtained for each of the other types of visual events.

Swallowtails

A *swallowtail* occurs along a viewline when this line has a contact of order 4 with the surface.

For each real solution in t of

$$\begin{cases} 3r'(t)r''(t) + r(t)r'''(t) = 0 \\ r''(t) > 0 \\ low \leq t \leq up \end{cases}$$

there is an hyperboloid of revolution whose intersection with the $z - x$ plane is the hyperbola with the equation:

$$x^2 = r^2(t) + (z-t)^2(r'^2(t) + r(t)r''(t)) + 2(z-t)r(t)r'(t).$$

In this case we see that the generating lines are skew with respect to the symmetry axis.

Using Bezout's theorem, we see that the visual events locus for *swallowtails* is the union of at most $2d - 3$ hyperboloids of revolution.

Tangent crossings originating from 2 occluding contours

Such a visual event occurs along a line that is tangent to the surafce at two

distinct points, where the normal vector direction is the same. This may happen only for lines that lie in a common plane with the symmetry axis.
For each real solution in (t_1, t_2) of

$$\begin{cases} r'(t_1) = \epsilon r'(t_2) \\ \epsilon r(t_2) - r(t_1) = (t_2 - t_1) r'(t_1) \\ low \leq t_1 \leq up \\ low \leq t_2 \leq up \\ t1 \neq t_2 \end{cases}$$

with $\epsilon = -1$ or $+1$, there is a cone whose intersection with the $z - x$ plane is the degenerate conic (that is the union of 2 lines) with equation:

$$x^2 = (r(t_1) + r'(t_1)(z - t_1))^2.$$

The visual events locus is then the union of at most $\frac{d^2-d}{2}$ cones.

Triple points originating from 3 occluding contours
Such a type of visual event happens along a line that is tangent to the surface at 3 distinct points. Such lines will in general be skew with respect to the symmetry axis.
For each real solution in (t_1, t_2, t_3) of

$$\begin{cases} \dfrac{r^2(t_1) - r^2(t_2)}{t_2 - t_1} + r(t_1) r'(t_1) + r(t_2) r'(t_2) = 0 \\ \dfrac{r^2(t_1) - r^2(t_3)}{t_3 - t_1} + r(t_1) r'(t_1) + r(t_3) r'(t_3) = 0 \\ \dfrac{r(t_2) r'(t_2) - r(t_1) r'(t_1)}{t_2 - t_1} - \dfrac{r(t_3) r'(t_3) - r(t_1) r'(t_1)}{t_3 - t_1} = 0 \\ (r(t_1) + r'(t_1)(t_2 - t_1))^2 - r^2(t_2) \leq 0 \\ (r(t_1) + r'(t_1)(t_3 - t_1))^2 - r^2(t_3) \leq 0 \\ low \leq t_1 \leq up \\ low \leq t_2 \leq up \\ low \leq t_3 \leq up \\ (t_1 - t_2)(t_1 - t_3)(t_2 - t_3) \neq 0 \end{cases}$$

there is an hyperboloid of revolution whose intersection with the $z - x$ plane is the hyperbola with equation:

$$x^2 = (z - t_1)^2 \frac{(r(t_2) r'(t_2) - r(t_1) r'(t_1))}{t_2 - t_1} + r^2(t_1) + 2(z - t_1) r(t_1) r'(t_1).$$

The visual events locus of triple points is then the union of at most

$$\frac{8d^3 - 16d^2 + 10d - 2}{6}$$

hyperboloids.

Example

The smooth surface of revolution obtained with the generating polynomial curve $r(t) = t^2 + 1$ along the whole t axis has 2 types of visual events:

a) *Swallowtails*: Its visual event locus is the hyperbola: $x^2 = 2z^2 + 1$.

b) *Tangent crossings originating from 2 occluding contours*: Its visual event locus is the degenerate conic: $x^2 - 4z^2 = 0$

Visibility of the visual events

When computing an aspect graph, the *visibility* is taken into account usually at the very last step: The visible contours are computed for a sample viewpoint in each cell of the tesselated viewspace, and adjacent cells in which the visible contours have the same topology are merged.

Here, we may take benefit of the rotational symmetry of the visual events locus and take the visibility into account from the beginning in order to avoid a lot of useless computations. As we have just seen, each type of visual event gives rise to a finite number of lines, and the visual event locus is the union of the cones or hyperboloids that we get by rotating these lines around the symmetry axis. For each such line, it is easy to check from where in viewspace the visual event is visible: We only have to compute where the line is intersecting the surface of revolution. Hence, by taking the visibility into account, we get a global visual event locus which is the union of a finite number of cones and hyperboloids, but also possibly of subsets of cones and of hyperboloids of revolution that are bounded by one or two planes that are perpendicular to the symmetry axis.

3 Complexity of the aspect graph

Theorem 1 *Given a generating curve described by an algebraic parametrization of degree d, the number of nodes in the aspect graph is upper-bounded by* $\frac{1}{2}C^2(d)$, *where* $C(d) = \frac{8}{3}d^3 + \frac{35}{3}d^2 + \frac{43}{3}d + \frac{31}{3}$.

Proof: For each type of visual event, we may use Bezout's theorem in order to upper-bound the number of conics (hyperbolas and degenerate conics). It can be shown in this way that the degree of the global visual events locus is upper-bounded by $C(d) = \frac{8}{3}d^3 + \frac{35}{3}d^2 + \frac{43}{3}d + \frac{31}{3}$. Note that we have included the generating curve and the z axis in the global visual events locus because we also need these two curves for the decomposition of the viewspace.

We may now use Thom-Milnor's results ([Mil], [T], [BCR]): the number of connected components of a semialgebraic set of $\mathbb{R}^n$ that is defined by one strict polynomial inequality, the degree of the polynomial being at most p, is upper-bounded by p^n. Usingthe fact that the degree of the polynomial characterizing the visual events locus is upper-bounded by $C(d)$, and $n = 2$ because we are in the plane, and also the symmetry of the viewplane with respect to the z axis, we have shown that the number of nodes in the aspect graph is upper-bounded by $\frac{1}{2}C^2(d)$, where $C(d) = \frac{8}{3}d^3 + \frac{35}{3}d^2 + \frac{43}{3}d + \frac{31}{3}$. (Note that $\frac{1}{2}C^2(d)$ is strictly increasing for $d > 0$.)

Remark
We have in fact proved the following: If N is the total degree of the visual events locus, the number of nodes in the aspect graph is upper-bounded by $\frac{1}{2}N^2$. In the worst case, N is $\frac{8}{3}d^3 + \frac{35}{3}d^2 + \frac{43}{3}d + \frac{31}{3}$. In all the examples we know, this number is much less. We do not have for the moment any lower bounds results. We may generalize this result when the generating curve of the surface of revolution is the union of s elementary parametrized curves of degree d. In this case we have to take the local events for each of the s elementary curves, look for bilocal events for all the combinations of two surface patches and/or edges, and look for triple points for all the combinations of three surface patches and/or edges. The degree of the visual events locus is then at most $O(s^6d^6)$. Applying again Thom-Milnor's bounds, we hence get the following.

Theorem 2 *Given a generating curve, union of s algebraic parametrizations of degree d, the number of nodes in the aspect graph is upper-bounded by $O(s^6d^6)$.*

Note that these upper-bounds should be compared with these established in [R2], [RV] for *general algebraic surfaces*. We showed in [RV] that the number of nodes of the aspect graph of a general algebraic surface of degree d (compact or not) is upper-bounded by $O(d^{12})$ under orthographic projection and $O(d^{18})$ under perspective projection.

4 Exact computation of the aspect graph

When the surface boundaries are semialgebraic (as it will be the case for any "CAD" object), the occluding contours, the edges, the visual events locus, the image contours and the visible contours are all *semialgebraic sets*! Remember that a semialgebraic set is a subset of some euclidean space $\mathbb{R}^n$ that is defined by a finite number of polynomial equalities and inequalities ([BCR]). The mathematical "objects" that we have to deal with here being all semialgebraic, we may achieve all the steps to compute aspect graphs using exact methods of *real algebraic geometry* (see [HRR] for a survey). To compute the *exact* aspect graph of bodies of revolution under perspective projection, we use:

- Sturm's theorem and its generalization, to compute exactly the number of real solutions of a system of univariate polynomial equalities and inequalities ([CR]).
- Hermite's method to compute exactly the number of real solutions of a system of polynomial equalities and inequalities in the multivariate case ([PRS], [P]).

- Thom's lemma to code uniquely and to sort out the real roots of a polynomial ([CR]).

The steps to compute the aspect graph are the following:

Characterize the visual events loci algebraically
We have shown in section 2 that the hyperbolas and the degenerate conics that build up the visual events loci are all given by the *real solutions* of univariate or multivariate polynomial systems involving equations and inequalities. Sturm's methods and Hermite's method give the means to count the *exact* number of these *real* solutions. Using Groebner bases and Thom's lemma, we may characterize exactly each of the visual events conics. Eventually, we may use Hermite's method again in order to check where the visual event is visible.

Compute sample viewpoints
Once we have the global visual events locus, Hermite's method may be used again in order to get a rational sample viewpoint in each connected component of the global visual events locus's complementary in viewspace.

Compute the exact topology of the aspects
For a fixed sample viewpoint $x = o_1$, $z = o_3$ (with o_1 and o_3 rational numbers), the image of the occluding contours and of the edges may be parametrized respectively (with $i = 1$ for the occluding contours, and $i = 2, 3$ for the 2 edges) by:

$$\begin{cases} A(t) = \dfrac{y_i\sqrt{(o_1^2 + o_3^2)}}{o_1 x_i + o_3 z_i - o_1^2 - o_3^2} \\ B(t) = \dfrac{o_1 z_i - o_3 x_i}{o_1 x_i + o_3 z_i - o_1^2 - o_3^2} \end{cases}$$

where A, B are the coordinates in the retinal plane and with

$$\begin{cases} x_1 = \dfrac{r(t)(r(t) + r'(t)(o_3 - t))}{o_1} \\ y_1 = \epsilon r(t)\sqrt{\dfrac{o_1^2 - (r(t) + r'(t)(o_3 - t))^2}{o_1^2}} \\ z_1 = t \end{cases}$$

with $\epsilon = -1$ or $+1$, $low \leq t \leq up$ and $o_1^2 - (r(t) + r'(t)(o_3 - t))^2 \geq 0$
This last constraint shows that the parametrization of the image of the *occluding contours* is defined for t belonging to a finite number of intervals.

$$\begin{cases} x_i = r(e)cos(t) \\ y_i = r(e)sin(t) \\ z_i = e \end{cases}$$

with $i = 1$ (resp. 2) for $e = low$ (resp. $e = up$).

The exact topology of the aspects may be computed by:

a) characterizing uniquely and sorting out the t values of the points corresponding to the borders of the intervals on which the occluding contours are defined and the different types of singular points (cusps, semi-cusps, ...).
b) sorting out the B values of the intersections of the contours with the B axis (which gives the means to establish exactly what are the inclusion/noninclusion properties of the contours that do not intersect).
c) Checking which segments of the image contour are visible.

These three steps may again be performed *exactly* by using the same methods we described above. Note that the symmetry of the surface gives the means to compute the exact topology of the aspects in a more economical way than in the case of a general surface.

Example
Figures 8 to 14 show some of the *aspects* of the body of revolution generated by $r(t) = \frac{124}{1000}t^3 - \frac{872}{1000}t + \frac{1145}{1000}$ for t varying between $\frac{-5}{2}$ and $\frac{5}{2}$ (figure 7). The theoretical upper-bound of the number of aspects is $\frac{1}{2}C^2(3) = \frac{231^2}{2} = 26681$. By using Sturm's and Hermite's methods however we obtain an upper-bound of $\frac{51^2}{2} = 1301$ aspects if we only take the *equalities* characterizing the different visual events. This upper-bound becomes $\frac{35^2}{2} = 613$ if we include the *inequalities* as well. This is still an upper-bound on the number of aspects that does not take the visibility into account, so that their actual number should still be lower. We see that this upper-bound of 613 aspects is already quite lower than the theoretical one of 26681.

Acknowledgment. We would like to thank Joachim Rieger of the Computer Science Department of the University of Hamburg for sending us his most recent unpublished results on piecewise-smooth algebraic surfaces.

References

[Ar] Arnol'd V. I.: Singularities of systems of rays. Russian Math. Surveys 38(2) pp .87–176. 1983.
[BCR] Bochnak J., Coste M., Roy M.-F.: Géométrie algébrique réelle. Ergebnisse des Mathematik.Vol.12. Springer Verlag. 1987.
[CR] Coste M., Roy M.-F.: Thom's lemma, the coding of real algebraic numbers and the topology of semialgebraic sets. J. of Symbolic Computation 5 (pp. 121–129). 1988.
[EB1] Eggert D., Bowyer K.: Computing the orthographic projection aspect graph for solids of revolution. Pattern Recognition Letters. 11 pp. 751–763. 1990.

[EB2] Eggert D., Bowyer K.: Computing the perspective projection aspect graph of solids of revolution. IEEE Trans. on PAMI. 1992.

[HRR] Heintz J., Recio T., Roy M.-F.: Algorithms in real algebraic geometry and applications to computational geometry. Dimacs series in discrete math. and theor. computer science. AMS ACM 6 pp. 137 164. 1991.

[K1] Koenderink J. J., Van Doorn A. J.: The singularities of the visual mapping. Biological cybernetics 24 pp. 51–59. 1976.

[K2] Koenderink J. J., Van Doorn A. J.: The internal representation of solid shape with respect to vision. Biological cybernetics 32 pp. 211 216. 1979.

[KP] Kriegman D. J., Ponce J.: Computing exact aspect graphs of curved objects: solids of revolution. Int. Journal of Comp. Vision 5:2 pp. 119 135. 1990.

[Mil] Milnor J.: On the Betti numbers of real algebraic varieties. Proc. AMS 15 pp. 275–280. 1964.

[P] Pedersen P.: Counting real zeros. Thesis, Courant Institute. New-York University. 1991.

[Pe] Petitjean S.: On the enumerative geometry of aspect graphs. Proceedings of MEGA'94, Santander, April 1994.

[PRS] Pedersen P., Roy M-F., Szpirglas A.: Counting real zeros in the multivariate case. Computational algebraic geometry. F. Eyssette. A. Galligo Editors. Birkhäuser. pp. 203–224. (Proceedings of the conference MEGA 92 in Nice.) 1993.

[R1] Rieger J. H.: Global bifurcation sets and stable projections of non-singular algebraic surfaces. Int. Journal of Computer Vision. 7:3. pp. 171 194. 1992.

[R2] Rieger J. H.: Computing view graphs of algebraic surfaces. Journal of Symbolic Computation Vol. 16 Number 3, pp. 259 272. 1993.

[R3] Rieger J. H.: On the complexity and computation of view graphs of piecewise smooth algebraic surfaces. Technical Report FBI-HH-M-228/93. Dept. of Computer Science, Univ. of Hamburg. 1993.

[RV] Roy M.-F., Van Effelterre T.: Aspect graphs of algebraic surfaces. Proceedings of the 1993 ISSAC conference, Kiev. 1993.

[T] Thom R.: Sur l'homologie des variétés réelles. In: Differential and combinatorial topology 255-265. Princeton University Press. 1965.

[V1] Van Effelterre T.: Graphes d'aspect d'assemblages de quadriques pour la reconnaissance visuelle d'objets. Proceedings of the "Neuvième congrès de Reconnaissance des Formes et Intelligence Artificielle" (Paris). 1994.

[V2] Van Effelterre T.: Aspect Graphs for Visual Recognition of 3-D Objects. To appear in a special issue of Perception: "Perceptual organization and object recognition".

[VVO1] Van Effelterre T., Van Gool L., Oosterlinck A.: Visual recognition of CAD-made objects with aspect graphs. Proc. of the 1992 IEEE Int. Symposium on Intelligent Control (Glasgow). 1992.

[VVO2] Van Effelterre T., Van Gool L., Oosterlinck A.: Construction of exact aspect graphs of Computer-Aided-Design objects with computer algebra. Proc. of the 1992 Seminar on Studies in Computer Algebra for Industry (SCAFI 1992 Brussels, to appear). 1992.

[W] Whitney H.: On singularities of mappings of Euclidean spaces. I. Mappings of the plane into the plane. Ann. of Math. 62 pp. 374–410. 1955.

M.-F. Roy (`costeroy@univ-rennes1.fr`)

T. Van Effelterre (`vaneffel@emmy.univ-rennes1.fr`)

IRMAR, University of Rennes I, 35042 Rennes Cedex (France).

Appendix: The Figures

Figures 1, 2, 3, 4, 5 and 6

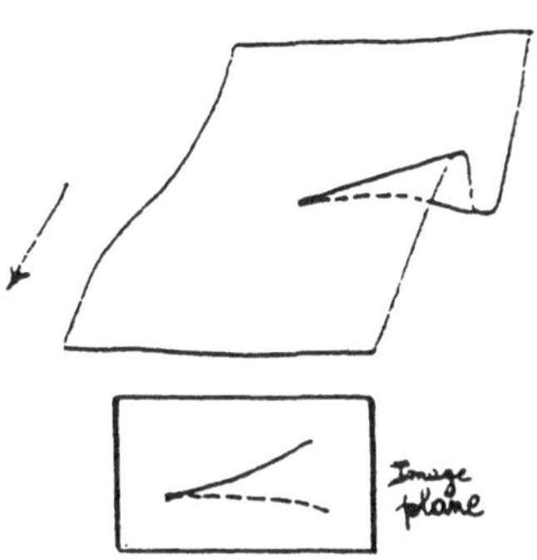

Figure 1

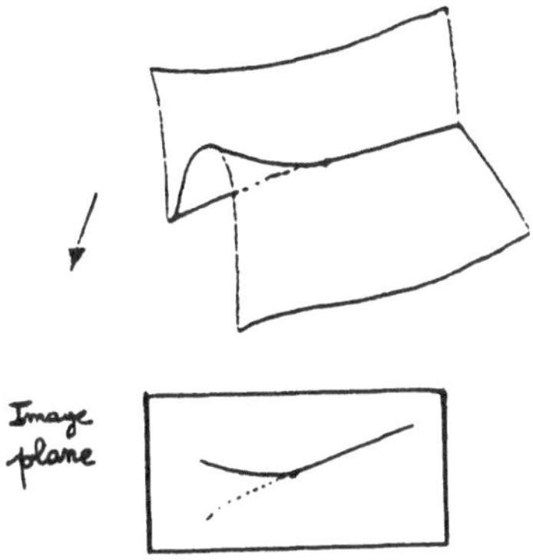

Figure 4

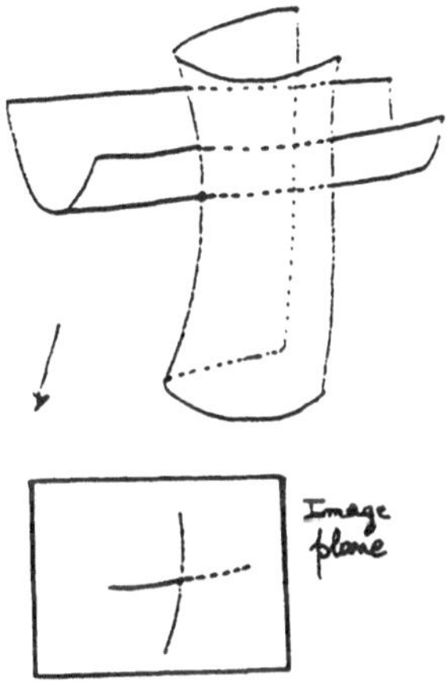

Figure 2

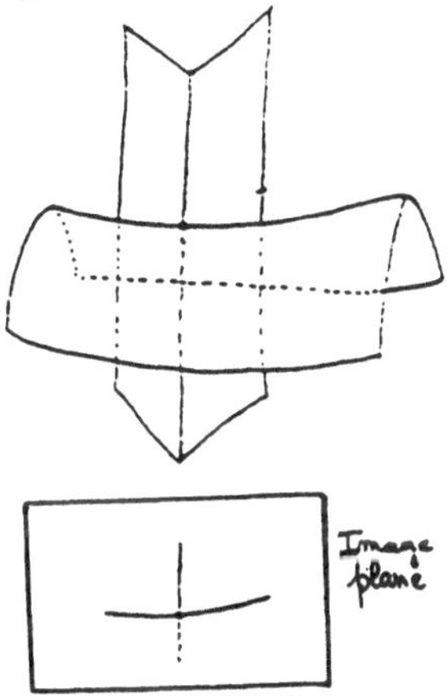

Figure 5

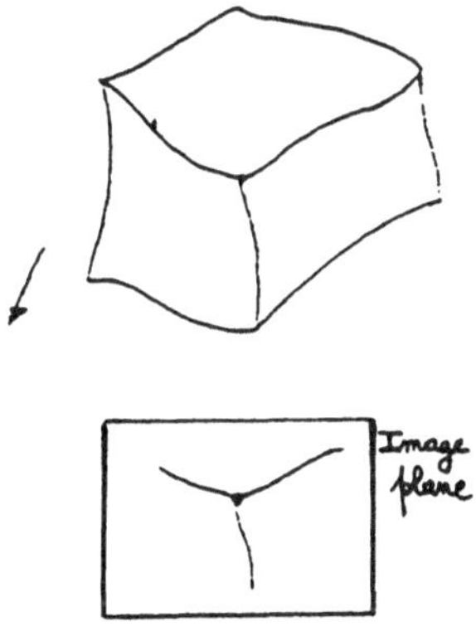

Figure 3

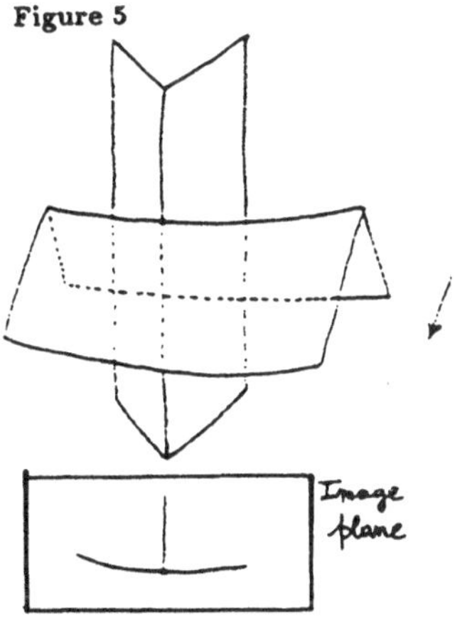

Figure 6

Figures 7, 8, 9, 10, 11, 12, 13 and 14

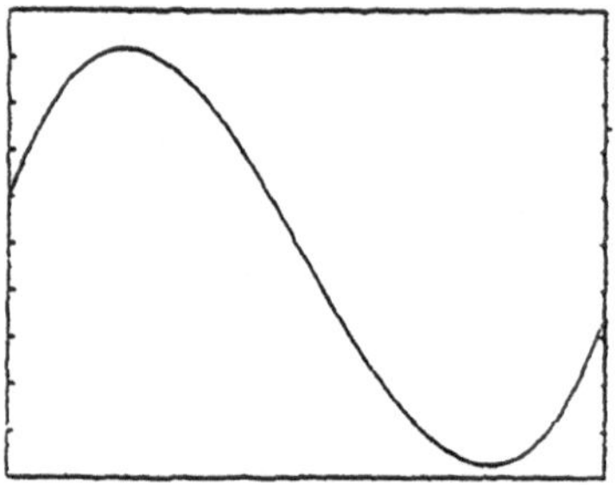

Figure 7

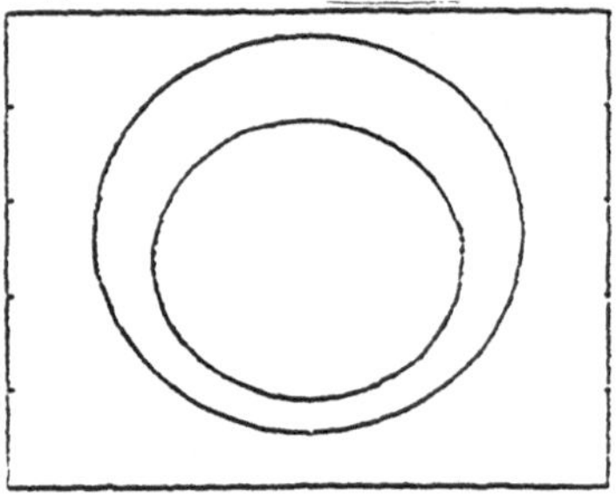

Figure 8

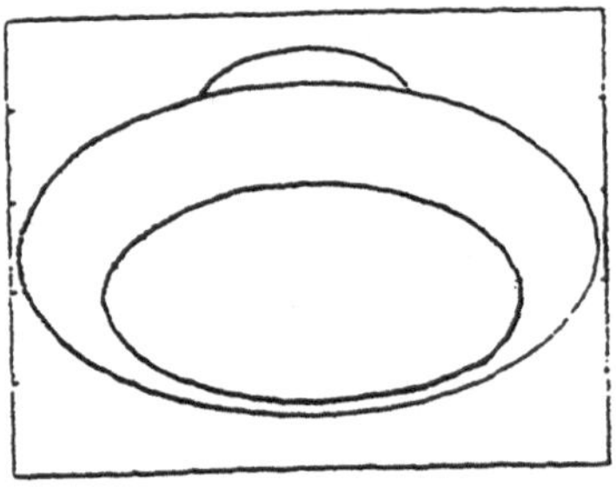

Figure 9

Figure 10

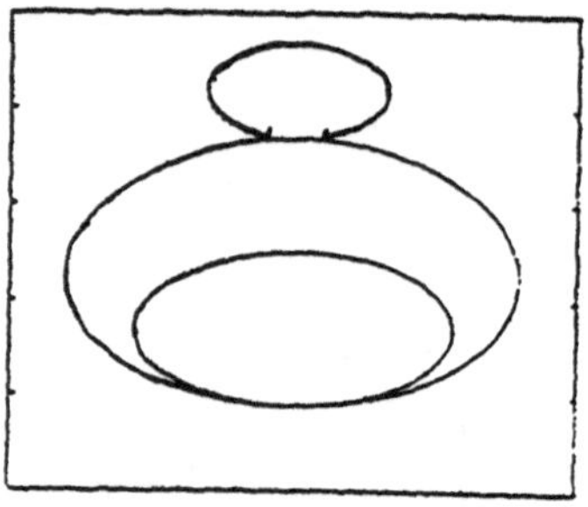

Figure 11

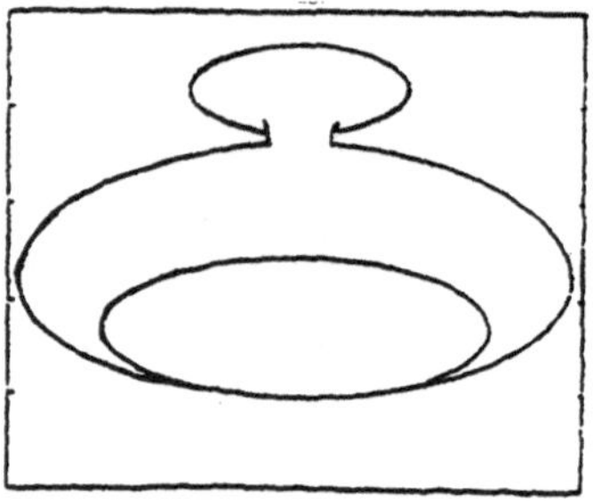

Figure 12

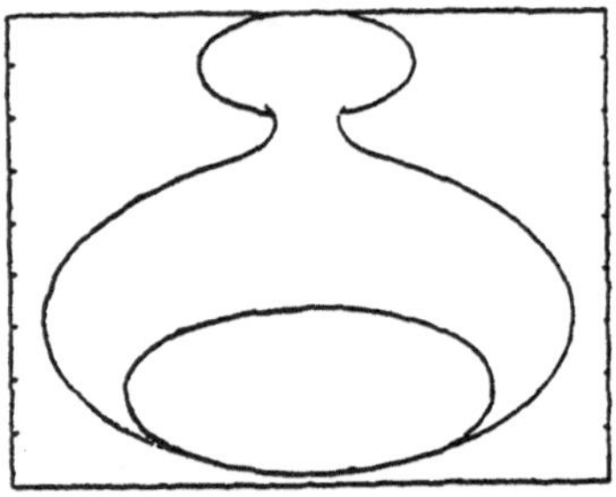

Figure 13

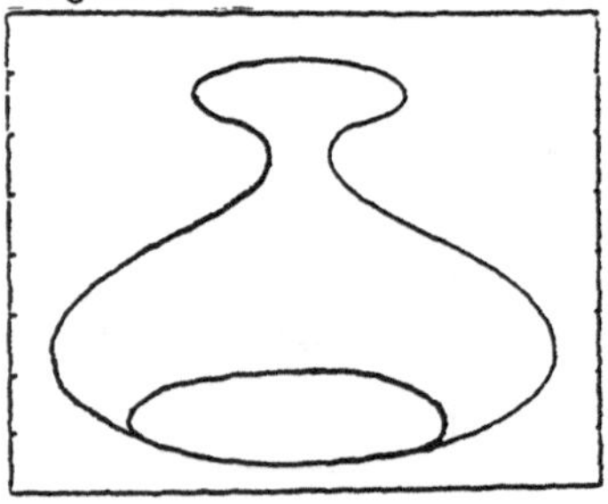

Figure 14

Progress in Mathematics, Vol. 143, © 1996 Birkhäuser Verlag Basel/Switzerland

Computational conformal geometry

M. Seppälä

1 Introduction

This paper describes some of the activities and results of the EC HCM network "Computational Conformal Geometry".[1] This network is a continuation of the Science Plan project "Computational Problems in the Theory of Riemann Surfaces and Algebraic Curves".[2]

Numerical uniformization and the correspondence between Riemann surfaces and their period matrices have been central themes in the above-mentioned projects. In this note I will concentrate on those topics.

The starting point is the fact that the following objects form equivalent categories:

1. compact Riemann surfaces
2. projective algebraic curves
3. Jacobian varieties of compact Riemann surfaces

It is our aim to be able to study that equivalence of categories in a concrete computational fashion, i.e., we would like to be able to compute:

1. equations, as a plane algebraic curve, for compact Riemann surfaces
2. discontinuous groups of Möbius transformations uniformizing a given algebraic curve
3. period matrices for compact Riemann surfaces or for algebraic (plane) curves
4. an algebraic curve or a Riemann surface having a given matrix as its period matrix

The first two items constitute the problem of *numerical uniformization*. In certain special cases we can solve *all* the above problems. Real algebraic plane curves (of a low genus) with maximal number of components of the real part constitute typically such a special case to which our methods can be applied.

It has been, around the turn of the century and early this century, the desire of a generation of mathematicians to carry out the above computations. In spite of intensive work carried out at that time (see e.g., [1], [2], [6], [5]),

[1]Human Capital and Mobility plan contract # CHRX-CT93-0408, 1993-1996.
[2]Science plan contract # SC1* CT91-0716, 1991-1993.

practical computations have been possible only in some particular cases, like that of the Klein's Riemann surface of genus 3, which has $168 = 84(3-1)$ automorphisms (see e.g., [4]). That is the largest possible group of automorphisms of genus 3 Riemann surfaces.

The methods developed early this century became quickly far too complex for numerical computations by hand. Today we can, however, use the same methods, write programs based on them, and carry the complex calculations through by computers. This is a part of what we have done.

2 Uniformization

In order to understand the classical methods developed early this century it is necessary to return to the origins of the uniformization problem and consider it in the way it was first presented.

We begin with an affine plane curve C defined by a polynomial P:

$$C = \{(x, y) \in \mathbb{C}^2 \mid P(x, y) = 0\}.$$

Classical uniformization starts from the observations that normally one cannot solve — as a univalent function — from the equation $P(x, y) = 0$, x in terms of y or vice versa. If you do that you get normally always a multivalued function.

It is, on the other hand, always possible to find a domain $\Omega \subset \mathbb{C}$ and a mapping

$$\pi : \Omega \to \mathbb{C}^2, \ \pi(z) = (\pi_x(z), \pi_y(z))$$

such that for all $z \in \Omega$

$$P(\pi_x(z), \pi_y(z)) = 0 \text{ i.e., } \pi(\Omega) \subset C.$$

The mapping $\pi : \Omega \to C$ can, furthermore, be taken to be surjective. So that the mapping π is a *one-valued* or *uniform* parametrization of the algebraic curve defined by the polynomial P. The construction of the mapping π, or rather its multivalued inverse, was the original *uniformization problem.*

Numerical uniformization problem can thus be formulated as the following: *Given a polynomial P defining a plane algebraic curve find:*

- *a domain $\Omega \subset \mathbb{C}$ and*
- *a mapping $\pi : \Omega \to \mathbb{C}^2$ which parametrizes the algebraic curve defined by the polynomial P.*

In this classical form we can solve the numerical uniformization problem in the case of hyperelliptic curves using methods of Myrberg ([6], see also [5]).

Modern uniformization of Riemann surfaces is presented in the following way.

Let X be a Riemann surface. Excluding certain special cases (sphere, annulus, torus etc.) can always find a discrete group G of Möbius transformations, acting in the unit disk $D \subset \mathbb{C}$ such that

$$X = D/G.$$

This solves the classical uniformization problem by setting $\Omega = D$ and defining $\pi : \Omega \to X = D/G$ as the projection.

3 Numerical uniformization of hyperelliptic curves

Assume that a hyperelliptic curve C of genus p is defined by the equation

$$y^2 = (x - \lambda_1)(x - \lambda_2) \cdots (x - \lambda_{2p+2}).$$

Assume further that the points λ_j are numbered in such a way that the euclidean straight arcs I_k joining λ_{2k-1} to λ_{2k}, $k = 1, \ldots, p+1$, are disjoint.

The usual topological construction for the curve C is the following:

- take first two copies of the Riemann sphere
- cut them both open along the arcs I_k
- identify the boundary components of the first Riemann sphere, that was cut open, with the boundary components of the second one in such a way that you get a double cover of the Riemann sphere ramified at the points λ_j.

The above topological construction for the curve C will be useful in understanding the following construction. We call the arcs I_k *slots* and the iterative process that we are going to describe involves *opening up the slots* I_k (and an infinite number of new slots that will be formed in this process).

The input data is the set of points

$$\lambda_1, \lambda_2, \ldots, \lambda_{2p+2}$$

which determine the *slots* $I_1, I_2, \ldots, I_{p+1}$.

Our iterative process will open the slots, one at a time, and create new slots at each step.

Set

$$\lambda_1^1 = \frac{\lambda_1 + \lambda_2}{2} - \frac{\lambda_1 - \lambda_2}{4}, \ \lambda_2^1 = \frac{\lambda_1 + \lambda_2}{2} + \frac{\lambda_1 - \lambda_2}{4}.$$

Consider first the domain $\hat{\mathbb{C}} \setminus I_1$. Solving x_1 in terms of x, the equation

$$\frac{x - \lambda_1}{x - \lambda_2} = \left(\frac{x_1 - \lambda_1^1}{x_1 - \lambda_2^1} \right)^2 \tag{1}$$

defines, by a suitable choice of the sign of the square root, a one-to-one mapping

$$\phi_1 : \hat{\mathbb{C}} \setminus I_1 \to \hat{\mathbb{C}} \setminus \overline{D}_1,$$

where $\overline{D}_1$ is a closed disk whose boundary passes through the points λ_1^1 and λ_2^1 and which has $(\lambda_1 + \lambda_2)/2$ as center.

The other alternative for the sign of the square root, when solving x_1 in terms of x from equation (1), gives a one-to-one mapping

$$\phi_1' : \hat{\mathbb{C}} \setminus I_1 \to D_1.$$

Let g_1 be the elliptic rotation, by the angle π, of the extended complex plane with the points λ_1^1 and λ_2^1 as fixed points. Then

$$\phi_1' = g_1 \circ \phi_1.$$

The above process has opened up the slot I_1 and modified all the other slots I_k which will now be replaced by arcs connecting $\phi_1(\lambda_{2k-1})$ to $\phi_1(\lambda_{2k})$, $k = 2, \ldots, p+1$.

We get, furthermore, new slots which connect the points $\phi_1'(\lambda_{2k-1})$ to $\phi_1'(\lambda_{2k})$.

At the second iterative step we start with the above defined $2p$ slots. Use the notation $\lambda_n^1 = \phi_1(\lambda_n)$ for $n = 3, 4, \ldots, 2p+2$. Repeat the above procedure replacing λ_1 and λ_2 by λ_3^1 and λ_4^1. Denote the corresponding mappings ϕ by ϕ_2 and ϕ_2' and use them to define the points λ_n^2 by setting

$$\begin{aligned}
\lambda_1^2 &= \phi_2(\lambda_1^1) \\
\lambda_2^2 &= \phi_2(\lambda_2^1) \\
\lambda_3^2 &= \frac{\lambda_3^1 - \lambda_4^1}{2} - \frac{\lambda_3^1 - \lambda_3^1}{4} \\
\lambda_4^2 &= \frac{\lambda_3^1 - \lambda_4^1}{2} + \frac{\lambda_3^1 - \lambda_3^1}{4} \\
\lambda_n^2 &= \phi_2(\lambda_n^1),\ n = 5, 6, \ldots
\end{aligned}$$

We iterate this process and form, by opening up each slot I_k (and all the new slots), the series λ_k^n for each index $k = 1, 2, \ldots, 2p+2$. The definition of the constants λ_k^n is such that all the above series converge. Let

$$\lim_{n \to \infty} \lambda_k^n = \lambda_k^\infty,$$

and let g_k^∞ be the rotation of the extended complex plane by the angle π around the points λ_{2k-1}^∞ and λ_{2k}^∞.

Let G be the group of Möbius transformations generated by all the products $g_k^\infty \circ g_l^\infty$. It turns out that this group G is a Kleinian group. Let Ω be the domain of discontinuity of G. Then

$$C = \Omega / G$$

is a uniformization for the hyperelliptic curve C in the classical sense. This is the construction of Myrberg (cf. [6]). This argument gives a solution for the classical problem of numerical uniformisation in the case of hyperelliptic curves and will be explained in detail in a forthcoming paper by K.-D. Semmler and M. Seppälä.

4 Period matrices

Consider Riemann surfaces X of genus $p > 1$. Let us first recall that the space $D^1(X)$ of holomorphic differentials of X is a p-dimensional complex vector space.

Let $\alpha_1, \beta_1, \alpha_2, \beta_2, \ldots, \alpha_p, \beta_p$ be a canonical homology base for X. Canonical means here that the curves α_j and β_j intersect in such a way that their intersection number is one and that there are no other intersections.

We can choose a normalized base $(\omega_1, \ldots, \omega_p)$ for holomorphic differentials of X in such a way that

$$\int_{\alpha_j} \omega_i = \delta_{ij} = \begin{cases} 1, & \text{if } i = j \\ 0, & \text{if } i \neq j. \end{cases}$$

A period matrix P_X for the Riemann surface X is then simply

$$P_X = \left(\int_{\beta_i} \omega_j \right). \tag{2}$$

When computing a period matrix for a given Riemann surface one has, therefore, to solve the following two problems:

1. Find a base $(\omega_1, \ldots, \omega_p)$ for holomorphic differentials on X.

2. Find a canonical homology base $(\alpha_1, \beta_1, \ldots, \alpha_p, \beta_p)$.

Once these problems have been solved it is a matter of numerical integration to compute the integrals (2).

Normally a Riemann surface X will be defined by giving its Fuchsian group, i.e., by representing $X = D/G$. When such a representation is given one can often read a base for the homology directly from the generators of the group G. Thus the second problem is easy for compact Riemann surfaces.

On the other hand, for general a compact Riemann surface $X = D/G$ there are no direct ways to give a base for holomorphic differentials. For such Riemann surfaces the usual Poincaré series *do not* converge and cannot therefore be used to construct differentials.

For complex algebraic (plane) curves one can, on the other hand, construct a base for holomorphic differentials directly by Noether's algorithm. Therefore the first problem regarding the computation of a period matrix is straightforward for curves. Problems arise from finding a homology base for curves. We do not know any algorithm that would work in the general case.

4.1 Real algebraic curves

Projective real algebraic curves C (with real points) are compact Riemann surfaces X together with an antiholomorphic involution $\sigma : X \to X$ having fixed points. Computation of a period matrix for such Riemann surfaces is easier and can be dealt with more straightforward methods than that of general compact Riemann surfaces.

Robert Silhol has developed methods that allow one to compute (good) approximations for period matrices of real curves with maximal number of real components ([9]). This is possible since from the equation of a real genus p plane curve, with maximal number of components of the real part, one can already find $p+1$ cycles that can be used to form a homology base. These cycles can then be completed to a full homology base and used to compute an approximation for the period matrix.

A related question is to develop programs that can be used to test whether two given (plane) curves are isomorphic or not. In low genus cases ($p = 2$ or 3) this can be settled by comparing their period matrices using their normal forms as presented by Robert Silhol ([8]).

4.2 Symmetric Riemann surfaces

Viewing real algebraic curves as Riemann surfaces one can compute approximations for their period matrices in the following way. Let X be a compact genus p, $p > 1$, Riemann surface with an antiholomorphic involution $\sigma : X \to X$.

Let Y be a component of $X \setminus X_\sigma$, where X_σ is the fixed point set of $\sigma : X \to X$. For technical simplicity assume that X_σ decomposes X into two components, i.e., that $Y \cap \sigma(Y) = \emptyset$. The case of non dividing real part can also be treated with the present method ([7]).

Let $Y = D/G$, where G is now a Fuchsian group *of the second kind* acting in the unit disk D.

The domain of discontinuity of G, $\Omega(G)$, contains the open unit disk and $X = \Omega(G)/G$. The mapping $\sigma : X \to X$ is induced by the reflection in the unit circle.

Since the group G is of the second kind (and since the quotient $\Omega(G)/G$ is compact), G is freely generated. Let the hyperbolic Möbius transformations $g_1, g_2, \dots, g_p$, mapping the unit disk onto itself, form a free set of generators for G.

For $j = 1, 2, \dots, p$ let

$$h_j(z) = \frac{\bar{\zeta}_j}{1 - \bar{\zeta}_j z}, \quad \zeta_j = g_j(0), \quad 1 \le j \le p. \tag{3}$$

Consider the Poincaré series

$$\Theta^m(h_j)(z) = \sum_{g \in G} h_j(g(z)) \left(\frac{\partial g(z)}{\partial z} \right). \tag{4}$$

We have ([3, Corollary 1, page 208]):

Lemma 1 $\{\Theta(h_j) \mid 1 \leq j \leq p\}$ *is a basis for holomorphic differentials on* $X = \Omega(G)/G$.

Any closed curve α, $\alpha : [0,1] \to Y$, on the Riemann surface $Y = D/G$ can be lifted to a curve $\alpha : [0,1] \to D$ on the unit disk. Using this notation we have the following result for the integrals of the base differentials of Theorem 1 ([7, Lemma 4]):

Lemma 2 *The integral of a base differential* $\Theta(h_j)$ *along* α *can be computed by the convergent series:*

$$\int_\alpha \Theta(h_j)dz = \sum_{g\in G} \log\left(\frac{1-\bar{\zeta}_j g(\alpha(0))}{1-\bar{\zeta}_j g(\alpha(1))}\right).$$

Important point to be observed here is that the above algorithm for computing periods of differentials is essentially *transcendental*. Even if we start with precise data, i.e., even if the coefficients of the generating Möbius transformations are precise rational or algebraic numbers, only floating-point approximations for the periods can be computed (with the present techniques). The same applies to the numerical uniformization of hyperelliptic curves.

5 Input device

When doing computations with Riemann surfaces, one has to define the Riemann surface in question for computations. That is done by giving the generators of a Möbius group corresponding to the Riemann surface to be studied.

This is, however, not always easy. It is a famous open problem to give sufficient conditions that guarantee that a group generated by a given set of Möébius transformations, mapping the unit disk onto itself, is discontinuous.

For this purpose K.-D. Semmler, in collaboration with Loris Renggli and Tero Venetjoki, has built an input device that allows one to build Riemann surfaces by gluing elementary parts together. These elementary parts are:

- Y-pieces or pairs of pants (i.e., spheres with three holes).
- Q-pieces or torii with one hole.

The input device is based on Maskit's combination theorems and allows one to build Fuchsian groups corresponding to compact Riemann surfaces of any given topological type. Furthermore, a homology base for the Riemann surface in question can be read directly from the generators. In a sense the input device will always generate, by default, a *random* Riemann surface of a given topological type. In the applications that we have in mind, we usually are not interested, for instance, in the period matrix or the equation of some particular Riemann surface. We are rather more interested in seeing what happens to the equation or to the period matrix of a Riemann surface when the Riemann surface itself is *deformed.* For that reason the input device has the capacity to deform Riemann surfaces by making certain geodesic curves shorter or performing unlimited Fenchel-Nielsen twists along the geodesics.

This graphical input device will be a central part of a set of programs that will be developed within the present HCM project and which can be used to study Riemann surfaces and algebraic curves numerically.

References

[1] H. F. Baker. *Abel's Theorem and the Allied Theory including the Theory of Theta Functions.* Cambridge University Press, Cambridge, 1897.

[2] H. F. Baker. *Multiply Periodic Functions.* Cambridge University Press, Cambridge, 1907.

[3] C. J. Earle and A. Marden. On Poincaré Series with Application to h^p Spaces of Bordered Riemann Surfaces. *Ill. J. Math.*, 13: pp. 202–219, 1969.

[4] Clifford J. Earle. H. e. Rauch, Function Theorist. In Isaac Chavel and Hersel M. Farkas, editors, *Differential Geometry and Complex Analysis*, pp. 15–31. Springer-Verlag, Berlin–Heidelberg–New York, 1985.

[5] Fernanda Esser. Die algebraische Uniformisierung mit numerischen Beispielen nach Myrberg. Manuscript probably written in the 1930's, exact date not known.

[6] P. J. Myrberg. Über die Numerische Ausführung der Uniformisierung. *Acta Soc. Sci. Fenn.*, XLVIII(7): pp. 1–53, 1920.

[7] Mika Seppälä. Computation of period matrices of real algebraic curves. *Dicrete Comput Geom*, 11: pp. 65–81, 1994.

[8] Robert Silhol. Normal forms for period matrices of real algebraic curves of genus 2 and 3. *J. Pure and Applied Algebra*, 87: pp. 79–92, 1993.

[9] Robert Silhol. Numerical approximation of period matrices of real algebraic curves, 1994. To appear.

M. Seppälä (`mika.seppala@helsinki.fi`)

Dept. of Mathematics, P.O. Box 4, University of Helsinki, FIN–00014 (Finland)

Progress in Mathematics, Vol. 143, © 1996 Birkhäuser Verlag Basel/Switzerland

An algorithm and bounds for the real effective Nullstellensatz in one variable

H. Warou

Introduction

In this paper we deal with the algorithm of construction of an effective positivstellensatz given in [Lom1], for the particular case of a family of univariate polynomials with coefficients in a real closed field.

We study in detail the mixed and generalized Taylor formulas, which are essential tools for constructing algebraic identities for the real effective Nullstellensatz. In particular we give more general and straightforward proof for mixed Taylor formulas, we prove that the coefficients which appear in both the mixed and generalized Taylor formulas are integers and we establish some related upper bounds.

We also show that a good control on the "glueing" procedure leads to a bound on the degree by $8rd^2$ in the final algebraic identity, where d is a bound on the degrees of the input polynomials and r is a bound on the number of real roots of these polynomials and their successive derivatives.

1 Ground tools

1.1 Incompatible system and strong implication

A *strict sign condition* is one of the following two: > 0, < 0. We denote them by -1 and $+1$ respectively. A *generalized sign condition* is one of the elements of $\{<0, \leq 0, =0, \geq 0, \neq 0, >0\}$. When we replace the strict sign condition < 0 (respectively > 0) by the generalized sign condition ≤ 0 (respectively ≥ 0) we say that the sign condition is relaxed.

We consider $\mathbf{R}$ a real closed field and $\mathbf{R}[\mathbf{X}]$ the polynomial ring $\mathbf{R}[X_1, \ldots, X_n]$. Let F be a finite subset of $\mathbf{R}[\mathbf{X}]$. We denote by: F^2 the set of squares of non-zero elements of F, $\mathcal{M}(F)$ the *multiplicative monoid* generated by $F \cup \{1\}$, $\mathcal{C}(F)$ the *positive cone* generated by F (the additive monoid generated by the elements of type $p.P.Q^2$ where p is positive in $\mathbf{R}$, P in $\mathcal{M}(F)$ and Q in $\mathbf{R}[\mathbf{X}]$) and $\mathcal{I}(F)$ the *ideal* generated by F.

Definition 1: *Consider a system $\mathbb{H} = [F_>, F_\geq, F_=, F_\neq]$ of generalized sign conditions, consisting of four finite subsets of $\mathbf{R}[\mathbf{X}]$. We say that the system $\mathbb{H}$ is strongly incompatible in $\mathbf{R}$ if we have in $\mathbf{R}[\mathbf{X}]$ an equality of the following type:*

$$S + P + Z = 0 \tag{1}$$

with $S \in \mathcal{M}(F_>^2 \cup F_\neq^2)$, $P \in \mathcal{C}(F_> \cup F_\geq)$ and $Z \in \mathcal{I}(F_=)$.

Notation 2: The strong incompatibility of a system $\mathbb{H}$ of generalized sign conditions is denoted by $\downarrow \mathbb{H} \downarrow$.

Remark 3: It is clear that a strong incompatibility is a very strong form of incompatibility. In particular, it implies that it is impossible to give the indicated signs to the polynomials in $\mathbb{H}$. The impossibility of a system of generalized sign conditions is constructively equivalent to its formulation in form of various implications: For example, that the system $[P=0, Q=0]$ is strongly incompatible in $\mathbf{R}$ is equivalent to

$$\forall x_1, \dots, x_n \in \mathbf{R} \quad P(x_1, \dots, x_n) = 0 \Longrightarrow Q(x_1, \dots, x_n) \neq 0.$$

We shall speak thus of *strong incompatibility, strong implication, or strong evidence*, meaning always implicitly a strong incompatibility.

Notation 4:

1) Let τ be a generalized sign condition. We use the following notation for strong implication:
$^*([S_1 > 0, \dots, S_i > 0, P_1 \geq 0, \dots, P_j \geq 0, Z_1 = 0, \dots, Z_k = 0, N_1 \neq 0, \dots, N_h \neq 0] \Longrightarrow Q\ \tau)^*$.

2) Let us denote by $\mathbb{H}$ the left-hand side in 1) and by $\mathbb{H}'$ a system of generalized sign conditions $Q_1\ \tau_1, \dots, Q_k\ \tau_k$. We then write: $^*(\mathbb{H} \Longrightarrow \mathbb{H}')^*$ to mean $^*(\mathbb{H} \Longrightarrow Q_j\ \tau_j\ )^*$ with $1 \leq j \leq k$.

The different variants of the real positivstellensatz are consequences of the following general theorem (see [BCR] ch.4):

Theorem 5: *Let $\mathbb{H} = [F_>, F_\geq, F_=, F_\neq]$ a be system of generalized sign conditions on polynomials of $\mathbf{R}[\mathbf{X}]$ and A be the semi-algebraic set of $\mathbf{R}^n$ defined by:*

$$\mathbf{x} \in A \quad \Longleftrightarrow \quad \begin{array}{ll} \forall\, f \in F_>,\ f(\mathbf{x}) > 0 & \forall\, g \in F_\geq,\ g(\mathbf{x}) \geq 0 \\ \forall\, h \in F_=,\ h(\mathbf{x}) = 0 & \forall\, q \in F_\neq,\ q(\mathbf{x}) \neq 0 \end{array}$$

The two following conditions are equivalent:

- *$\mathbb{H}$ is strongly incompatible in $\mathbf{R}$.*
- *A is empty in $\mathbf{R}^n$.*

1.2 A bound on the degree of a strong incompatibilityconstructed from other strong incompatibilities

We introduce now some results that will be useful for the rest of this note, concerning the manipulation of strong incompatibilities and strong implications (see [Lom2]).

Definition 6: *We call degree of a strong incompatibility the maximum degree of polynomials that appear in the corresponding algebraic identity.*

For example, if we have a strong incompatibility: $\downarrow [A > 0, B > 0, C \geq 0, D \geq 0, E = 0, G = 0] \downarrow$ characterized by the following algebraic identity:

$$A^2.B^6 + C.\sum_{i=1}^{h} p_i.P_i^2 + A.B.D.\sum_{j=1}^{k} q_i.Q_j^2 + E.U + G.V = 0$$

then the degree of this strong incompatibility is the maximum of $deg(A^2.B^6)$, $deg(C.P_i^2)$ $(1 \leq i \leq h)$, $deg(A.B.D.Q_j^2)$ $(1 \leq j \leq k)$ and $deg(E.U)$, $d(G.V)$.

In the following proposition we give some precisions on the degree of some basic constructions of strong incompatibilities (cf [Lom1]).

Proposition 7: *Let $\mathbb{H}$ be a system of generalized sign conditions on polynomials of $\mathbf{R}[\mathbf{X}]$ and $Q \in \mathbf{R}[\mathbf{X}]$. Then*

(i) *If $\downarrow [\mathbb{H}, Q \leq 0] \downarrow$ with the degree δ and $\downarrow [\mathbb{H}, Q \geq 0] \downarrow$ with the degree δ', then one has $\downarrow \mathbb{H} \downarrow$ with the degree bounded by $\delta + \delta'$.*

(ii) *If $\downarrow [\mathbb{H}, Q < 0] \downarrow$ with the degree δ and $\downarrow [\mathbb{H}, Q > 0] \downarrow$ with the degree δ', then one has $\downarrow [\mathbb{H}, Q \neq 0] \downarrow$ with the degree bounded by $\delta + \delta'$.*

(iii) *If $\downarrow [\mathbb{H}, Q \neq 0] \downarrow$ and $\downarrow [\mathbb{H}, Q = 0] \downarrow$, then one has $\downarrow \mathbb{H} \downarrow$. Moreover if Q has the degree q, $\downarrow [\mathbb{H}, Q \neq 0] \downarrow$ has the degree δ with $Q^{2m}S$ in the monoid part and $\downarrow [\mathbb{H}, Q = 0] \downarrow$ has the degree δ', then one has the strong incompatibility $\downarrow \mathbb{H} \downarrow$ with the degree bounded by $\delta + 2m\delta' - 2mq$.*

Proof: We give only the proof for the most intricate case (iii). Call $F_>$, $F_\geq$, $F_=$ and $F_{\neq}$ the four finite subsets of $\mathbf{R}[\mathbf{X}]$ containing in $\mathbb{H}$.

The hypothesis $\downarrow [\mathbb{H}, Q \neq 0] \downarrow$ corresponds to the identity:

$$Q^{2m}S_1 + P_1 + Z_1 = 0 \tag{1}$$

with

$$S_1 \in \mathcal{M}(F_>^2 \cup F_{\neq 0}^2), P_1 \in \mathcal{C}(F_> \cup F_\geq), Z_1 \in \mathcal{I}(F_=)$$

Likewise, the second hypothesis means we have an equality:

$$S_2 + P_2 + Z_2Q + Z_3 = 0 \tag{2}$$

with

$$S_2 \in \mathcal{M}(F_>^2 \cup F_{\neq 0}^2), P_2 \in \mathcal{C}(F_> \cup F_\geq), Z_1 \in \mathcal{I}(F_=).$$

In (1) we have $deg(S_1) \leq \delta - 2mq$ and in (2) $deg(Z_2) \leq \delta' - q$. We rewrite the equality (2) as

$$S_2 + P_2 + Z_3 = -Z_2Q,$$

and we take the both sides to the power $2m$, so we obtain the equality

$$S_3 + P_3 + Z_4 = Q^{2m} Z_5, \tag{3}$$

where $deg(Z_5) \leq 2m(\delta' - q)$. And next we multiply (1) by Z_5 and (3) by S_1 and get

$$\underbrace{S_1S_3 + P_3S_1 + Z_4S_1}_{\deg \leq 2m\delta' + (\delta - 2mq)} - \underbrace{P_1Z_5 - Z_1Z_5}_{\deg \leq \delta + 2m(\delta' - q)} = 0.$$

2 Mixed and generalized Taylor formulas

The mixed and generalized Taylor formulas are an important tool for the constructions of algebraic identities leading to Henri Lombardi's proof of the real effective Nullstellensatz. The aim of this section is to extend the mixed Taylor formulas to differentiable functions, which allows us to prove by a straightforward way the algebraic theorem given in [Lom1]. Next we study the algebraic identities called generalized Taylor formulas. These formulas generalize the mixed Taylor formulas in the polynomial case.

2.1 Mixed Taylor formulas

Notations:
Let $\varepsilon = [\epsilon_1, \epsilon_2, \ldots, \epsilon_n]$ $(n > 1)$ be a n-tuple of strict sign conditions with $\epsilon_1 = 1$. We denote by

- $\varepsilon_k = [\epsilon_1, \ldots, \epsilon_k]$ the k-tuple formed by the k first elements of ε;
- $\xi_{n-1} := \frac{1}{2}(1 + \epsilon_{n-1}\epsilon_n)$ (i.e., 1 if $\epsilon_{n-1}\epsilon_n = 1$, 0 if $\epsilon_{n-1}\epsilon_n = -1$);
- $\mathrm{R}_{\varepsilon_n}$ the polynomial of $\mathbb{Q}[X]$ strictly positive on $]0,1[$ defined by the induction:

$$\mathrm{R}_{\varepsilon_1}(t) := 1$$

$$\mathrm{R}_{\varepsilon_n}(t) := (-1)^{\xi_{n-1}}(n-1)\int_{\xi_{n-1}}^{t} \mathrm{R}_{\varepsilon_{n-1}}(x)dx;$$

- c_{ε_k} the positive rational number defined by $c_{\varepsilon_k} = k\int_0^1 \mathrm{R}_{\varepsilon_k}(t)dt.$

Moreover, we denote the differential operator $\dfrac{d^k}{k!dx^k}$ by D_k.

Theorem and Definition 8: *If u and v are two reals with $u < v$, we let $\Delta = u - x$ for $u \leq x \leq v$ and $y_k = \begin{cases} u & if\ \ \epsilon_k\epsilon_{k+1} = 1 \\ x & otherwise \end{cases}$. Then for each function $f : [u,v] \longrightarrow \mathbb{R}$ of class C^{n+1}, $n \geq 0$, there exist 2^n mixed Taylor formulas and all the possible sign combinations occur. More precisely, let $\varepsilon = [\epsilon_1, \epsilon_2, \ldots, \epsilon_{n+1}]$*

be a $(n+1)$*-tuple of strict sign conditions with* $\epsilon_1 = 1$*, then one has the following equality*

$$\begin{aligned} f(x) &= f(u) + \sum_{k=1}^{n} \epsilon_k c_{\varepsilon_k} \Delta^k \mathrm{D}_k(f)(y_k) \\ &\quad + \epsilon_{n+1}(n+1)\Delta^{n+1} \int_0^1 \mathrm{R}_{\varepsilon_{n+1}}(t) \mathrm{D}_{n+1}(f)(u+t\Delta)dt. \end{aligned} \tag{E_1}$$

The previous formula is called the mixed Taylor formula for f *associated to the combination* ε.

Proof: We show (E_1) by induction on n. The formula (E_1) is immediate for $n = 0$. Indeed for f of C^1 we have:

$$f(x) - f(u) = \Delta \int_0^1 f'(u+t\Delta)dt.$$

Suppose that the formula (E_1) holds for f of class C^n for $n \geq 1$ on $[u, v]$. Then for f of class C^{n+1} we have the following two cases.

1rst case: $\epsilon_n \epsilon_{n+1} = -1$

In this case $y_n = x$. Integrating by parts yields

$$\mathrm{R} = \epsilon_n \frac{\Delta^n}{(n-1)!} \int_0^1 \mathrm{R}_{\varepsilon_n}(t) f^{(n)}(u+t\Delta)dt$$

which is equal by induction to

$$f(x) - f(u) - \sum_{k=1}^{n-1} \epsilon_k c_{\varepsilon_k} \Delta^k \mathrm{D}_k(f)(y_k)$$

and we obtain

$$\mathrm{R} = \epsilon_n \frac{\Delta^n}{(n-1)!} [\mathrm{Q}(t) f^{(n)}(u+t\Delta)]_0^1 - \epsilon_n \frac{\Delta^{n+1}}{(n-1)!} \int_0^1 \mathrm{Q}(t) f^{(n+1)}(u+t\Delta)dt$$

where $\mathrm{Q}(t) = \int_0^t \mathrm{R}_{\varepsilon_n}(x)dx = \frac{1}{n}\mathrm{P}_{\varepsilon_{n+1}}(t)$. Consequently we have

$$\mathrm{R} = \epsilon_n c_{\varepsilon_n} \Delta^n \mathrm{D}_n(f)(y_n) + \epsilon_{n+1} \frac{\Delta^{n+1}}{n!} \int_0^1 \mathrm{R}_{\varepsilon_{n+1}}(t) f^{(n+1)}(u+t\Delta)dt$$

with $c_{\varepsilon_{n+1}} = n \int_0^1 \mathrm{R}_{\varepsilon_n}(t)dt = \mathrm{R}_{\varepsilon_{n+1}}(1)$.

2nd case: Similar computations. □

Proposition 9: *Let n be an integer ≥ 1. Then each $\mathrm{R}_{\varepsilon_n}$ has integer coefficients.*

Proof: For $n = 2$ we have $\mathrm{R}_{\varepsilon_2}(t) = (-1)^{\xi_1}(t - \xi_1)$. Suppose now inductively that $\mathrm{R}_{\varepsilon_n}$ $n \geq 2$, has integer coefficients. Write

$$\mathrm{R}_{\varepsilon_{n+1}}(t) = \sum_{i=0}^{n} a_i t^i$$

$$\mathrm{R}^{(k)}_{\varepsilon_{n+1}}(t) = \sum_{i=k}^{n} i(i-1)\cdots(i-k+1)a_i t^{i-k}$$

Using the definition it is easy to see that

$$\mathrm{R}^{(k)}_{\varepsilon_{n+1}}(t) = k!\binom{n}{k}(-1)^{\left(\sum_{j=n-k+1}^{n}\xi_j\right)}\mathrm{R}_{\varepsilon_{n-k+1}}(t)$$

and by identification it follows that $(1 \leq k \leq n)$

$$a_k = \binom{n}{k}(-1)^{\left(\sum_{j=n-k+1}^{n}\xi_j\right)}\mathrm{R}_{\varepsilon_{n-k+1}}(0).$$

One deduces that $a_k \in \mathbb{Z}$, $k = 1, \ldots, n$ and $\mathrm{R}_{\varepsilon_{n+1}} \in \mathbb{Z}[t]$ noting that $a_0 = 0$ or $a_0 = -(a_1 + a_2 + \cdots + a_n)$ according to $\epsilon_n\epsilon_{n+1} = -1$ or 1. □

Corollary 10: (polynomial mixed Taylor formulas)
Let $\mathbf{A}$ be a commutative ring. One considers two variables U and V and one lets $\Delta = U - V$. Let $\varepsilon = [\epsilon_1, \epsilon_2, \ldots, \epsilon_n]$ be a n-tuple of strict sign conditions with $\epsilon_1 = 1$, then the mixed Taylor formula associated to the combination ε for a polynomial $P \in \mathbf{A}[X]$ of degree $n \geq 1$ is an algebraic identity of the following form:

$$P(U) - P(V) = \sum_{k=1}^{n-1} \epsilon_k c_{\varepsilon_k} \Delta^k \mathrm{D}_k(P)(Y_k) + \epsilon_n c_{\varepsilon_n} \Delta^n \mathrm{D}_n(P) \qquad (E_2)$$

where c_{ε_k} are positive integers and $Y_k = \begin{cases} V & \text{if } \epsilon_k\epsilon_{k+1} = 1 \\ U & \text{otherwise.} \end{cases}$

Proof: Since it concerns algebraic identities in variables U, V and the coefficients of the polynomial P, it is enough to show the corollary for the ring $\mathbf{A}$=$\mathbb{Z}$, and this clear from theorem 8. □

Proposition 11: *Let $\varepsilon = [\epsilon_1, \epsilon_2, \ldots, \epsilon_n]$ $(n \geq 1)$ be a n-tuple of strict sign conditions with $\epsilon_1 = 1$. Then the integers c_{ε_k} $(k = 1, \ldots, n)$ satisfy the following inequalities:*

$$\leq c_{\varepsilon_k} \leq k!, \qquad c_{\varepsilon_k} \leq c_{\varepsilon_{k+1}} \leq (k+1)c_{\varepsilon_k}. \qquad (E_3)$$

Proof (see [War]). In this proof, we show that if u_n is the maximum value for c_{ε_n}, then

$$\sum_{n=0}^{+\infty} \frac{u_n}{n!} t^n = \frac{1+\sin(t)}{\mathrm{cost}(t)}$$

with $(u_0 = 1)$. So, for n odd, u_n is an Euler number. □

Proposition 12: *Let $\varepsilon = [\epsilon_1, \epsilon_2, \ldots, \epsilon_n]$ be a n-tuple of strict sign conditions with $\epsilon_1 = 1$. Then the coefficients c_{ε_k} satisfy:*

$$c_{\varepsilon_1} = 1, \text{ and } \epsilon_k c_{\varepsilon_k} = 1 - \sum_{i=1}^{k-1} \epsilon_i (1-\xi_i) c_{\varepsilon_i} \binom{k}{i} \qquad (E_4)$$

for $i = 1, \ldots, k-1$, $k = 2, \ldots, n$ and the computation of $[c_{\varepsilon_1}, c_{\varepsilon_2}, \ldots, c_{\varepsilon_n}]$ takes time $O(n^4 [\log(n)]^2)$ using classical arithmetic.

Proof: The equality (E_4) is obtained taking the particular case $P = X^k$ in the previous corollary 10. By the same equality one deduces that the computation of c_{ε_n} takes $O(n^2)$ arithmetic operations. And by the inequalty (E_3) the size of the integers and the coefficients c_{ε_k} $(k \leq n-1)$ previously computed is bounded by $n\log(n)$, so using classical arithmetic operations, our total computation is in $O(n^4 [\log(n)]^2)$. □

2.2 Generalized Taylor formulas

Let P be polynomial of degree n with coefficients in a commutative ring and $\sigma_0, \sigma_1, \ldots, \sigma_n$ a list of strict sign conditions. We denote by

$$\mathbb{H}(X) = \left[P(X)\ \tau_0, P'(X)\ \tau_1, \ldots, P^{(d)}(X)\ \tau_n\right]$$

and

$$\mathbb{H}'(X) = \left[P(X)\ \tau_0', P'(X)\ \tau_1', \ldots, P^{(d)}\ \tau_n\right]$$

the system of generalized sign conditions obtained from $\mathbb{H}(X)$ by relaxing all inequalities except the last one. Thom's lemma claims (among other things) that:

$$[\mathbb{H}'(U), \mathbb{H}'(V),\ U < Z < V] \Rightarrow \mathbb{H}(Z). \qquad (\star)$$

The generalized Taylor formulas will be a way of expressing this geometric fact by particular algebraic identities, as we shall see later on examples.

Notation 13: Let $\varepsilon = [\epsilon_0, \ldots, \epsilon_n]$ be $(n+1)$-tuple of strict sign conditions with $\epsilon_0 = 1$. We denote by

- $\varepsilon^{(k)} = [1, \epsilon_k \epsilon_{k+1}, \ldots, \epsilon_k \epsilon_{n+1}]$

- λ: 1 if $\epsilon_1 = 1$, 2 if $\epsilon_1 = -1$
- $\Delta_\lambda = \begin{cases} \Delta_1 & \text{if } \epsilon_1 > 0 \\ \Delta_2 & \text{otherwise} \end{cases}$

Theorem 14: *Let* $\mathbf{A}$ *be a commutative ring and* $P \in \mathbf{A}[X]$ *a polynomial of degree* $n \geq 1$. *One considers two new variables* U *and* V *and one lets* $\Delta_1 = X - U$, $\Delta_2 = V - X$. *Then there exists an equality of the following type:*

$$P(X) = P(Y_0) + \sum_{k=1}^{n-1} \epsilon_k \mathrm{D}_k(P)(Y_k)\mathrm{H}_{k,\varepsilon}(\Delta_1, \Delta_2) + \epsilon_n \mathrm{D}_n(P)\mathrm{H}_{n,\varepsilon}(\Delta_1, \Delta_2) \quad (E_5)$$

where

$$Y_k = \begin{cases} U & if\ \epsilon_k\epsilon_{k+1} = 1 \\ V & otherwise \end{cases}$$

and $\mathrm{H}_{k,\varepsilon}$ *is a non-zero homogeneous polynomial of degree* k *with nonnegative integer coefficients, given by the inductive relation:*

$$\mathrm{H}_{1,[1,\epsilon_1]} = \Delta_\lambda$$

$$\mathrm{H}_{n,\varepsilon}(\Delta_1, \Delta_2) = \sum_{k=1}^{n-1} \frac{1 - \epsilon_1\epsilon_k\epsilon_{k+1}}{2} \binom{n}{k} c_{\varepsilon'_k} \Delta_\lambda^k \mathrm{H}_{n-k,\varepsilon^{(k)}}(\Delta_1, \Delta_2) + \ c_{\varepsilon'_n} \Delta_\lambda^n$$

with $\varepsilon' = \left[\epsilon_1^k \epsilon_k\right]_{k=1,\ldots,n}$. *The equality* (E_5) *is called the generalized Taylor formula for* P *associated to the combination* ε.

Proof: By induction on the degree n of P. If $n = 1$ the theorem is easy. Suppose that the equality (E_5) holds for a degree $n - 1 \geq 0$. Let $\varepsilon = [\epsilon_0, \epsilon_2, \ldots, \epsilon_n]$ be a $(n+1)$-tuple of strict sign condtions with $\epsilon_0 = 1$ and P be a polynomial of degree n. We distinguish two cases:

Case 1: $\epsilon_1 = -1$. We write the mixed Taylor formula corresponding to the combination $\varsigma = [\sigma_k]$ where $\sigma_k = (-1)^k \epsilon_k$, $k = 1, \ldots, n$:

$$\begin{aligned} P(X) &= P(V) + \sum_{k=1}^{n-1} \sigma_k c_{\varsigma_k} (X - V)^k \mathrm{D}_k(P)(Z_k) + \sigma_n c_{\varsigma_n} (X - V)^n \mathrm{D}_n(P) \\ &= P(V) + \sum_{k=1}^{n-1} \epsilon_k c_{\varsigma_k} \Delta_2^k \mathrm{D}_k(P)(Z_k) + \epsilon_n c_{\varsigma_n} \Delta_2^n \mathrm{D}_n(P) \end{aligned}$$

with

$$Z_k = \begin{cases} V & \text{if } \epsilon_k\epsilon_{k+1} = -1 \\ X & \text{otherwise.} \end{cases}$$

Then either $Z_k = V, k = 1, \ldots, n-1$ and then the theorem is proved, or there are two sets I and J such that $I \cup J = \{1, \ldots, n\}$, $I \cap J = \emptyset$, $J \neq \emptyset$ and

$$\begin{aligned} P(X) &= P(V) + \sum_{k\in I} \epsilon_k c_{\varsigma_k} \Delta_2^k \mathrm{D}_k(P)(V) \\ &\quad + \sum_{k\in J} \epsilon_k c_{\varsigma_k} \Delta_2^k \mathrm{D}_k(P)(X) + \epsilon_n c_{\varsigma_n} \Delta_2^n \mathrm{D}_n(P). \end{aligned}$$

Let

$$Q(X) = \sum_{k\in J} \epsilon_k c_{\varsigma_k} \Delta_2^k \mathrm{D}_k(P)(X).$$

By induction one has:

$$\begin{aligned}\mathrm{D}_k(P)(X) &= \mathrm{D}_k(P)(Y_k) + \epsilon_k \sum_{j=1}^{n-k} \epsilon_{k+j} \mathrm{D}_j[\mathrm{D}_k(P)](Y_{k+j}) \mathrm{H}_{j,\varepsilon^{(k)}}(\Delta_1, \Delta_2) \\ &= \mathrm{D}_k(P)(Y_k) + \epsilon_k \sum_{m=k+1}^{n} \epsilon_m \binom{m}{m-k} \mathrm{D}_m(P)(Y_m) \mathrm{H}_{m-k,\varepsilon^{(k)}}(\Delta_1, \Delta_2)\end{aligned}$$

with $\mathrm{H}_{m-k,\varepsilon^{(k)}}$ homogeneous, with positive integer coefficients. of degree $m-k$ and $Y_m = U$ or V. If j_0 is the smallest index in J then

$$\begin{aligned}Q(X) &= \sum_{k\in J} \sum_{j=k}^{n} \epsilon_j \mathrm{D}_j(P)(Y_j) \mathrm{G}_{j,\varepsilon^{(k)}}(\Delta_1, \Delta_2) \\ &= \sum_{l=j_0}^{n} \epsilon_l \mathrm{D}_l(P)(Y_l) \mathrm{G}_{l,\varepsilon}(\Delta_1, \Delta_2)\end{aligned}$$

with $\mathrm{G}_{j,\varepsilon^{(k)}} = \binom{j}{j-k} c_{\varsigma_k} \Delta_2^k \mathrm{H}_{j-k,\varepsilon^{(k)}}$, $j \geq k$, and $\mathrm{G}_{l,\varepsilon} = \sum_{k\in J} \mathrm{G}_{l,\varepsilon^{(k)}}$. So one deduces

$$P(X) = P(V) + \sum_{k\in I} \epsilon_k \mathrm{D}_k(P)(V) c_{\varepsilon_k} \Delta_2^k + \sum_{l=j_0}^{n} \epsilon_l D_l(P)(Y_l) \mathrm{G}_{l,\varepsilon}(\Delta_1, \Delta_2).$$

One remarks now that if $l \in I$ then $\epsilon_l \epsilon_{l+1} = -1$ and by induction $Y_l = V$.
Case 2: $\epsilon_1 = 1$, similar reasoning with $\sigma_k = \epsilon_k$. □

Remark 15: For a degree n, we have 2^n generalized Taylor formulas and all the possible sign combinations do appear.

Proposition 16: *Let $\varepsilon = [\epsilon_0, \dots, \epsilon_n]$ be a combination of strict sign conditions. Then the sum of coefficients of the homogeneous polynomial $\mathrm{H}_{k,\varepsilon}$, $k = 1, \dots, n$ is bounded by:*

$$\frac{n!}{2}\left(\left[1+\sqrt{2}\left(\frac{2}{\pi}\right)^{\frac{3}{2}}\right]^n + \left[1-\sqrt{2}\left(\frac{2}{\pi}\right)^{\frac{3}{2}}\right]^n\right) \leq \frac{n!}{2}\left((1,75)^n + 1\right).$$

Proof: The proof can be deduced easily by from proposition 11 and theorem 14. □

2.3 Some explicit examples of generalized Taylor formulas

We consider two variables U and V, and we let $\Delta_1 = X-U$ and $\Delta_2 = V-X$. Let P be a polynomial with coefficients in a commutative ring $\mathbf{A}$. If $deg(P) = 4$, one has sixteen generalized Taylor formulas: The following eight and the symmetric ones obtained by exchanging U and V:

$$P(X)=P(U)+P'(U).\Delta_1+P^{(2)}(U).\left[\tfrac{1}{2}{\Delta_1}^2\right]+P^{(3)}(U).\left[\tfrac{1}{6}{\Delta_1}^3\right]+P^{(4)}.\left[\tfrac{1}{24}{\Delta_1}^4\right],$$

$$P(X)=P(U)+P'(U).\Delta_1+P^{(2)}(U).\left[\tfrac{1}{2}{\Delta_1}^2\right]+P^{(3)}(V).\left[\tfrac{1}{6}{\Delta_1}^3\right]-P^{(4)}.\left[\tfrac{1}{6}{\Delta_1}^3\Delta_2+\tfrac{1}{8}{\Delta_1}^4\right],$$

$$\begin{aligned}P(X)=&P(U)+P'(U).\Delta_1+\\&P^{(2)}(V).\left[\tfrac{1}{2}{\Delta_1}^2\right]-P^{(3)}(V).\left[\tfrac{1}{2}{\Delta_1}^2\Delta_2+\tfrac{1}{3}{\Delta_1}^3\right]+P^{(4)}.\left[\tfrac{1}{4}{\Delta_1}^2{\Delta_2}^2+\tfrac{1}{3}{\Delta_1}^3\Delta_2+\tfrac{1}{8}{\Delta_1}^4\right],\end{aligned}$$

$$\begin{aligned}P(X)=&P(U)+P'(U).\Delta_1+\\&P^{(2)}(V).\left[\tfrac{1}{2}{\Delta_1}^2\right]-P^{(3)}(U).\left[\tfrac{1}{3}{\Delta_1}^3+\tfrac{1}{2}{\Delta_1}^2\Delta_2\right]-P^{(4)}.\left[\tfrac{1}{2}{\Delta_1}^3\Delta_2+\tfrac{1}{4}{\Delta_1}^2{\Delta_2}^2+\tfrac{5}{24}{\Delta_1}^4\right],\end{aligned}$$

$$\begin{aligned}P(X)=&P(U)+P'(V).\Delta_1-P^{(2)}(V).\left[\tfrac{1}{2}{\Delta_1}^2+\Delta_1\Delta_2\right]+\\&P^{(3)}(U).\left[\tfrac{1}{6}{\Delta_1}^3+\tfrac{1}{2}\Delta_1{\Delta_2}^2+\tfrac{1}{2}{\Delta_1}^2\Delta_2\right]+P^{(4)}.\left[\tfrac{3}{4}{\Delta_1}^2{\Delta_2}^2+\tfrac{1}{3}\Delta_1{\Delta_2}^3+\tfrac{1}{2}{\Delta_1}^3\Delta_2+\tfrac{1}{8}{\Delta_1}^4\right],\end{aligned}$$

$$\begin{aligned}P(X)=&P(U)+P'(V).\Delta_1-P^{(2)}(V).\left[\Delta_1\Delta_2+\tfrac{1}{2}{\Delta_1}^2\right]+\\&P^{(3)}(V).\left[\tfrac{1}{2}\Delta_1{\Delta_1}^2+\tfrac{1}{2}{\Delta_1}^2\Delta_2+\tfrac{1}{6}{\Delta_1}^3\right]-P^{(4)}.\left[\tfrac{1}{6}\Delta_1{\Delta_2}^3+\tfrac{1}{4}{\Delta_1}^2{\Delta_2}^2+\tfrac{1}{6}{\Delta_1}^3\Delta_2+\tfrac{1}{24}{\Delta_1}^4\right],\end{aligned}$$

$$\begin{aligned}P(X)=&P(U)+P'(V).\Delta_1-P^{(2)}(U).\left[\tfrac{1}{2}{\Delta_1}^2+\Delta_1\Delta_2\right]-\\&P^{(3)}(V).\left[\tfrac{1}{2}\Delta_1{\Delta_2}^2+{\Delta_1}^2\Delta_2+\tfrac{1}{3}{\Delta_1}^3\right]+P^{(4)}.\left[{\Delta_1}^2{\Delta_2}^2+\tfrac{5}{6}{\Delta_1}^3\Delta_2+\tfrac{1}{3}\Delta_1{\Delta_2}^3+\tfrac{5}{24}{\Delta_1}^4\right],\end{aligned}$$

$$\begin{aligned}P(X)=&P(U)+P'(V).\Delta_1-P^{(2)}(U).\left[\tfrac{1}{2}{\Delta_1}^2+\Delta_1\Delta_2\right]-\\&P^{(3)}(U).\left[\tfrac{1}{3}{\Delta_1}^3+{\Delta_1}^2\Delta_2+\tfrac{1}{2}\Delta_1{\Delta_2}^2\right]-P^{(4)}.\left[\tfrac{1}{2}{\Delta_1}^3\Delta_2+\tfrac{1}{2}{\Delta_1}^2{\Delta_2}^2+\tfrac{1}{6}\Delta_1{\Delta_2}^3+\tfrac{1}{8}{\Delta_1}^4\right].\end{aligned}$$

Assume that U and V give the same relaxed sign condition $\sigma'_0, \sigma'_1, \sigma'_2, \sigma'_3$ to $P, P', P^{(2)}, P^{(3)}$, respectively, and the same strict sign condition σ_4 to $P^{(4)}$, and that $U < X < V$. Then one of the sixteen generalized Taylor formulas is strong evidence showing that $P(X)$ has the strict sign condition σ_0. For example, If σ'_0 is ≥ 0, σ'_1: ≥ 0, σ'_2: ≤ 0, σ_3: ≤ 0, σ_4: < 0, the last generalized Taylor formula can be reread

$$\begin{aligned}-P(X)+P(U)+P'(V).\Delta_1-P^{(2)}(U).\left[\frac{1}{2}{\Delta_1}^2+\Delta_1\Delta_2\right]-P^{(3)}(U).\left[\frac{1}{3}{\Delta_1}^3\right.\\ \left.+{\Delta_1}^2\Delta_2+\frac{1}{2}\Delta_1{\Delta_2}^2\right]-P^{(4)}.\left[\frac{1}{2}{\Delta_1}^3\Delta_2+\frac{1}{2}{\Delta_1}^2{\Delta_2}^2+\frac{1}{6}\Delta_1{\Delta_2}^3+\frac{1}{8}{\Delta_1}^4\right].\end{aligned}$$

The eight generalized Taylor formulas for a degree 3 are obtained from the previous sixteen ones for a degree 4, by deleting the terms with $P^{(4)}$.

3 Algorithm for the construction of the identity providing the Positivstellensatz

Let $\mathbf{D}$ be an ordered domain, $\mathbf{K}$ its fraction field, and $\mathbf{R}$ the real closure of $\mathbf{K}$. Let $F_1, \ldots, F_k \in \mathbf{D}[X]$. We shall call the complete tableau of signs for the F_i

denoted by $\mathcal{T} = \mathcal{T}_k[F_1, \ldots, F_k]$ the data of the number N of distinct real roots in $\mathbf{R}$: $\zeta_1 < \zeta_2 < \cdots < \zeta_N$ of F_j, $(j = 1, \ldots, k)$ and a tableau with k rows and $2N + 1$ columns giving the sign of each F_j at each zero ζ_i and on each open interval $]-\infty, \zeta_1[$, $]\zeta_j, \zeta_{j+1}[$, $]\zeta_N, +\infty[$, $j = 1, \ldots, N-1$.

Note that the tableau does not provide the value of roots but simply their Thom code.

Proposition 17: *Let (P_j) be a family in $\mathbf{D}[X]$ stable by derivation. Let (ζ_i) be the family of real roots of P_j in $\mathbf{R}$. Then one can set up the complete tableau of signs $\mathcal{T}$ for the family (P_j) using only the signs of $P_j(\zeta_i)$.*

Proof: It is an easy consequence of Thom's lemma. □

Let $L = [\mathrm{P}_1, \ldots, \mathrm{P}_s]$ be a list of polynomials of $\mathbf{D}[\mathrm{X}]$, $\varsigma = (\sigma_i)_{i=1,\ldots,s}$ be a n-tuple of generalized sign conditions, and $\mathbb{H} = [\mathrm{P}_1\sigma_1 0, \ldots, \mathrm{P}_s\sigma_s 0]$ be the system of generalized sign conditions corresponding to L and ς. One considers the family $\mathcal{P}$ generated by L and by the operation $P \mapsto P'$, $\zeta_1 < \zeta_2 < \cdots < \zeta_r$ the real roots of polynomials in $\mathcal{P}$, and $\mathcal{T}$ the corresponding complete tableau of signs.

Lemma 18: *If one has*

a) *the incompatibilities*

$$\downarrow [\mathbb{H}, \mathrm{X} < \zeta_1] \downarrow, \downarrow [\mathbb{H}, \zeta_1 < \mathrm{X} < \zeta_2] \downarrow, \ldots, \downarrow [\mathbb{H}, \zeta_r < \mathrm{X}] \downarrow$$

with degree bounded by δ, and

b) *the incompatibilities*

$$\downarrow [\mathbb{H}, \mathrm{X} = \zeta_i] \downarrow \quad (1 \leq i \leq r)$$

with degree bounded by δ', then one can construct $\downarrow \mathbb{H} \downarrow$ with a degree bounded by $2r\delta\delta'$.

Proof: We denote by $\mathbb{E}_1 = \downarrow [\mathbb{H}, \mathrm{X} < \zeta_1] \downarrow$, $\mathbb{E}_k = \downarrow [\mathbb{H}, \zeta_{k-1} < \mathrm{X} < \zeta_k] \downarrow$ for $k = 2, \ldots, r$, $\mathbb{E}_{r+1} = \downarrow [\mathbb{H}, \zeta_r < \mathrm{X}] \downarrow$ and $Q_k = X - \zeta_k$, $k = 1, \ldots, r$. Applying r-times the proposition 7 (ii), one constructs $\downarrow [\mathbb{H}, Q_1 \neq 0, Q_2 \neq 0, \ldots, Q_r \neq 0] \downarrow$ with a degree bounded by $(r+1)\delta$ according to the following diagram:

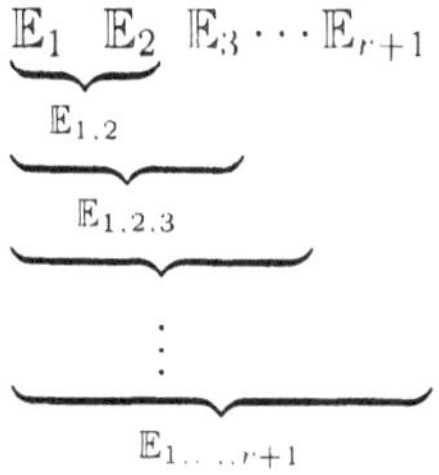

The incompatibility $\mathbb{E}_{1,2} =\downarrow [\mathbb{H}, Q_1 \neq 0, Q_2 < 0] \downarrow$ is obtained by applying from proposition 7 (ii) to $\mathbb{E}_1$ and to $\mathbb{E}_2$. The incompatibility

$$\mathbb{E}_{1,2,3} =\downarrow [\mathbb{H}, Q_1 \neq 0, Q_2 \neq 0, Q_3 < 0] \downarrow$$

is obtained by applying proposition 7 (ii) to $\mathbb{E}_{1,2}$ and to $\mathbb{E}_3$ as previously. Iterating this process until $\mathbb{E}_{r+1}$ provides us with the incompatibility $\mathbb{E}_{1,\ldots,r+1} :=\downarrow [\mathbb{H}, Q_1 \neq 0, Q_2 \neq 0, \ldots, Q_r \neq 0] \downarrow$. We apply next r-times the proposition 7 (iii) with $q = 1$ and, noticing that the exponent of Q_i is bounded by 2δ at each step, one deduces that the degree increases at most by $2\delta(\delta' - 1)$. Finally one gets the bound $(r+1)\delta + 2r\delta(\delta' - 1)$. □

Theorem 19: (Real effective positivstellensatz in one variable)
Let **D** *be an ordered domain,* **K** *its fraction field, and* **R** *the real closure of* **K**. *Let L be a list of polynomials of $\mathbf{D}[X]$ of degree at most d; r be the number of real roots in* **R** *of the family generated by L; and the operation $P \mapsto P'$ and $\mathbb{H}(X)$ be a system of generalized sign conditions on elements of L. Then, either $\mathbb{H}(x)$ is possible in* **R** *or, $\mathbb{H}(x)$ is impossible in* **R** *and then $\downarrow \mathbb{H} \downarrow$ in* **R**. *Moreover, the degree of $\downarrow \mathbb{H} \downarrow$ is bounded by $8rd^2$.*

Proof: Let $\mathcal{T}$ be the complete tableau of signs for the family generated by L and the operation $P \mapsto P'$ and $\zeta_1 < \zeta_2 < \cdots < \zeta_r$ be the ordered list of finite points of $\mathcal{T}$. We set $\zeta_0 = -\infty$ and $\zeta_{r+1} = +\infty$. Assume that $\mathbb{H}(x)$ is impossible in **R**. Then theorem 14 gives a strong incompatibility in **R**

$$\downarrow [\mathbb{H}, \zeta_i < \mathrm{X} < \zeta_{i+1}] \downarrow$$

for $i = 0, \ldots, r$ with degree bounded by $2d$. On the other hand, we have easily a strong incompatibility

$$\downarrow [\mathbb{H}, \mathrm{X} = \zeta_i] \downarrow$$

for $i = 1, \ldots, r$ with degree bounded by $2d$. We conclude applying lemma 18. □

4 Some precisions about implementation and cost of the algorithms in Axiom

Given $L = [P_1, \ldots, P_s]$ a list of univariate polynomials of $\mathbf{D}[X]$, we denote by $\mathcal{P}$ the family of the polynomials of L and their derivatives, and by $\zeta_1 < \zeta_2 < \cdots, < \zeta_r$ the sorted list of real roots of the polynomials in $\mathcal{P}$. The procedure *TS* gives us the complete tableau of signs $\mathcal{T}$ for the family $\mathcal{P}$.

Procedure TS:

Its input is a list L of polynomials with coefficients in **D**.
Its output is the complete tableau of signs for the polynomials in L and their derivatives (in a suitable coded form).

Remark: We use Roy's algorithm SI (simultaneous inequalities) [see RS] in the procedure TS to obtain the complete signs tableau. The procedure TS performs then n times the procedure SI.

Given a polynomial $P = \sum_{i=0}^{n} a_i X^i \in \mathbb{Z}[X]$, we recall that its *norm* is defined as $\left(\sum_{i=0}^{n} a_i^2\right)^{\frac{1}{2}}$ and its *size* $\| P \|$ is defined as the logarithm of its norm.

Proposition 20: *Let* $\mathbf{D} = \mathbb{Z}$, $[P_1, \ldots, P_s]$ *be a list of polynomials with integer coefficients,* d *the maximum of their degree,* t *the maximum of their size, and* r *the number of real roots of* $P_1, \ldots, P_s$ *and their successive derivatives. The procedure TS runs in time* $O(s^2 d^6 r (\log r)^3 t^2)$.

Proof: TS performs at most sd times the procedure SI. Each call of SI takes $O(sd^5 (\log r)^3 t^2)$, so we get the stated total computation time by multiplication by sd. □

In the algebraic identity which provides the positivstellensatz, we introduce for each ζ_i a variable x_i which represents it. The algebraic identity computed can be represented in compact form. When we use the algorithm described in proposition 7 (iii), we take to the power $2k$ ($k \geq 0$) an algebraic identity of the following type:

$$S + P + N = Y$$

which can be written

$$S^{2k} + \left[(S+P)^{2k} - S^{2k}\right] + \left[(S+P+N)^{2k} - (S+P)^{2k}\right] = Y^{2k}.$$

In the right side of the previous equality, we get $S_1 + P_1 + N_1$, with S_1, P_1 and N_1 in, respectively, the monoid, the cone, and the ideal. Instead of expanding P_1 in sum of squares with positive weights, we can use the following notation:

$$P_1 = diffp_1(S, P, 2k).SP + diffp_2(S, P, 2k)$$

where $diffp_1$ and $diffp_2$ are sums of squares which are easy to put in the explicit form.

Also, to avoid expanding N_1, we shall write:

$$N_1 = diffp_0(S, P, N, 2k).N.$$

Our algorithm produces, from the procedure TS, a straight-line program which gives the desired identity. The evaluation of the last instruction of this straight-line program is an algebraic expression in the variables X and $P_j(x_i)$, which is a suitable incompatibility as far as that the x_i are replaced by the ζ_i. It follows that this expression is an identity with coefficients in $\mathbb{R}_{alg}$ (the real closure of $\mathbb{Q}$). Whenever $P_j(\zeta_i) = 0$, we have substituted the term $P_j(x_i)$ by 0 in the expression which has been especially simplified. The final identity is hence an equality in $\mathbb{Z}[\mathrm{X}, (x_i)] / \mathbf{I}$ where $\mathbf{I}$ the ideal generated by $P_j(x_i)$ corresponding to the zero value of $P_j(\zeta_i)$. In practice, the computation time by straight-line program of the algebraic identity is negligible when compare to the time of the computation of the complete tableau of signs.

Example: Let us consider the following polynomial

$$P = -7X^3 + 5X^2 + 2X + 3.$$

We have the following complete tableau of signs for P and its successives derivatives:

		x_1		x_2		x_3		x_4	
$P^{(3)}$	$-$	$-$	$-$	$-$	$-$	$-$	$-$	$-$	$-$
$P^{(2)}$	$+$	$+$	$+$	0	$-$	$-$	$-$	$-$	$-$
P'	$-$	0	$+$	$+$	$+$	0	$-$	$-$	$-$
P	$+$	$+$	$+$	$+$	$+$	$+$	$+$	0	$-$

The system $\downarrow \mathbb{H} = [P \leq 0, P' \geq 0] \downarrow$ is impossible in $\mathbb{Q}_{alg}$. We denote:

- $p_{i,j} = |P^{(i)}(x_j)|$ for $i = 0, \ldots, 2$, $j = 1, \ldots, 4$
- $\Delta_k = X - x_k$ for $k = 1, \ldots, 4$
- $R_{i,j}$ the polynomial such that: $P^{(i)}(X) = P^{(i)}(x_j) + \Delta_j R_{i,j}$ $i = 0, 1, j = 1, \ldots, 4$
- $T_0 = -P(X)$, $T_1 = P'(X)$

The strong incompatibility of $\mathbb{H}$ on the open intervals $]-\infty, x_1[$, and $]x_i, x_{i+1}[$ $(i = 1, 2, 3)$, $]x_4, +\infty[$ is given respectively by the following generalized Taylor formulas:

$$q_1 + T_1 - p_{2,1}\Delta_1 = 0$$
$$p_{0,1} + T_0 + \Delta_1 q_{2,1} - \Delta_2 q_{2,2} = 0$$
$$p_{0,2} + T_0 + \Delta_2 q_{3,1} - \Delta_3 q_{3,2} = 0$$
$$q_4 + T_1 + p_{2,3}\Delta_3 = 0$$
$$p_{1,4} + T_1 + q_5 + p_{2,4}\Delta_4 = 0$$

where

$$q_1 = 21\Delta_1^2$$
$$q_{2,1} = 14\Delta_1^2,\ q_{2,2} = 21\Delta_1^2$$
$$q_{3,1} = 21\Delta_3^2 + 14\Delta_2^2,\ q_{3,2} = 42\Delta_2^2$$
$$q_4 = 21\Delta_3^2$$
$$q_5 = 21\Delta_4^2.$$

The identity which provides the strong incompatibility of $\downarrow \mathbb{H} \downarrow$ is given by the equality:

$$A + B + C = 0$$

where

$$A = 441 p_{0,1}^3 p_{0,2} p_{0,3}^2 \quad . \quad C = 0.$$

and B is expanded in the following straight-line program:

$$\begin{aligned}
B_1 &= q_1 p_{0.1} \\
B_2 &= (q_1 + T_1)T_0 + p_{0.1}T_1 + p_{2.1}q_{2.1}\Delta_1^2 \\
B_3 &= q_{2,2}(q_1 + T_1) \\
B_4 &= p_{0,2}B_1 \\
B_5 &= (B_1 + B_2)T_0 + p_{0,2}B_2 + q_{3,1}\Delta_2^2 B_3 \\
B_6 &= q_{3,2}(B_1 + B_2) \\
B_7 &= (B_4 + B_5)T_1 + q_4 B_5 + p_{2,3}\Delta_3^2 B_6 \\
B_8 &= 21 p_{0,1}p_{0,2}q_4 \\
B_9 &= p_{0,1}T_0 B_8 diffp_1(p_{0,1}, T_0, 2) + B_8 diffp_2(p_{0,1}, T_0, 2) + R_{0,1}^2 B_7 \\
B_{10} &= 441 p_{0,1}^3 p_{0,2} \\
B &= p_{0,3}T_0 B_{10} diffp_1(p_{0,3}, T_0, 2) + B_{10} diffp_2(p_{0,3}, T_0, 2) + R_{0,3}^2 B_9
\end{aligned}$$

References

[BCR] J. Bochnak, M. Coste, M.-F. Roy, *Géométrie algébrique réelle*. A series of Modern Surveys in Mathematics 11, Springer-Verlag, 1987.

[Lom1] H. Lombardi, *Effective real Nullstellensatz and variants*. Effective Methods in Algebraic Geometry, Progress in Mathematics 94, pp. 263–288, Birkhäuser, 1991.

[Lom2] H. Lombardi, *Une borne sur les dégrés pour le théorème des zéros réel effectif*, Lecture Notes in Mathematics 1524, eds. M. Coste, L. Mahe, M.-F. Roy, pp. 323–345, 1991.

[RS] M.-F. Roy, A. Szpirglass, *Complexity of computation on real algebraic numbers*, Journal of Symbolic Computation 10, pp. 39–51, 1990.

[CLGR] F. Cucker, L. González-Vega, F. Rosello *On algorithms for real algebraic plane curves*, Effective Methods in Algebraic Geometry, Progress in Mathematics 94, pp. 63–87, Birkhäuser, 1991.

[War] H. Warou "Thèse", Univ. de Rennes1, 1994.

H. Warou (warou@emmy.univ-rennes1.fr)
IRMAR, University of Rennes I, 35042 Rennes Cedex (France).

Progress in Mathematics, Vol. 143, © 1996 Birkhäuser Verlag Basel/Switzerland

Solving zero-dimensional involutive systems

A. Yu. Zharkov

1 Introduction

In our recent paper [1], the notion of involutive bases of polynomial ideals was introduced and an algorithm for computing involutive bases was presented. The improved form of this algorithm together with the proof of its correctness in the zero-dimensional case is given in [2]. In the positive-dimensional case, a linear change of variables is generally required for constructing involutive bases defined in our sense. It turns out that when the involutive basis exists (without change of variables), it can be computed considerably faster by our algorithm than the minimal standard basis by Buchberger's algorithm [3]. On the other hand, an involutive basis computed in the total-degree term ordering often looks more complicated than the corresponding minimal standard basis. The reason is that the involutive basis of a zero-dimensional ideal is nothing but a standard basis enlarged to an "overdetermined" linear algebraic system in monomials irreducible modulo this ideal. From this fact, some interesting properties of involutive bases may be deduced, and a simple method for solving zero-dimensional systems may be constructed.

In the second section of this paper, we recall the notions of the Janet normal form and involutive bases of polynomial ideals. We present also a new version of algorithm *Invbase* for computing involutive bases with further improvements in comparison with that presented in [2]. In the third section we systematically investigate the structure of zero-dimensional involutive bases. In the fourth section we describe the method of converting the total-degree involutive basis to the pure lexicographical reduced standard basis in case the shape lemma holds [7].

2 Involutive bases

Throughout, we use the following notation.

K arbitrary field of characteristic zero
$K[x_1, \dots, x_n]$ polynomial ring over K and f, g, h polynomials in such ring
F, G, H finite subsets in $K[x_1, \dots, x_n]$
card(F) number of elements (cardinality) of F
u, v, w terms in polynomials (without coefficients from K)
$\deg(u, x_i)$ degree of u in variable x_i and $\deg(u)$ total degree of u
$cf(f, u)$ the coefficient of u in f and (F) the ideal generated by F.

Let variables x_i be ordered as $x_1 < \cdots < x_n$ and let $<_T$ be some admissible term ordering. Denote

$lt(f)$– leading term in f with respect to $<_T$
$lc(f) = cf(f, lt(f))$
$lt(F) = \{lt(f) \mid f \in F\}$
$\deg(F) = \max\{\deg(lt(f)) \mid f \in F\}$.

Definition 1. Variable x_i is *multiplicative* for the term u if its index i is not greater than the index of the lowest variable in u. Otherwise x_i is *non-multiplicative* for u.

For a given polynomial g, denote by Nonmult(g) a set of non-multiplicative variables for $lt(g)$.

Definition 2. The *class of a term* u (symbolically class(u)) is the index of the lowest variable contained in u with a non-zero power. The class of a unit term (containing all variables in zero powers) is defined to be $n+1$, where n is the number of variables. The *class of a polynomial* g (symbolically class(g)) is the class of $lt(g)$.

Denote $u \cdot v$ by $u \times v$ if all variables in v are multiplicative for u or if $\deg(v) = 0$. Write also $g \cdot u = g \times u$ if $lt(g) \cdot u = lt(g) \times u$.

Definition 3. The term u is called a *Janet divisor* for the term w if there exists a term v such that $w = u \times v$ (symbolically $u \mid_J w$).

Definition 4. The polynomial f *reduces to* h *modulo* G *in the sense of Janet* if there exist $g \in G$ and u such that $lt(g) \cdot u \equiv lt(g) \times u,\ a \equiv cf(f, lt(g) \times u) \neq 0$, and $h = f - a \cdot g \times u$. The polynomial f is given in *Janet normal form* modulo G if for each term in f there are no Janet divisors in $lt(G)$. The polynomial h is a *Janet reduced form* of f modulo G (symbolically $h = NF_J(f, G)$) if there exists a chain of Janet reductions from f to h and h is given in Janet normal form modulo G.

In contrast to the Janet normal form, we denote by $NF(f, G)$ a usual normal form of f modulo G. An algorithm for computing NF_J may be obtained from one for computing NF [3] replacing usual division of terms by Janet division.

Definition 5. G is *autoreduced* (in the sense of Janet) if $\forall_{g,g' \in G, g \neq g'} \neg(lt(g) \mid_J lt(g'))$. G is *completely autoreduced* if $\forall_{g \in G}\ NF_J(g, G \setminus \{g\}) = g$.

Denote by Autoreduce(F) a function that for a given F computes G which is autoreduced and $(F) = (G)$. An algorithm for computing Autoreduce may be obtained from the well-known algorithm Reduce *All* [3], replacing usual NF by NF_J.

Definition 6 [1]. G is an *involutive basis* if it is autoreduced and

$$\forall_{g\in G}\,\forall_{x\in \mathrm{Nonmult}(g)}\; NF_J(g\cdot x,\, G) = 0. \tag{1}$$

Some general properties of the involutive bases established in [1, 2] are summarized below.

Properties of involutive bases [1, 2].

- If G is involutive then $\forall_{f\in(G)}\; NF_J(f, G) = 0$.
- If G is involutive then $\forall_h\; NF_J(h, G) = NF(h, G)$.
- Any involutive basis is a standard basis, generally not minimal.
- Let G be an involutive basis and $G_{\min}$ be a minimal standard basis of (G). Then, for each term u there exist $g \in G$ such that $lt(g) \mid_J u$ if and only if there exist $g' \in G_{\min}$ such that $lt(g') \mid u$.
- If G, H are involutive bases and $(G) = (H)$, then $lt(G) = lt(H)$. Moreover, if G, H are completely autoreduced, then $G = H$ up to multiplication of polynomials by non-zero elements of K.
- If G is involutive then for each $g \in G$ and for each term u such that $u \mid_J lt(g)$ and $\mathrm{class}(u) > \mathrm{class}(g)$ there exist exactly one $g' \in G$ such that $\mathrm{class}(u) = \mathrm{class}(g')$ and $u \mid_J lt(g')$.
- Let G be involutive. The dimension of (G) is k if and only if $T(x_i) \cap lt(G) = \emptyset$ for $1 \le i \le k$ and $T(x_j) \cap lt(G) \ne \emptyset$ for $k+1 \le j \le n$ where for any subset $S \subseteq \{x_1, \ldots, x_n\}$ we let $T(S)$ be the set of all terms of variables in S.
- For any zero-dimensional ideal there exist an involutive basis.
- For any positive-dimensional ideal an invertible linear change of variables may be found such that involutive basis of a given ideal does exist in terms of the new variables.

Below, an improved version of algorithm *Invbase* for constructing involutive basis G of the ideal generated by a given set F is presented (see [4]).

Algorithm 1 ($G = \mathit{Invbase}(F)$).
Input: $F = \{f \mid f \in K[x_1, \ldots, x_n]\}$
Output: G - involutive basis of (F)
$G :=$ Autoreduce(F);
$L := lt(G)$;
while $L \ne \emptyset$ **do**

$g :=$ element of G with minimal $lt(g) \in L$;
$L := L \setminus \{lt(g)\}$;
$L_1 := lt(G)$;
for each x **in** Nonmult(g) **do**
  $G :=$ Autoreduce$(G \cup \{g \cdot x\})$;
$L_2 := lt(G)$;
$L := (L \cap L_2) \cup (L_2 \setminus L_1)$;

Algorithm 1 may be obtained from the algorithm *Invbase* of the paper [2] by avoiding the so-called repeated prolongations. That is, in each step of the **while** loop we add a product $g \cdot x$ to the current basis only if such product with the same $lt(g)$ and x has not been already considered in the previous steps. Using a noetherian argument, it can be proved that the repeated prolongations in fact reduce to zero; a detailed proof is to be given elsewhere. Our computational experience shows that avoiding such zero-reduced prolongations construction gives a considerable speed-up.

3 Structure of zero-dimensional involutive bases

Throughout this section, G refers to an involutive basis of a *zero-dimensional* ideal in some admissible term-ordering $<_T$. Now we are beginning to study some properties of zero-dimensional involutive bases.

Theorem 1. For any term v such that $\deg(v) \geq \deg(G)$, there exist $g \in G$ such that $lt(g) \mid_J v$.

Proof. Since G is an involutive basis, and consequently a standard basis, of a zero-dimensional ideal, then for each $1 \leq i \leq n$ there exist $g_i^* \in G$ such that $lt(g_i^*) = x_i^{d_i}$ and $d_i > 0$ [3]. Assume for contradiction that there exists a term u such that $\deg(u) \geq \deg(G)$ and u has no Janet divisors in $lt(G)$. Consider two alternative cases. First assume that there exist i such that $\deg(u, x_i) \geq d_i$. In this case u may be represented as $u = v \cdot lt(g_i^*)$. Since the polynomial $v \cdot g_i^* \in (G)$, its leading term u has a Janet divisor in $lt(G)$, which contradicts our assumption. Another possibility is $\deg(u, x_i) < d_i$ for each $i = 1, \ldots, n$. Let $u' = u \mid_{x_1=1}$. Since $u' \cdot g_1^* \in (G)$, the term $u' \times x_1^{d_1}$ has a Janet divisor, say v, in $lt(G)$. The latter should have the form $v = u' \times x_1^p$ where $\deg(u, x_1) < p \leq d_1$ because otherwise v would be a Janet divisor for u. Hence $\deg(u) < \deg(v) \leq \deg(G)$ which contradicts the fact that $\deg(u) > \deg(G)$. □

Let U be a set of all irreducible terms (in the sense of Janet) modulo G. By the properties of involutive bases (see above), U is nothing but the set of all irreducible terms (in the usual sense) modulo a standard basis. Since (G) is zero-dimensional, U is finite and $D \equiv \text{card}(U)$ is the number of roots of (G)

counting their multiplicities [3]. Let us denote

$$U_i = \{u \in U \mid \text{class}(u) > i\}, \quad D_i = \text{card}(U_i), \quad i = 1, \dots, n.$$

We remember that class(1) $= n + 1$, and therefore $U_n = \{1\}$. It is natural to set $U_0 = U$ and $D_0 = D$. Evidently $U_{i+1} \subseteq U_i$, and hence $D_{i+1} \leq D_i$. Denote also

$$G_i = \{g \in G \mid \text{class}(g) = i\}.$$

Now we are ready to state the following interesting property of zero-dimensional involutive bases.

Theorem 2. Let G be an involutive basis of a zero-dimensional ideal. and let G_i, U_i be defined as above. Then for all $i = 1, 2, \dots, n$ and for all $g \in G_i$, $lt(g) \mid_{x_i=1} \in U_i$. Conversely, for all $i = 1, 2, \dots, n$ and for each $u \in U_i$ there exists exactly one $g \in G_i$ such that $u = lt(g) \mid_{x_i=1}$.

Proof. Let $g \in G_i$, that is, $lt(g) = u \times x_i^k$, class$(u) > i$. Since G is autoreduced, u has no Janet divisors in $lt(G)$. Therefore, $u \equiv lt(g) \mid_{x_i=1} \in U_i$.
Conversely, let $u \in U_i$. Consider $v = u \times x_i^N$ such that $\deg(v) \geq \deg(G)$. From theorem 1 it follows that there exist $g \in G$ such that $lt(g) \mid_J v$, that is, $lt(g) = u \times x_i^k$. Since u is irreducible modulo G, $k > 0$, and hence $g \in G_i$. Since any monomial has no more than one Janet divisor in $lt(G)$ (see [1], proposition 1), g is determined uniquely. □

Theorem 3. The number of elements in G is

$$\text{card}(G) = \sum_{i=1}^{n} D_i = \sum_{i=1}^{n} i \cdot N_{i+1}$$

where D_i is defined as above and N_j is a number of all terms of the class j irreducible modulo G.

Proof. Since $G = \cup_{i=1}^{n} G_i$ and $G_i \cap G_j = \emptyset$ for $i \neq j$, then card$(G) = \sum_{i=1}^{n}$ card(G_i). By theorem 2, card$(G_i) = D_i$. Hence card$(G) = \sum_{i=1}^{n} D_i$. Taking in account the recurrence relations $D_{i-1} = D_i + N_i$, $i = n, n-1, \dots, 1$, we obtain that card$(G) = \sum_{i=1}^{n} i \cdot N_{i+1}$. □

Corollary 1. Let G_{inv} be an involutive basis of a zero-dimensional ideal and G_{min} be a corresponding minimal standard basis. Then the following chain of inequalities obviously holds:

$$\text{card}(G_{\text{min}}) \leq \text{card}(G_{\text{inv}}) \leq 1 + (n-1)D_1 \leq nD_1 \leq nD.$$

Theorem 4. Let G be an involutive basis of zero-dimensional ideal. Then

$$D_{i-1} = \sum_{g \in G_i} \deg(g, x_i).$$

Proof. By theorem 2,

$$lt(G_i) = \{u_k \cdot x_i^{d_k} \mid u_k \in U_i,\ d_k > 0,\ k = 1, \dots, D_i\}, \quad i = 1, 2, \dots, n,$$

where $u_l \neq u_m$ for $l \neq m$. It is easy to observe that U_{i-1} is a union of D_i disjoint sets $\{u_k \cdot x_i^j \mid 0 \le j < d_k\}$ where $k = 1, \dots, D_i$. From this it immediately follows $D_{i-1} = \sum_{k=1}^{D_i} d_k$, which proves the theorem. □

Corollary 2. Let G be an involutive basis of zero-dimensional ideal. The number of roots of (G) counting their multiplicities is

$$D = \sum_{g \in G} \deg(g, x_1).$$

4 Conversion to lexicographical standard basis

In this section we propose a method for converting an involutive basis of a zero-dimensional ideal in any admissible (normally, total degree) term ordering to the pure lexicographical minimal standard basis. We describe the theoretical foundations of the method as well as a version of the corresponding algorithm. Our method is based on the following property of zero-dimensional involutive bases resulting from the theory developed in the previous section.

Theorem 5. Let G be a completely autoreduced involutive basis of a zero-dimensional ideal, and let G_i, U_i, D_i be as defined above theorem 2. Then G_i is nothing but a system of D_i linear algebraic equations in $D_i - 1$ unknowns $u \in U_i \setminus \{1\}$ over $K[x_1, \dots, x_i]$. These equations are linearly independent over $K[x_1, \dots, x_i]$.

Proof. Since G is *completely* autoreduced, any term w in the reductum of any polynomial $g \in G_i$ should have the form $w = u \times v$ where $u \in U_i$. Indeed, if u were not an element of U_i, the term w could not be irreducible in the sense of Janet. On the other hand, by theorem 2, for each $g \in G_i$ its leading term has the form $lt(g) = u \times x_i^k$, $k > 0$, $u \in U_i$ (including $u = 1$), which gives a one-to-one correspondence between the sets U_i and $lt(G_i)$. Thus, G_i is evidently a linear algeraic system of D_i equations in $D_i - 1$ unknowns $u \in U_i \setminus \{1\}$ over $K[x_1, \dots, x_i]$.

Assume for a contradiction that these equations are not linearly independent over $K[x_1, \dots, x_i]$, that is, $c_1 \cdot g_1 + c_2 \cdot g_2 + \dots + c_k \cdot g_k = 0$ where $c_j \in K[x_1, \dots, x_i]$, $g_j \in G_i$, and, say, $c_1 \neq 0$. Since $\mathrm{class}(g_j) = i$, we may write

$$lt(g_1) \times lt(c_1) = \max\{lt(g_2) \times lt(c_2), \dots, lt(g_k) \times lt(c_k)\}$$

where by max is meant the maximal term in the sense of $<_T$ ordering. Let $lt(g_2) \times lt(c_2)$ be such a term. Then the term $lt(g_1) \times lt(c_1)$ has 2 different

Janet divisors in G_i: the term $lt(g_1)$ and the term $lt(g_2)$. This contradicts the fact that G_i is autoreduced in the sense of Janet (see [1], proposition 1). □

An immediate consequence of theorem 5 is an algorithm for isolating the lowest variable x_1. Indeed, assuming $i = 1$ in theorem 5 and denoting D_1 by N, we see that G_1 is nothing else but the set of components of the vector $A(x_1) \cdot \mathbf{u}$ where $A(x_1)$ is a square $N \times N$ matrix whose elements are univariate polynomials in x_1 and $\mathbf{u}$ is a vector with D_1 components $u_i \in U_1$ arranged so that $u_i > u_j$ for $i < j$ with respect to the pure lexicographic ordering (note that $u_N = 1$). By theorem 5, the elements of G_1 are linearly independent over $K(x_1)$, and hence $det\, A(x_1) \neq 0$. Polynomial matrix $A(x_1)$ may be transformed to the equivalent *upper triangular form*

$$B(x_1) = \|\, b_{ij}(x_1)\, \|, \quad b_{ij}(x_1) = 0\ (i > j), \quad i, j = 1, \ldots, N$$

by means of the *left elementary operations* (see [5], chapter VI, theorem 1):

1. Multiplication of the row by a non-zero number.
2. Addition to some row another row multiplied by any polynomial in x_1.
3. Permutation of two rows.

Applying the algorithm described in [5], one can find the following representation of the matrix $A(x_1)$:

$$A(x_1) = Q(x_1) \cdot B(x_1), \quad det\, Q(x_1) = c \neq 0, \quad c \in K$$

where $Q(x_1)$ is a square $N \times N$ matrix whose elements are polynomials in x_1. From this relation and inequality $det\, A(x_1) \neq 0$, it follows $detB(x_1) \neq 0$ that implies $b_{ii}(x_1) \neq 0$ for $i = 1, \ldots, N$. Let $\tilde{G}_1$ be a set of components of the vector $B(x_1) \cdot \mathbf{u}$. Since the left elementary operations correspond to the equivalent transformations of the polynomial set G_1, we have $(\tilde{G}_1) = (G_1)$ (we shall refer to $\tilde{G}_1$ as a *triangular set* equivalent to G_1). Taking into account that $u_N = 1$, we have $b_{NN}(x_1) \in (G_1)$. Below it will be proved that $b_{NN}(x_1)$ is just the lowest element of the lexicographical standard basis of (G).

The algorithm for constructing the triangular set $\tilde{G}_1$ [5] may be formally described in the following way. Let $<_T$ be any admissible (normally, total-degree) term ordering, let $<_L$ be the pure lexicographical term ordering with the same order of variables, and let G be completely autoreduced zero-dimensional involutive basis in $<_T$ ordering.

Algorithm 2.

Input: $G_1 = \{g \in G \mid \text{class}(g) = 1\}$.

Output: $\tilde{G}_1$ - a triangular basis with respect to $<_L$ such that $(\tilde{G}_1) = (G_1)$.

1. Fix $<_L$ term ordering and rearrange G_1 with respect to $<_L$.
2. $\tilde{G}_1 := \text{Reduce}(G_1, 1)$.

The function Reduce(F, i) in algorithm 2 computes an autoreduced form of F in terms of the so-called i-division. We say that the term u is an i-divisor of the term v iff $u \mid v$ and $u \mid_{x_1=\cdots=x_i=1} = v \mid_{x_1=\cdots=x_i=1}$. The algorithm for computing Reduce$(\ldots, i)$ may be obtained from the well-known algorithm Reduce*All* [3] by replacing usual division by i-division in the normal form algorithm.

As it is shown above, the minimal with respect to $<_L$ element of $\tilde{G}_1$ is an equation in the single variable x_1. To prove that it is just the minimal element of the corresponding minimal lexicographical standard basis, we need the following theorem.

Theorem 6. Let G be a completely autoreduced zero-dimensional involutive basis with respect to $<_T$ ordering and $G_{1,\ldots,i} = \{g \in G \mid \text{class}(g) \leq i\}$. Let H be the minimal standard basis of (G) with respect to $<_L$ ordering and $H_i = H \cap K[x_1 \ldots x_i]$. Then for each $i = 1, \ldots, n$ and for each $h \in H_i$, the equality $NF_J(h, G_{1\ldots i}) = 0$ holds where NF_J is computed with respect to $<_T$ ordering.

Proof. For fixed $i = 1, \ldots, n$, we let $P(U_i)$ be the set of all finite sums of the form

$$P(U_i) = \{\sum_{j,k} \alpha_{jk} \cdot u_j \times v_{jk} \mid \alpha_{jk} \in K,\ u_j \in U_i\}.$$

Evidently, any $f \in P(U_i)$ is in Janet normal form modulo $G \setminus G_{1,\ldots,i}$. Since G is completely autoreduced, and since for any term $u \in U_i$ all its Janet divisors also lie in U_i, from theorem 5 it follows that $G_{1,\ldots,i} \subset P(U_i)$. Note that H_i is also a subset of $P(U_i)$ (with all $u_j = 1$). Consider any $h \in H_i$. From $h \in (G)$ it follows that $NF_J(h, G) = 0$. We have to prove that in fact a stronger condition holds, namely $NF_J(h, G_{1,\ldots,i}) = 0$. First we claim that any $f \in P(U_i)$ may be reduced by means of polynomials from $G_{1,\ldots,i}$ and not from $G \setminus G_{1,\ldots,i}$. Indeed, otherwise the terms in f could not have the form $u \times v (u \in U_i)$ since u would not be irreducible modulo G. So, as $h \in P(U_i) \cap (G)$, then $NF_J(h, G_{1,\ldots,i}) = 0$. □

Corollary 3. $(H_i) \subseteq (G_{1,\ldots,i})$ for each $i = 1, \ldots, n$. □

Corollary 4. The minimal element of H coincides with the minimal element of $\tilde{G}_1$ computed by algorithm 2.

Proof. Let $\tilde{g}_1 \in K[x_1]$ be the minimal element of $\tilde{G}_1$ and $h_1 \in K[x_1]$ be the minimal element of H. As above, we let G_1 be subset of involutive basis G containing all the elements of class 1. From algorithm 2 it follows that each $g_i \in G_1$ is a linear combination of $\tilde{g}_j \in \tilde{G}_1$ $(i, j = 1, \ldots, D_1)$ with coefficients in $K[x_1]$. Because of theorem 6, $NF_J(h_1, G_1) = 0$ (with respect to $<_T$), and hence

$$h_1 = \sum_{i=1}^{D_1} c_i(x_1) \cdot g_i = \sum_{j=1}^{D_1} \tilde{c}_j(x_1) \cdot \tilde{g}_j, \quad c_i(x_1), \tilde{c}_j(x_1) \in K[x_1].$$

Taking into account the triangular structure of $\tilde{G}_1$ mentioned above algorithm 2, we conclude that $\tilde{c}_j(x_1) = 0$ for $j = 2, \ldots, n$, and so $h_1 = \tilde{c}_1(x_1) \cdot \tilde{g}_1$. On the other hand, since H is a standard basis of (G) in $<_L$ ordering and $\tilde{g}_1 \in K[x_1] \cap (G)$, the equality $NF(\tilde{g}_1, \{h_1\}) = 0$ (with respect to $<_L$) holds, i.e., $\tilde{g}_1 = c(x_1) \cdot h_1$ where $c(x_1) \in K[x_1]$. Hence $\tilde{g}_1 = h_1$ (up to multiplication by non-zero element of K). □

In most cases, applying algorithm 2 gives not only the minimal equation but the whole lexicographical standard basis. So, the following theorem holds.

Theorem 7. Let G, H be as in theorem 6, $<_T$ be the total degree, and $<_L$ be the pure lexicographical term ordering. Suppose card$(H) = n$ and $\{x_2, \ldots, x_n\} \subset lt(H)$. Assume moreover that $(G) \cap E = \emptyset$ where $E \subset K[x_1, \ldots, x_n]$ is the set of all linear forms in $x_1, \ldots, x_n$. Then, after removing redundant elements, $\tilde{G}_1$ computed by algorithm 2 coincides with H.

Proof. For each $i = 2, \ldots, n$, let h_i be the element of H such that $lt(h_i) = x_i$, and $\tilde{g}_i$ be the element of $\tilde{G}_1$ such that $lt(\tilde{g}_i)\,|_{x_1=1} = x_i$ where the leading terms are defined with respect to $<_L$. Since $<_T$ is the total degree ordering, the assumption $(G) \cap E = \emptyset$ implies $\{x_2, \ldots, x_n\} \subset U_1$. Consequently, $H \subset P(U_1)$ and so $NF_J(h_i, G_1) = 0$ for each $i = 2, \ldots, n$. Repeating the same reasonings as in the proof of corollary 4, we obtain

$$h_i = \sum_{j=1}^{D_1} c_j \cdot \tilde{g}_j, \quad \tilde{g}_j \in \tilde{G}_1, \quad c_j \in K[x_1], \quad i = 2, \ldots, n.$$

Taking into account the triangular form of $\tilde{G}_1$ and considering successively each $i = 2, \ldots, n$, one can easily observe that the only possibility is $h_i = \tilde{g}_i$ for each i. Together with corollary 4 this proves the theorem. □

Note that the form of H supposed in theorem 7 is known to happen for zero-dimensional radicals in generic position [6] and, more generally, for the sets of curvilinear points in generic position [7]. It means that algorithm 2 computes the whole lexicographical standard basis for the *most* zero-dimensional ideals and may be considered as an alternative to the well-known FGLM-technique [8].

A "natural" generalization of algorithm 2 for the arbitrary zero-dimensional ideals is given below.

Algorithm 3 ($H = Invlex(G)$).
Input: G - zero-dimensional involutive basis with respect to $<_T$
Output: H - triangular basis of (G) with respect to $<_L$
$H := \emptyset$;
for $i := 1 : n$ **do**
  $H :=$ Reduce$(H \cup G_i, i)$;
  if $lt(H) \cap T(x_k) \neq \emptyset$ for all $k \in \{1, .., n\}$ **then go to** *exit*;
exit: $H :=$ Remred(H);

The function Reduce$(\ldots, i)$ is computed with respect to $<_L$ ordering by using i-division. The function Remred removes in a given set all redundant elements, i.e., those elements whose leading terms are the multiples (in the usual sense) of the leading terms of other elements.

Algorithm 3 had been tested on many examples, mainly for non-radical ideals. Almost always the output set H is just the minimal lexicographical standard basis. However, there are some examples (i.e., the so-called cyclic root problems with 5 and 6 variables) for which the result of algorithm 3 is not a standard basis, though it is very closed to it. Thus if the output set H does not have the form of theorem 7, one should apply Buchberger's algorithm with respect to $<_L$-ordering with H as input. Since H is closed to a standard basis or coincides with it, this computation is rather fast.

The algorithm *Invlex* has been implemented in the computer algebra system REDUCE. By using this implementation, the author has computed the lexicographical standard basis for the famous polynomial system by K. Rimey [9] unsolvable during the last 10 years by any computer algebra tools and considered as hopeless. The computation took about 10 hours on the computer ALPHA/DEC with 50 Mb memory.

Acknowledgements. The author is grateful to J. Apel, Yu. Blinkov, V. Gerdt, and T. Mora for useful discussions.

References

[1] Zharkov, A. Yu., Blinkov, Yu. A. Involution Approach to Solving Systems of Algebraic Equations. Proceedings of the IMACS'93, pp. 11–16, 1993.

[2] Zharkov A. Yu., Blinkov Yu. A. Involutive Bases of Zero-Dimensional Ideals. To appear in Journal Symbolic Computation.

[3] Buchberger B. Gröbner bases: An Algorithmic Method in Polynomial Ideal Theory. In (Bose N.K., ed.) Recent Trends in Multidimensional System Theory, Reidel, 1985.

[4] Zharkov A. Yu., Blinkov Yu. A. Involutive Systems of Algebraic Equations. To appear in Russian journal "Programmirovanie" (in Russian).

[5] Gantmacher R. F. *Theory of Matrices.* Moscow, "Nauka", 1988 (in Russian).

[6] Gianni P., Mora T. Algebraic Solution of Systems of Polynomial Equations using Gröbner Bases. Lecture Notes in Computer Science, Springer 356, pp. 247–257, 1989.

[7] Mora T. The shape of the Shape Lemma. To appear.

[8] Faugère, J. C., Gianni, P., Lazard, D. and Mora. T. Efficient Computation of Zero-Dimensional Gröbner Bases by Change of Ordering. Technical Report LITP 89-52, 1989.

[9] Rimey K. A System of Polynomial Equations and a Solution by an Unusual Method. ACM-SIGSAM Bull. 18, pp. 30–32, 1984.

A. Yu. Zharkov[†]

Saratov University, Astrakhanskaya 83, Saratov 410071 (Russia).

(†) We regret to inform that the author died shortly before the Mega Conference

Progress in Mathematics

Edited by:

H. Bass
Columbia University
New York
10027
U.S.A.

J. Oesterlé
Dépt. de Mathématiques
Université de Paris VI
4, Place Jussieu
75230 Paris Cedex 05, France

A. Weinstein
Dept. of Mathematics
University of CaliforniaNY
Berkeley, CA 94720
U.S.A.

Progress in Mathematics *is a series of books intended for professional mathematicians and scientists, encompassing all areas of pure mathematics. This distinguished series, which began in 1979, includes authored monographs, and edited collections of papers on important research developments as well as expositions of particular subject areas.*

We encourage preparation of manuscripts in such form of TeX for delivery in camera-ready copy which leads to rapid publication, or in electronic form for interfacing with laser printers or typesetters.

Proposals should be sent directly to the editors or to: Birkhäuser Boston, 675 Massachusetts Avenue, Cambridge, MA 02139, U.S.A.

1 GROSS. Quadratic Forms in Infinite-Dimensional Vector Spaces
2 PHAM. Singularités des Systèmes Differentiels de Gauss-Manin
4 AUPETIT. Complex Approximation
5 HELGASON. The Radon Transform
6 LION/VERGNE. The Weil representation Maslov index and Theta series
7 HIRSCHOWITZ. Vector bundles and differential equations
10 KATOK. Ergodic Theory and Dynamical Systems I
11 BALSLEY. 18th Scandinavian Congress of Mathematicians
12 BERTIN. Séminaire de Théorie de Nombres, Paris 79-80
13 HELGASON. Topics in Harmonic Analysis on Homogeneous Spaces
14 HANO. Manifolds and Lie Groups
15 VOGAN JR. Representations of Real Reductive Lie Groups
16 GRIFFITHS/MORGAN. Rational Homotopy Theory and Differential Forms
17 VOVSI. Triangular Products of Group Representations and Their Applications
18 FRESNEL/VAN DER PUT. Géometrie Analytique Rigide et Applications
19 ODA. Periods of Hilbert Modular Surfaces
20 STEVENS.Arithmetic on Modular Curves
21 KATOK. Ergodic Theory and Dynamical Systems II
22 BERTIN. Séminaire de Théorie de Nombres, Paris 80-81
23 WEIL. Adeles and Algebraic Groups
24 LE BARZ. Enumerative Geometry and Classical Algebraic Geometry
25 GRIFFITHS. Exterior Differential Systems and the Calculus of Variations
27 BROCKETT. Differential Geometric Control Theory
28 MUMFORD. Tata Lectures on Theta I
29 FRIEDMANN. The Birational Geometry of Degenerations
30 YANO/KON. Submanifolds of Kaehlerian and Sasakian Manifolds
31 BERTRAND. Approximations Diophantiennes et Nombres Transcendant
32 BROOKS. Differential Geometry
33 ZUILY. Uniqueness and Non-Uniqueness in the Cauchy Problem
34 KASHIWARA. Systems of Microdifferential Equations
35/36 ARTIN/TATE. Vol. 1 Arithmetic. Vol. 2 Geometry
37 BOUTET. Mathématique et Physique
38 BERTIN. Séminaire de Théorie de Nombres, Paris 81-82
39 UENO. Classification of Algebraic and Analytic Manifolds
40 TROMBI. Representation Theory of Reductive Groups
41 STANLEY. Combinatorics and Commutative Algebra
42 JOUANOLOU. Théorèmes de Bertini et Applications

43 MUMFORD. Tata Lectures on Theta II
45 BISMUT. Large Deviations and the Malliavin Calculus
47 TATE. Les Conjectures de Stark sur les Fonctions L d'Artin en s=0
48 FRÖHLICH. Classgroups and Hermitian Modules
49 SCHLICHTKRULL. Hyperfunctions and Harmonic Analysis on Symetric Spaces
50 BOREL ET AL. Intersection Cohomology
51 Séminaire de Théorie de Nombres, Paris 82-83
52 GASQUI/GOLDSCHMIDT. Déformations Infinitésimales desStructures Conformes Plates
53 LAURENT. Théorie de la 2ième Micro-localisation dans le Domaine Complexe
54 VERDIER. Module des Fibres Stables sur les Courbes Algébriques
55 EICHLER/ZAGIER. The Theory of Jacobi Forms
56 SHIFFMAN/SOMMESE. Vanishing Theorems on Complex Manifolds
57 RIESEL. Prime Numbers and Computer Methods for Factorization
58 HELFFER/NOURRIGAT. Hypoellipticité Maximale pour des Opera-teurs Polynomes de Champs de Vecteurs
59 GOLDSTEIN. Séminaire de Théorie de Nombres, Paris 83-84
60 ARBARELLO. Geometry Today
62 GUILLOU. A la Recherche de la Topologie Perdue
63 GOLDSTEIN. Séminaire de Théorie des Nombres, Paris 84-85
64 MYUNG. Malcev-Admissible Algebras
65 GRUBB. Functional Calculus of Pseudo-Differential Boundary Problems
66 CASSOU-NOGUES/TAYLOR. Elliptic Functions and Rings of Integers
67 HOWE. Discrete Groups in Geometry and Analysis
68 ROBERT. Autour de l'Approximation Semi-Classique
69 FARAUT/HARZALLAH. Analyse Harmonique: Fonctions Speciales et Distributions Invariantes
70 YAGER. Analytic Number Theory and Diophantine Problems
71 GOLDSTEIN. Séminaire de Théorie de Nombres, Paris 85-86
72 VAISMAN. Symplectic Geometry and Secondary Characteristic Classes
73 MOLINO. Riemannian Foliations
74 HENKIN/LEITERER. Andreotti-Grauert Theory by Integral Formulas
75 GOLDSTEIN. Séminaire de Théorie de Nombres, Paris 86-87
76 COSSEC/DOLGACHEV. Enriques Surfaces I
77 REYSSAT. Quelques Aspects des Surfaces de Riemann
78 BORHO/BRYLINSKI/MCPHERSON. Nilpotent Orbits, Primitive Ideals, and Characteristic Classes
79 MCKENZIE/VALERIOTE. The Structure of Decidable Locally Finite Varieties
80 KRAFT/ SCHWARZ/PETRIE (Eds). Topological Methods in Algebraic Transformation Groups
81 GOLDSTEIN. Séminaire de Théorie des Nombres, Paris 87–88
82 DUFLO/PEDERSEN/VERGNE (Eds). The Orbit Method in Representation Theory
83 GHYS/DE LA HARPE (Eds). Sur les Groupes Hyperboliques d'après M. Gromov
84 ARAKI/KADISON (Eds). Mappings of Operator Algebras
85 BERNDT/DIAMOND/HALBERSTAM/HILDEBRAND (Eds). Analytic Number Theory
89 VAN DER GEER/OORT/STEENBRINK (Eds). Arithmetic Algebraic Geometry
90 SRINIVAS. Algebraic K-Theory
91 GOLDSTEIN. Séminaire de Théorie des Nombres, Paris 1988-89
92 CONNES/DUFLO/JOSEPH/RENTSCHLER. Operator Algebras, Unitary Representa tions, Enveloping Algebras, and Invariant Theory. A Collection of Articles in Honor of the 65th Birthday of Jacques Dixmier
93 AUDIN. The Topology of Torus Actions on Symplectic Manifolds
94 MORA/TRAVERSO (Eds). Effective Methods in Algebraic Geometry
95 MICHLER/RINGEL (Eds). Represen-tation Theory of Finite Groups and Finite–Dimensional Algebras
96 MALGRANGE. Equations Différen-tielles à Coefficients Polynomiaux
97 MUMFORD/NORMAN/NORI. Tata Lectures on Theta III
98 GODBILLON. Feuilletages, Etudes géométriques
99 DONATO/DUVAL/ELHADAD/ TUYNMAN. Symplectic Geometry and Mathematical Physics. A Collection of Articles in Honor of J.-M. Souriau

100 TAYLOR. Pseudodifferential Operators and Nonlinear PDE
101 BARKER/SALLY. Harmonic Analysis on Reductive Groups
102 DAVID. Séminaire de Théorie des Nombres, Paris 1989-90
103 ANGER/PORTENIER. Radon Integrals
104 ADAMS/BARBASCH/VOGAN. The Langlands Classification and Irreducible Characters for Real Reductive Groups
105 TIRAO/WALLACH. New Developments in Lie Theory and Their Applications
106 BUSER. Geometry and Spectra of Compact Riemann Surfaces
107 BRYLINSKI. Loop Spaces,Characteristic Classes and Geometric Quantization
108 DAVID. Séminaire de Théorie des Nombres, Paris 1990-91
109 EYSSETTE/GALLIGO. Computational Algebraic Geometry
110 LUSZTIG. Introduction to Quantum Groups
111 SCHWARZ. Morse Homology
112 DONG/LEPOWSKY. Generalized Vertex-Algebras and Relative Vertex Operators
113 MOEGLIN/WALDSPURGER. Décomposition Spectrale et Series d'Eisenstein
114 BERENSTEIN/GAY/VIDRAS/YGER. Residue Currents and Bezout Identities
115 BABELON/CARTIER/KOSMANN-SCHWARZBACH (Eds). Integrable Systems. The Verdier Memorial Conference
116 DAVID (Ed.). Séminaire de Théorie des Nombres, Paris, 1991-1992
117 AUDIN/LAFONTAINE (Eds). Holomorphic Curves in Symplectic Geometry
118 VAISMAN. Lectures on the Geometry of Poisson Manifolds
119 JOSEPH/MIGNOT/MURAT/PRUM/RENTSCHLER (Eds) First European Congress of Mathematics (Paris, July 6-10, 1992). Volume I: Invited Lectures (Part 1)
120 JOSEPH/MIGNOT/MURAT/PRUM/RENTSCHLER (Eds). First European Congress of Mathematics (Paris, July 6-10, 1992). Volume II: Invited Lectures (Part 2)
121 JOSEPH/MIGNOT/MURAT/PRUM/RENTSCHLER (Eds). First European Congress of Mathematics (Paris, July 6-10, 1992). Volume III: Round Tables
122 GUILLEMIN. Moment Maps and Combinatorial Invariants of Hamiltonian T^n-spaces
123 BRYLINSKI/BRYLINSKI/GUILLEMIN/KAC (Eds). Lie Theory and Geometry: In Honor of Bertram Kostant
124 AEBISCHER/BORER/KÄLIN/LEUENBERGER/REIMANN. Symplectic Geometry. An Introduction based on the Seminar in Bern, 1992
125 LUBOTZKY. Discrete Groups, Expanding Graphs and Invariant Measures
126 RIESEL. Prime Numbers and Computer Methods for Factorization
127 HÖRMANDER. Notions of Convexity
128 SCHMIDT. Dynamical Systems of Algebraic Origin
129 DIJKGRAAF/FABER/VAN DER GEER (Eds). The Moduli Space of Curves
130 DUISTERMAAT. Fourier Integral Operators
131 GINDIKIN/LEPOWSKY/WILSON. Functional Analysis on the Eve of the 21st Century in Honor of the 80th Birthday of I.M. Gelfand. Vol. I
132 GINDIKIN/LEPOWSKY/WILSON. Functional Analysis on the Eve of the 21st Century in Honor of the 80th Birthday of I.M. Gelfand. Vol. II
131/132 Set Vols I+II
133 HOFER/TAUBES/WEINSTEIN/ZEHNDER (Eds). The Floer Memorial Volume
134 CAMPILLO LOPEZ/ NARVAEZ MACARRO (Eds). Algebraic Geometry and Singularities
135 Amrein/Boutet de Monvel/Georgescu. C_0-Groups, Commutator Methods and Spectral Theory of N-Body Hamiltonians
136 BROTO/CASACUBERTA/MISLIN (Eds). Algebraic Topology. New Trends in Localization and Periodicity
137 VIGNÉRAS. Représentations l-modulaires d'un groupe réductif p-adique avec $l \neq p$
140 KNAPP. Lie Groups Beyond an Introduction
141 CABANES (Ed.). Finite Reductive Groups: Related Structures and Representations
142 MONK. Cardinal Invariants on Boolean Algebras